CYBERCARTOGRAPHY: THEORY AND PRACTICE

Cybercartography: Theory and Practice

Edited by

D. R. FRASER TAYLOR

Geomatics and Cartographic Research Centre (GCRC) and
Distinguished Research Professor in International Affairs,
Geography and Environmental Studies
Carleton University, Ottawa, Ontario, Canada

ELSEVIER

Amsterdam – Boston – Heidelberg – London – New York – Oxford
Paris – San Diego – San Francisco – Singapore – Sydney – Tokyo

Elsevier
Radarweg 29, PO Box 211, 1000 AE Amsterdam, The Netherlands
The Boulevard, Langford Lane, Kidlington, Oxford OX5 1GB, UK

First edition 2005
Reprinted 2007

Copyright © 2005 Elsevier BV. All rights reserved

No part of this publication may be reproduced, stored in a retrieval system or transmitted in any form or by any means electronic, mechanical, photocopying, recording or otherwise without the prior written permission of the publisher

Permissions may be sought directly from Elsevier's Science & Technology Rights Department in Oxford, UK: phone (+44) (0) 1865 843830; fax (+44) (0) 1865 853333; email: permissions@elsevier.com. Alternatively you can submit your request online by visiting the Elsevier web site at http://elsevier.com/locate/permissions, and selecting *Obtaining permission to use Elsevier material*

Notice
No responsibility is assumed by the publisher for any injury and/or damage to persons or property as a matter of products liability, negligence or otherwise, or from any use or operation of any methods, products, instructions or ideas contained in the material herein. Because of rapid advances in the medical sciences, in particular, independent verification of diagnoses and drug dosages should be made

British Library Cataloguing in Publication Data
A catalogue record for this book is available from the British Library

Library of Congress Cataloging-in-Publication Data
A catalog record for this book is available from the Library of Congress

ISBN–13: 978-0-444-51629-9
ISBN–10: 0-444-51629-8

For information on all Elsevier publications
visit our website at books.elsevier.com

Printed and bound in *The Netherlands*

07 08 09 10 10 9 8 7 6 5 4 3 2

Working together to grow
libraries in developing countries

www.elsevier.com | www.bookaid.org | www.sabre.org

ELSEVIER BOOK AID International Sabre Foundation

Contents

Preface

This book on the *Theory and Practice of Cybercartography* presents current thinking on an emerging new paradigm. Since its formal introduction to the International Cartographic Association conference in Stockholm, Sweden in 1997, most concentrated research on cybercartography has been carried out in two main research centres: the Geomatics and Cartographic Research Centre (GCRC) at Carleton University in Ottawa, Canada and Centro de Investigación en Geografia y Geomática "Ing. J. L. Tamayo" (CentroGeo) in Mexico City. GCRC is an Organized Research Unit in the Department of Geography and Environmental Studies at Carleton University and CentroGeo was established as a Consejo Nacional de Ciencia y Technologica (CONACYT) Research Centre in 1999.

In November 2002, the Geomatics and Cartographic Research Centre (GCRC) was awarded a major research grant by the Social Sciences and Humanities Research Council of Canada (SSHRC) for a research project entitled Cybercartography and the New Economy (CANE). The award of $2.56 million was under an innovative SSHRC program entitled the Initiative on the New Economy (INE). Implementation began in January 2003. The research project runs for 4 years and will be completed by December 2006.

The research team involved is a multidisciplinary one and, in addition to GCRC, a number of other research laboratories at Carleton University are involved including the Human Oriented Technology Laboratory (HOTLab); The Institute for Comparative Studies in Literature, Art and Culture (ICSLAC); and the Centre for Applied Cognitive Research (CACR) which includes the Cognitive Science group and the Navigational and Situational Awareness and Information Group. These research laboratories are all members of the Human Oriented Technology group established to draw together research on this topic at Carleton. This group, led by Dr. Gitte Lindgaard, Director of the Human Oriented Technology Laboratory, was awarded a major infrastructure grant of $4.56 million by the Canadian Foundation for Innovation (CFI) to create a Human Computer Interaction (HCI) Institute which provides a new building and sophisticated computer technology to carry out research, including that related to the CANE project. The new building is scheduled for completion in 2005.

The book reflects progress by researchers in both Ottawa and Mexico City in both the theory and practice of cybercartography.

In addition to the research teams at Carleton and CentroGeo, chapters have been contributed to the book by three key members of the International Cartographic Association's Commission on Maps and the Internet. One chapter has also been written by the research team which produced the Georgia Basin Digital Laboratory.

It is difficult to fully capture research on cybercartography using only a fixed print medium so a CD-ROM has been used which includes links to various Web sites to capture the more dynamic and interactive elements of the paradigm. This cybercartographic product was created with the help of CentroGeo and reflects the considerable expertise in producing cybercartographic atlases in CD-ROM format which has been developed by researchers in Mexico.

This book is a result of a multidisciplinary team effort and has benefited from the input of partners from government, industry and non-governmental organizations which are part of the CANE project. These research team members and organizations are acknowledged in the next section of the book and more details on the nature of their contribution and the structure of both the book and the project are given in Chapter 1.

Special thanks go to Tracey P. Lauriault, a PhD candidate in the Department of Geography and Environmental Studies at Carleton University, who acted as Managing Editor for this volume. Without her efforts the book would not have come to fruition.

Cybercartography is a new paradigm and this volume marks the first major study of its many elements. Maps and mapping were central to the Age of Exploration. It is argued that maps and mapping in the new form of cybercartography are equally important to the information era. Cybercartography has the potential to help us navigate and better understand the sea of information threatening to drown us all as computer based information increases exponentially on almost a daily basis.

D. R. Fraser Taylor
Ottawa, December 2004

Acknowledgments

The editor, who directs the Geomatics and Cartographic Research Centre (GCRC) at Carleton University and is Principal Investigator for the Cybercartography and the New Economy (CANE) project, would like to acknowledge a large number of people who helped make this book possible. First, thanks are due to the authors who have contributed to the book including Tracey P. Lauriault, the managing editor. The contribution of Dra Carmen Reyes, Director General of CentroGeo and her colleagues in Mexico has been of special importance. The Cybercartography and the New Economy project research team at Carleton University consists of the following individuals and research groups. Not all have written chapters for this book but their contributions to the research have been significant.

Human Oriented Technology Laboratory (HOTLab) including Professors Gitte Lindgaard (Director), Richard Dillon (deceased), Robert Biddle, Avi Parush; Postdoctoral Fellows Claire Dormann, Leo Ferres and Kate Oakley; PhD students Sheila Narasimham, Maria Rasouli, Shelley Roberts, Bruce Tsuji and Patricia Trbovich; MA students Michelle Gauthier, Shamima Khan, Karen Philp, Elizabeth Whitworth, Aida Hadziomerovic, Adam Bronsther (graduated) and Greg Dunn (graduated).

Geomatics and Cartographic Research Centre (GCRC) including Professor D. R. Fraser Taylor; Postdoctoral Fellows Sebastien Caquard and Diana DeStefano; PhD students Brian G. Eddy, Tracey P. Lauriault and Peter L. Pulsifer; MA students Samantha Baulch (graduated), Glenn Brauen, Xiuxia Liu, Kenneth Pawliw, Barry Dale Powell, Birgit Woods and Yuchai Zhou; and Christine Earl, an Instructor in the Department of Geography and Environmental Studies.

Centre for Applied Cognitive Research (CACR) including Professors Jo-Anne LeFevre (Director) and Chris Herdman; PhD students Aryn Pyke and Matthew Rutledge-Taylor; and MA student Lisa Hagen.

Institute for Comparative Studies in Literature, Art and Culture (ICSLAC) including Professors Brian Greenspan and Paul Theberge, PhD students Christopher Eaket; Robert Evans and Esther Post; and MA students Megan Graham (graduated), Kristopher Moran (graduated) and Benjamin Wright.

The Norman Paterson School of International Affairs (NPSIA) including Professor Michael Hart and MA students Ilka Guttler, Oksana Pidufala, and Francis MacDonnell (graduated).

A number of individuals also act as leaders for interdisciplinary research clusters organized around specific research and production challenges. These clusters include faculty, postdoctoral fellows, students and technical staff. In December 2004 cluster leaders included Brian Eddy, Dr. Sebastien Caquard, Dr. Claire Dormann, Amos Hayes, Tracey P. Lauriault, Professor Avi Parush, Peter L. Pulsifer, and Bruce Tsuji. Research clusters can include individuals not formally associated with the CANE project but interested in the research topic.

The project also benefits from the input of a number of international research collaborators including Dr. Michael Peterson of the University of Omaha, Nebraska; Dr. Georg Gartner of the University of Vienna and Dr. William Cartwright of the Royal Melbourne Institute of Technology, Australia; and Dr. Robert McCann of NASA Ames Research Centre, USA Three of these individuals have contributed chapters to the book and Dr. Peterson spent several months at Carleton University working with the GCRC research team.

The Cybercartography and the New Economy Project also has a number of formal partners including the GeoAccess Division and National Atlas Program of Natural Resources Canada, the Geography and Trade divisions of Statistics Canada, the Canadian Department of Foreign Affairs and International Trade, the Geomatics Industry Association of Canada, the Orbital Media Group, Telecom Italia, the Scientific Committee for Antarctic Research (SCAR), the Canadian National Institute for the Blind (CNIB) Library, the Canadian Committee for Antarctic Research (CCAR) of the Canadian Polar Commission (CPC), the International Cartographic Association's (ICA) Commissions on Maps and the Internet on Mountain Cartography.

In addition to the formal relationships listed above a number of additional relationships have developed with Students on Ice, Innotive Corporation and the DM Solutions Group. The partnership with SCAR has involved special relationships with the British Antarctic Survey and with Antarctic research organizations in a number of countries including Argentina, Australia, China, Germany and USA. Dr. Daniel Vergani and Zulma Stanganelli of CENPAT in Argentina have made a significant contribution to the Cybercartographic Atlas of Antarctica.

Management and Administration includes Barbara J. George, Project Manager and YamHean Kong, Administrative Assistant. Technical Support includes Amos Hayes, network specialist and GCRC laboratory manager, consultant Jean-Pierre Fiset of ClassOne Technologies Inc., and Jeff McKenna of DM Solutions Group.

The project Advisory Board includes:

- Dr. Robert O'Neil, Director (Advisory Board Chair), GeoAccess Division, Natural Resources Canada
- Dr. Gordon Deecker, Director, Geography Division, Statistics Canada
- Dr. Norman Vinson, Research Officer, Interactive Information, National Research Council of Canada
- Dr. Olav Loken, Secretary, Canadian Committee for Antarctic Research
- Ronald C. Drews, President and CEO, the Orbital Media Group
- Dr. Ed Kennedy, Managing Director, Canadian GeoProject Centre
- Dr. Lise Paquet, Associate Dean (Research), Faculty of Arts and Social Science, Carleton University
- Dr. Fabrizio Davide, Telecom Italia, S.p.A.
- Mr. John Manning, Scientific Committee for Antarctic Research
- Dr. Hilary Young and Dr. Gordana Krcevinac (ex officio)

The research reported in this book would not have been possible without the major funding support from the Social Sciences and Humanities Research Council (SSHRC) of Canada Initiative on the New Economy (INE) program which is gratefully acknowledged. Funding has also been provided by the Canadian Department of Foreign Affairs and International Trade, Carleton University, the Canadian Committee for Antarctic Research (CCAR), Scientific Committee for Antarctic Research (SCAR) and Agencia para la Promocion de las Cooperation Cientifica y Tecnologia of Argentina.

The production of this book has been a team effort and this human knowledge network has been of inestimable value. All errors and omissions in the book remain the responsibility of the editor.

D. R. Fraser Taylor
Ottawa
December 2004

Contributors

Regina Araujo de Almeida (Vasconcellos)
Departamento de Geografia
Faculdade de Filosofia Letras e Ciências Humanas
Universidade de São Paulo
Av. Prof. Lineu Prestes, 338
Cidade Universitária
05508900 Sao Paulo, SP
Brasil

Samantha Baulch
The Delphi Group
428 Gilmour St.
Ottawa, (ON) K2P OR8
Canada

Boyan Brodaric
Geological Survey of Canada
Natural Resources Canada
234B-615 Booth Street
Ottawa (ON) K1A 0E9
Canada

Adam Bronsther
Human Oriented Technology Lab (HOTLab),
Department of Psychology,
210 Social Science and Research Building
1125 Colonel By Drive
Carleton University
Ottawa (ON) K1S 5B6
Canada

Allison Brown
Professor and Director, Teaching and Learning
School of Electrical and Computer Engineering
Royal Melbourne Institute of Technology (RMIT) University,
P.O. Box 2476V, Melbourne, Victoria
Australia, 3001

Sébastien Caquard
Geomatics and Cartographic Research Centre (GCRC)
Department of Geography and Environmental Studies
Carleton University
1125 Colonel By Drive
Ottawa (ON) K1S 5B6
Canada

William E. Cartwright
School of Mathematical and Geospatial Science
Royal Melbourne Institute of Technology (RMIT) University
RMIT City Campus
Building 12, Level 11, Swanston Street
Melbourne, VIC 3000
Australia

Shannon Denny
Geological Survey of Canada
Natural Resources Canada
101-605 Robson Street
Vancouver (BC) V6B 5J3
Canada

Brian G. Eddy
Geomatics and Cartographic Research Centre (GCRC)
Department of Geography and Environmental Studies
Carleton University
1125 Colonel By Drive
Ottawa (ON) K1S 5B6
Canada

Georg Gartner
Research Group Cartography
Department of Geoinformation and Cartography
Vienna University of Technology, Gusshausstr. 30,
1040 Wien Austria

Ryan Grant
Institute for Resources, Environment and Sustainability
University of British Columbia
2nd floor, 1924 West Mall
Vancouver (BC) V6T 1Z2
Canada

Brian Greenspan
Institute for Comparative Studies in Literature, Art and Culture (ICSLAC)
and the Department of English Language and Literature
Carleton University
201 St. Patrick's Building
1125 Colonel By Drive
Ottawa (ON) K1S 5B6
Canada

Rob Harrap
Queen's University GIS Laboratory
Mackintosh-Corry Hall Room E223
Queen's University
Kingston (ON) K7L 3N6
Canada

Murray Journeay
Geological Survey of Canada
Natural Resources Canada
101-605 Robson Street
Vancouver (BC) V6B 5J3
Canada

Tracey P. Lauriault
Geomatics and Cartographic Research Centre (GCRC)
Department of Geography and Environmental Studies
Carleton University
1125 Colonel By Drive
Ottawa (ON) K1S 5B6
Canada

Gitte Lindgaard
Professor, Natural Sciences and Engineering Research Council (NSERC)
Industry Chair User-Centred Design, and
Director Human Oriented Technology Lab (HOTLab),

Department of Psychology
210 Social Science and Research Building
1125 Colonel By Drive
Carleton University
Ottawa (ON) K1S 5B6
Canada

Elvia Martínez
Centro de Investigación en Geografía y Geomática
J.L. Tamayo (CentroGeo)
Contoy No. 137
Col. Lomas de Padierna
Delegación Tlalpan
México D.F. C.P. 14770

Ronald Douglas Macdonald
Kawartha Pine Ridge District School Board
Lakefield District Intermediate Secondary School
Lakefield (ON) K0L 2HO
Canada

Dr. Mark Monmonier
Distinguished Research Professor of Geography
Department of Geography and
Maxwell School of Citizenship and Public Affairs
Syracuse University
Department of Geography
144 Eggers Hall
Syracuse University
Syracuse (NY) 13244-1020
USA

Avi Parush
Professor of Psychology
Human Oriented Technology Lab (HOTLab)
Department of Psychology
210 Social Science and Research Building
1125 Colonel By Drive
Carleton University
Ottawa (ON) K1S 5B6
Canada

Michael P. Peterson
Department of Geography / Geology
University of Nebraska at Omaha
Omaha, Nebraska NE 68182-0199
USA

Peter L. Pulsifer
Geomatics and Cartographic Research Centre (GCRC)
Department of Geography and Environmental Studies
Carleton University
1125 Colonel By Drive
Ottawa (ON) K1S 5B6
Canada

María del Carmen Reyes
Directora General
Centro de Investigación en Geografía y Geomática
J.L. Tamayo (CentroGeo)
Contoy No. 137
Col. Lomas de Padierna
Delegación Tlalpan
México D.F. C.P. 14770

Shelley Roberts
Human Oriented Technology Lab (HOTLab)
Department of Psychology
210 Social Science and Research Building
1125 Colonel By Drive
Carleton University
Ottawa (ON) K1S 5B6
Canada

Sonia Talwar
Department of Geography
University of British Columbia
217-1984 West Mall
Vancouver (BC) V6T 1Z2
and Geological Survey of Canada
Natural Resources Canada
101-605 Robson Street
Vancouver (BC) V6B 5J3
Canada

D. R. Fraser Taylor
Geomatics and Cartographic Research Centre (GCRC) and
Distinguished Research Professor in International Affairs
Geography and Environmental Studies
Geomatics and Cartographic Research Centre
Department of Geography and Environmental Studies
Carleton University
1125 Colonel By Drive
Ottawa (ON) K1S 5B6
Canada

Paul Théberge
Canada Research Chair in Technological Mediations of Culture
Institute for Comparative Studies in Literature, Art and Culture (ICSLAC)
and Music
201 St. Patrick's Building
1125 Colonel By Drive
Ottawa (ON) K1S 5B6
Canada

Patricia Leean Trbovich
Human Oriented Technology Lab (HOTLab)
Department of Psychology
210 Social Science and Research Building
1125 Colonel By Drive
Carleton University
Ottawa (ON) K1S 5B6
Canada

Bruce Tsuji
Human Oriented Technology Lab (HOTLab)
Department of Psychology
210 Social Science and Research Building
1125 Colonel By Drive
Carleton University
Ottawa (ON) K1S 5B6
Canada

Joost Van Ulden
Geological Survey of Canada
Natural Resources Canada
101-605 Robson Street
Vancouver (BC) V6B 5J3
Canada

CHAPTER 1

The Theory and Practice of Cybercartography: An Introduction

D. R. FRASER TAYLOR

Geomatics and Cartographic Research Centre (GCRC) and Distinguished Research Professor in International Affairs, Geography and Environmental Studies, Carleton University, Ottawa, Ontario, Canada

Abstract

The concept of cybercartography was introduced in 1997 in the keynote address entitled *Maps and Mapping in the Information Era,* presented to the International Cartographic Conference in Sweden. The central argument made was that if cartography was to play a more important role in the information era, then a new paradigm was required. An initial version of cybercartography was introduced as that paradigm. This chapter describes developments in cybercartography since that time and, in particular, the developments emerging from the Cybercartography and the New Economy research project, funded for a 4 year period, by the Social Sciences and Humanities Research Council of Canada. Many of the chapters in this book grew out of this research project. Parallel developments in the theory and practice of cybercartography are taking place in a very different cultural context at Mexico by CentroGeo, a new and dynamic research agency, which is part of the Consejo Nacional de Ciencia y Technologica (CONACYT) network. The significant differences between cybercartography and Geographic Information Systems (GIS) and the related concept of Geographical Information Science will be discussed, and this chapter will establish the context for the rest of this book.

1. Introduction

The term "cybercartography" was introduced in June 1997 in the keynote address to the 18th International Cartographic Conference in Stockholm, Sweden (Taylor 1997). The title of that presentation was "Maps and Mapping in the Information Era." It was argued that mapping as a process and the map, both as a concept and a product, could become increasingly important to the information era but that this would require a change in the thinking of cartographers and an increased awareness of the opportunities with which the discipline and profession was presented. The key elements required to achieve this were identified as imagination, foresight, and effort. It was argued, "We must move away from narrow 'technological' normative and formalistic approaches to cartography to a more holistic approach where both mapping as a process and the map as a product are expanded." (Taylor 1997) As Cartwright observes later in this volume (Chapter 14), this was not a sudden shift in thought on the part of the author, but an evolution of thinking over time, including an article in 1991 entitled "A Conceptual Basis for Cartography: New Directions for the Information Era" (Taylor 1991) and one in 1994 on "Cartography for Knowledge, Action and Development: Retrospective and Perspective" (Taylor 1994). The author sees the paradigm of cybercartography not as a sudden and dramatic shift from past ideas and practice, but as an evolutionary and integrative process which incorporates important elements from the past, redefines others, and introduces new ideas and approaches to both cartographic practice and theory. Theory and practice are closely linked in a series of ongoing and iterative feedback loops. The concept of cybercartography, for example, was strongly influenced by a product, The *Canadian Geographic Explorer*, produced in 1996 by the multimedia firm IQMedia (now superseded by the Orbital Media Group) (Taylor 1997). Continuous interaction between research and production is a feature of the cybercartographic paradigm. Theory and practice are not two discrete processes. Cybercartography is a new way of looking at the role of cartography in the 21st century and, in particular, at cartography's contribution to the new knowledge-based economy. Cybercartography is an evolving concept and concentrated research on the paradigm is a very recent phenomenon and, as several authors in this volume indicate, there are still many research questions to be answered.

The departure point for this book is the chapter entitled "The Concept of Cybercartography" (Taylor 2003) in the important book on *Maps and the Internet,* edited by Michael Peterson (Peterson 2003). All the authors in this volume read this chapter. Seven major elements of cybercartography were outlined in that chapter:

- Cybercartography is multisensory using vision, hearing, touch, and eventually, smell and taste;
- Cybercartography uses multimedia formats and new telecommunication technologies, such as the World Wide Web;
- Cybercartography is highly interactive and engages the user in new ways;
- Cybercartography is applied to a wide range of topics of interest to the society, not only to location finding and the physical environment;
- Cybercartography is not a stand-alone product like the traditional map, but part of an information/analytical package;
- Cybercartography is compiled by teams of individuals from different disciplines; and
- Cybercartography involves new research partnerships among academia, government, civil society, and the private sector. (Taylor 2003).

Each of the chapters in this book considers one or more of these seven elements, expands on them, and introduces new ideas and understanding about both theory and practice. In addition, the book builds on the cybercartographic-integrated framework of content, context, and contact (Eddy 2002; Taylor 2003), which considers the interactions between and among the seven elements outlined above and the holistic nature of the cybercartographic paradigm.

2. Why Cybercartography?

Why do we need a new term? A major reason is to reassert and demonstrate the importance of maps and mapping and the centrality and utility of cartography in the information era. Harmon has argued that the impulse to map is an innate human characteristic, "I sense that humans have an urge to map and this mapping instinct, like our opposable thumbs, is part of what makes us human. I map, therefore I am." (Harmon 2004: 10, 11) She comments:

> Mapmaking fulfills one of our deepest desires: understanding the world around us and our place in it. But maps need not show just continents and oceans: there are maps to heaven and hell; to happiness and despair; maps of moods, matrimony and mythological places. There are maps to popular culture, from Gulliver's Island to Gilligan's Island; speculative maps of the world before it was known; and maps to secret places known only to the map-maker. Artists' maps show another kind of uncharted realm: the imagination. What all these maps have in common is their creator's willingness to venture beyond the boundaries of geography or convention.

These observations are not new although they are eloquently stated but, they do reemphasize the importance of imagination in cartography. For example, a centerpiece of the National Broadcasting Company's (NBC) television coverage of the 2004 US presidential election was a large choropleth map painted on the surface of the ice rink at the Rockefeller Center in New York with the results being colored in by individuals using paint spray guns. This was complemented by huge colored streamers acting as bar charts, indicating the number of electoral votes won by the candidates. The National Broadcasting Company was using imaginative analog cartography. At the same time, the American Broadcasting System was utilizing the *iBrowser* software, developed by Innotive Systems of Toronto, to scroll across the electoral map of the United States and derive trends in statistical fashion in real-time from each map, as the results were reported (Innotive Systems works closely with the Orbital Media Group, which is a partner in the Cybercartography and the New Economy project). This was an imaginative form of computer cartography including several cybercartographic elements. There is no shortage of imaginative cartography, but most of it is not produced by cartographers. However, the dominant paradigm in recent years for cartographers has been cartography as a science. Cybercartography has a strong scientific component but sees cartography as both an art and a science, and has a qualitative as well as a quantitative element.

3. Cybercartography and Geographic Information Systems

Geographic Information Systems (GIS) have grown in importance over the last four decades and Goodchild (1992) introduced the term Geographic Information Science, arguing that GIS was the basis of a new science. A new University Consortium for Geographic Information Sciences has since been established. The Third International Conference on Geographic Information Sciences was held in October 2004 (http://www.giscience.org). Although GIScience is interdisciplinary, it is still primarily a scientific paradigm as its name suggests. For many GIScientists the map is often seen only as an input and/or output device for a GIS and as a result some have argued that cartography is of limited relevance in the GIS era despite the fact that GIS was initially known as analytical cartography.

There are significant differences between cybercartography and both GIS and geographic information science. Cybercartography values and includes the analytical strengths of GIS, but differs from it in a number of ways. Elvia Martinez and Carmen Reyes in this volume (Chapters 4, 5, and 6) argue that cybercartography owes much to concepts of cybernetics and, in particular, second-wave socio-cybernetics, and that cybercartographic atlases are primarily social products. The "cyber" in cybercartography does not only refer to delivery mechanisms in cyberspace but has theoretical linkages with cybernetics. Geographic Information Systems and GIScience are more in line with Norbert Weiner's concepts of a scientifically objective first-order cybernetic system where the observer is not seen as part of the process. Cybercartographic atlases are produced together with the societies in which they are designed and are interactive in both technological and societal sense. In Mexico cybercartographic atlases are instrumental in creating social action by communities on environmental issues. This process is qualitatively different from simply consulting the intended users in order to determine their needs, which many GIS systems do. In cybercartography, the user is an integral part of the creative process. This places the User Needs Assessment (UNA) techniques of human factors psychologists in a wider societal context. Applying Wilber's integral theory to cybercartography (Eddy 2002; Eddy and Taylor, Chapters 3 and 22 in this volume) leads to similar conclusions. For Reyes and Martinez, researchers producing cybercartographic products are participants in an active social process, not "objective" and "scientific" observers.

Dangermond (2004) has argued that GIS is "the language of geography" and utilizes this term as part of the marketing strategy for his company Earth Systems Research Institute (ESRI). Given the different theoretical perspectives on geography it is unlikely that there is any one "language of geography" but, a good case can be made that GIS is one important language given the pervasive nature of GIS today. Geographic Information Systems are primarily a visual form of communication. Cybercartography includes the visual but adds other senses, especially sound and touch, and methodologies, such as multimedia that creates a new form of communication with society. The communication paradigm for cartography, which dominated the 1970s and early 1980s (Taylor 1986) is being revisited, but from new perspectives. Cybercartography is developing a new multisensory language for cartography, which is being extended to include smell and taste. Geographic Information Systems as "the language of geography" are exclusive, and primarily scientific in nature. Cybercartography as a new language for cartography is inclusive, more subjective, and integrative. It includes the scientific and positivistic elements of GIS, but goes well beyond these limited concepts.

GIS and GIScience are products of the specialization of knowledge, which has dominated thinking for at least the last five decades. Cybercartography is holistic and returns to an earlier, more integrative approach which incorporates ideas from the humanities, the social sciences, and the physical sciences to help deal with the challenges of knowledge integration, which some have argued as one of the major challenges of the information era (Evans 2002; Johnston 2002; Rock 2002). Over a century ago human geographers of the French school, such as Vidal de la Blache (1903) were using geography and cartography to integrate knowledge by including regions, such as the French "pays" as the "container" for integrating knowledge of the physical environment with "genre de vie," which covered all aspects of society, economy, language, and culture. Later geographers such as Taylor (1936) were arguing for geography as a holistic, integrative discipline, and Sauer (1956) made an eloquent case for the centrality of the map to geography. The technologies available to cybercartography allow a much more extensive integration of knowledge. They also allow new forms on interconnectivity, interaction, and integration which makes cybercartography different from conventional cartography (Eddy and Taylor, Chapter 3). In integrating this knowledge, cybercartography recognizes that there may well be different interpretations and views of each situation being mapped. The "objective" view of GIS is not the only view. Caquard argues later in this volume (Chapter 12) that cybercartography can be used to challenge what they call "the false objectivity of maps." Cybercartography allows the presentation of several views of the same "reality" and in a variety of different forms, in addition to the map. The map is the central organizing principle of cybercartography, but users can choose to access information in a variety of other forms and can choose to ignore the map as a presentation method, if they so wish. Cybercartography allows different voices to be heard both literally and figuratively.

Both GIS and cybercartography involve new forms of interaction with computer databases. Geographic Information Systems tend to concentrate on the technological elements of this interaction. Cybercartography makes use of these important developments, but builds on them to consider how users interact with, and navigate through, virtual space from a wider perspective. Human factors psychology, cognitive psychology, and studies in language and literature, such as those on hypertext, are used to give new insight into these important processes and to develop new products in an interactive fashion.

Some of the individual elements of cybercartography appear in the literature of many disciplines, including cartography itself, but the whole is very much more than the sum of the parts. Cybercartography is attempting a new synthesis, is developing new cooperative processes, and is creating a new language for communication and societal involvement. In Mexico, for example, building on the ideas of Luscombe, CentroGeo is using the Strabo technique to help resolve environmental conflict (Luscombe and Reyes 2004). The language of cybercartography is proving to be great value in this respect. Environmental specialists from different disciplines and perspectives use cybercartographic images interactively to arrive at consensus. Cartography has tended to be undervalued and, in the minds of many, is seen as a subset of GIS, or as a simple illustrative or location-finding communication tool. Cybercartography is a new paradigm for cartography and the term is adopted to reflect this. The chapters in this book illustrate the significance and vibrancy of this new approach, as well as some of its limitations and the important research questions, which remain to be answered.

4. The Cybercartography and the New Economy Project

In 2002 the Social Sciences and Humanities Research Council of Canada (SSHRC) announced a new program for funded research, the Initiative on the New Economy (INE) (http://www.sshrc.ca). A multidisciplinary research team from cartography, film studies, geography, international trade, comparative studies in literature, language and culture, and music and psychology at Carleton University in Ottawa, made application to the INE Collaborative Research Initiatives under this program to develop a new foundational paradigm for cybercartography. In the application it was argued that the development of this new paradigm would constitute a fundamental contribution to the New Economy by making the increasing volume of information available from databases, more accessible, understandable and useful to the general public, decision makers, and researchers in a wide variety of disciplines. To illustrate this it was proposed to produce two new innovative products – the Cybercartographic Atlas of Antarctica and a Cybercartographic Atlas of Canada's Trade with the world. A full version of the project proposal can be found at http://gcrc.carleton.ca. The application for funding for the Cybercartography and the New Economy (CANE) project was successful and received a $2.56 million grant from SSHRC over a period of 4 years. Full implementation of the project began in January 2003.

The team at Carleton were all members of the Human-Oriented Technology group, established to draw together research from existing research units which had been in existence for some time. The group obtained a major infrastructure grant of $4.64 million from the Canadian Foundation for Innovation (CFI) to create a Human Computer Institute, which provides new laboratory space and sophisticated computer technology to carry out research including that related to the Cybercartography and the New Economy project.

The core research activities of the Cybercartography and the New Economy project are based at Carleton, but are supplemented by researchers from other institutions including three key members of the International Cartographic Association's Maps and the Internet Commission. These are Michael P. Peterson of the University of Nebraska, Omaha (USA), William Cartwright of the Royal Melbourne Institute of Technology (RMIT) (Australia), and Georg Gartner of the University of Vienna (Austria). Peterson and Gartner are co-chairs of the ICA Commission facilitating research contributions from another 60 members worldwide.

Two major partners for the research are the National Atlas Program of the GeoAccess Division of Natural Resources, Canada and both the Geography and the International Trade divisions of Statistics, Canada. Researchers from these groups are actively involved in the Cybercartography and the New Economy project research. Both the Canadian Committee for Antarctic Research (CCAR) and the Scientific Committee for Antarctic Research (SCAR) are partners and contribute knowledge to the project. Students on Ice, an organization which takes students on field trips to Antarctica, is an important partner.

A key researcher has been identified to coordinate the research team for the project in each of the nations participating in the creation of the Antarctic Atlas through the Scientific Committee for Antarctic Research. These nations include Argentina, Australia, Belgium, Canada, Chile, China, Germany, Poland, the United Kingdom, and the United States.

The research also has linkages with industry through the Geomatics Industry Association of Canada (GIAC) and individual member firms such as the DM Solutions Group and ESRI, as well as the multimedia firm Orbital Media Group. Of special importance is the work of the Open Geospatial Consortium (OGC) and project researchers participating actively in the work of OGC. The Cybercartographic Atlas of Antarctica, in particular, is implementing OGC specifications on interoperability and semantic ontologies.

Archiving digital data and ensuring adequate metadata descriptions and authentication is a major research challenge. Much of our cartographic digital heritage over the last three decades has been lost due to the lack of attention given to preserving ephemeral data. In map preservation terms the last few decades have some similarities with The Dark Ages. All we have are a few fragments of the original computer-produced maps and only written descriptions of others. Some have completely disappeared and cannot be retrieved. The Atlas of Antarctica is a case study for the InterPARES project, led by Dr Luciana Duranti of the University of British Columbia, which brings the expertise of the world's leading archivists to research on these important topics. The Cybercartography and the New Economy project is developing new metadata descriptions for multimedia and multisensory products and is actively considering how cybercartographic products can be archived. Here we are building on the work of the International Standards Organization (ISO) as well as OGC.

Graduate students are an integral part of the Cybercartography and the New Economy project, both in terms of carrying out research and of helping to organize and manage research. There are 11 individuals at the PhD level including three from geography, Brian Eddy, Tracey P. Lauriault, and Peter L. Pulsifer who are also members of the project management team. Seventeen Master's students have been involved in the project so far. The active research involvement of students is illustrated by their contributions to several chapters in this book.

Several postdoctoral fellows are involved with the project including Sebastien Caquard (Geography) and Diana DeStefano (Cognitive Science). Three postdoctoral fellows in human factors psychology, Claire Dormann, Kate Oakley, and Leo Ferres, are also pursuing research related to the project.

The project is supported by a strong technical and administrative team. It is utilizing a Web services approach with OGC specifications, but includes the use of both commercially available and customized software, specific to the needs of the project. Of special importance is the design of new graphic interfaces and modified games for use in Cybercartography and the New Economy research production. Amos Hayes and J. P. Fiset are key members of the technical support group and Barbara George, Project Manager, and YamHean Kong, Administrative Assistant, provide management and administrative support.

The original management structure of the project is shown in Fig. 1.1. The core research activities took place in the four key laboratories involved: the Geomatics and Cartographic Research Centre (GCRC), The Human Oriented Technology Laboratory (HOTLAB), the Institute for Comparative Studies in Literature, Art and Culture (ICSLAC), and the Centre for Applied Cognitive Research (CACR), which includes the Cognitive Science group and the Navigational and Situational Awareness and Informational Group (NSAVIS).

In May 2004, this structure was supplemented by a new matrix structure developed around specific themes to help make interdisciplinary interaction more effective. Interdisciplinary teamwork is an important part of the paradigm of cybercartography. The research clusters included groups looking at Atlas content, cybercartographic theory, information visualization, narratives, and games. These will be described in more detail in Chapter 8 which will focus attention more strongly on key interdisciplinary research questions. Whereas the laboratory structure is fixed, the research clusters will change over time as new questions emerge. A major objective is to create transdisciplinary knowledge defined as "new knowledge" emerging from interdisciplinary interaction.

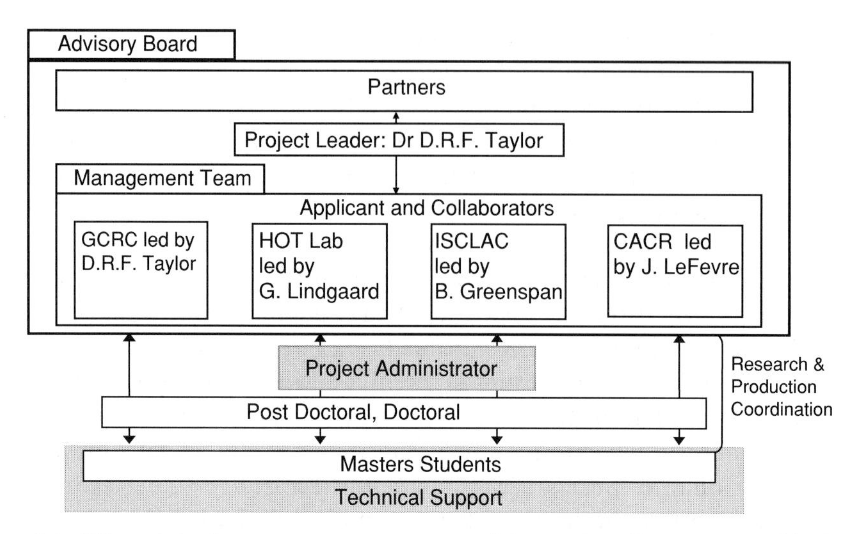

FIG. 1.1. The original management structure of the project. Image created by Peter Pulsifer and submitted as part of the funding proposal, 2002.

As is shown in Fig. 1.1, the project benefits from the advice of an eight-person Advisory Board drawn from government, industry, and academia. The project is funded until late 2006.

5. The Structure of the Book

After this introductory chapter, the book discusses the legacy of existing cartographic products in a chapter by Mark Monmonier. Monmonier argues that important elements of cartographic design are being ignored in the creation of new online products. This is followed by a chapter on the integral and inclusive nature of the cybercartographic paradigm, building on Ken Wilber's integral theory, in which a comprehensive Cybercartographic Human Interface (CHI) model is introduced. The CHI model integrates models from critical theory and science studies, information and organizational management, systems engineering, ecological interface design, and visualization and cartography. It builds on and expands the initial Cybercartographic Integrated Framework of content, context, and contact (Eddy 2002; Taylor 2003). Three comprehensive chapters written by Carmen Reyes and Elvia Martinez review the important work being carried out at CentroGeo in Mexico. CentroGeo is the only other research centre, other than the Geomatics and Cartographic Research Centre (GCRC) at Carleton University in Ottawa, carrying out research on the theory and practice of cybercartography and the importance of social context, systems theory, modeling, and second-wave cybernetics being the key elements in their approach. CentroGeo has produced a number of cybercartographic atlases in CD format and cybercartography owes much to the outstanding work done by Mexican researchers. CentroGeo is one of the important network of research centers funded by the Consejo Nacional de Ciencia y Tecnologica (CONACYT).

Chapter 7 introduces the concept of geographic and cartographic mediation to frame the emerging processes of integrating and representing geographic information in a distributed and interoperable way. This further enhances the integrative nature of cybercartography and our understanding of semantic ontologies. The topic of the challenges of research integration and collaboration, which includes consideration of some of the more important but underresearched elements of cybercartography, is discussed in Chapter 8. The theory of interdisciplinarity and partnership is much easier than the implementation, and this chapter is an excellent example of the interaction between theory and practice in building the paradigm of cybercartography. Chapters 3, 7, and 8 are important contributions to the paradigm of cybercartography by PhD students in the Geomatics and Cartographic Research Centre (GCRC). Chapters 9–11 illustrate the contributions which psychology has to make to cybercartography including the important topic of human interaction with, and navigation through, cyberspace. Understanding the needs, wants, and capabilities of users is critical. Multimedia and multisensory techniques in cybercartography clearly have to be used with great care. A technological "bells and whistles" approach devoid of good theory and practice is inappropriate and can lead to increased confusion. Engaging the user through "edutainment," for example, does not always lead to meaningful learning. Chapters 12 and 13 look at the contribution of art and literature to cybercartography, drawing on theories from the humanities which bring some very different and informative perspectives and open up new opportunities for a more critical cybercartography, expanding the field of critical cartography and GIS in general. Chapters 14–16 deal with the design and production of new cartographic artifacts and products, including public map displays and the use of cartography in the rapidly expanding field of mobile devices. There is an interesting contrast between Monmonier's comment in Chapter 2 that, small screen size is the "Achilles heel" of cybercartography in technical terms and the innovative use of multimedia techniques to overcome even smaller screen sizes in Gartner's "telecartography." These chapters also make an important contribution to the design challenges identified by Monmonier. Chapters 17 and 18 look at the use of sound and touch, two key elements for communication in cybercartography. Chapters 19–22 describe how cybercartographic theory is being put into practice by describing the production of the Georgia Basin Digital Library, the Cybercartographic Atlas of Antarctica, Cybercartography for Education with the Antarctic Atlas as a case study, and the Cybercartographic Atlas of Canada's Trade with the world. The book concludes with a consideration of some of the remaining challenges for the future of cybercartography.

6. Conclusion

In an article on cybercartography in Information Highways, Roy (2004) describes cybercartographers as the "new explorers," contrasting the work that the Cybercartography and the New Economy project is doing in exploring and mapping information space with that of the early explorers such as Samuel Champlain and others in Canada. It took many years for the efforts of the early explorers of Canada to bear fruit, and they often found themselves going in wrong directions and having to retrace their steps. Cybercartographic exploration is in the very early stages. The purpose of this book is to indicate the present scenario and to stimulate further discussions. Cybercartography is a work in progress. We are not presenting a complete guide map for the reader to follow but a set of directional signposts which will hopefully stimulate further exploration.

References

Dangermond, J. (2004) *Keynote address to ESRI Users Conference*, San Diego accessed 5 September from www.esri.com

Eddy, B. G. (2002) *Cybercartographic Integrated Framework*, presentation to the Research Workshop on Cooperation between Carleton University and the National Atlas of Canada, 15 April.

Evans, J. R. (2002) *Universities and Government: Centre Stage in Innovation and the New Economy*, presentation to the 8th International Forum on the New Economy, La Conférence de Montréal, June.

Goodchild, M. F. (1992) "Geographical information science", *International Journal of Geographical Information Systems*, Vol. 6, pp. 31–47.

Harmon, K. (2004) *You are Here: Personal Geographies and Maps of the Imagination*, Princeton Academic Press, New York.

Johnston, D. (2002) *Innovation and Development Strategies*, presentation to the 8th International Forum on the New Economy, La Conférence de Montréal, June.

Luscombe, B. W. and C. Reyes (2004) *Building Consensus in Environmental Decision-Making–A Methodology Integrating GIS Tools and Structured Communication*, National Association of Environmental Professionals.

Peterson, M. (2003) *Maps and the Internet*, Elsevier, Amsterdam.

Rock, A. (2002) *Canada's Innovation Strategy*, presentation to the 8th International Forum on the New Economy, La Conférence de Montréal, June.

Roy, V. (2004) "The new explorers: how cybercartography is mapping the information age", *Information Highways,* Vol. 2, No. 5, July–August.

Sauer, C. O. (1956) "The education of a geographer", *Annals, Association of American Geographers*, Vol. 46, pp. 287–299.

Third International Conference on Geographic Information Sciences (2004) Washington, Accessed September 2004 from http://www.geoscience.org

Taylor, D. R. F. (ed.) (1986) "Graphic communication and design in contemporary cartography", *Progress in Contemporary Cartography,* Series Vol. 2, John Wiley and Sons, Chichester.

Taylor, D. R. F. (1991) "A conceptual basis for cartography: new directions for the information era", *The Cartographic Journal*, Vol. 28, No. 2, pp. 213–216.

Taylor, D. R. F. (1994) "Cartography for knowledge, action and development: retrospective and perspective", *The Cartographic Journal*, Vol. 31, No. 1, pp. 52–55.

Taylor, D. R. F. (1997) *Maps and Mapping in the Information Era*, Keynote address to the 18th ICA Conference, Stockholm, Sweden, in Ottoson, L. (ed.), *Proceedings*, Vol. 1, Swedish Cartographic Society, Gavle, pp. 1–10.

Taylor, D. R. F. (2003) "The concept of cybercartography", Chapter 26 in Peterson, M. P. (ed.), *Maps and the Internet*, Elsevier, Amsterdam, pp. 405–420.

Taylor, T. G. (1936) *Environment and Nation: Geographical Factors in the Cultural and Political History of Europe*, University of Toronto Press, Toronto.

Vidal de la Blache, P. (1903) *Tableau de la Geographie de la France*, Librarie Jules Talandier, reprinted 1979, Paris.

CHAPTER 2

POMP and Circumstance: Plain Old Map Products in a Cybercartographic World

MARK MONMONIER

Department of Geography, Maxwell School of Citizenship and Public Affairs, Syracuse University, Syracuse, New York, USA

Abstract

The analogy between traditional analog telephone service and traditional paper maps fosters a useful exploration of the importance of standards and the emergence of hybrid products facilitated by new technology but bound by inertia or circumstance to past practices. Compared to telecommunications, in which protocols of all types have been indispensable, few universal standards have emerged for mapping. Although the increased demand for interoperability and flexible exchange formats is impelling a greater standardization of geospatial data, the "standards" discourse of recent decades has largely ignored graphic quality and visual effectiveness. Particularly ominous for cybercartography is a failure to adopt the graphic logic embodied in prominent exemplars like the US Bureau of the Census's recent statistical atlas of racial diversity. Available in both paper and electronic formats, the *Diversity* atlas illustrates a number of highly effective conventional practices likely to survive the transition to cybercartography. By contrast, the Census Bureau's American *FactFinde*r Web site suggests that authors of cybercartographic products readily ignore well-established, more traditional practices, as well as promising innovations that could make map use more engaging and informative. Since governments and other organizations that establish standards for geospatial data are likely to remain focused on data quality and interoperability, it is up to educators and authors of critical reviews to raise the public's awareness and expectations.

1. Introduction

Cell phones, broadband two-way fiber-optic cable, and similar innovations in electronic communications so radically altered the way we consume information and converse over long distances that engineers and regulators needed a word to represent the older technology upon which average citizens still relied. The result was the acronym POTS, meaning plain old telephone service, and referring specifically to analog communications over the historical two-wire, public switched network, once synonymous in the United States with the "Bell System." Most residential customers with a hardwire telephone connection have POTS, even those with enhancements like call-waiting, cordless telephones, and low-speed, dial-up Internet connections. Inspired by POTS, a clever wordsmith coined a companion acronym, PANS, for "pretty amazing new stuff" such as the digital alternative to POTS known as Integrated Services Digital Network (ISDN). Although, industry observers are waiting patiently for ISDN or some other form of PANS like Digital Subscriber Line (DSL) or Voice-over-IP (VoIP) phone service to replace POTS, conventional telephony appears amazingly hardy (Taylor 2002).

There is a cartographic parallel here, insofar as cybercartography (Taylor 2003) poses a similar threat to plain old map products (POMP), which includes conventional cartographic products ranging from wall maps, printed topographic quadrangle maps, and paper atlases to newspaper maps, sketch maps, and cartographic placemats. Although, replacement of POMP by cyberequivalents might not be as swift or complete as the eventual substitution of PANS for POTS, a radical restructuring of the way we receive, manipulate, and display geospatial information seems inevitable. For instance, the lowly spring-roller wall map will no doubt surrender its wall space to the large flat-panel display that can offer multiple views, cartographic and otherwise, without the constraint of a single, monotonous, and often ossifying map projection. Well-heeled commercial and educational establishments that are the prime market for wall maps will eagerly embrace an electronic upgrade, promoted perhaps by an apocryphal scenario of how an accommodating office manager can pander to visiting dignitaries with a straight-laced Mercator map for Old Money yachtists, a droopy Gall-Peters worldview for Third World sympathizers from UNESCO, or a well-rounded Robinson for the cartographically savvy. By contrast, simple ad hoc sketch maps drawn on napkins or scraps of paper might seem less easily dislodged were it not for the electronic drawing tablets basic to handheld computers and an attractive option on some newer laptops. Perhaps the single most pervasive impetus for cybercartography is the map's uncanny ability, starting with the humble line printer in the late 1950s, to piggyback on the circumstance of whatever electronic displays become available – hence the second term in this essay's title (Tobler 1959).

In addition to exploring briefly the similarities and differences between POTS and POMP, this chapter examines cybercartography's likely reliance on conventions and techniques inherited from POMP. In particular, it looks at the role of standards in data collection and graphic design, examines recent applications of cybercartography and conventional mapping by the US Bureau of the Census, and questions the validity of design principles and the graphic understanding of software developers. How effectively, it asks, are software developers taking advantage of existing knowledge? What principles and practices of conventional cartography can enhance making and using maps in a cybercartographic world? And what new strategies are needed to enhance the effectiveness of cybermaps?

2. Cartography and Telecommunications Compared

The analogy between mapping and telecommunications is far from perfect. Although both technologies depend on networks that require regulation and standardization, uniform protocols are clearly more basic in telecommunications, and for a very good reason without mandated consistency in the signals carried over the network, as well as design standards for every piece of equipment connected to the network, a system that includes over a thousand independent telephone companies would be inefficient if not dysfunctional (Flaherty 2000). While developments during the 1990s illustrated the side-by-side compatibility of analog and digital networks for mobile telephony, design and performance standards allowed these parallel systems to interface efficiently and effectively with each other as well as with the traditional, locationally fixed POTS (Arensman 1999). Most important, a combination of government regulation and voluntary industry standards guarantee that no two customers in North America have the same ten-digit phone number. Equally significant, negotiated tariffs and technical standards allow worldwide direct-dial calling. What's more, technical standards combined with technological innovation have fostered intense competition among service providers, and telecommunications monopolies, whether government sanctioned or state-owned, have become less common than in past decades, thanks to the federal judiciary in the

United States (Laffont and Tirole 2000) and the Thatcher Government in Britain (Moon *et al.* 1986).

In cartography, by contrast, state-run mapping organizations dominate geodesy and topographic mapping, as well as some comparatively specialized applications like cadastral mapping and meteorological cartography. This dominance persists despite occasionally impressive collaborations with the private sector. Although the number of joint ventures between government mapmakers and their commercial partners will no doubt increase, the degree of privatization found in telecommunications is unlikely because of the key roles maps play in national defense and public administration (Monmonier 1985; Cloud 2002). Private-sector contractors have taken over much of the work, but investment in cartographic infrastructure is largely a government responsibility. Indeed, some of the more impressive partnerships link federal mapping with the cartographic activities of regional and local governments (National Research Council 2001).

Compared with the technical standards developed in the United States communications industry, American cartographic standards have been amazingly but understandably flexible, at least until recent concern about interoperability and flexible data-exchange formats (Sondheim *et al.* 1999). The emergence of ESRI-compatible formats as a de facto exchange format reflect the enormous collective market share represented by Arc/Info, ArcView, Arc/GIS, and their successors, offspring, and corporate-partner, plug-in cousins.

Data quality is another matter. Instead of focusing on a universal minimum level of spatial precision, mapmakers subscribe to an operational notion that data quality is akin to "fitness for use" supported by a "truth in labeling" paradigm (Veregin 1999). This ad hoc standard is readily apparent in the diversity of street centerline files, ranging from raw TIGER (Topographically Integrated Geographically Encoded Referencing) files with a bad case of the jaggies to subsequent refinements by data vendors eager to turn a profit by cleaning up the aesthetic and informational deficiencies of a database that was quite adequate for the Census Bureau's intended use at the time. Local utilities and highway departments dissatisfied with the Census Bureau standard have added to the diversity by developing their own databases from scratch (Cooke 1995).

Perhaps the most historically significant cartographic protocol in the United States is the National Map Accuracy Standards, devised in the early 1940s by the Bureau of the Budget to promote consistency among federal topographic maps (Marsden 1960; Thompson 1960; Monmonier 2002). Separate standards addressing horizontal and vertical accuracy are administered as a "pass–fail" test for individual quadrangle maps. Each test involves a selection of approximately 20 "well-defined" points and the comparison of the distance between their exact and mapped positions with a context-dependent tolerance. For horizontal accuracy the tolerance is 1/30 inch [0.85 mm] for maps published at scales greater than 1:20,000 and 1/50 inch [0.51 mm] for scales of 1:20,000 and smaller. (For a 1:24,000 topographic map, the latter tolerance represents a horizontal error of 40 ft [12.2 m].) A map passes the test if the horizontal distance between exact and mapped positions exceeds the tolerance for no more than 10% of the points tested. Similarly, for vertical accuracy no more than 10% of the checkpoints may exceed a tolerance of half a contour interval. With no additional credit for deviations well below the tolerance and no declared penalty for a grossly deficient error – which presumably the mapmaker would be required to fix – the National Map Accuracy Standards are emblematic of a state-run mapping industry focused on producing printed maps and committed to at least minimal consistency in data and basic symbology.

Cartographers with an appreciation of quantitative analysis and quality control have recognized the shortcomings of pass–fail testing for nearly half a century (Thompson 1960). A suggested replacement was the root mean square error (RMSE), defined in the maptesting context as the square root of the arithmetic mean of the squared deviations, that is,

$$\mathrm{RMSE} = \left[\sum x_i/n\right]^{1/2},$$

where $x_1, x_2, x_3, \ldots, x_n$ are the deviations or "errors" at n test points. If map errors are assumed to be normally distributed, the critical RMSE for a 1:24,000 scale topographic would be 24 ft [7.3 m] – the average error for a set of checkpoints 10% of which register errors greater than the 40 ft [12.2-m] tolerance. By contrast, mapmakers have avoided formal tests of graphic effectiveness, pass–fail or otherwise, except perhaps for minimal standards for the legibility of labels (Robinson *et al.* 1978).

The most revealing similarity between telecommunications and mapping is the disconnect between the data and graphic design. Standardization in telephony stops abruptly once the caller punches in a number or speaks into the microphone. Because the configuration of "the instrument" has no inherent effect on interoperability, agencies responsible for telecommunications protocols have accorded designers considerable leeway to develop convenient or dysfunctional features. The shape of the handset or the layout of the dial or keypad became minimally standardized through a combination of monopolized manufacturing and user expectations. Similarly, the discourse on cartographic standards has focused almost exclusively on geospatial data and basic maps intended largely as data. Like the humble telephone instrument, the derived map is at the mercy of the designer's skill and probity and the public's experience and expectations. And in the same way that democratic governments seldom meddle in art and literature, they are reluctant to set standards for the visual effectiveness and graphic excellence of map displays. What's more, it seems unlikely that cybercartography will blur this distinction between data and display.

3. GPS, Census Cartography, and the Persistence of Paper Maps

Although rooted in the numerical procedures of late twentieth century conventional cartography, statistical notions of geometric error are highly relevant to cybercartographic applications based on global positioning (Lange and Gilbert 1999). This connection was especially apparent in the 5-year, $200-plus million Census Bureau contract awarded in 2002 to Harris Corporation for development of a nationwide street centerline file with a 7.6-m [25-ft] spatial accuracy requirement, based on a 95% circular error (US Bureau of the Census 2001, 2002). The contract supports the Census Bureau's Master Address File (MAF)/TIGER Accuracy Improvement Project with a realignment of the TIGER database designed to serve field enumerators using mobile GPS-equipped computers. The "95 CE" spatial accuracy standard not only gives Census Bureau geographers 95% confidence that the correct position of a point is within 7.6 m of the position stated in the realigned TIGER database, but also allows a field enumerator to record horizontal positions that are 100% certain of being associated with the intended residential structure, 100% certain ofbeing recorded on the correct side of the street, and 100% percent certain of being related correctly to legal and other boundaries and to neighboring structures (US Bureau of the Census 2003).

Mobile computers with map displays and built-in GPS receivers represent a significant advance along the road from POMP to cybercartography, even for a government agency like the US Census Bureau, which pioneered imageless demographic data collection in the late 1960s map with address coding guides, dual independent map encoding (DIME) files, and machine-readable mail out/mail back questionnaires. Although census planners and supervisors might still use large-format printed maps in orchestrating their 2010 enumeration, it is now technologically feasible and probably cost effective to dispense with these conventional map products altogether.

Far less likely is the agency's abandonment of paper maps in its publications program, even though the Internet will continue to play an increasingly important role in the dissemination of census data and census maps. In other words, paper maps can be enormously appealing and user friendly, and thus a valuable vehicle for a federal agency rich in information worth sharing with power elites in government and academia. A prominent recent example is the colorful, intriguing and painstakingly designed atlas *Mapping Census 2000: The Geography of US Diversity* (Brewer and Suchan 2001). This atlas's careful development, heavy promotion, and wide distribution attest to the continued value of conventional paper maps even when cartographers with ready access to Internet distribution, dynamic software, and CD-ROM replicating equipment can develop some "pretty amazing new stuff."

In the same way that systems supporting digital telephony conferred significant improvements on conventional analog telephone service, so too most conventional paper maps are hybrid products, influenced or expedited in important ways by cybercartography. For a present-day Census Bureau atlas these links are apparent not only in the electronic collection and collation of the raw enumeration data, but also in the computational software used to aggregate data, calculate rates, experiment with design and zlayout, and generate prepress negatives for color printing. Perhaps the strongest case for labeling *Mapping Census 2000* a hybrid is the Census Bureau's use of the Internet to distribute on demand its constituent sheets as either Portable Document Format (PDF) or Postscript files. In this sense, the *Diversity* atlas highlights the fact that many instances of cybercartography are transitional, hybrid products rather than pure, fully electronic displays. And while better examples of hybridism exist, the *Diversity* atlas dramatically illustrates that cybercartography is not only analogous to PANS but also heavily influenced by the telecommunications revolution.

4. Cybercartography and the Conventions of POMP

This section attempts to define conditions under which cybercartography reflects or relies on rules, theories, and other pragmatics devised for conventional cartography. A related task is to identify design conventions for statistical maps so fundamental that they apply universally (or nearly so) throughout cybercartography and to differentiate these practices from rules or traditions that can readily be ignored in the cybercartographic realm. This exercise sheds light on both the revolutionary nature of cybercartography and the need for map authors trained in conventional mapping to abandon old customs and adopt new ones.

The *Diversity* atlas is a case in point. Published in both conventional and electronic modes but designed and promoted as a paper atlas, *Mapping Census 2000* reflects a cybercartography tightly bound to established practices. In particular, it honors standard interpretations of Jacques Bertin's

(1983) concept of "visual variables," notably by relying largely on value (rather than hue) for portraying differences in intensity and occasionally on size for showing differences in magnitude. Adherence to these rules is far from slavish though. The atlas consists principally of choropleth maps, some of which have diverging or partial–spectral color schemes that highlight departures from the national mean, while others rely on single-hue sequential schemes to show differences in numbers of persons with a specific racial heritage (Brewer 1997).

The authors' use of choropleth maps to portray magnitude data as well as intensity data is inherently controversial insofar as it breaches the "bigger means larger" rule for magnitudes like population counts and the "darker means more intense" rule for area-independent rates and ratios (Dent 1999). There are mitigating factors though. As the authors observe in their introduction,

> Choropleth maps are well-suited to showing derived values such as percent or density. They are less appropriate for representing total numbers of people. They are, however, used for totals in the atlas for county identification and county-to-county comparisons with other mapped data. As an alternative to choropleth mapping, the distribution map with symbols proportional to total number of people is included to give a better sense of the overall distribution of population. (Brewer and Suchan 2001: 2)

Titled *Population Distribution, 2000*, this "alternative" map uses graduated squares to portray differences in total population among the nation's roughly 3,200 counties. Although the display provides a usefully vivid illustration of marked differences in density, it also demonstrates the weakness of theoretically appropriate graduated-point symbols in heavily concentrated areas like the Megalopolitan Corridor between Southern New Hampshire and Washington, D.C. (Fig. 2.1). Overlapping squares that completely obscure county boundaries in this portion of the map make it difficult, if not impossible, to relate individual symbols to the counties they represent. The authors are quite correct in violating a rule in conflict with the Need for Clarity, one of two overriding principles in map design. (The other is the Need for Informative Contrast.) As in any design arena, rules can be dangerous if the designer ignores situations in which the rules do not apply, break down, or conflict with higher objectives.

Despite the latter caveat, the principles of statistical mapping exemplified in the *Diversity* atlas are sufficiently fundamental to survive, and indeed flourish, in cybercartography. Even though choropleth maps are sometimes necessary for reasons naïve purists might reject, viewers readily appreciate the inherent logic of the "bigger means more" strategy for decoding graduated circles and squares, and the "darker means more intense" metaphor for light-to-dark sequences of area symbols on choropleth maps. Three

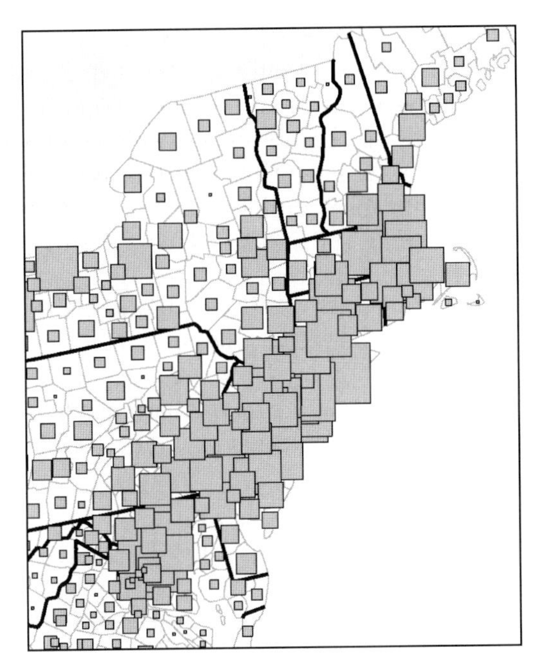

FIG. 2.1. High population densities along part the east coast of the United States yield overlapping graduated squares that obscure county boundaries in the *Diversity* atlas's map of "Population Distribution, 2000." From Brewer, Cynthia A. and Trudy A. Suchan (2001) *Mapping Census 2000: The Geography of U.S. Diversity*, U.S. Census Bureau, Census Special Reports, Series CENSR/01–1, U.S.Government Printing Office, Washington, D.C., pp. 13. Available online at http://www.census.gov/prod/ 2001pubs/censr01-1.pdf

additional principles provide a conceptual basis for diverging or partial–spectral color schemes by linking the notions "darker means more deviant," "different hues signify different types," and "more pronounced shifts in hue reflect more pronounced shifts in rate."

A comparison of two maps illustrates these latter concepts. On the map titled *Percent of Population, 2000: One or More Races Including Asian*, six categories ranging from "50.0 to 61.6" at the upper end to "0.0 to 0.9" at the lower end rely on a progressively lighter series of colors running from a medium blue for the highest category through various shades of green to a light yellow-beige for the lowest category. On the facing page, the map titled *Percent Change, 1990 to 2000: One or More Races Including Asian,* employs a six-color diverging scheme that shifts abruptly from purple to orange between the fourth highest category "0.0 to 48.2" and the fifth highest category "–10.0 to –0.1" – a break underscored by the label "No change." Especially prominent is the visual contrast between the strong purple of the highest category, labeled "Gain of 200 percent or more," and the prominent orange of the lowest category, labeled "Loss of more than 10 percent." Understandably, the most common symbol on the map (white) is not a part of this six-color diverging scheme: as noted in the key, counties colored white had "fewer than 100 people" self-identified as wholly or partly Asian. The graphic logic embodied in this assignment reflects a sixth principle equally valid for statistical cybercartography: "white or neutral gray identifies a nonparticipating area or one with no data or a statistically questionable rate." It is a wordy concept and, however logical its metaphor, the association must be pointed out in the map key.

Another asset of the *Diversity* atlas is its inherently meaningful class breaks. Although internally homogeneous categories are highly recommended for choropleth maps, map authors as well as developers of software for selecting optimal class intervals too often ignore the map viewer's interest in knowing whether a place is above or below the national or regional mean (Armstrong *et al.* 2003). Interest in relative performance is especially important for demographic maps, which beg questions like "Which places have higher than average concentrations of Asians?" Or "Which counties grew faster than the country as a whole?" It is, of course, essential that these breaks should also be highlighted in the map key (Fig. 2.2). Moreover, category breaks can foster useful comparisons of demographic groups as on the *Diversity* atlas's map showing the percentage of African–Americans under age 18; in addition to the break labeled "US

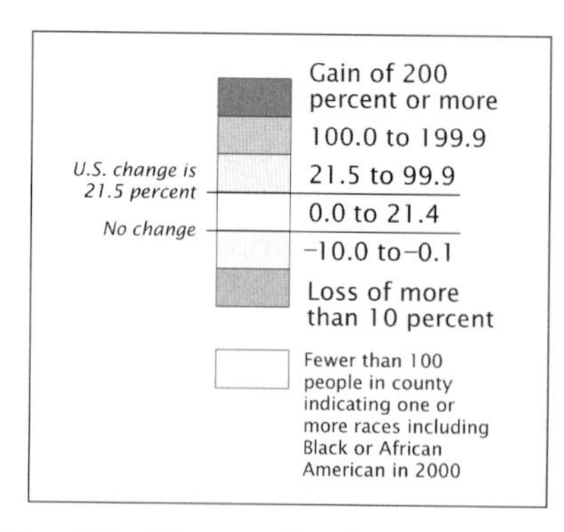

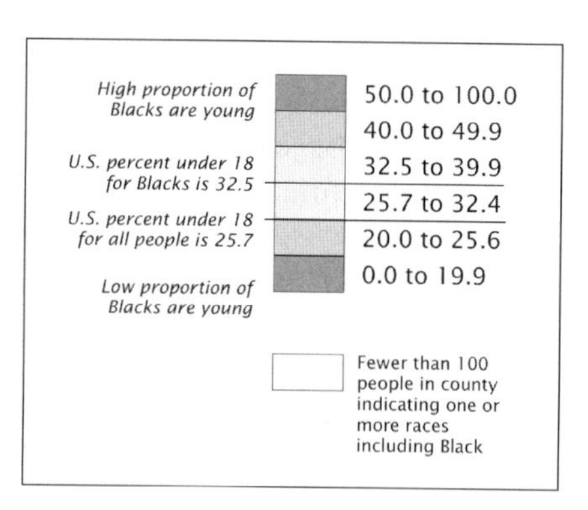

FIG. 2.2. The map key identifies inherently meaningful breaks for the Diversity atlas's maps of "Percentage Change, 1990 to 2000: One or More Races Including Black or African–American" (right) and "Percentage Under 18, 2000: One or More Races Including Black or African–American" (left). From Brewer, A. Cynthia A. and Trudy A. Suchan, *Mapping Census 2000: The Geography of U.S. Diversity*, U.S. Census Bureau, Census Special Reports, Series CENSR/01-1, U.S. Government Printing Office, Washington, D.C., 2001, p. 44 (left excerpt) and 46 (right excerpt). Available online at http//www.census.gov/prod/2001pubs/ censr01-1.pdf

percent under 18 for Blacks is 32.5," a second inherently meaningful break labeled "US percent under 18 for all people is 25.7" helps map viewers identify counties in which persons under 18 are less common in the local African–American population than nationally in the population at large. Another type of inherently meaningful break conveniently identified in the *Diversity* atlas is zero, especially useful on rate-of-change maps for differentiating increases from decreases.

An additional positive feature of the *Diversity* atlas is the use of complementary maps with state and county units. Every map sheet has a small inset map in each of its four corners. In addition to county-unit maps covering outlying states and territories (Alaska at the upper left, Hawaii at the lower left, and Puerto Rico and the Virgin Islands at the lower right), a state-unit map (at the upper right) affords a complementary state-level summary, especially useful to viewers eager to compare, for example, California, New York, and Texas. Although the county-level map in the center of the sheet affords a comparatively detailed picture of geographic variation, the state-level map promotes comparisons of these larger politically meaningful units – comparisons particularly meaningful to senators, governors, and state legislatures, as well as to journalists and scholars eager to generalize. The dual importance of detailed and aggregated views and the difficulty of visually aggregating from counties to states are strong arguments for two complementary levels of aggregation.

Electronic versions of the atlas's maps accentuate the limitations of coarse display devices that are, for the time being at least, the Achilles heel of cybercartography. There is no comparison in crispness and color fidelity between the atlas pages, transferred by a four-color press from 180-lines-per-inch color separated negatives onto smooth, coated paper with a satin finish and trimmed to pages 8.5 in. [21.6 cm] tall by 11 in. [27.9 cm] wide, and the corresponding PDF images displayed on my laptop's 12 in. × 9 in. (30.5 cm × 22.9 cm) color LCD screen, configured as 1,400 pixels across and 1,050 pixels down. With Acrobat Reader's 136% full page, "fit in window" view, I can barely read the anti-aliased fine print describing source data or labeling inherently meaningful breaks. For easier reading of small print or careful examination of local detail, I must position Acrobat Reader's magnifying glass icon over an appropriate center and click the mouse button four times for a 400% enlargement. The type is now clear and moderately crisp, colored area symbols are more easily differentiated, and county boundaries, though still noticeably jaggy when compared to their gracefully rendered counterparts on the printed page, are less aesthetically objectionable than at 136%. But, if imperfect centering cuts off a nearby area of interest or one needs to consult the map key, it is necessary to drop back to a lesser level of magnification or use Acrobat Reader's hand tool to recenter the image. Needless to say, image quality and ease of viewing are markedly worse on less expensive laptop's stingy 9.7 in. × 7.3 in. (24.6 cm × 18.5 cm) screen with the resolution frozen at 800 by 600 pixels. I can, of course, print the PDF image on my 600 dpi color ink-jet printer, but lackluster colors and type with noticeably jagged edges merely underscore electronic technology's inadequacies in coping with cartographic artwork designed for more aesthetically competent media.

5. Cybercartography and Dynamic Display

Cybercartography is, of course, far more than the use of zooming and panning to overcome the inadequacies of screen resolution. Richer modes of interactivity promise unprecedented levels of exploratory map analysis as well as ready access to timely or informatively customized spatial data. For instance, a cursor-driven rollover function can enrich the viewer's understanding by providing exact data values or added background facts for specific areas or features. And a cybercartography that lets users toggle among several supported languages and their respective renderings of toponyms can overcome the cost and design impediments of multilingual mapping. For place names, based on local languages – a useful feature if the viewer intends to travel in the area – sound clips can help users cope with a baffling orthography or unfamiliar pronunciations.

A promising example is the active map (Fig. 2.3) offering a native speaker's enunciation of 34 place names in Nisga'a Lands, a semi-autonomous aboriginal territory in the northwestern part of British Columbia. Each large red dot is a hot button linked to the *British Columbia Geographic Names Information System* online database. A mouse click summons an information screen that describes the place's location, offers a brief history of its name, and provides an "Audio Accompaniment" hot button that returns a small RealAudio file, launches the RealOne Player, and speaks otherwise arcane names like "Ksi Xts' at'kw" three times. Alas!, these 34 names seem to be the only ones on the BC Web site with audio clips. However accessible and intriguing its geography, this disappointingly sparse prototype of audio maps underscores the need to compile and validate an enormous amount of new, not easily certified geographic data. Because pronunciation is no less prone to uncertainty than numerical estimation, fussy map users would require multiple audio examples illustrating variation in a name's locally recognizable pronunciation.

The Internet abounds with maps suggesting that cybercartography is not approaching its full potential. A case in point is the US Census

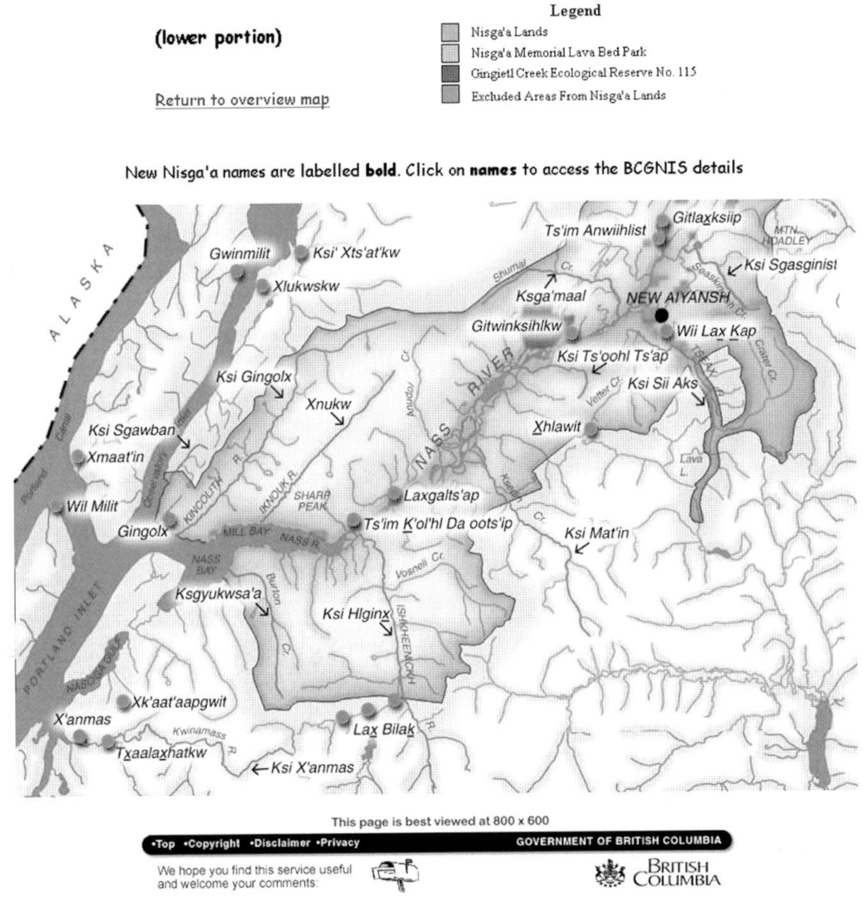

FIG. 2.3. Active map for the lower portion of the Nisga'a Treaty Area, British Columbia. Circular hot buttons linked to the British Columbia Geographical Names Information Systems lead to audio clips on which native speakers pronounce the respective names. This is from Nisga'a Lands Names, British Columbia Ministry of Sustainable Resource Management, http://srmwww.gov.bc.ca/bcnames/g2 nl.htm (accessed 30 December 2003).

Bureau's American *FactFinder* Web site (*factfinder.census.gov*), which offers an impressive menu of reference and thematic maps, delivered promptly in their initial rendering (at least to users with a high-speed, broadband Internet connection). For users eager to explore the data more fully, this early success experience can be misleading, thanks to a computationally challenged map server and a one-size-fits-all graphic interface. Especially frustrating was the small size of the county-subdivision map of New York State showing the "*Percent of Workers 16 Years and Over*

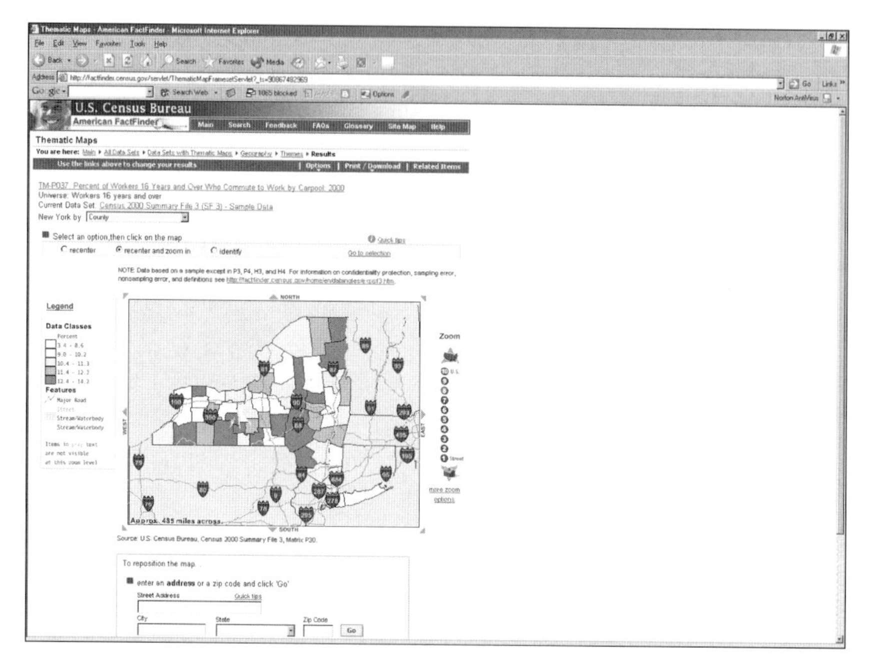

FIG. 2.4. Map server's inability to expand the size of its maps seriously limits the detail and flexibility of online maps. From the US Bureau of the Census, American FactFinder, http://factfinder.census.gov

Who Commute to Work by Carpool: 2000." Although my laptop can display 1,400 pixels across and 1,050 pixels down, *FactFinder* delivered a skimpy 500 by 375 GIF image, which fills a scant 13% of the screen (Fig. 2.4). (The map key, to its left, is a separate 125 × 257 GIF image.) For a more detailed view I had to zoom in and pan. While the resulting GIF images are markedly less elegant than the PDF files I downloaded for the *Diversity* atlas, the zoomed-in *FactFinder* maps at least reduced in size and number the superfluously obnoxious Interstate highway shields that obscured more than a dozen towns on the smaller-scale, statewide map.

Although the interface offers a wide variety of map themes and permits adjustments of display parameters, viewers cannot replicate the highly functional design of maps in the *Diversity* atlas. A mouse click on the blue hypertext word "Legend" summons a menu with three options. The viewer dissatisfied with the default number of classes can request between 2 and 7 categories, change the partial–spectral color system from a scheme favoring green to one based on orange, blue, magenta, violet, or gray, and switch the classing method from the default natural-breaks algorithm to equal intervals, quantiles, or "user defined." Tutorial screens describe the rationale for each classing scheme and offer an illustrative example. As the tutorial for user-defined categories suggests, "if the median income value for your county is $24,045, you could make a map showing other counties in your state with higher or lower median incomes by making a 3 class map" that assigns counties with the same median income to the intermediate class. Nowhere does the tutorial mention far broader inherently meaningful breaks like the national or state median, values of which the Web Site could easily supply. And nowhere does the system offer the user the option of filtering out counties with insignificantly low numbers of commuting workers. These limitations are puzzling insofar as the developers of the American *FactFinder* Web site no doubt had access to the *Diversity* atlas and its designers, Cindy Brewer and Trudy Suchan. In other words, cybercartography will fall short of its full potential until it can deliver maps as large, detailed, aesthetically appealing, and logically designed as PDF versions of the best conventional printed maps.

Programmer intransigence is not limited to Web cartography. For several years in the late 1990s I served on a multidisciplinary board of experts recruited to advise Microsoft Corporation on the design and content of Encarta World Atlas. It was an impressive product, and having said so, I don't mind telling a tale out of school, involving place names. Microsoft's technicians had implemented a highly efficient algorithm for displaying toponyms rapidly and without overlap, and they enriched the product by adding the names and coordinates of all populated places in the US Geological Survey's Geographic Names Information System (GNIS). Since the labeling algorithm was given no guidance about a place's relative importance, a minor crossroads would often eclipse the name of a town with 20,000 inhabitants. What's more, a slight recentering of the map could precipitate a wholly different mix of names, especially in regions rich in populated places. I suggested matching the list of named places with a list of Post Offices – a quick and dirty way to identify and keep in check many less significant toponyms – but my recommendation was consistently ignored. As I recall, meaningful naming was apparently less important than the remote possibility that a video snippet or audio clip would somehow prove offensive.

Although cybercartography demands new paradigms, cybermaps can benefit significantly from strategies described over a decade ago for making dynamic maps more engaging and broadly informative. Particularly promising are geographic brushing, graphic scripts, and user profiles (Monmonier 1992). A simple form of geographic brushing (Monmonier 1989) is the dynamic two-category map with a movable cut-point, which the user can shift back and forth along the number line as he or she explores the distribution (Fig. 2.5). A graphic script could present viewers with a

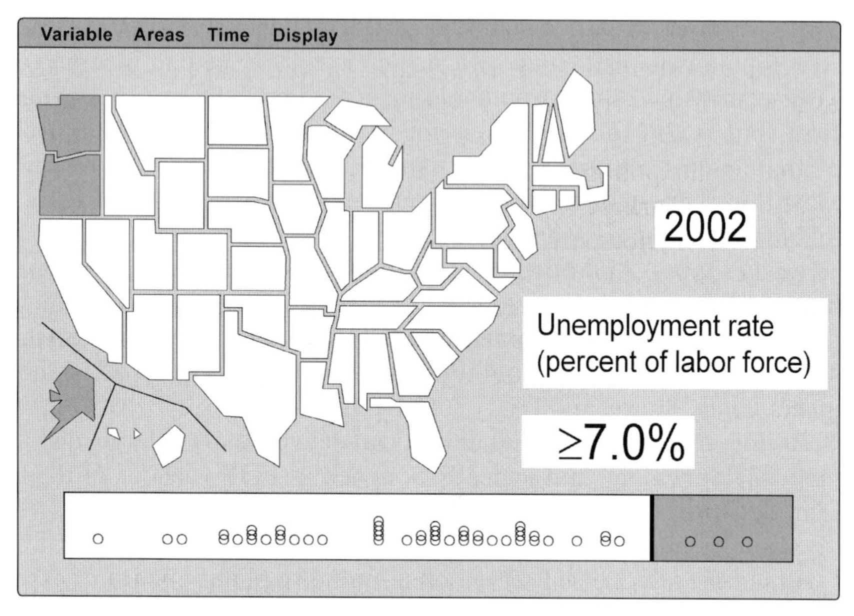

FIG. 2.5. A dynamic two-category choropleth map invites exploration of data with a movable cut-point, the current value of which (7.0%) is displayed in the numerical window above the right end of the number-line scatterplot. In this view data values above the 7.0% cut-point are highlighted on the map as well as on the number line.

logical sequence of maps either designed by human experts or fabricated by expert-system software to illustrate important multivariate relationships, as well as revealing interpretations of a single geospatial distribution. A typical script might offer a logical sequence of complementary views composed by varying category breaks and level of aggregation as well as by juxtaposing maps offering revealing reexpressions of the data or presenting intensity and magnitude measurements of the same phenomenon (DiBiase *et al.* 1992). And a user profile could automatically trigger the customized "my county compared to the rest of the state" display suggested in the *FactFinder* tutorial, as well as accommodate idiosyncratic viewer preferences such as a preferred color scheme (especially helpful for those with color impairments), or a sequence of zoomed views highlighting places where the viewer has lived or traveled. Since the graphics might need an explanation, an author-recorded or computer-generated voice-over could usefully explain the strategy or point out salient details.

Although graphic scripts authored by alleged experts might take on a persuasive power, multiple graphic scripts could more than compensate by presenting viewers with diverse viewpoints. Moreover, instead of serving as the final word, a script might serve as a pump primer, designed to instigate serious exploration of a dataset through data mining as well as critical questioning of the data themselves (Andrienko *et al.* 2001). Revealing expert scripts might, for example, explore the consequences of uncertainty in the data or the political connotations, if any, of diverse map projections (Mulcahy and Clarke 2001; Aerts *et al.* 2003). At the risk of appearing sacrilegious, I cannot resist adding that scripting by experts might even yield one or more socially sensitive, ethically grounded displays constructed to address the question "What kind of map would Jesus make?" I say "one or more" because it is plausible that different well-informed and well-meaning experts could claim the moral high ground for markedly different maps.

6. Conclusion

Except for one-off printouts of downloaded maps, plain old map products (POMP) are likely to decline in number, if not prominence. Even so, high-end audiences like the users of *Mapping Census 2000* will continue to justify precision printing and attentive design. For more mundane applications in which immediate response is of paramount importance, especially GPS navigation systems and Internet street maps, users willingly tolerate aesthetically crude displays that might once have been shunned or ridiculed. A troubling paradox of cybercartography is that dynamic display, however powerful a tool for analysis and fact-finding, might well lower standards of graphic excellence.

Momentarily at least, bandwidth and the computational power of map-servers are prime impediments to a more fulfilling and aesthetically admirable cybercartography. No less significant is the vision of cartographic administrators who control map-oriented Web sites and the knowledge and prowess of the contractors and vendors who supply them. Although the Internet and high-interaction computing have revolutionized map use, those in charge ought not overlook proven design principles like those listed in Table 2.1, or newer dynamic strategies like those listed in Table 2.2, introduced in the literature over a decade ago but largely ignored by software developers. Since governments and other organizations that establish standards for geospatial data are likely to remain focused on data quality and interoperability, it is up to educators and authors of critical reviews to heighten public awareness of graphic quality.

TABLE 2.1
Conventional Design Strategies Worth Implementing in Cybercartography

'Bigger means more' (graduated circles and squares for count data)
'Darker means more intense' (light-to-dark sequences on choropleth maps)
'Darker means more deviant' (diverging color schemes)
'Different hues signify different types' (diverging color schemes, qualitative data)
'More pronounced shifts in hue reflect more pronounced shifts in rate' (choropleth maps)
'White or neutral gray identifies a nonparticipating area or one with no data or a statistically questionable rate'
Inherently meaningful breaks (choropleth maps)
Complementary maps at different levels of aggregation (choropleth maps)

TABLE 2.2
Dynamic Design Strategies Awating Implementation in Cybercartography

Geographic brushing, including interactive two-category choropleth maps
Graphic scripts
User profiles
Multiple expert designs
Recorded or machine-generated voice-overs

References

Aerts, J. C. J. H., K. C. Clarke and A. D. Keuper (2003) "Testing popular visualization techniques for representing model uncertainty", *Cartography and Geographic Information Science*, Vol. 30, pp. 249–261.

Andrienko, N., G. Andrienko, A. Savinov, H. Voss and D. Wettschereck (2001) "Exploratory analysis of spatial data using interactive maps and data mining", *Cartography and Geographic Information Science*, Vol. 28, pp. 151–165.

Arensman, R. (1999) "Analog renaissance", *Electronic Business*, Vol. 25, No. 12, pp 104–106.

Armstrong, M., N. Xiao and D. A. Bennett (2003) "Using genetic algorithms to create multicriteria class intervals for choropleth maps", *Annals of the Association of American Geographers*, Vol. 93, pp. 595–623.

Bertin, J. (1983) *Semiology of Graphics: Diagrams, Networks, Maps* (translated by William J. Berg), University of Wisconsin Press, Madison, pp. 60–97.

Brewer, C. A. (1997) "Spectral schemes: controversial colour use on maps", *Cartography and Geographic Information Systems*, Vol. 24, No. 4, pp. 203–220.

Brewer, C. A. and T. A. Suchan (2001) *Mapping Census 2000: The Geography of US Diversity*, US Census Bureau, Census Special Reports, Series CENSR/01-1, US Government Printing Office, Washington, D.C.

Cloud, J. (2002) "American cartographic transformations during the Cold War", *Cartography and Geographic Information Science*, Vol. 29, pp. 261–282.

Cooke, D. F. (1995) "Sharing street centerline spatial databases", in Onsrud, Harlan J. and Gerard Rushton (eds.), *Sharing Geographic Information*, Centre for Urban Policy Research, New Brunswick, NJ, pp. 363–376.

Dent, B. D. (1999) *Cartography: Thematic Map Design*, 5th edn, WCB/McGraw-Hill, Boston, pp. 140–142.

DiBiase, D., A. M. MacEachren; J. B. Krygier and C. Reeves (1992) "Animation and the role of map design" in *Scientific Visualization, Cartography and Geographic Information Systems*, Vol. 19, pp. 201–214, 265–266.

Flaherty, J. (2000) "Where phone service is way above average, and competitive", *New York Times*, 9 March, C1.

Government of British Columbia, *BC Geographical Names Web site*, http://srmwww.gov.bc.ca/bcnames/

Laffont, J. J. and J. Tirole (2000) *Competition in Telecommunications*, MIT Press, Cambridge, Massachusetts.

Lange, A. and C. Gilbert (1999) "Using GPS for GIS data capture", in Paul A. Longley, Michael F. Goodchild, David J. Maguire, and David W. Rhind (eds.), *Geographic Information Systems, Vol. 1: Principles and Technical Issues*, John Wiley and Sons, New York, pp. 467–476.

Marsden, L. E. (1960) "How the national map accuracy standards were developed", *Surveying and Mapping*, Vol. 20, pp. 427–439.

Microsoft Corporation, *Encarta Home Page*. http://www.microsoft.com/products/encarta/default.mspx

Monmonier, M. (1985), *Technological Transition in Cartography*, University of Wisconsin Press, Madison.

Monmonier, M. (1989) "Geographic brushing: enhancing exploratory analysis of the scatterplot matrix", *Geographical Analysis*, Vol. 21, pp. 81–84.

Monmonier, M. (1992) "Authoring graphic scripts: experiences and principles", *Cartography and Geographic Information Systems*, Vol. 19, pp. 247–260, 272.

Monmonier, M. (2002) "Targets, tolerances, and testing: taking aim at cartographic accuracy", *Mercator's World*, Vol. 7, No. 4, pp. 52–54.

Moon, J., J. J. Richardson, and P. Smart (1986) "The privatisation of British telecom: a case study of the extended process of legislation", *European Journal of Political Research*, Vol. 14, pp. 339–355.

Mulcahy, K. A. and K. C. Clarke (2001) "Symbolization of map projection distortion: a review", *Cartography and Geographic Information Science*, Vol. 28, pp. 167–181.

National Research Council (2001) *National Spatial Data Infrastructure Partnership Programs: Rethinking the Focus*, National Academy Press, Washington, DC.

Robinson, A., R. Sale and J. Morrison (1978) *Elements of Cartography*, 4th edn, John Wiley and Sons, New York, pp. 284–285.

Sondheim, M., K. Gardels, and K. Buehler (1999) "GIS interoperability", in Paul A. Longley, Michael F. Goodchild, David J. Maguire and David W. Rhind (eds.), *Geographic Information Systems*, Vol. 1, *Principles and Technical Issues*, John Wiley and Sons, New York, pp. 347–358.

Taylor, D. R. F. (2003) "The concept of cybercartography", in Peterson, M. P. (eds.), *Maps and the Internet*, Elsevier Science, Oxford, pp. 405–420.

Taylor, P. (2002) "Get me a line online: conventional phone services have remained unrivalled by net-based offerings - until now", *Financial Times*, December 24, p. 7.

Thompson, M. M. (1960) "A current view of the national map accuracy standards", *Surveying and Mapping*, Vol. 20, pp. 449–457.

Tobler, W. R. (1959) "Automation and cartography", *Geographical Review*, Vol. 49, pp. 526–534.

US Bureau of the Census, *FactFinder Home Page*, http://factfinder.census.gov/home/ saff/main.html?_lang=en

US Bureau of the Census (2001) *Master Address File/Topologically Integrated Geographic Encoding and Referencing (MAF/TIGER) Accuracy Improvement Project*, Solicitation no. 52-SOBC-2-00005, issued 16 November (2001) online at http://www.census.gov/geo/mod/ rfpfeb5.doc, accessed 26 December (2003).

US Bureau of the Census (2002) *Harris Corporation Awarded Contract for US Census Bureau's MAF/TIGER Accuracy Improvement Project*, 26 June (2002) accessed 26 December (2003) from http://wwwS.census.gov/geo/mod/ maftiger.html

US Bureau of the Census (2003) *Background Information on the MAF/TIGER Accuracy Improvement Project (MTAIP) and Governmental Partnerships,* accessed 26 December (2003) from http://www.census.gov/geo/mod/ backgrnd.html

USES, Geological Survey's Geographic Names Information System (GNIS), http://geonames.usgs.gov/

Veregin, H. (1999) "Data quality parameters", in Paul A. Longley, Michael F. Goodchild, David J. Maguire and David W. Rhind (eds.), *Geographic Information Systems*, Vol. 1: *Principles and Technical Issues*, John Wiley and Sons, New York, pp.177–189.

CHAPTER 3

Exploring the Concept of Cybercartography Using the Holonic Tenets of Integral Theory*

BRIAN G. EDDY

Geomatics and Cartographic Research Centre (GCRC)
Department of Geography and Environmental Studies,
Carleton University,
Ottawa, Ontario, Canada

D. R. FRASER TAYLOR

Geomatics and Cartographic Research Centre (GCRC)
and Distinguished Research Professor in International Affairs,
Geography and Environmental Studies,
Carleton University,
Ottawa, Ontario, Canada

Abstract

The concept of cybercartography draws on theory from a variety of disciplines including the physical and human sciences, geomatics sciences and technology, collaborative research, and the humanities. Integration of theory across disciplines is required in such approach. This chapter examines cybercartography through a transdisciplinary lens of "integral theory" which uses a set of tenets describing "whole–part" relationships, or "holons" as a basis for theoretical integration and synthesis. Central to this approach is an examination of what we mean by "information," in particular, with its relationship to data, knowledge, and meaning. Extending conventional user–computer

*This chapter is based on a PhD comprehensive paper presented as part of the lead author's PhD program at Carleton University, Ottawa (ON), Canada

oriented concepts, a "Cybercartographic Human Interface" (CHI) model is proposed.

1. Introduction – Cybercartography in the Information Era

An important aspect of cybercartography is how people and society can learn more effectively about the world using geospatial information. In Chapter 1 it was argued that, although cybercartography is different from Geographic Information System (GIS), it includes GIS as an integral part of the cybercartographic paradigm. This chapter considers cybercartography in relationship with the emerging trends in geomatics and geographical information processing, and related technologies and infrastructures. The important historical relationship between geography and cartography is extended by situating these disciplines along a stream of information flow. Such an examination requires a closer look at some of the characteristics of "information."

From an information perspective, we view the role of geography as the theory and practice of studying the world and the generation of numerous types of geographical information. Cartography looks at ways of presenting information about the world – primarily in the form of maps augmented by other forms – with the aim to communicate geographical information to an intended user base. This connection is summarized in Fig. 3.1, which situates the geographer and cartographer as "facilitators" of information flow between the world and the user base (Fig. 3.1).

One of the main distinctions of cybercartography is how cyber technology affects the dynamics of this general information flow pattern. As Peterson (1995) argues, the conventional singular, linear, monological flow is now

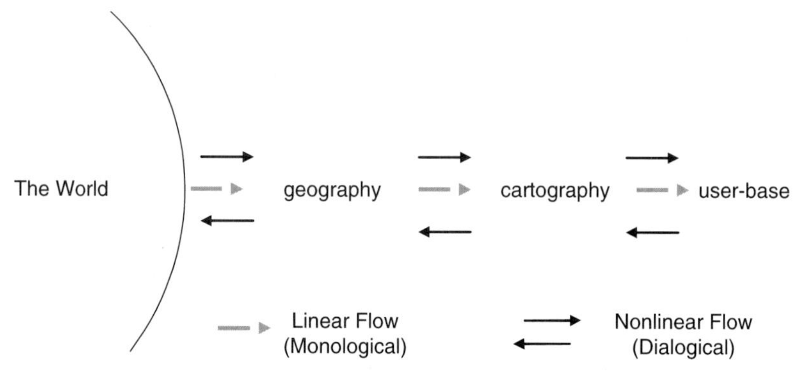

FIG. 3.1. Geography, cartography, and information flow.

seen as nonlinear and dialogical. Users as learners also become part of the process of mapmaking, and the conventional roles of geographers and cartographers, as "construction workers" of information about the world, also take the additional role of learner in the process. This interaction is part of the socio-cybernetic feedback process discussed in Chapter 1 and by Martinez and Reyes later in this volume (Chapters 4–6).

One aim of cybercartography is to provide better information about the world, and this is now both enhanced and challenged by rapidly emerging cyber technologies. There are more maps being generated today than ever in history, and as argued in Chapter 1, many of these maps are not being formally created by cartographers. The result has been an increase in the quantity of maps and geographical information on the Web, but says little of the "quality" of information and understanding about the world attained by users. There is certainly greater access to data and knowledge, but one must also ask questions as to whether current practices and modes of delivery are helping to generate better meaning and understanding of the world. How might cybermaps serve to augment "meaning" as a result of dialogical information processing, and how does cybercartography enhance this process?

One way of approaching this question is to examine cybercartography from an information theory perspective. In doing so, we focus on the complexities of the information flow among the various levels and dimensions among geographers, cartographers, and other participants in the mapmaking process. Examining the characteristics of information flow presents several challenges. What is "information"? Is information something different than data, knowledge, facts, or interpretations? How is information presented in the form of a map different than that of a table, a narrative, or a video?

One of the difficulties in exploring these questions is that, there does not seem to be a consensus on what "information" actually is. Dictionaries often quote colloquial and circular reasoning (e.g. Lexicon 1987), and philosophical examinations, while highlighting important aspects that must be taken into account do not lead to a clear definition of the term (Winder *et al.* 1997; Adams 2001). It is argued that an understanding of the various concepts of information is required in order to adequately characterize the important dialogical mapmaking process. What is presented here is intended to initiate a dialogue with respect to how geographers, cartographers, and people from many other disciplines place themselves in the emerging information age or information economy, and more specifically within the emerging cybercartographic paradigm.

Conventionally, geographers and cartographers may not have regarded themselves as information processors or information professionals, with the exception of those who specialize in GIS and geomatics. A scan of various geography, GIS, and cartography textbooks indicates that the concept of information is rarely, if ever, examined (Freeman 1949; Shaw 1965; Lock 1976; Ullman 1980; Fisher 1982; Haggett 1983; Szegö 1987; Tomlin 1990; Wood and Fels 1992; Medyckyj-Scott and Hearnshaw 1993; Robinson 1995; Small and Witherick 1995; De Blij and Muller 1997; ESRI Inc. 1997; Leung 1997; Burrough and MacDonnell 1998; Raper 2000; Clarke 2001; Longley *et al.* 2001; Chrisman 2002). In few cases, concepts such as "data," "information," and "knowledge," if not simply taken for granted, are often defined vaguely, and in a circular fashion. If cybercartography is to be part of an "information/analytical package" (Taylor 2003), then these relationships must be more fully considered.

The need for better understanding of information cannot be understated. Confusion over terminology creates misunderstanding and confusion. It is, therefore, necessary to explore a working model of information that may be used to better suit commonly used terms, such as data, knowledge, facts, interpretations, and meanings and their relative influence in the cybercartographic process. Such a model would need to be built from an integration of the approaches, taken to the concept of information by a variety of disciplines. Drawing from multiple disciplines can be challenging, both in practical and theoretical terms, as Lauriault and Taylor discuss later in this volume (Chapter 8). How multiple disciplines or perspectives intersect or interact can take on a number of forms, including multidisciplinarity, cross-disciplinarity to interdisciplinarity and transdisciplinarity (these concepts are defined in Chapter 8 of this volume).

This chapter explores the concept of information in cybercartography by using "Integral Theory," which offers itself as a "transdisciplinary" framework for integration and synthesis across disciplines. The chapter first introduces some of the basic elements of integral theory, then examines how the concept of information can be explained, and then examines information theories from a variety of disciplines to explore how these may be integrated. Finally, this chapter presents a synthesis of author's analysis in the form of a Cybercartographic Human Interface (CHI) model, a working model that includes consideration of the most important elements required in considering information flow in cybercartographic applications.

2. Integral Theory and the Holonic Tenets

Integral Theory was developed by Ken Wilber (1995, 1996, 2000) during several decades of research on ecology and consciousness studies. A comprehensive understanding of this subject necessarily involves reconciling numerous fields of knowledge from the physical and social sciences, psychology and anthropology, cultural studies, philosophy, aesthetic, and the world's spiritual traditions. Integral theory is both voluminous and complex, and a full consideration of its depth is beyond the scope of this chapter. For an outline of the application of integral theory to geography, the reader is referred to Eddy (2005). This section presents some basic elements of integral theory in relation to its application to cybercartography.

Wilber argues that the first step necessary in integrating knowledge from different fields is to consider each as having some important contributions to make in its own right, and that all knowledge domains "emerge" as a result of the necessity of perspective. In the integral approach, integrating knowledge across disciplinary boundaries starts with examining lines of general agreement and focus, which may sometimes reveal patterns that individual perspectives might not have otherwise captured on their own. Using Arthur Koestler's (1967) conceptualization of "holons" (or whole/parts) and "holarchies," Integral Theory examines where potential interfaces occur by presenting views of the world (ontologies), and knowledge of the world (epistemologies), as series of nested "whole–part" relationships. It suggests that anything we study is viewed in whole–part relational terms, and is also understood and expressed in whole–part terms. Koestler originally proposed a set of general rules or guidelines that characterize whole–part relationships. Wilber (1995) built upon these, added a few of his own, and summarized them as the "holonic tenets" (Table 3.1).

To complement the 20 tenets, Fig. 3.2 illustrates a simplified model of the base theoretical framework used in integral theory. The following sections will examine information from an ecological perspective, and this version of the framework highlights some of the required elements. Specifically, in terms of holonic elements, this model emphasizes the very broad holonic relationships among matter, life, and mind maintaining that all information contains some combination of these basic elements. According to Wilber, "there are also different 'dimensions' of these elements that must be taken into account." They are summarized in Fig. 3.2 as "quadrants" that correspond with the intersections among: (1) individual and (2) collective dimensions, combined with (3) exterior (physical/biophysical, societal), and (4) interior (experiential/cultural) dimensions or aspects of reality. The four quadrants are sometimes used to differentiate intention and behavior in individuals (the interior and exterior dimensions of the upper quadrants), and culture and society in the collective sphere (the interior and exterior dimensions in the lower quadrants) (Fig. 3.2).

TABLE 3.1
The Holonic Tenets

1. Reality is composed, not of 'parts,' nor of 'wholes,' but of holons (whole/parts)
2. Holons display 4 fundamental capacities:
   (a) Self-preservation (wholeness/agency)
   (b) Self-adaptation (partness/communion)
   (c) Self-transcendence
   (d) Self-dissolution
3. Holons emerge
4. Holons emerge holarchically
5. Each emergent holon transcends but includes its predecessors
6. The lower sets the possibilities of the higher; the higher sets the probabilities of the lower
7. The number of levels which a hierarchy comprises determines whether it is 'shallow' or 'deep'; and the number of holons on any given level we shall call its span
8. Each successive level of evolution produces greater depth and less span
9. Destroy any type of holon, and you will destroy all of the holons above it and none of the holons below it
10. Holarchies co-evolve
11. The micro is in relational exchange with the macro at all levels of its depth
12. Evolution has directionality:
    (a) Increasing complexity
    (b) Increasing differentiation/integration
    (c) Increasing organization/structuration
    (d) Increasing relative autonomy
    (e) Increasing telos

This main template of Integral Theory is referred to as an "All-Quadrants, All-Levels" (AQAL) framework. The quadrants are used to situate different aspects of our perceived reality, as embedded in our languages across cultures, and are also used to help situate many methods for studying the world. There are several ways of studying and understanding the intentions and behaviors of individuals, culture, and society, because each

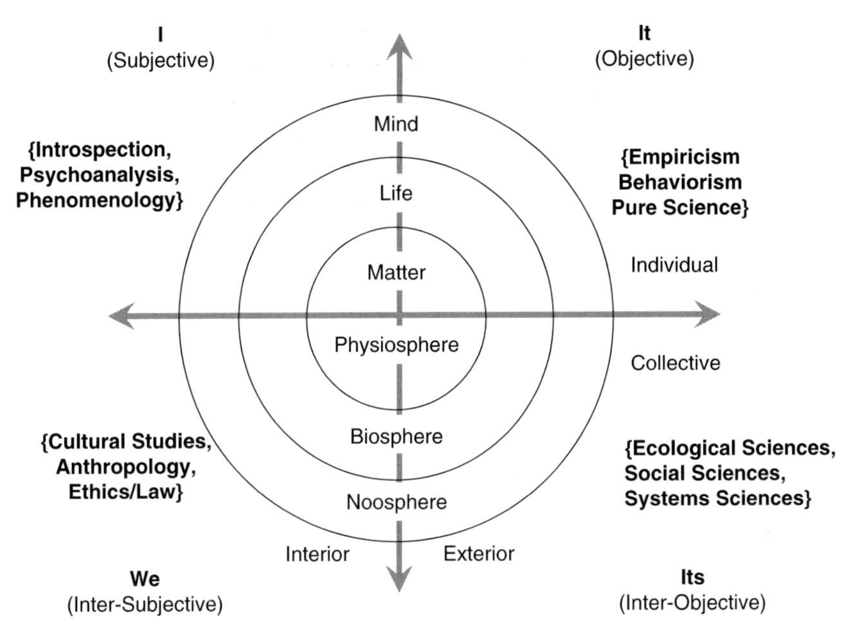

FIG. 3.2. All-Quadrants, All-Levels (AQAL) framework used in Integral Theory. (Modified after Wilber, 1995, Figure 5.1, pg. 198.)

of these dimensions occupies distinctively different domains that require specific methods of inquiry. In the upper right quadrant, we have empirical and behavioral sciences that study individual, physical, and biophysical exterior realities. In the lower right quadrant, we can draw from ecological, social, and systems sciences while examining physical things in their collective dimension. In the upper left quadrant, we make room for disciplines that consider the inner person, from phenomenology to psychoanalytic theory, to aesthetics or perspectives. In the lower left quadrant, the study of collective interior spaces falls within disciplines, such as cultural studies, cultural anthropology, ethics, law, linguistics, and various approaches to collaborative inquiry.

Wilber describes the AQAL model in common cross-cultural language as representing ways of orienting 1st, 2nd, and 3rd person perspectives corresponding with I, We, and It. In other areas, he summarizes the AQAL model in terms of Art/Aesthetics (I), Morals/Ethics (We), and Science (It/Its), or as Self, Culture, and Nature (see Wilber 1996 for an elaboration on these parallels). It is this facet of integral theory that provides a transdisciplinary framework and language to fill various information gaps drawing on various disciplines as required. Integral theory suggests that in order to be comprehensive, it is important to take "All-Quadrants, All-Levels" into account as best as possible. If any quadrant or level is missing from an approach, then we are likely leaving out something very important – or "integral" – to the reality we are studying. These basic theoretical elements of the integral map will be expanded upon in later sections. Now, the chapter focuses on how geographical information, and its representation and communication can be examined within the integral framework.

3. Understanding Information – An Integral Perspective

An integral approach takes as broad a perspective as possible, taking into account all quadrants and all levels (to which other elements may be added). It is assumed that information is intrinsically holonic. Every piece of information, or any information event or element can be said to be holonic, and its emergence follows the holonic tenets. For example, a letter (as a symbol) is a part of a word, which is part of a sentence, which is part of a paragraph, and so on. In geography, place is part of a region, which is part of a nation, which is part of the world. But we are not only concerned with the things we see visually or physically. There is also an "ephemeral" dimension of our perceived reality that sets the "context" within which information "content" is situated. This suggests that information be placed in a holonic "content–context" relation (Fig. 3.3).

In one sense, the "content" can be touched, seen, or sensed because it is "bounded" in some way. The "context," however is ephemeral, fluid, or "unbounded." Some of the most significant challenges in information processing occur at the "interface" between the content and context, or what is considered here as various points of "contact." The three elements of content, context, and contact (the 3-Cs) are integral to the concept of cybercartography as originally suggested by Eddy (2002) and elaborated in Taylor (2003). To elaborate their intrinsic characteristics and further explore their potential relevance to cybercartography, it is necessary to look at different "types" of holons as treated in integral theory.

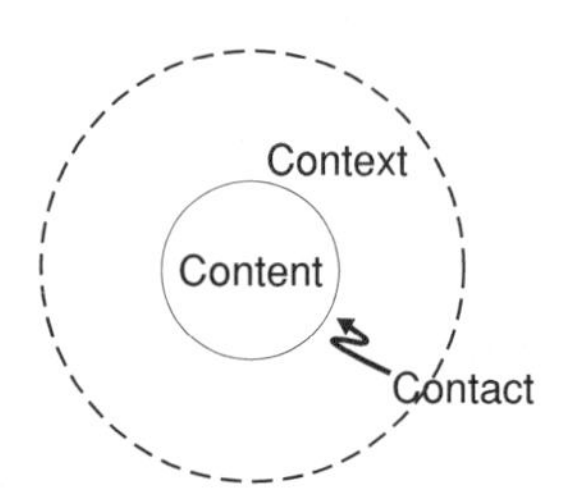

FIG. 3.3. Content, context, and contact as a holonic relation.

4. Holon Typology

Although integral theory considers everything to be holonic, not all holons share the same characteristics. Each AQAL matrix is considered to be some mixed combination of holonic interaction among three general holonic typologies. The integral terms used include: (1) Sentient and Non-sentient, (2) Individual and Social, and (3) Artifact and Heap. Sentient holons are those that emerge above the matter–life boundary, generally corresponding to "perceptive, living creatures." Non-sentient holons are those that occur below the matter–life boundary. From an ecological perspective, one of the distinctive differences between sentient and non-sentient holons is that sentient holons (living creatures) have the capacity for "perception" and "adaptation," and in the process, have developed additional capacities for both sensing and registering environmental conditions, and changing environmental conditions as part of the adaptation process (e.g. see Wicken 1987 for an elaboration of this ecological perspective).

The distinction between individual and social holons, in this context, is that an individual holon has a "dominant monad," and has some form of discernible physical boundary (i.e. an animal or a person) (Wilber 2002). A social holon does not have a dominant monad, but rather, is characterized by a "regnant-nexus" of intersubjective (internal) and interobjective (external) interactivity among holons (e.g. a flock of geese or human societies). Artifacts can be either sentient or non-sentient, or "hybrid" holons that are created by sentient holons to provide some form of extended functional utility (e.g. a beaver dam, an automobile, an information system, or biological weapons). By contrast, a "heap" is any holon that is progressively undergoing self-dissolution (Tenet 2d), losing its intended functional integrity, or innate drive for higher telos. Human holonic artifacts are often used to mark eras in the development of human societies from hunter–gatherer, to horticultural, to agrarian, to industrial, to information societies (in the lower-right quadrant of the AQAL framework). In all cases, artifacts are created to serve or extend some human function, and in many ways have augmented our ability to adapt to environments that would not otherwise be possible. Here, we focus specifically on the use of "Information and Communication Media" (ICM) as material holonic artifacts of human information processing and communication.

5. Information and Communication Media (ICM)

Information and Communication Media (ICM) emerged in various forms throughout history, from stone writings and pictures, to the development of languages and script, to books and libraries, to radio and television, to advanced computer networks. Each new form builds upon the preceding forms, which is stated in the holonic tenets as the lower (or previous) sets the possibilities of the higher (or latter). Today's computer and telecommunication networks are the most advanced form of ICM development, not only affecting how information is being captured about us and our environment, but also the rate and quality of information that is communicated with each other. Whereas most artifacts of the industrial era serve to function as extensions of human needs for "mechanical" work (corresponding with matter and the body in Fig. 3.2), ICM extend the function of 'mental' work (corresponding with the mind or the noosphere in Fig. 3.2) (Stonier 1997). From an integral perspective, such a distinction marks a boundary between the body and the mind. Wilber has argued that this distinction marks a shift from the modern industrial era, to the post-modern information era affecting economy, society, and culture in all aspects. Our use of ICM is essentially "cybernetic," and this has significant implications for the fields of geography, cartography, and geomatics.

ICM serve as an extended storage medium for both short-term and long-term memory, and partly a communication medium that significantly reduces the space, time, and effort required for transmission of information among people. They also function as new environmental phenomena unto themselves, enhancing human imagination, and further reinforcing cognitive development and other aspects of human potential, complexity, and interaction. The emergence of modern and postmodern forms of ICM raises a number of important questions. Although mechanical artifacts serve well to extend human mechanical work functions, to what extent do ICM serve their intended function? Can the fullness of the human mind and experience be replicated in, or by, ICM? To what extent do ICM contribute and redirect the human mind and experience (from alternative pathways that might have otherwise occurred)?

These types of questions are central to various philosophical debates in information theory, such as whether artificial intelligence will ever supersede human intelligence (depending, of course, on how we define intelligence). Although such questions are not primary concern here in this chapter, some aspects of these questions are relevant for developing a working model of information, and especially for addressing human expectations of ICM in Human–Computer Interaction (HCI). This is discussed in later sections of the chapter, but this chapter now focuses on an examination of how integral theory supports the view that ICM are constructed to mimic, in many ways, the human mind.

6. Vertical Levels of Information Processing

If we extend the ecological view of information processing, all "incoming and outgoing" multisensory, cognitive, and behavioral activity can be considered as a composite of information processing. All humans and all living creatures are continually processing information received from their environments, and react and behave in response to the information they receive. Figure 3.4 summarizes several basic elements involved in information processing from a vertical (or individual) holonic perspective (modified after Eddy *et al.* 2003). In transdisciplinary terms – we look at how our common language references notions, such as in/out, up/down, top/bottom, higher/lower, or before/after in terms of holonic relations, among information processing elements (Fig. 3.4).

Figure 3.4 illustrates some of these basic relationships in terms of data, knowledge, and meaning. The primary upward flow of information is regarded as "data," which is most commonly defined as raw, sensory-based, factual information. Different types of data (not limited to quantitative or scientific data) can be captured from each of the four quadrants by using various methods, such as those outlined in Fig. 3.2. Retention of all past and current data reconciled in various cognitive capacities, is

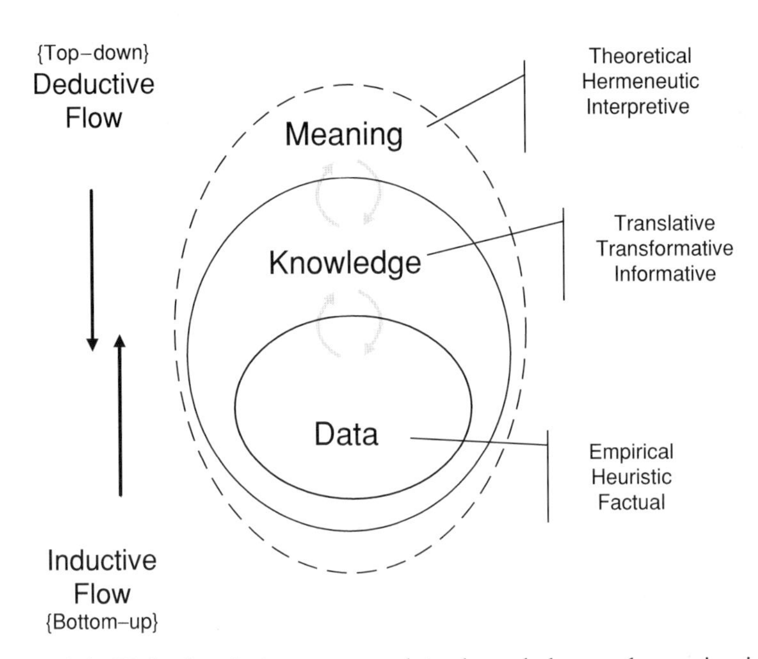

FIG. 3.4. Holonic relations among data, knowledge, and meaning in information. (Modified after Eddy *et al.* (2003).)

considered as "knowledge." Since all data is considered here as incoming (from any of the four quadrants), it is illustrated here as the "bottom–up" flow (from a holonic perspective). Data that have been retained and previously reconciled with experience as knowledge, exists a priori to any current incoming data stream. Prior knowledge works "upon" incoming data, and therefore is seen to have a "top–down" influence on the incoming data stream. Updating of knowledge by new data is considered here as posteriori affect. In this view, "deductive" processes are defined as prior knowledge working upon data (top–down), whereas "inductive" processes are data working on, and updating knowledge (bottom–up). All human cognitive functioning involves some complex mixture of both inductive and deductive processing, using corresponding data and knowledge, respectively (see Copi 1982 on inductive and deductive reasoning).

There is also a third general level, or capacity, in all information that goes beyond both data and knowledge that must also be taken into account. For the purposes of this analysis, we will refer to this as the existential level of "Meaning." Meaning can be used in a definitive sense to denote or ascribe meanings of symbols within a semiotic stream, and these are necessary for our cognitive capacities to function. More significantly, it is used here in a connotative sense, wherein it is that which gives resonance and reconciliation to both incoming data and prior knowledge "relative to a given existential context" of experience (as a composite of all three levels). Since knowledge supercedes data, meaning supercedes both knowledge and data, and therefore depends upon knowledge and data. At the level of meaning, deductive flow is extended by one level, and the inductive flow (from knowledge to meaning) is also extended by one level.

This view outlines the depth or "all-levels" aspect of information flow and processing. Data is primarily sensory as all forms of information have material states of transfer. For example, reading of this text involves transfer of photons from the page to the eye. This sensory process is then extended through neurological capacities of the body through complex biochemical and biophysical processing, and then into the "mind." In the collective dimensions, information is embedded in corresponding physiospheric, biospheric, and noospheric forms. The emerging Internet, mass communications, and the media, may be considered as artifacts of a global noosphere, providing us with a new level of infrastructure on which the emerging information economy will function.

In considering Tenet 8, where "each level produces greater depth, and less span," we see the reflection of many ways of looking at the hierarchical dimension of information flow (a heterarchical dimension is discussed below). What this suggests, if looked at as a holonic relation, is that there are always more data elements than there are knowledge structures to act upon them. Subsequently, there are always many knowledge structures that are combined to form any given meaning. Looking at it from the top down, we see that any given meaning requires a synthesis of many knowledge elements, which in turn requires synthesis of many more data elements. Applying this model in the collective dimensions gives rise to a plurality of information processing pathways and contexts.

7. Horizontal Information Transfer and Exchange

The AQAL model illustrates a critically important aspect that, all singular (individual) information processing contexts do not occur in isolation. There is also a corresponding collective or social dimension that is simultaneously taking place. Whereas the above singular processing streams were treated holarchically as "bottom–up" and "top–down" processes in the vertical dimension of holarchy (the "hierarchical" component), the social, ephemeral dimension is treated as the horizontal dimension of inside–out and outside–in relations, referred to here as the "heterarchical" component. Although the general levels and information processing structures (of data, knowledge, and meaning) are available in all individuals, the specific contents and contexts of all individual holons are different due to unique circumstances in space, time, and place (the information environment). More complexity is added by the need to reconcile horizontal information exchange, invoked by differences of experience and perspective among individuals.

Integral theory uses the concept of "involution" as a top–down (or from higher to lower) pull on evolutionary processes (as derived from the philosophical perspectives of Teilhard de Chardin 1969). This is to say evolution (through bottom–up, inside–out holonic relations) proceeds by way of involution (top–down, outside–in holonic relations). More simply, in order for something to evolve (as an agency, or a whole unto itself), it must simultaneously "involve" (as in a communion, or a part of a social holon). Evolution can be seen as that which "ascends" hierarchically, and involution can be seen as that which "descends" heterarchically through horizontal exchanges among "agencies-in-communion." The convoluted mixing of individual contexts and circumstances are constituting the horizontal or relational aspect, as summarized by Tenet 11. Content–context relations can also be examined by using this tenet, wherein they carry intrinsic levels of depth and extrinsic extensions of span across social holonic space, and serve to qualify information for its authenticity (depth) and legitimacy (span). At this point, it is helpful to incorporate an understanding of "entropy" as it may relate to information processing.

8. Information Entropy

If individual contexts could be quantified, considering all dimensions of space, time, circumstance, or "place" of any group of individual contexts, then the number of "possibilities" of content–context relations expand exponentially, as the number of individuals involved increased together with the complexity of the levels within which information is processed. Here, the concept of "entropy" is useful as means to convey the intrinsic aspect of these relational possibilities. Wicken (1987) examines entropy as applied to information in the context of the evolution and emergence of complex living, self-adaptive systems. As Wicken's analysis reveals, "information entropy" is not always (nor necessarily) synonymous with disorder, randomness, or noise, but rather pertains to the "number of possibilities" available to any given micro–macro (or content–context) relationship. Our adaptation of this view is that entropy is not limited to measure complexity of information "content," but rather information "activity" will contain corresponding levels of complexity in relation to the depth and span of content–context relationships.

A generalized relationship can be stated as "higher" the information processing through the three general levels of data, knowledge, and meaning (or depth), and "wider" the number of potential contexts to which some information potentially applies or is used (the span), higher is the entropy of information. In this analysis, we are concerned with different orders of information entropy, associated with primary (data-oriented), secondary (knowledge-oriented), and tertiary (meaning-oriented) levels of a cybercartographic processing environment. This general understanding is used as a central feature in examining the concept of cybercartography from an information processing perspective. Essentially, we view cybercartography as aiming to address the higher entropy "meaning-oriented" levels of information processing and use in society, where the same content (map or map elements) can have virtually unlimited numbers of potential contexts (users, interpretations, or meanings derived from the content). This level necessarily transcends and includes the many sciences, technologies, and disciplines engaged in the data and knowledge levels of an information process. It is, therefore, useful to construct a synthetic model of information processing that helps to differentiate these levels and domains and bring clarity to the types of integration necessary in any given cybercartographic environment. Before doing so, this chapter summarizes some of the main elements from this integral analysis.

9. Summary of Integral Elements

The preceding sections introduced some of the basic elements of Integral Theory and examined how it can be applied to improve the understanding of "information," in particular, within the context of geographical information and communication in cybercartography. This integral perspective leads to the following general statements:

- Information is intrinsically holonic, in both individual and collective capacities, and with both interior (personal/cultural) and exterior (expressive/societal) forms.
- Any complete information package includes three general levels of "data," "knowledge," and "meaning." The intersection of these three levels becomes increasingly complex as the number of individuals that an information package aims to serve increases, corresponding to higher degrees of information entropy.
- Information is processed in sequential, hierarchical, and evolutionary patterns in response to nonsequential, heterarchical, and involutionary needs and demands.
- ICM aim to mimic human information processing capacities, but in themselves are not (presently) "sentient" processing environments. Human–Computer Interaction (HCI) is essentially an interaction between sentient and non-sentient ecological entities (people and technology), and this has a significant influence on how we approach human–computer interaction.
- All information processing occurs within broad cultural–societal contexts, within which individuals are situated, and therefore have differences in perspective and user context.
- Information systems and maps "emerge" in response to needs within the context of human affairs.
- Information artifacts may emerge as some combination of bottom-up (or science/technology "push/supply") activities and top–down (or market "pull/demand") requirements. The aim of cybercartography is to find an appropriate or optimal balance within this dynamic, nonlinear interaction. Here the perspective of second order socio-cybernetics as outlined in Chapter 1, and by Martinez and Reyes later in this volume is informative (Chapters 4–6).

Now this chapter turns to look at how these elements can be used to synthesize a number of multidisciplinary and interdisciplinary perspectives on information and cartographic theory.

10. Integration and Synthesis

This analysis highlights four models in Fig. 3.5 that cover the disciplines of Critical Theory and Science Studies (Latour 1999), Information and Organizational Management (Reeve and Petch 1999), Systems Engineering and Ecological Interface Design (Rasmussen 1999), and Visualization and Cartography (Lindholm and Sarjakowski 1994) (Fig. 3.5).

Latour's work (and others like it) provides an important starting point to allow questions to be asked with respect to the societal needs and circumstances in which cybercartographic applications emerge. His relationship between "dictae" and "modis" is considered as one form of content–context relationship: the dictae representing the content of information, and the modis, the societal circumstance within which the content emerges.

Extending this model to fit more closely with a systems–environment level, another form of content–context relation is evident in Ecological Interface Design (EID). More immediate environment of a given information system can be considered a "micro-modis" nested within a broader societal context. Encompassing a holistic approach to system design by considering the electromechanical constraints of a system and the human environment, EID makes use of an "abstraction hierarchy." Bottom–up and top–down interactive flows are considered in relating low-level, physical objects, and specific functions and processes (considered here as non-sentient ICM and their information contents), to high level, more abstract process-oriented environments (sentient human environments). The organizational pyramid model presented by Reeve and Petch (1999) extends many of these aspects of information processing to an organizational scale. As with the EID model, their model also emphasizes top-down and bottom-up dimensions of information flow.

Lindholm and Sarjakoski (1994) present interactive flows more explicitly as parallel distinctions between the individual user (a sentient processing holon) and an information system (a non-sentient holon). Their model reveals a more explicit linkage between system components and information processing stages as mirroring similar information processes and capacities in individuals. In this scheme, primary measurements and observations about the world, as represented in raw data, extend the human sensory capacity to an ICM form. Structured data and models/theories

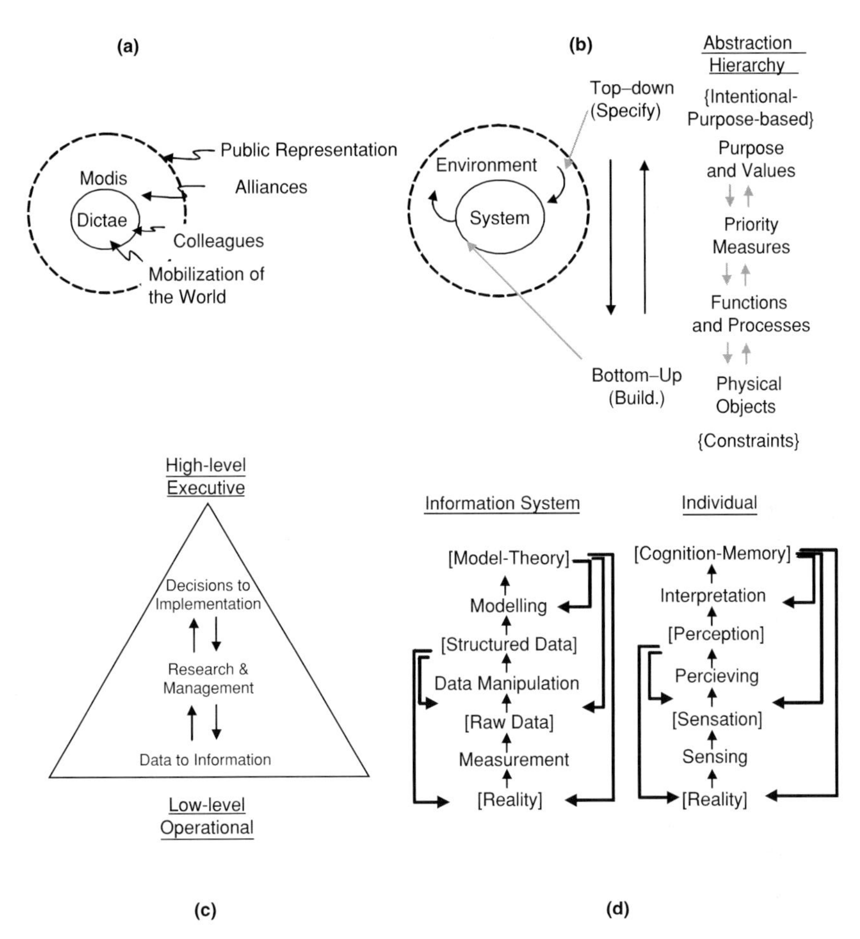

FIG.3.5. Multidisciplinary perspectives on information flow and use adapted from (a) Latour (1999) Science Studies, (b) Rasmussen (1999), Ecological Interface Design, (c) Reeve and Petch (1999), Information and Organizational Theory, and (d) Lindholm and Sarjakoski (1994), Visualization and Cartography.

embedded in ICM extend higher level human cognitive functions. Note also their explicit reference to top–down and bottom–up information flow and feedback functions.

The inclusion of these particular models highlights a number of general aspects that we believe should be considered as a minimum in any cybercartographic application. We used these four models to highlight different scales as follows:

- ***Societal Level*** – How and why cybercartographic applications and maps are used in society, and to draw from various schools of critical theory, science studies, and postmodern studies of the emerging information society, including socio-cybernetics.

- ***Organization Level*** – How information is used at an individual organizational scale, and the relations among mandated organizations, all of which are nested within broader societal contexts.
- ***Information Processing Level*** – How information is structured, stored, processed, and managed internally within a cybercartographic environment in ways that mimic human information processing, nested within a variety of organizational contexts.
- ***Infrastructure Level*** – Designing, engineering, and maintaining systems to adapt to ever changing human environments; again, nested within and supporting all of the above levels.

To provide additional support for this approach, we propose a synthesis of these models that can be used as a guideline or "map" for finding our way through a cybercartographic environment. This constitutes one of the directional signposts for the paradigm of cybercartography as discussed in Chapter 1.

11. A Cybercartographic Human Interface (CHI) Model

A synthetic "Cybercartographic Human Interface" (CHI) model is illustrated in Fig. 3.6. This model combines some of the main elements of each model presented above, and follows the principles derived from an integral perspective on information flow and use. First, we outline three broad domains for consideration: 1) A Cyber Domain (or Cybercartographic or Content Processing Domain), 2) A Human Domain, and 3) An Interface Domain (Fig. 3.6)

Any ICM used in a cybercartographic context is considered to be a Cybercartographic Domain. The societal, organizational, and operational environments in which applications are used is considered the Human Domain, and the various points of direct interaction between people and the cybercartographic domain comprise the Interface Domain. In this model, we prefer to make an important distinction between "Human" and "User." Many computer-user models focus on normative approaches to understand user needs and behavior at the immediate interface level of human–computer interaction. Some of these approaches are described in subsequent chapters in this book such as that by Lindgaard and Brown (Chapter 9), Roberts *et al.* (Chapter 10), and Trbovich *et al.* (Chapter 11). Such studies are of great value to cybercartography, but we believe it is important to consider users in a broader societal or 'human' context

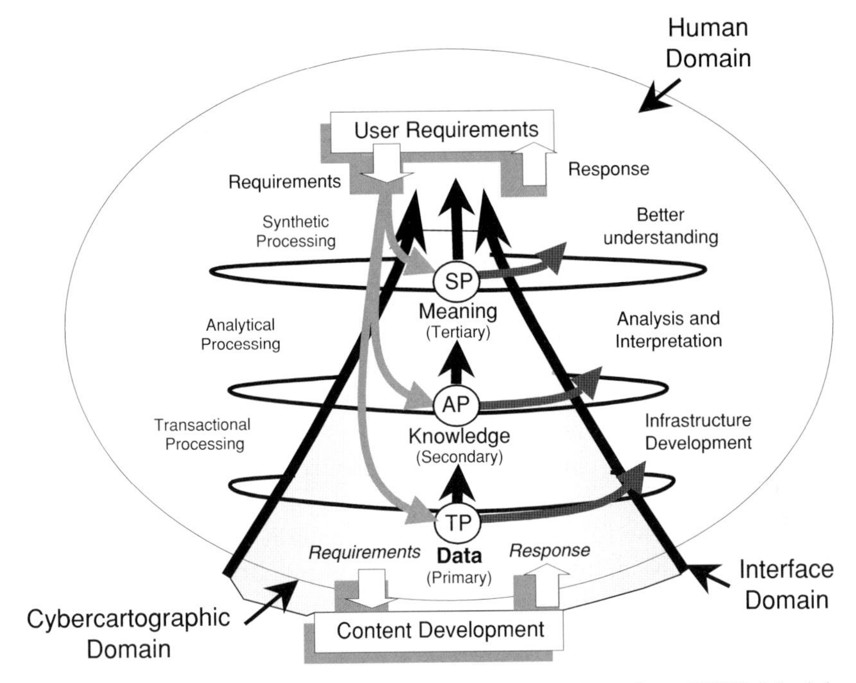

FIG. 3.6. A Synthetic Cybercartographic Human Interface (CHI) Model. Note: TP, AP, and SP refer to interfaces appropriate for transactional processing, analytical processing, and synthetic processing, respectively. See text and Table 3.2 for additional information.

(considering methods, such as critical and social theory of information, or broader philosophical and aesthetic perspectives). A socio-cybernetic approach to cybercartography as described by Martinez and Reyes in this volume (Chapters 4–6) is an example of the value of this broader societal and cultural context in creating cybercartographic products.

In addition, no distinction is made between "providers" and "users" in this model because of the dialogical dynamics involved in cybercartography. We view cybercartographic applications as constructed, maintained, and used by people who fill a variety of roles and levels of expertise at different times in the process. A programmer, for example, may eventually be a user of an application, and similarly, a high school student may add content to an application, or embed one application within their own. More significantly, it is important to differentiate and place emphasis on the human-sentient environment within which non-sentient ICM emerge. The characteristics of each of these three domains are fundamentally different, and often diametrically opposite in many ways. Some of the more significant challenges occur at the "interface" of the cyber and human domains. Interface designers must take into consideration both scientific and technological aspects of the cyber domain, as well as broader human and cultural context for which the content is intended.

Figure 3.3 depicts both vertical (hierarchical) and horizontal (heterarchical) dimensions as a series of successively nested content–context relationships along different levels of information processing. The three broad levels considered here (in Fig. 3.6) are: data-oriented (primary), knowledge-oriented (secondary), and meaning-oriented (tertiary) application environments. Elements from Lindholm and Sarjakoski's model, or other models of information processing may be used to provide additional detail. It is believed that most cybercartographic applications aim to support tertiary level requirements, but unlike conventional maps and cartographic systems, that remain detached from their primary and secondary information sources. A cybercartographic environment provides the possibility for maintaining access to multiple levels of information, supporting different levels of use for general public or general purpose use, to policy and scientific analysis, and to experts and organizations who collect and maintain primary information sources. The triangular information flow depicting the cyber domain also captures Taylor's (2003) three dimensions involved in cartographic visualization and communication, whereby the principal elements of cybercartography intersect in different ways at each level. The funnel in Fig. 3.6 may also be used to symbolize organizational and information pyramids by condensing larger volumes of primary data into smaller volumes of synthetic information packages at the tertiary level. An important differentiation is also made with respect to the different types of interfaces that correspond to different levels of information processing. Interfaces that are used by the general public or for educational purposes, have different requirements and operational environments than interfaces used in analytical or transactional levels. Table 3.2 summarizes additional characteristics for these three general levels.

Primary content that flows upward to a secondary level is often subject to the application of statistical or theoretical models, or at least, selection and processing criteria stipulated in part by some specific needs. This is equivalent to what Pulsifer has identified as "geographic" mediation in the cybercartographic process (Pulsifer and Taylor, Chapter 7). It is the deductive element of prior knowledge working upon primary data, which when implemented in a standardized spatial data infrastructure, involves selecting aspects of geographic data that are required for analysis and presentation. The retrieved data stream, in turn, updates the secondary level of knowledge, and this process is often iterative and heuristic. Secondary level content is considered to be primarily objective-based information set in a knowledge-oriented context.

TABLE 3.2

Characteristics of Cybercartographic Human Interface Levels

	Primary	Secondary	Tertiary
Inputs	**Internal:** Recorded observations and measurements that are of primary importance to the CUI context	**Internal:** Any output produced on the primary level, the selection of which is framed within the context of a given theoretical framework	**Internal:** Any output produced on the secondary level, the selection and presentation of which is framed according to its higher level meaning in a given context
	External: Outputs of other domains that are similar or related to the primary level of the internal CUI domain	**External:** Outputs of other domains that are similar or related to the secondary level of the internal CUI domain	**External:** Outputs of other domains that are similar or related to the tertiary level of the internal CUI domain
Processing and Mediation	**Transactional Processing:** Primarily those associated with validating and authenticating primary data, structuring, organizing, and management according to working data models	**Analytical Processing:** Analysis of patterns in primary data according to various theoretical and knowledge-based frameworks constructed from prior experience	**Synthetic Processing:** Reporting and presenting secondary information (and where necessary primary also), for specific contexts and intended use. Tertiary products provide meaning to an immediate context of decisions relating to real-world affairs
Outputs	**Internal:** Extracted data elements that are fed upward to the secondary level	**Internal:** Secondary, knowledge-based products that are fed upward to the tertiary level	**Internal:** Different types of secondary content that are meaningful to a particular context
	External: Outputs of primary data elements to other domains (for primary, secondary, or tertiary level input)	**External:** Secondary outputs to other domains (for primary, secondary, or tertiary level input)	**External:** Tertiary outputs to other domains (for primary, secondary, or tertiary level input)

(Continued)

TABLE 3.2
Characteristics of Cybercartographic Human Interface Levels (Continued)

	Primary	Secondary	Tertiary
Infrastructure	Generally more complex matrix of data recording instruments, with VLDB (very large database) capacities, restrictions, and highly specialized operations	Generally more localized matrix of analytical processing tools, with moderate (often temporary) storage capacities, less restricted, but still may be highly specialized operations depending on the types of analyses	Generally tools that are most commonly used for broad public consumption, and compatibility within a full range of possible end-user environments (Web browsers, office desktops, CD-ROM, personal communication devices, multimedia)
Types of Interfaces	Very technical and highly specialized depending on the primary data, online transactional processing (OLTP)	More analytical oriented tools, expert systems, or on-line analytical processing (OLAP)	More product/report oriented tools that allow ease of use, general purpose interfaces, regarded here as online synthetic processing (OLSP)
Visualization, Abstraction and Cartography	Primary control, calibration, abstraction as data symbolization and modeling, visualization, and representation of primary data. Cartography as a component of transactional processing/map-based data capture	Multiple forms of higher abstraction and scientific/analytical visualization, geographic and non-geographic, cartography as internal to GIS science	Visualization and abstraction, as part of a realization process and enacting of real-world context and use. Broader, and more immediate domain of cybercartography. GIS science as internal to cybercartography
Communication and User Base	Complex and specialized and often restricted and highly secured communication networks, limited technical user base	Localized and somewhat less restricted communication networks, limited scientific/analytical user base, often in clusters of specialized communities	Broad communication networks, multimedia, both public and private, potentially very large user base (i.e. public internet environments, mass media)
Examples	Primary geospatial data collection (field to computer, GIS/GPS input, remote sensing capture and calibration), different forms of geographic and non-geographic transactional processing	Specialized GIS modeling and applications (e.g. forestry, resources, environmental, human demographic modeling), policy and scenario analysis. Nongeographic knowledge-mapping (e.g. AQAL map)	Educational, mass communication, public education and information consumption. Day-to-day pragmatic use. Simplified, intuitive and easy-to-use cybermaps

Different theories and models can be applied to the same data, which often yield different results and interpretations, and this signifies the introduction of an increasing element of subjectivity in the processing stream. The choice of model or analysis is partly dependent upon the intended or desired application of the information output at the tertiary level. Tertiary level information aims to contextualize secondary content through the addition of "meaning" for specific contexts. The transition from secondary to tertiary level representation and synthesis is equivalent to Pulsifer's "cartographic" mediation (as discussed in Chapter 7 in this volume). This is often a very challenging boundary in the development of any information system, because tertiary level information carries higher degrees of entropy. But it is also the level to which aesthetic creativity and innovation can be applied more freely to make information more appealing and meaningful for an intended user base. As argued by Taylor in Chapter 1, cybercartography is both an art and a science and is qualitative as well as quantitative, and all of these factors need to be taken into account in interface design and content development.

It is important to note that Fig. 3.6 is a simplified and somewhat idealistic representation of a singular CHI context, such as an individual organization or a specific cybercartographic application. In reality, no organization or geographical information environment exists in isolation. Each CHI context is at least partly dependent upon information flow from other CHI contexts (domains in other organizations). This is similar to the importance in differentiating individual and social (or collective) holons with respect to their mutual interdependencies. In the collective domain, different CHI models necessarily interact on different levels. Although primary, secondary, and tertiary levels can be differentiated internally within a given CHI, they are "relational" in the collective dimension. Secondary level content from one organization may serve as a primary input to another, or tertiary level content from one domain may serve as a secondary input to another CUI domain (Table 3.2).

12. Conclusion

Although cybercartography appears primarily oriented around tertiary level applications, all such tertiary level applications "transcend and include" primary and secondary issues and environments in some way. The construction of a cybercartographic atlas, for example, must take into account the CHI environments and types of information processing that need to occur on the primary and secondary levels. It is the potential for this interconnectivity in both vertical (intradomain) and horizontal (interdomain) dimensions that marks a significant distinction between conventional cartography and cybercartography. Different types of interfaces and corresponding infrastructures are usually encountered at these different levels. If connections to these levels are retained in a cybercartographic environment, these need to be considered in the vertical integrative dimension. The concept of "intraoperability" in the vertical stream must also be considered with "interoperability" issues in the horizontal exchange dimension. There are not only different infrastructures, but also contrasting human domains, which from an operational standpoint are often "culturally" different. Different levels and types of decision making need to permit information to flow from one level to another, and inside and outside of any given domain. The CHI model may also be used to examine visualization, abstraction, and cartographic practices associated within different levels. These suggestions indicate that a variety of cartographic practices and different forms of visualization are to be expected in these environments.

If there is something unique that integral theory offers to cybercartography, it may be its panoptic view of where we find ourselves at the pioneering edge of the emerging information age in laying the foundations for new forms of visualization and communication about the world. Wilber draws an approximate contour associating formal-operational cognitive capacities with modern rationalism, and he uses the term "vision-logic" to denote higher and deeper level cognitive capacities that transcend (and include) rationalism and modernity into postmodern forms of representation, discourse, epistemology, expression, and being. Vision-logic is essentially "transrational," and therefore, more apt to better contextualize levels of meaning beyond strictly factual or codified knowledge, and is presently emerging as the next wave of human consciousness. Formalisms, as represented in various texts, narratives, statistics, and conventional maps, all have limited capacity to respond to the increasing need to put very large amounts of information into forms that are meaningful to society. This creates an opportunity for cybercartography to advance itself as an important tool for this emerging vision-logic capacity, as a means to help integrate, synthesize, visualize, and communicate very complex information in ways that go much beyond the conventional use of the geographic map, thereby giving maps and mapping a new level of application and meaning in the information era.

Acknowledgments

This chapter is based on research completed by the first author as part of a PhD comprehensive examination in the Department of Geography and Environmental

Studies at Carleton University. We are therefore grateful to the members of the examination committee, Fran Klodawsky, Ken Torrance, and Iain Wallace for their constructive and critical review of the integral theoretical framework. We are also grateful to Donna Williams at the Atlas of Canada, Michael P. Peterson, as well as members of the CANE project for their review and suggestions on earlier versions and presentations. We thank Patricia Trbovich for bringing our attention to the Ecological Interface Design (EID) literature. The material contained in this chapter remains the responsibility of the authors.

References

Adams, F. (2001) "Information theory (definition of)", in Audi, R. (ed.) *The Cambridge Dictionary of Philosophy*, Cambridge University Press, Cambridge, pp. 435–437.

Burrough, P. A. and R. A. MacDonnell (1998) *Principles of Geographical Information Systems*, Oxford University Press, New York.

Chrisman, N. (2002) *Exploring Geographic Information Systems*, Wiley, New York.

Clarke, K. C. (2001) *Getting Started with Geographic Information Systems*, Prentice Hall, Upper Saddle River, New Jersey.

Copi, I. M. (1982) *An Introduction to Logic*, Macmillan, New York.

De Blij, J. and P. O Muller (1997) *Geography: Realms, Regions and Concepts*, Wiley, New York.

Eddy, B. G. (2002) *Cybercartographic Integrated Research Framework*. Presentation to Research Workshop on Cooperation between Carleton University and The National Atlas of Canada, Carleton University, Ottawa, 15 April.

Eddy, B. G., G. F. Bonham-Carter and C. W Jefferson (2003) "Mineral potential analyzed and mapped at multiple scales: a modified fuzzy logic method using digital geology", *Geological Association of Canada (GAC): Special Volume on GIS Applications in the Earth Sciences* (in press).

Eddy, B. G. (2005) "Integral geography: Space, Place and Perspective", *World Futures: Journal of General Evolution*, 61: 151–163.

ESRI Inc. (1997) *Understanding GIS: the Arc/Info Method*, Wiley, New York.

Fisher, H. T. (1982) *Mapping Information: The Graphic Display of Quantitative Information*, Abt Books, Cambridge, Massachusetts.

Freeman, O. W. (1949) *Essentials of Geography*, McGraw-Hill, New York.

Haggett, P. (1983) *Geography: A Modern Synthesis*, Harper & Row, New York.

Koestler, A. (1967) *The Ghost in the Machine*, MacMillan, New York.

Latour, B. (1999) *Pandora's Hope: Essays on the Reality of Science Studies*, Harvard University Press, Cambridge, MA.

Leung, Y. (1997) *Intelligent Spatial Decision Support Systems*, Springer-Verlag, New York.

Lexicon (1987) *The New Lexicon Webster's Dictionary of the English Language*, Lexicon Publications, Inc.

Lindholm, M. and T. Sarjakoski (1994) "Designing a visualization user interface", in MacEachren, A. M. and D. R. F. Taylor (eds.), *Visualization in Modern Cartography*, Elsevier Science, Oxford, pp. 167–184.

Lock, C. B. (1976) *Geography and Cartography: A Reference Handbook*, Bingley, London.

Longley, P. A., M. F. Goodchild, D. J. MaGuire and D. W. Rhind (2001) *Geographic Information Systems and Science*, Wiley, Chichester, New York.

Medyckyj-Scott, D. and H. M. Hearnshaw (1993) *Human Factors in Geographical Information Systems*, Belhaven Press, London.

Peterson, M. P. (1995) "Maps and the Changing Medium", in Peterson, M. P. (ed.) *Interactive and Animated Cartography*, Prentice Hall, Englewood Cliffs, NJ, pp. 3–9.

Raper, J. (2000) *Multidimensional Geographic Information Science*, CRC Press, London, New York.

Rasmussen, J. (1999) "Ecological interface design for reliable human-machine systems," *International Journal of Aviation Psychology*, Vol. 9, pp. 203–223.

Reeve, D. E. and J. R. Petch (1999) *GIS, Organisations and People: A Socio-Technical Approach*, Taylor & Francis, London, Philadelphia.

Robinson, A. H. (1995) *Elements of Cartography*, Wiley, New York.

Shaw, E. B. (1965) *Fundamentals of Geography*, Wiley, New York.

Small, R. J. and M. Witherick (1995) *A Modern Dictionary of Geography*, Edward Arnold, London.

Stonier, T. (1997) *Information and Meaning: An Evolutionary Perspective*, Springer-Verlag, New York.

Szegö, J. (1987) *Human Cartography, Swedish Council for Building Research*, Stockholm, Sweden.

Taylor, D. R. F. (2003) "The concept of cybercartography", in Peterson, M. P. (ed.), *Maps and the Internet*, Elsevier, Cambridge, pp. 403–418.

Teilhard de Chardin, P. (1969) *The Phenomenon of Man*, Harper Torchbooks, New York.

Tomlin, C. D. (1990) *Geographic Information Systems and Cartographic Modelling*, Prentice Hall, Englewood Cliffs, New Jersey.

Ullman, E. L. (1980) *Geography as Spatial Interaction*, University of Washington Press, Seattle.

Wicken, J. S. (1987) *Evolution, Thermodynamics, and Information: Extending the Darwinian Program*, Oxford University Press, Oxford.

Wilber, K. (1995) *Sex, Ecology, Spirituality: The Spirit of Evolution, The Collected Works of Ken Wilber,* Shambhala, Boston.

Wilber, K. (1996) *A Brief History of Everything*, Shambhala, Boston.

Wilber, K. (2000) "Waves, streams, states, and self: further considerations for an integral theory of consciousness," *Journal of Consciousness Studies*, Vol. 7, No. 11–12, pp. 146–176.

Wilber, K. (in prep) (2002). *Excerpts A-G - Volume 2 of the Kosmos Trilogy,* Shambhala, Boston; accessed 15 January 2005 from http://wilber.shambhala.com/

Winder, R., S. K. Probert and I. A. Beeson (1997) *Philosophical Aspects of Information Systems,* Taylor and Francis, London.

Wood, D. and J. Fels (1992) *The Power of Maps,* Guilford Press, New York.

CHAPTER 4

Cybercartography from a Modeling Perspective

MARÍA DEL CARMEN REYES

General Director, Centro de Investigación en Geografía y Geomática J. L. Tamayo. (CentroGeo), México, D. F.

Abstract

Modeling processes have always been present in the design, production, and use of maps. Geometric language, since the times of the Greeks, has played an essential role in the representation of the geographic landscape. In a similar way as geometry emerged from observations more than 20 centuries ago, the genesis of cybercartography is found on empirical work developed in the last seven years. Such work, besides the traditional elements of cartography, encompasses a wide variety of information forms, languages and media, and uses technological advancements to integrate in a holistic, dynamic and interactive way multiple disciplines. The revolutionary and rapid advances in information and communication technology can be identified as the driving force in the emergence of a new paradigm in cartography, which is transforming the manner in which a map is conceived, produced, and used. This chapter deals with the construction of a theoretical and methodological framework for cybercartography by exploring: Modeling (which is inherent in Cartography), systems theory (which provides a holistic approach to formalize knowledge derived from the empirical work), and cybernetics (as the encompassing conceptual framework which allows approaching the processes of communication, control, interaction and feedback inherent in cybercartography). Finally, the modeling of Cybercartographic atlases using a qualitative approach is presented as a multidimensional system and the use of these atlases is viewed as the unfolding of a virtual spiral, which represents a qualitative dynamic model of the interaction among society, technology, information, and knowledge in chapter 23.

1. Introduction

The invention of the printing press made possible the reproduction of Ptolemy's Geography in the fifteenth century and in the sixteenth century the first atlas was published. Moreover, various hard copy media are historically reported for the creation of maps, such as wall paintings, papyrus, and clay tablets among others (O'Connor and Robertson 2002:1).

Cartography had compiled until the twentieth century a considerable amount of knowledge and techniques. However, with the advent of telecommunications and the revolution in information technology of the last century, new resources were offered to the mapmaker. Software and hardware technologies were in the market ready to be applied by mapmakers. A new language emerged; digital maps, computer and automated cartography, analytical cartography, cartographic information systems, multimedia cartography, Web cartography and cybercartography have become part of the jargon of the cartographic community.

These new technologies are relatively recent, and there is a lack of maturity in the new ways of approaching cartography. There is also a lack of conceptual clarity of the differences between the various terms mentioned above and as a consequence, there is a lack of a well established theoretical framework.

Two different approaches can be identified in the last 50 years towards the application of new technologies in cartography. The first is the conviction that cartography in essence has not changed and that information technology has only modified the manner in which maps are produced. In that regard Clarke (1995: 3–4) wrote, "cartography is a set of skills and a body of theory, and the theory remains the same independent of what particular technology one happens to use to make any particular map."

On the other hand, several authors such as Taylor (1997, 2003 and in this volume) and Wood (2003) argue that new paradigms are emerging in the cartography of the twenty-first century. In the last 30 years, computer knowledge and technology have had a strong impact in the manner in which geographers use maps. Peucker (1972) expressed it in his breakthrough paper on *Computer Cartography*, argued that either the cartographic scientific community accepts the fact that there are new avenues of research leading to a whole new body of knowledge in cartography or that a new discipline is emerging. In either case there is a need for innovative theoretical and methodological frameworks.

In cybercartography, the interrelationship between knowledge and technology plays a central role in the development of this new scientific concept. As mentioned by Russell (1967: 23) both concepts and tools are intrinsic to science. Throughout the twentieth century, examples of the development of technologies based on scientific knowledge are found in many different disciplines. However, technology has also been throughout history, a driving force to open new avenues of theoretical research.

The revolutionary and rapid advances in information and communication technology can be identified as the driving force in the emergence of a new paradigm in cartography, which is transforming the manner in which a map is conceived, produced, and used. As mentioned by Taylor (2003), multimedia, virtual visualizations, and the Internet among other technological tools, are demanding new paradigms for cartography. Cybercartography is a novel concept that is expressed in experimental prototypes resulting from empirical research over the last seven years. There is currently a need to focus on epistemological and conceptual issues behind cybercartography in order to develop a common language amongst researchers and a theoretical framework that result in the scientific advancement of this discipline. This is a major objective of this book.

Empirical and experimental work is essential in the advancement of science. But if one wants to further the development of theoretical knowledge, methodological frameworks are required. Moreover, society demands that the products and services that are offered through high technology tools are effective in solving its problems. Such effectiveness can be achieved by the application of a scientific approach in designing and engineering of such products and services.

Since 1998, following the initial ideas of cybercartography proposed by Dr Taylor, a research team in a Mexican institution (Centro de Investigación en Geografía y Geomática Ing. J. L. Tamayo, CentroGeo) started experimenting with the concept by developing several prototypes of "cybercartographic atlases" (Chapter 6). From 1998 to 2003 the main focus was on empirical work that included an intensive involvement with society, modeling, production processes, and technological development.

Four main points were considered essential in reporting the results of this first stage of research project:

- to explore the theoretical and methodological foundations of cybercartography;
- to sketch a basic framework;

- to analyze the interaction between cybercartography and society; and
- to document the results obtained through the empirical work.

Within this book, these questions are considered in three chapters that are intimately interrelated and complement each other (Chapters 4, 5, and 6). In this chapter, cybercartography is approached from a modeling perspective in order to advance the construction of theoretical and methodological frameworks for this innovative concept.

A conceptual approach to cybercartography poses important challenges due to the different and complex ways in which the geospatial landscape is represented. However, for these representations to be sound, they have to capture the holistic nature of the landscape and reflect the main processes or forces that influence significant changes on it. This is why in the Section 2 of this chapter modeling, cybernetics, and systems theory are reviewed in an effort to explore the building blocks that may support the scientific stature of cybercartography (Fig. 4.1).

In the second part of the chapter a systems approach is undertaken to identify the components and interrelationships among the different geospatial subsystems and models involved in cybercartography. Then, the genesis and development of cybercartography is visualized as a feedback process, represented by an "unfolding virtual helix," where societal demands, technology, information and knowledge are emerging and interacting in the process of generating scientific results, which impact on society.

Finally, the methodological approach used for the design, production, and use of the prototypes of cybercartographic atlases is briefly described in order to offer the reader the essential elements of the empirical results obtained in the past five years at CentroGeo. The methodology is more fully explained in Section 4 of this chapter.

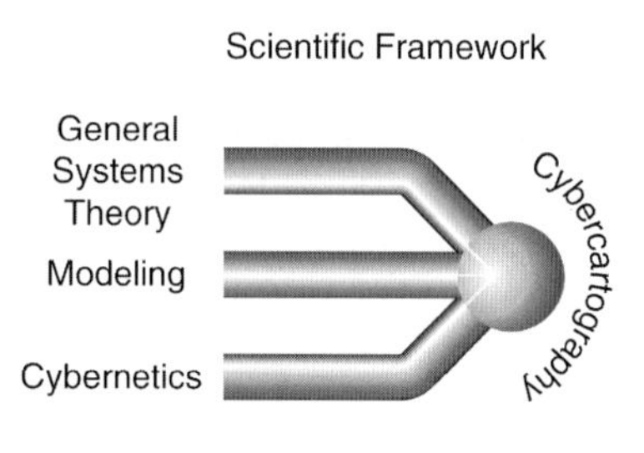

FIG. 4.1.

The working hypothesis of this chapter is that cybercartography is part of an empirical discipline. The manner in which the concept was first proposed by Taylor (1997) was inspired by the observations of results from experimental work (Canadian Explorer, China Cyberatlas) and their interaction with different societal groups. The major observations in this chapter are based on the results from empirical work at CentroGeo within the Mexican and Latin American context.

CentroGeo was established in 1999. The following research group had already worked together with Dr Fraser Taylor during the time he was exploring the cybercartography concept. During the last several years the research groups at the Geomatics and Cartographic Research Center (GCRC) at Carleton University and CentroGeo have worked collaboratively on cybercartography, as well as on a number of other research projects.

In terms of scientific practice, we want to draw to the attention of the reader that the initial developments of cybernetics resulted from 10 years of joint research between Norbert Weiner and with Arturo Rosenbleuth (a Mexican scientist from the National Cardiology Institute of Mexico), who had established a working group on scientific methodology as Wiener acknowledges in his book (Weiner 1967: 11). They both agreed at that time, that it was precisely at the borderlines of the different disciplines that fruitful avenues of research could be found.

2. Modeling, Cybernetics, and Systems Theory

Modeling, cybernetics, and systems theory are strongly related and in this first stage of the development of cybercartography, can be identified as the main "building blocks" of a scientific paradigm of cybercartography. Historically, modeling has ancient origins in mathematics while cybernetics and systems theory emerged in the twentieth century.

In the following paragraphs, concepts taken from these approaches that describe and explain cybercartography are presented and discussed. Modeling is inherent in cartography and systems theory provides a holistic approach, which is crucial in formalizing knowledge derived from the empirical work. The cybernetic character can be approached from the processes of communication, control, interaction and feedback inherent in cybercartography a term first proposed by Taylor in 1997.

2.1. Modeling

Modeling is a well established concept in the cartographic cognition process. In general terms, a model is a representation of an abstract or of an empirically observed situation. Mathematical modeling is an integral part of various scientific disciplines. In fact, cartography and mathematics have been intimately related since the time of the Greeks. Maps are visual models of the geographical landscape where a geometrical language has been used for centuries to represent geospatial information.

Around 325 BC Euclid established the axiomatic system for Geometry; Lobachevsky, in 1829, demonstrated that "the truth of Euclidean geometry could only be established by observation, not by reasoning" (Russell 1962: 61). In other words, geometry is an "empirical science based on observations" (Russell 1962: 61).

This statement is better understood by analyzing it from a historical perspective, in the manner in which the Euclidean axiomatic system was developed. According to Reyes (2005, 2:3), the first evidence of known mathematical knowledge was found in Geometry. The Babylonians and Egyptians discovered geometrical relationships such as the Pythagorean theorem and many others that are still used in engineering and other sciences. The manner in which these geometrical discoveries were made was empirical and responded to the need to resolve practical problems. Some of the results obtained were erroneous. Such is the case of the method to calculate the area of a quadrilateral with sides a, b, c, and d that happens to be incorrect [$K = (a + c)(b + d)/4$].

Many other examples of erroneous geometrical knowledge can be found in the history of mathematical development. It was not until Thales of Miletus around the year AD 624, that a systematic formalization of geometry was explored. It was Thales who made the first deductive demonstration. When Euclid wrote the "Elements" 300 years later, he incorporated hundreds of logical propositions that are deductively interrelated. After Euclid, for a proposition to be true in mathematics, it has to be either an axiom or deducted from other true propositions. This is the origin of the axiomatic system that has supported the development of mathematics for more than 20 centuries.

Once knowledge is formalized in either mathematical or other terms, a reader might deduce wrongly that the frameworks have emanated from pure reasoning processes, rather than from empirical observations.

Historically and from a modeling perspective, in the first process of abstraction, drawings and drafts were used to represent the geographical landscape. Lines, points, and areas have been basic geometric elements in map building since the time of the Greeks. As mentioned by O'Connor and Robertson around 250 BC. Eratosthenes besides measuring the circumference of the earth with great accuracy, used a grid to locate position of places on the earth.

Different authors discussed the way to construct the geographical grid and Ptolemy used coordinates to establish the location of major places. Similarly, the problem of the projection of a sphere onto the plane was undertaken by Ptolemy around AD 140. As is well known, map projections became a central problem in cartography and this still remains of great interest to mathematicians (Feeman 2002).

To synthesize, the three pillars of geometrical knowledge in mapping essential to our theoretical building blocks are: geometric figures, coordinate systems, and sphere–plane transformations. These mathematical concepts have been part of the modeling process in cartography for at least 20 centuries.

From a different perspective, Board (1967) examines maps as models and explains the modeling cycle involved in map making where a process of selection, reduction, data processing, and cartographic design are involved. The main actors in this process are the mapmaker, the map-user and the geographer. He also presents the first approach to discuss maps in terms of general communication systems where elements, such as the source, transmitter, receiver, and destination are analyzed. Other areas of mathematical research in formalizing cartography are error determination, formal description of map language, map projections, and algorithm development among others. This chapter discusses only those that are considered relevant at this point, in the process of advancing a conceptual basis for cybercartography.

It is important to note that the modeling process in cybercartography goes beyond mathematical cartography incorporating mathematical modeling in a broader sense and, as shall be discussed later, other related areas of knowledge such as cybernetics and system theory.

2.1.1. *Quantitative and Qualitative Modeling*

Quantitative and qualitative thinking have been present throughout the development of mathematics. Often in the solution of practical problems numbers are essential. However, in the context of the axiomatic system qualitative thinking is unavoidable. The best example of the application of quantitative mathematical approaches is found in physics but, even there, as Feynman (1999: 18) comments describing his work on quantum electrodynamics, it is sometimes necessary to get a qualitative idea of how a phenomenon works before obtaining a quantitative solution.

On the other hand, as mentioned by Bissell and Dillon (2000: 3), engineers have usually applied existing mathematical models and the application of physics in engineering is characterized by numerical modeling and as such is essentially quantitative. In other areas of knowledge, such as sociology or psychology, statistical methods which are essentially quantitative have prevailed.

As mentioned by Harley (1991), through cartography, humans substitute the geographical landscape with an analogy model of space that has allowed an intellectual dominion over the world and has given power to people and nations. In cartography, topographic mapping as conceived in the last 20 centuries has been essentially a quantitative modeling exercise, although, qualitative elements such as the relative position of geospatial features is in many cases implicit. In addition, in thematic mapping qualitative elements are more emphasized in the representation of features, such as land use, vegetation or ecological units.

In participatory mapmaking and map use, such as the Strabo Technique (Luscombe 1986), the process is essentially qualitative. For example, the Strabo Technique which is built upon the Delphi Technique (Linstone and Turoff 1975) incorporates the geospatial dimension in such a way that a diversity of experts involved in the participatory process can build consensus and define and deal with questions by using a geospatial approach. Qualitative research has also been applied in the conceptual design and implementation of Geographic Information Systems (GIS) (Reyes 1997). In other areas of geographical research mathematical qualitative modeling is encountered, as exemplified by the application of fuzzy sets, graph theory and topology in spatial analysis (Reyes 1986).

While taking a modeling perspective in cybercartography, one encounters that both approaches (quantitative and qualitative) are considered valuable. Cybercartography encompasses traditional mapping and as such inherits quantitative modeling in geometry and mathematical cartography.

As Taylor (2003: 403) argues, "organizational, institutional, and community processes will be very important functions of cartography in the twenty-first century." Moreover, geospatial relations and processes are also expressed through natural language, music and videos among others. In this sense, qualitative modeling becomes an indispensable resource in building cybercartographic atlases and a general systems theory approach appears to be a suitable route for our purposes.

2.2. Cybernetics

Norbert Wiener (1894–1964), a mathematician, founded the discipline of cybernetics. In 1950 he wrote the book *The Human Use of Human Beings, Cybernetics and Society*. He mentions, that the word cybernetics is derived from the Greek kybernetes (which means steersman) and shares its etymological root with the word "governing" to which Plato gave the meaning of "the art of science." Wiener introduced the word cybernetics in scientific language to describe the science of communication and control in the organism, the machine, humans, and society. In this regard, Wiener's main thesis was:

> Society can only be understood through a study of the messages and the communication facilities which belong to it; and that in the future development of these messages and communication facilities, messages between man and machines, between machines and man, and between machine and machine, are destined to play an ever increasing part (1950: 25).

Wiener clearly identified the relevance of the study of communications and the role of information in human life. The explicit process of receiving information and using it to adapt to the environment allows the human race to live effectively in its own context.

He clearly expressed the concept of feedback by mentioning, "For any machine subject to a varied external environment to act effectively, it is necessary that information concerning the results of its own action be furnished to it as part of the information on which it must continue to act." (Wiener 1950: 35). His concern for "the organic responses of society itself" (Wiener 1950: 39) beyond his machine feedback process is of interest here.

Reading Wiener's book (1950), bearing in mind the results of empirical work in cybercartography, several very relevant issues of cybernetics can be identified as strongly related with the modeling process in this new concept.

- The role played by messages and information in the communication process.
- The awareness of the relevance of communication in the control of society.
- Through the concept of feedback, the implication of the explicit involvement of the "user" of the system within the system itself.
- The fact that all "sense organs" are involved in the communication processes.
- The recognition of the importance of the communication of visual images.
- The role played by information in measuring the relevance of content of a set of messages in a communication process.

In Section 3 of this chapter, each of these issues will be analyzed in the context of establishing modeling approaches in cybercartography. In the initial empirical stages of cybercartography, these concepts were not formally acknowledged but appear to have been intuitively applied in the creation of the first cybercartographic products and in the development of cybercartographic theory.

2.3. Systems Theory

The concept of system was introduced in scientific language by Von Bertalanffy (1950) in the middle of twentieth century. His intention was to propose a new vision that would introduce an effective way to consider problems from nature, which could transcend the limitations of reductionism (Ackoff 1974).

A system is defined as a set of two or more elements where each part can affect the behavior of the whole; the manner in which each part affects the whole depends at least on what another part is doing, and each subset of elements satisfies the previous properties. A system cannot be an ultimate element, it cannot be divided into independent parts and none of the parts of the system has an independent effect on the whole (Ackoff 1974: 13).

The systems paradigm can be viewed at three levels: the system as a "holistic structure," where the role or function of the parts is considered within the whole, the intrasystem that is concerned with subsystems or groups of subsystems and the suprasystem that makes explicit the fact that every system is contained in a larger system called environment or context (Fig. 4.2). Eddy and Taylor utilize integral theory which has some similarities to aspects of systems theory in discussing cybercartography in Chapter 3.

Modeling is a concept that emerged long before systems theory, and it has been extensively used in science with a reductionist approach. Physics is probably the best example of the use of mathematical modeling to simplify and represent phenomena in a very successful manner.

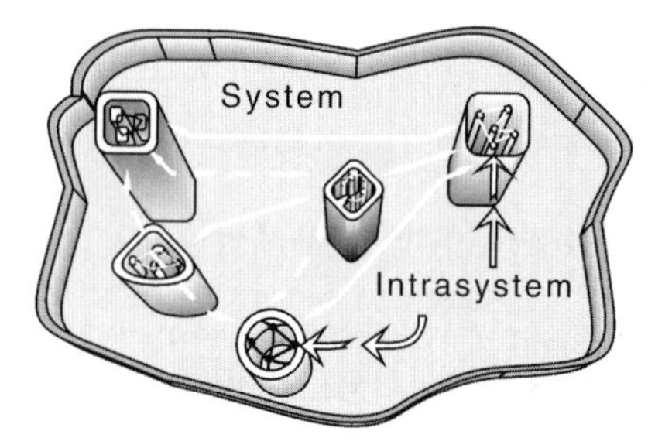

FIG. 4.2.

From the perspective of systems theory, a model can be viewed as an abstract relational structure analogous to a certain aspect of a system or to an empirically observed situation. In this sense, models are "scientific metaphors" significantly similar to the system they attempt to represent.

In the design of systems, modeling is implicit. The identification of the components and interrelationships among the subsystems requires a process of abstraction, and understanding of the interaction, functionality, and interrelationship among the subsystems.

The modeling process or the construction of a model involves the representation of the system or observed situation in terms of a conceptual model that describes its generic structure. The researcher builds a system of symbols and rules in such a way that its structure is homomorphic (in mathematical terms) to the phenomenon or process of interest; for that purpose it identifies the logical components, and the meaningful interactions of the phenomenon or processes observed and the structure of the model.

One of the main problems in modeling lies in the complexity of the situations that are to be represented since, in many disciplines nowadays, the amount and variety of interactions and variables is considerable and the rate of change of the processes involved is exponential. In order for a model to be relevant, it has to represent the key variables and interactions that characterize the main features, the dynamics and the structure of the system.

From this perspective, modeling can require a systems approach that allows different levels of structuring the observed situation and the selection of the intervening elements or agents. Modeling must also consider the structural and procedural interactions that derive from behavioral patterns in either a static or dynamic fashion.

The need to adopt a holistic approach, derived from the Theory of Systems, becomes evident if one recognizes the value of incorporating knowledge and approaches from different disciplines in the building of robust scientific models. Wilber's integral theory, discussed in Chapter 3 by Eddy and Taylor takes a similar approach.

2.4. Maps, Modeling, and Communication

A map can be viewed as an inseparable binomial phenomenon of modeling and communication. Is a map a model designed only for communication or can a map also be an input for another model? It is argued here that both "propositions" are true and that there might be other ways of adopting, applying or using a map. The fact that must be emphasized is that the intrinsic binomial characteristic of mapping is an initial assumption in the first step to formalize the concept of cybercartography.

As Wiener (1967: 101) comments "communication and language are also inseparable." The relevance of language issues in cybernetics are not only for people or living creatures but also, for communication between humans and machines or among machines themselves.

In the past, the language of maps was extensively studied by Peucker (1972:1), who mentions that "geographers use maps for the analysis, communication and storage of their findings and explores the relationship of the new languages of computers and its impact on the development of cartography." Keates (1996) in two chapters: *Theories and Models of Cartographic Communication* and *Communication, Map Use and Cartographic Objectives*, presents state of the art research approaches to the modeling of cartographic communication.

Among other interdisciplinary areas of knowledge, neurophysiology, anthropology, psychology, cognitive sciences, and linguistics have analyzed different aspects of the relationship between language and space (Bloom *et al.* 1994). For example, Bierwisch (1994) poses the question "which components of natural language accommodate spatial information, and how?" In this case natural language includes both writing and speech. For centuries geographers have used language as a means to communicate and represent spatial knowledge. Possible answers to this question are of major interest for cybercartography, since natural language is one of the main resources in building cybercartographic atlases. The term "geo-text" is suggested to describe the body of language relevant to geospatial knowledge and/or information, which is communicated through words.

On the other hand, Jackendoff (1994) raises the question of the relationships between natural language and knowledge, and spatial images and knowledge. Understanding these relationships requires much further research. Jackendoff remarks that the theory of spatial representation is less developed than that of the encoding of linguistic meaning. His emphasis on spatial representation, however, only takes into account visualization implying that the building of meaning relies only on the sense of sight. In addition he does not incorporate symbolic mathematical language into his model.

There is an explicit bridge between spatial representation and mathematics. The realms of geometry in its different expressions (Euclidean, projective, hyperbolic, etc.) (Baragar 2001) as well as topology (algebraic and surface) (Firby and Gardiner 1982) are examples of ways of encoding and communicating spatial information and knowledge through a symbolic language.

In cybercartography, communication takes place using diverse resources including cartographic, linguistic, mathematical, statistical, musical, and

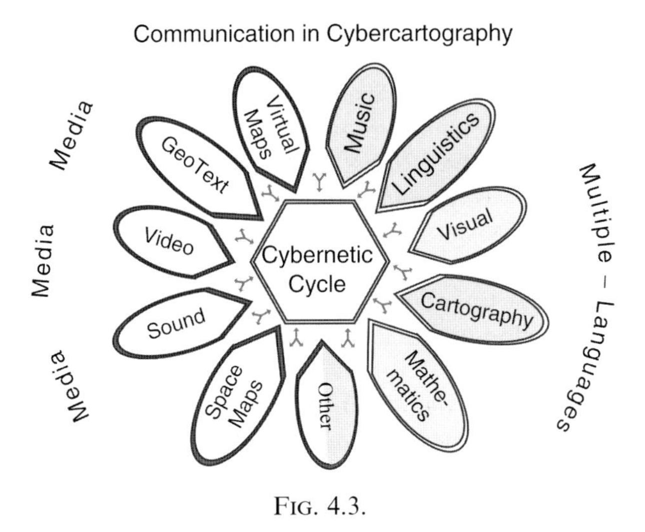

FIG. 4.3.

visual languages. Virtual maps, geo-text, videos, photographs, space-maps, satellite images, computer simulations, graphs, sound, and diagrams among others, have been used in the construction of cybercartographic atlases (Fig. 4.3). With them, the subject interacting with cybercartography engages only sight and hearing. Other senses, such as smell, touch and taste as elements of the communication process are being explored in the development of cybercartographic atlases (Taylor 2003). Whether there can be formal languages for each of these senses is still an open question.

The formalization of the language of cybercartography has still to be explored. The syntax and semantics of this metalanguage still need to be defined. The empirical work done up till now, gives some idea of possible avenues to explore. Following Jackendoff, the encoding process of geospatial knowledge and information in linguistics, visual, mathematical, and musical structures, as well as their interconnection would have to be modeled. In this regard a more holistic, approach based on systems, theory might point to the right direction for future research.

3. Modeling and Cybercartography

Taylor (2003) describes cybercartography as a new concept that responds to the impact of the advancements in information technology and telecommunications. Since 1997 when the concept was first proposed, community and organizational processes where cybercartography has had an impact have been accompanied by high-technological products (artifacts) that have been called cybercartographic atlases. As described in other chapters of this book, the stimulating manner in which different societal groups from different countries (Canada, China, Mexico) have embraced the empirical work of cybercartography, and the change in the manner of communicating and using geospatial information and knowledge clearly has pointed to possible new avenues of research that: (1) need to explore the concept from different theoretical and methodological perspectives and (2) advance the formalizing of the knowledge derived from 7 years of experimental work.

As a means of justifying the semantics of a concept named after two well-established disciplines (cybernetics and cartography) and the seven essential elements of cybercartography that Taylor (2003) mentions, one needs only to look at each of these disciplines to find a clear and relevant relationship between the naming of the concept and its scientific significance. For example, the multisensory and multidisciplinary characteristics of cybernetics were identified by Wiener from early on as discussed in Section 2.3. The feedback and interactivity processes are also cybernetic characteristics, as well as the interest in the human–machine and machine–machine communication processes. In summary the prefix "cyber," goes beyond the fact that cybercartographic atlases use computers and WEB technology, and instead points to possible theoretical frameworks.

On the other hand, in Taylor's characterization the understanding of mapping as a communication model is implicit, and as such, the binomial characteristic of modeling-communication in traditional cartography is fulfilled. Moreover, some of the key concepts presented by Wiener regarding messages, information and communication bring illumination to the meaning of cybercartography. Finally, the fact that the first cybercartographic prototypes are atlases, with a strong focus on using maps to communicate geospatial information, points to the essentially cartographic focus of this new concept.

Atlases have been part of cartography for many centuries and have included drawings, photographs, and geo-text and in multimedia cartography other elements such as videos and simulations. In other words, the selection of atlas as a conceptual point of departure, to generate products in cybercartography is not fortuitous.

It is argued here that embracing the concept of atlas in cybercartography has been a step in the right direction. It allowed a natural incorporation of different languages and resources such as multimedia cartography. On the other hand, the concept of "cyber map" within cybercartography still has to be more fully explored. Taking the concept of map as the sole point of departure in designing a Cybercartographic product does not seem a promising line of approach. However, if the definition of a cyber map as one that is designed, produced, and/or used through computers and the WEB is adopted, then it can be argued that this has been part of the evolution of computer cartography and does not seem to add any new conceptual element to the field despite its importance in building cybercartographic atlases.

The first of the following two sections considers the modeling of cybercartographic atlases using a qualitative approach, through the description of the components of a multidimensional system and the interrelationships and functionality of its dimensions. This is followed by a consideration of the processes involved in an exercise that uses cybercartographic atlases represented through a qualitative model.

3.1. Modeling Cybercartographic Atlases from a Systems Approach

In 1967 Wiener (15) described how cybernetics would emerge within systems theory so that a common theory would come up. In the realm of cybercartography the systems theory approach is essential as a modeling framework.

Cybercartographic atlases are multidimensional systems composed of three axes: models, knowledge representation, and communication. Conceptually, the applications can be approached systemically from each of these perspectives. For example, a cybercartographic atlas can be viewed as a model of models or a geospatial meta-model. Alternatively, the communication process is accomplished through messages and information organized in subsystems using multiple languages. Knowledge is incorporated through implicit models derived from the suprasystem that reflects the specific interests of the users (Fig. 4.4).

The suprasystem is the context where the results of the demand from society could have an impact. In cartographic information systems development, the suprasystem is usually considered the immediate user of the results. In cybercartography it is in the broader context, the "first order user" and is conceptualized as being part of the system. Whether the "second or third or *n*th order user" of the cybercartographic atlases is also part of the system still needs to be explored. This characteristic clearly responds to the argument by Taylor (2003) that there is a need for "new user-centered model" for cartographic applications. This is discussed further in a number of chapters in this book.

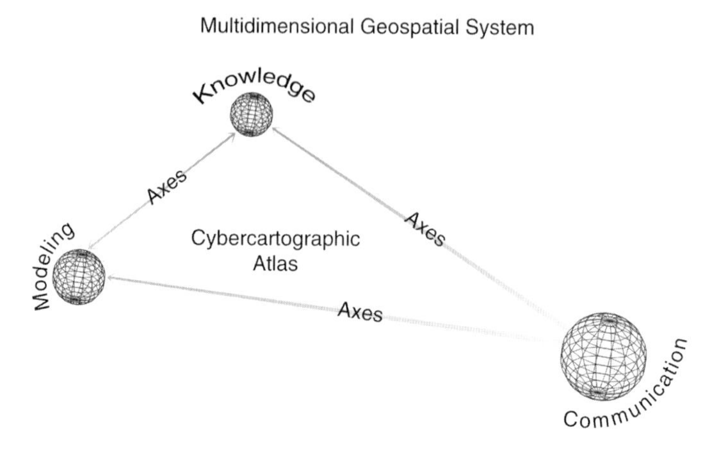

FIG. 4.4

3.1.1. *Communication Axes in Cybercartographic Atlases*

In the communication axes there are two types of subsystems: the structural subsystem and communication subsystem. The structural subsystem functions as a backbone that gives support to the user-prototype interaction in terms of geospatial information. Its main components include data bases, spatial data structures, geospatial data libraries, cartographic data banks, and WEB and graphic structures among others.

The communication aspect consists of a systemic combination of languages. Multimedia in itself has its own structure that incorporates video, photograph, text, sound, and music. As mentioned before, cartography has developed its own theoretical language frameworks.

In cybercartography the geospatial messages that are expressed through maps, graphs, images, diagrams, videos, photographs, text, sounds, and music (and potentially via touch and smell) have to be designed, integrated, and presented in such a way that the user receives geospatial information.

From a methodological point of view the use of multiple languages (visual, linguistics, musical, etc.), poses an important challenge. The arbitrary combination of geospatial messages does not necessarily produce a prototype cybercartographic atlas. Such a prototype emerges from the selection of information relevant to communicate whatever is needed, in order to reach the objective of the modeling research. This process of selection and modeling gives rise to a holistic product that articulates different languages and media in geospatial messages. Similarly, as in traditional cartography, the production process implies along with the use

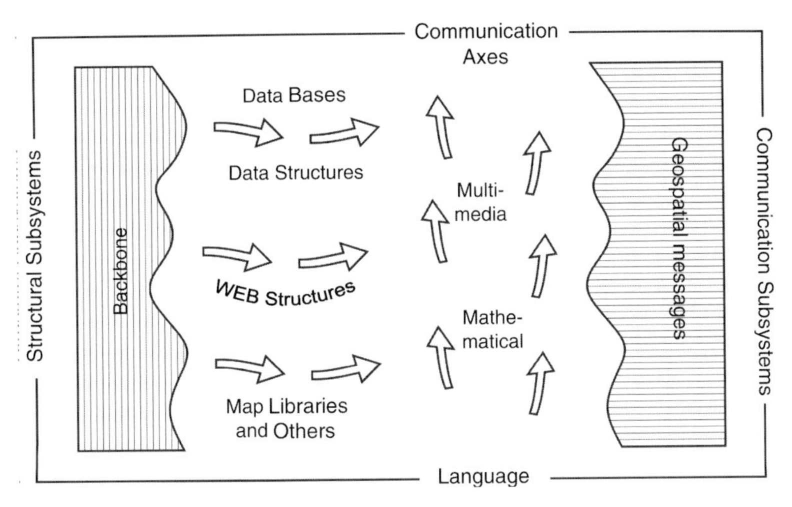

FIG. 4.5

of formal methodological frameworks, "artistic talent" that adds to the effectiveness of the communication processes (Fig. 4.5).

As explained in Chapter 6 (*Technology and Culture in Cybercartography*), from a technological point of view, although there are some elements in the commercial geomatics sector that can be used to develop cybercartography, there is a need for additional customized software development. Moreover, culture as in other cartographic contexts, plays a fundamental role in terms of approaches utilized in the production process.

3.1.2. Meta-Model Axes

A cybercartographic atlas is a model of models. Each one of the models has a purpose to represent the geographical landscape in one way or the other. Some of the models have been present in cartography for many centuries, some are derived from technological developments in computer science, and others are part of the thematic knowledge framework of the specific cybercartographic applications.

For example, cybercartographic atlas prototypes could include: cartographic digital models, virtual maps, digital terrain models, space-maps, relational data bases, topological data structures, raster models, geo-text models, icon models (photographs, videos, images), musical models (Lluis-Puebla 2004; Mazzola 2002), landscape ecology models, and geographical business models, among others. The computing science, geographical and communication models that can be used are numerous and in this respect, the development of cybercartography is just in its initial stages.

What characterizes the meta-model is the description of cybercartography in itself. The meta-model is qualitative and it is a meta-representation of the geographical landscape. Similarly, as in traditional mapping, the meta-model is part of the modeling-communication binomial (Fig. 4.6).

Each of the models represents geospatial entities, relationships, and processes. In the first part of this chapter, the manner in which these elements are treated in traditional cartography was explored. Depending on the characteristics and scope of the model, different conceptual resources are utilized. For example, in topological data structures, the entities are usually points, lines or polygons that are elements of computational structures (files, fields, lists, etc.). The essential relationship represented is "spatial contiguity relations."

Each of the models incorporated into an atlas plays distinct and complementary roles. The key factor is that the universe of geospatial entities, relationships and processes represented in all the models have to be integrated so that "they make sense together." In other words, in the meta-modeling there is a "storytelling" process involved explicitly. As will be argued in Section 4.2 of this chapter, the building of a cybercartographic atlas has similarities with the production process involved in the entertainment industry (movies, TV shows, or theatre). The right models have to be included so that the "story" is adequately presented to the user. This requires from the "director" of the production process knowledge, artistic talent, and a holistic approach.

But that is only half of the story. With new technological resources, cybercartographic atlases can be dynamic and open systems. Spatio-temporal

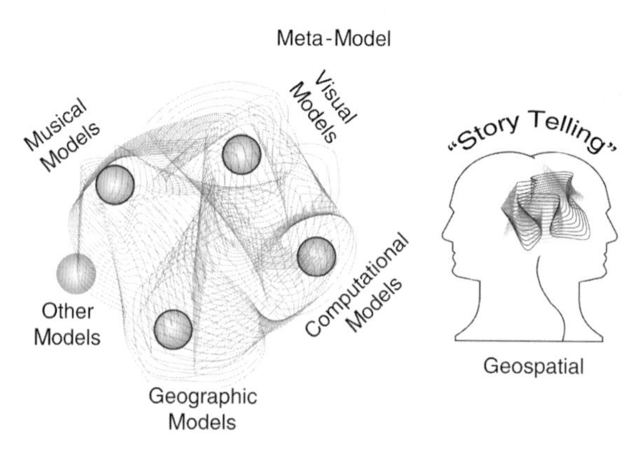

FIG. 4.6

interactivity can be incorporated so that the user can "navigate" the re-creation of spatial processes at a fixed point in time or through historical periods. The applications can also be designed so that the user can incorporate new models and expand the scope of the "storytelling" process.

3.1.3. Knowledge Axes

Cybercartography, like traditional cartography, undertakes specific thematic topics, which encompass the context where the services, products, or prototypes will be embedded. The universe of problems and environments where a geospatial analysis approach is of interest, is very wide across society. To exemplify the manner in which a theoretical framework is an integral component of cybercartography, some of the results of the empirical work of the cybercartographic atlases developed at CentroGeo are used as a reference.

As explained in Chapter 6, the Cybercartographic Atlas of Lake Chapala responded to the requirements of a nongovernmental organization for geospatial information and knowledge that could provide the community with a framework to solve the conflicts of interest around the management of a natural resource in environmental danger.

For this particular case, a landscape ecology approach was adopted. Its comprehensive and holistic concepts seemed appropriate in addressing such a complex problem. Several landscape ecology models were introduced throughout the prototype. For example, the cartographic information was organized using a diagram (Fig. 4.7) adapted from Zonneveld's (1989) conception and its systemic components are used as icons so that the user can have access to the interactive maps. The geo-texts are also conceptually framed and organized according to the theory of landscape ecology.

The final prototype is therefore comprised with a specific body of knowledge. As mentioned by Kuipers (1978), traditionally cartographers are interested in how "information presented as a map is intended to become knowledge in the head of the map-reader." In contrast, in cybercartographic atlases knowledge is incorporated through multithematic geographical models and multiple languages. There is, therefore, an explicit transfer of message–information–knowledge to the user. The meaning that the user derives from such knowledge and feeds back to the atlas will be explored in Chapter 5.

At first glance, it may be assumed that the knowledge embedded in cybercartographic atlases may be treated in a similar way as it is treated in traditional atlases. There is however an essential difference, in terms of cognitive modeling, the "reader" is conceived as being part of the process

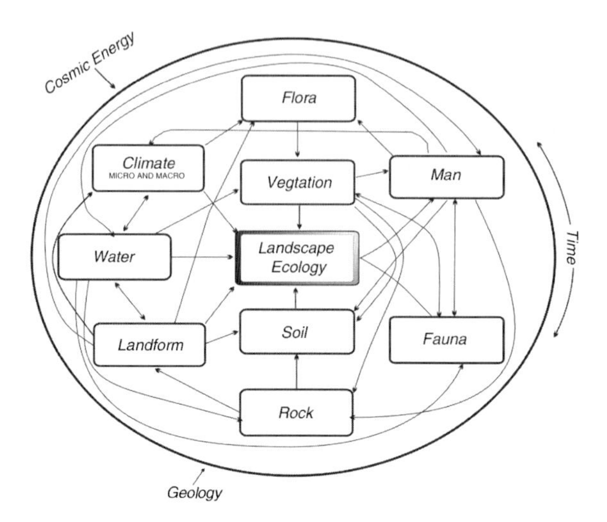

FIG. 4.7

so that his/her knowledge interacts and modifies the prototypes. According to the proposed model, new information and knowledge can be created through feedback processes and incorporated into the prototype

3.1.4. Systems Integration and Functionality in the Cybercartographic Atlases

There is conceptually a common "interaction space" among the communication, modeling, and knowledge axes. For example, while a cartographic database is part of a structural component that gives support to a geospatial communication process, it can also play a key role in the representation of a geospatial phenomenon. Formally, this process can be better understood by making an analogy with the construction of mathematics where there can be multiple representations of a concept (numerical, visual, and relational). Dubinsky and Harel (1992) describe mathematical functions using these three representation axes.

The proposed axes in modeling cybercartographic atlases aid in the systemic characterization of the artifact from different perspectives, and the understanding of roles played by various models, languages, and knowledge frameworks that are incorporated in the design of the prototypes.

While approaching the integration and functionality of cybercartographic atlases, the incorporation of technological elements is unavoidable. As mentioned by Winograd and Flores (1987) "All new technologies develop within the background of a tacit understanding of human nature and human work. The use of technology in turn leads to fundamental changes in what we do, and ultimately in what it is to be human."

Computer technology is essential in the production of cybercartographic atlases. Without the advances in geotechnology, multimedia, and WEB tools, it would be impossible to recreate the conceptual description of cybercartography. In the production process, the cognitive frameworks involved have to be interrelated through the use of technological tools.

Questions regarding the differences between cybercartographic atlases and other approaches naturally arise. For example, how is it different from a conventional paper atlas? Technology certainly plays a central role in the answer to such a question. Making an analogy between a collection of photographs accompanied with music and sound and the impact on the viewer of any of the movies currently playing in the nearest theatre, as well as the analysis of the evolution of "games" from a chess blackboard to computer video games could point to possible answers. The technology used in movie-making incorporates multiple languages and is a visual communication exercise while in video games interactivity is added as an essential element (Fig. 4.8).

In this regard, the interrelationship among the different systemic components, their complementarities, the synergetic effect of integrating various models and languages, the aesthetic elements required are all combined to produce a cybercartographic atlas. An explicit model of the manner in which the process results in the products desired, is still not in place. At this

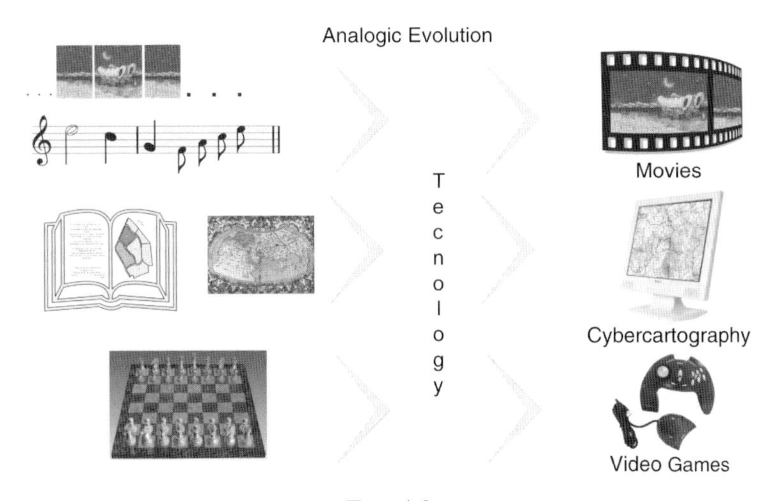

FIG. 4.8

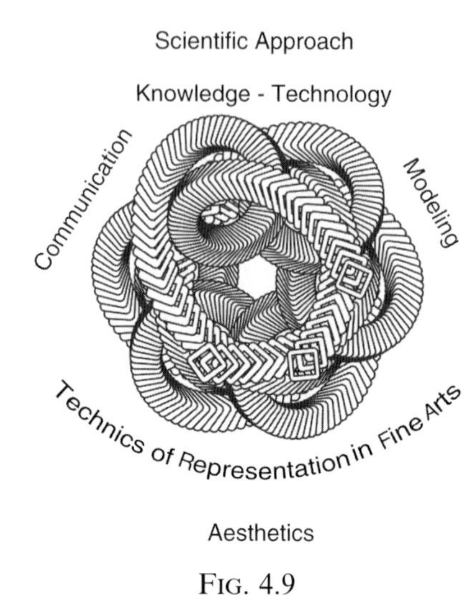

FIG. 4.9

point, the only certainty is the very enthusiastic manner in which different societal groups have embraced the empirical results (Fig. 4.9).

3.2. *Modeling Cybercartographic Processes: The Virtual Helix Model*

Process is essential in cybercartography (Taylor 2003). The previous section explored the modeling of cybercartographic atlases as products; however, the products as such have to be embedded into processes in order to complete the cybernetic cycle. In this section, cybercartographic processes are represented as a graphic image, which we called an "unfolding virtual helix."

Several theoretical considerations are required as precursors to a description of the virtual helix model, these include:

- The early vision of Wiener on the wide range of applications of cybernetics; for instance, in medicine, economics, sociology, and engineering (Haseloff 1970: 15);
- The awareness from early researchers in cybernetics, of its potential to study complex structured systems from the perspective of the transmission of information;
- The various types of circular relationships that can be found in cybercartography. For example, cybercartographic atlases provide information and knowledge, and through feedback new information and knowledge can be created;

- The nonlinear processes present in the design and implementation stages of the atlases;
- The different levels of interlocution of the feedback process in cybercartography; and
- The openness, adaptability, and high interactivity of cybercartographic products.

Information and knowledge have become key inputs for the functioning of modern society. In economically developed nations in particular, geospatial information and knowledge are represented in the form of cartography and geospatial data infrastructures have become strategic elements. The variety of applications is considerable and is constantly growing. Hence, the point of departure of the unfolding helix is societal demands. Societal demand creates the need to design and produce cartographic services and products that satisfy specific purposes. This demand comes from individuals, organizations, or communities.

Cybercartography is composed of geospatial messages. These messages are created using various simple and complex modeling and communication resources that respond to the needs of society for specific geospatial information and knowledge. The building of such geospatial messages is the next hallmark in the unfolding path.

Using an analogy to the message–information relationship mentioned by Wiener, geospatial information in cybercartography is the measure of the adequacy of content and structure of geospatial messages. The interactivity inherent in cybercartography allows the user to manipulate messages so that new geospatial information is created. This feedback process between the prototype and the users is an essential characteristic of cybernetics and allows the advancement of the helix trajectory (Fig. 4.10).

Usually, in each demand from society for geospatial information, there is implicitly or explicitly a knowledge framework. In some cases, a geospatial model has been explicitly identified and in others, the results are derived from the "spatial intuition" of the user. In any case, the geospatial message provided through cybercartography is structurally coupled with the spatial model of the user (see Chapter 5).

Empirical results indicate that in order to adequately communicate with the user, it is highly advisable to engage with them in a process of uncovering their geospatial mental model: to make explicit the spatial entities, relationships, or process of interest for their specific applications. In this way, the cybercartographic product will incorporate not only geospatial information but also knowledge, which will function as a "conceptual bridge" between the user and the prototype. Given the importance of

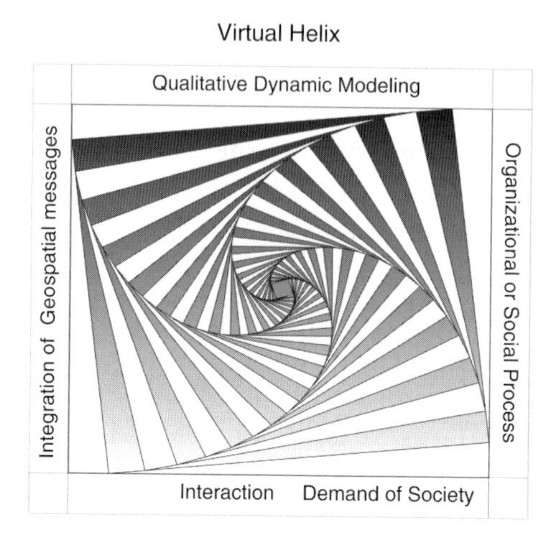

FIG. 4.10

these issues on the cybernetic cycle, they are broadly discussed in Chapter 5 (Martínez and Reyes).

Evidence from the incorporation of cybercartographic prototypes into public policy processes points to the potential of the impact that such an approach can have on society (Chapter 6). In fact, governmental decisions on environmental issues at high levels in the Mexican government were strongly influenced by the *Cybercartographic Atlas of the Lake of Chapala* in the year 2000 (a prototype developed at CentroGeo). Moreover, the perception of the community significantly changed with the incorporation of this prototype in the negotiations among different actors, regarding the application of resources to environmental programs of local governments.

The technological functions embedded in the atlases developed at CentroGeo were designed so that the user is able to incorporate new geospatial information and expand the geospatial models of the cybercartographic atlas. The current advances in geotechnology allow a level of interactivity between the user and the product that in the past could not possibly be conceived of, in the design of cartographic products.

Once a product is incorporated into organizational or social processes (public policy exercises, planning efforts, political interaction, consensus approaches, etc.), it becomes dynamic and "vivid" in the sense that it evolves according to the desires of the users. The current products certainly have limitations so that the scope of the impact on society, the creation of new demands for geospatial information or the advance in knowledge frameworks produce new demands on the designers of cybercartographic atlases (Rodríguez 2001).

Thus, the virtual helix is back at the initial point of departure: the demand from society for geospatial information and knowledge; but clearly, at a different level of content, scope and functionality as illustrated in Fig. 4.10. These processes can be repeated over and over so that the helix does not necessarily have an end point. That is the reason it is characterized as "unfolding".

As mentioned in Section 2.2, cybercartography encompasses several of the essential elements that Wiener used for the development of the conceptual framework of cybernetics. Since cybercartography is in an early stage of development its potential is likely to expand as new cybernetic qualities and characteristics are incorporated. For example, CentroGeo prototypes rely mostly on visual and auditory sense organs. The research group at Carleton University is experimenting with touch and smell with the ultimate aim of using all of the senses.

4. Methodological Framework

Since the seventies, the literature has stressed the need to design and use robust methodologies in order to successfully apply GIS. Since that time, the author has been involved in an intensive interaction with Mexican society on the need for geospatial analysis and information, and in suggesting technological solutions for a wide variety of decision-making, planning, and management problems relating to government and international projects. During the 90s, the complexity of the problems posed by society made it necessary to formalize a methodological framework to assure the success of applied projects where geospatial modeling was involved, particularly in cartographic and GIS. More than 25 years of experience in the design and implementation of geomatic projects provided the basic framework and conceptual guidelines to define a methodology for the design and implementation of the cybercartographic atlases developed at CentroGeo.

4.1. *A Methodology for the Creation of Cybercartographic Atlases*

The methodology used to design and construct cybercartographic atlases adopted a basic framework, which is in alinement with the central principles of cybercartography in terms of both its systemic and holistic approach, and the explicit incorporation of feedback processes supported by qualitative research practices.

The methodology involves two distinct but interrelated issues: (1) The structural methodology, that is, the approach that provides conceptual guidance for the cybernetic cycle of observation, modeling and feedback and (2) The content framework or conceptual elements used to design, steer, and build the thematic content of the model.

4.1.1. *The Structural Methodology*

The first step is the design of the meta-model which focuses on a number of different processes and these are:

(1) The understanding and study of societal requirements in terms of geospatial information, analysis, and knowledge;
(2) The behavior of the research group producing the Atlas, including what they observe, how they proceed, how they interact with relevant actors, how they identify relevant themes, and how they arrive at a synthesis;
(3) The assurance of the systemic coherence and the synergetic effect among the different languages and models involved, which is guided by the societal needs outlined above; and
(4) The geospatial messages and information transmitted through the communication process.

This 'structural methodology' is therefore essential to any cybercartographic exercise and its purpose is to provide a basic conceptual framework for the applied research and production process chosen.

The 'structural methodology' encompasses three conceptual levels: the meta-system, the system modeling process, and the technological solution.

4.1.1.1. *The Meta-System – Third Order*

In developing the meta-system, four main issues are considered: (1) the political, organizational, cultural, and social context in which the atlas will be inserted. At this stage a qualitative research approach is adopted in order to assure that the product is structurally coupled with community, organizational, institutional or individual processes; (2) The adoption of organizational approaches that guide the behavior of the research team during the design and production stages, such as collaborative approaches, team work, and the formation of multidisciplinary groups; (3) The user-requirement analysis; and (4) The specification of the overall geospatial modeling, analysis, and content frameworks.

Research in information systems now focuses on organizational and managerial issues; therefore and as a result, there is an increasing use of qualitative research in the design of such systems. "Qualitative research involves the use of qualitative data, such as interviews, documents, and participant observation, to understand and explain social phenomena. The field work notes and the experience of living there become an important addition to any other gathering technique that may be used" (Myers 1997). The qualitative approach is complementary to user-requirement analysis. In applying qualitative research methodologies emphasis is placed on an understanding of work setting, cultural, and political issues that could affect the implementation of the information system.

The author has applied this approach in a number of geospatial information and analysis projects in México and Latin America over the last 10 years including the *Electoral Federal Redistricting for the 1996 Presidential and Congressional elections in Mexico* (Reyes and Levi 2001: 161–171) and the *Corporate Geographic Information System for the Mexican Oil Company* (SICORI, PEMEX) (Jordá and Reyes 1993: 331–340). The success of the approach in these complex projects, dealing with high-profile topics, led to the adoption of a similar qualitative research approach for cybercartography.

The integration of "working groups" responsible for the atlas, as well as their organizational structure, have been identified as essential elements in assuring the success of cybercartographic projects due to its multidisciplinary and holistic character. There are many different studies reported in the literature on these issues (Tall 1991; Lauriault 2003; Lauriault and Taylor in Chapter 8). At CentroGeo multidisciplinary team work is the institution's organizational structure and has been important in the development of cybercartographic theory and practice as suggested by Taylor (2003).

User Needs Assessment (UNA) methodologies are widely used in the design of geospatial information and analysis solutions. Methodologies for this purpose are widely available in the literature, though the cultural characteristics of the societal groups where the cybercartographic project is embedded, certainly plays a central role in the application of the different approaches. The manner in which these issues have an impact on the methodological aspects of the design and production of CentroGeo's five cybercartographic atlases thoroughly discussed in Chapter 6. User needs approaches are also considered in several other chapters in this volume.

The definition of the overall geospatial modeling and analysis approach is a key factor in the success of system modeling. At this stage the identification of theoretical frameworks and the concomitant methodologies becomes necessary. Modeling approaches will depend on the specific purpose of the cybercartographic exercise. An exercise, focused on business geographics, will follow a different approach from one focused on environmental issues. For example, a Spatial Analysis Modeling focus, together with a Territorial Planning Structure could be adopted for a specific project.

The metasystem stage is a conceptualization process where there is an essential feedback among its different components and it is a fundamental input for the other two stages of the structural methodology.

4.1.1.2. System Modeling

The production of a cybercartographic atlas involves three interlinked dimensions: modeling, communication, and knowledge. Modeling is a transversal axis in the sense that it is present in all of the dimensions. The metasystem guides the models needed in the design of a cybercartographic atlas, such as computational, geographical, visual, and cartographic models. In some cases well established models are used. In others, new models must be developed. Given the variety of modeling types, a multi-disciplinary team work approach has to be adopted.

4.1.1.3. The Technological Solution

Once the design of the system's model is in place, the technological strategy that will meet the needs of the atlas must be identified. As mentioned above, technology plays a central role in the development of cybercartography. The designers have to take advantage of the information and communication technologies available, and in some cases, develop new tools. Chapter 6 analyzes the manner in which these powerful resources relate to and support the development of cybercartography.

4.1.2. The Content Framework

The content framework differs from the structural methodology as it focuses on the expert knowledge, derived from scientific disciplines that respond to the need for explanation and/or prediction of the situation posed by societal demands. This theoretical and methodological approach responds to specific problems or situations identified by the users, which provide essential guidance for the content of the cybercartographic atlas. For example, in the case of the Lake Chapala Atlas, the theoretical and methodological framework of landscape ecology and territorial planning theories were adopted in order to provide a tool that could be embedded into the public policy processes for the region (Fig. 4.11).

4.2. Application Aspects of the Methodological Approach

The conceptual guidance in the design and implementation of the atlases is supported by the structural methodology, which is intimately related with

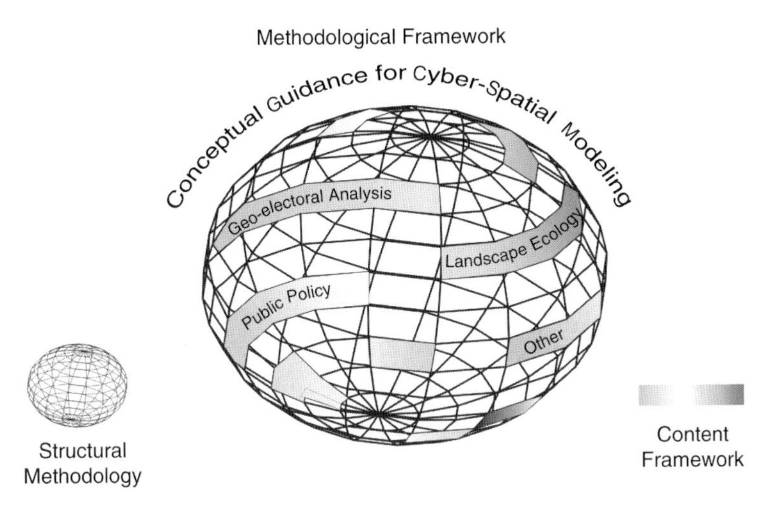

FIG. 4.11

the content framework. The cybercartographic atlas is conceived as a system consisting of a set of subsystems as described in Section 3.2. The identification, characterization, and functionality of the subsystems respond to the modeling approach adopted for specific problems or situations. For example in the Lake Chapala Atlas, each of the subsystems correspond to the basic components of a territorial planning exercise (description, diagnostic, prospective, management, feedback) and the computer menu to access information is based on a landscape ecology approach as shown in Fig. 4.7.

The three levels of modeling and design in the proposed methodology are strongly interrelated. Throughout the overall design process, these three levels are interacting; the analysis, knowledge, and information derived from the meta-system is used as an input for the system modeling process, and the technological solution responds to the needs established at the first two conceptual levels.

This design stage is not a linear process. The working team undertakes the methodological approach through a collaborative approach that allows the creative abilities of each of the members to be incorporated. For example, in the communication axes, the definition of the geospatial messages that incorporate specific elements, such as maps or geo-text, and the computational models involved, are the responsibility of a multidisciplinary group. The structural methodology points to the right language and media to be used, while the content framework gives guidance to the specific bodies of knowledge required and to the content of the geospatial messages.

The production stage uses procedures that have to fit the methodological framework. The design of the production phase and the specific manner in which these tasks are organized and accomplished are dependent upon the cultural characteristics of the working groups involved in the project. A first prototype is usually created that serves as the initial contact point for the feedback process in the cybernetic cycle.

The methodological conceptual framework discussed above served as a guide in the design, production, and implementation of CentroGeo's cybercartographic atlases. The results obtained since 1999 provided empirical evidence that the success of cybercartographic atlases, as tools in environmental and territorial planning projects, within Mexican and Latin American society is due to the adoption of this methodological framework.

As with any methodology, however, the approach is not a simple conceptual "recipe." When a group is established to undertake a project in cybercartography, the organizational structures should already be in place and the research team must have the knowledge, background, empirical "know how," leadership capabilities and artistic talent required if a successful result is to be obtained.

5. Conclusions

There are certainly many unanswered questions in this very early stage of conceptualizing and applying cybercartography. Are we really at a breakthrough in the development of cartography as Taylor suggests in Chapters 1 and 23? Is this new paradigm worth exploring? Which other empirical avenues of research should be tested?

The ideas behind cybercartography still need to be further developed, tested, strengthened, and broadened before key philosophical questions can be answered. It is simply too early to either abandon the concept or to try to establish a philosophical framework around it.

Since 1997 when the concept was first put forward by Taylor at least two research groups have embraced the idea. The empirical work derived from this first vision has had an impact on societal groups with distinct characteristics and in different parts of the world. The geospatial awareness of governments, international organizations, public policy specialists, and communities among others has evolved with the insertion of cybercartographic atlases to societal processes that include management, decision making, and planning. Geospatial awareness includes the development of national spatial data infrastructures in over 20 countries.

This book responds to the need to make explicit the knowledge acquired by the actors involved since 1997. There is a need to deepen the theoretical and methodological frameworks but also to pursue empirical research. The chapters in this book make an important contribution in this respect. There are a number of important avenues for further research:

- The concepts underlying the theoretical framework used in this chapter should be further explored. Only basic ideas on cybernetics and general systems theory were used as building blocks, and these need to be elaborated.
- The principle of undertaking cybercartography as a trilogy together with science and society should be retained and further explored.
- Existing cybercartographic atlases should be tested within the cybernetic cycle.
- The methodological framework should be refined, transmitted, and compared to other research and development groups so that cybercartographic atlases are enriched and applied in other contexts in terms of content themes, geographic regions, and societal groups (see chapter 23). Given the results of the educational Atlas of Lake Chapala and the Canadian Geographic Explorer (Taylor 1997), education seems to be a natural context for further work. This is explored and developed in several chapters of this book.
- The development of cartography has responded to the needs of society in different periods of history. Globalization, the inability of contemporary governments to solve even the basic needs of their populations and in general the complexity of the information society is demanding creativity from geo-scientists.
- O'Connor and Robertson (2002), argue that, cartography was from its onset part of the mainstream of mathematics. A new examination of the interrelationship between mathematics and cartography would be a fruitful area for research.
- Given that cybercartographic products are very adaptable, exploring collaborative research with artificial intelligence could lead to the development of intelligent atlases.
- New bridges are being built between different languages, such as those of space, speech, and music (Patel *et al.* 1998), and between the languages, music and mathematics. Taylor builds on these ideas and argues in Chapter 1 that a new language of cybercartography is emerging to complement GIS as "the LANGUAGE of geography."

The development of this language and its relationships with other languages needs to be explored. Of special interest in this respect is the relationship between geo-text and mathematics.

- Weinberger (1994) observes; "When we see an object such as a red ball, we do not experience the shape of the ball separately from its color. And when we hear a violin, we do not perceive its pitch separately from its timber. The notes in a chord are not heard as several individuals but rather in a more holistic fashion." In this sense there is much research to be done around the manner in which cybercartographic atlases are perceived and processed by the brain as a "whole," where several models and languages are combined to represent the geospatial landscape and to communicate spatial information and knowledge. The contribution of philosophy in this respect is explored in several chapters in this book.
- To explore the concept cybercartography through "atlases" has produced some important results, though other cartographic and/ or geomatics empirical and theoretical frameworks could also be developed around the essence of this innovative concept. The author proposed the concept of "Complex Solutions in Geomatics" (Reyes et al. 2005) and some empirical and theoretical work is already underway at CentroGeo.

In this chapter cybercartography has been approached from a modeling perspective. For that purpose three well-established scientific areas of knowledge (modeling, general systems theory, and cybernetics) were used as building blocks to explore possible theoretical and methodological frameworks. There are, however, other issues and unanswered questions that should be addressed from other perspectives. For example, within the cybernetic cycle, society plays a central role in the development of cybercartography. Social issues are involved in cybercartographic processes. This is certainly the case with collaborative research, the organizational and institutional contexts of research, and the adoption of technological innovations among others. Some of these issues are explored by Lauriault and Taylor in this volume and by Martinez and Reyes using some of the more recent developments in second generation cybernetic theories.

Acknowledgments

Fraser Taylor had the vision to create the original concept and promote a new avenue of research in cybercartography. The team of researchers, technologists, and technicians from CentroGeo were heavily involved in the empirical stage of the development of cybercartographic atlas prototypes and lead the production and technological groups. Yosu Rodriguez and members of Mexican society contributed to the design of prototypes by sharing their geospatial information and knowledge needs. Special thanks to Elvia Martínez, Araceli Reyes, Mauricio Santillana, Wayne Luscombe, and Tracey Lauriault for aiding in the revision of the chapter and to Carlos Payno his contribution in the design of the figures.

References

Ackoff, R. L. (1974) *Redesigning the Future, A Systems Approach to Societal Problems*, John Wiley & Sons, New York.

Baragar, A. (2001) *A Survey of Classical and Modern Geometries*, Prentice Hall, Inc.

Bierwisch, M. (1994) "How much space gets into language?" in Bloom, P., M. A. Peterson, L. Nadel and M. F. Garrett (eds.), *Language and Space, A Bradford Book,* The MIT Press, Cambridge Massachusetts, London, England.

Bissell, C. and C. Dillon (2000) "Telling tales: Models, stories and meanings", *For the Learning of Mathematics*, Vol. 20, No. 3, 3–11.

Bloom, P., M. A. Peterson, L. Nadel and M. F. Garrett (eds.) (1994) in *Language and Space, A Bradford Book*, The MIT Press, Cambridge Massachusetts, London, England.

Board, C. (1967) "Maps as models," Chapter 16, in Chorley R. J. and P. Haggett (eds.), *Models in Geography*, Methuen, London.

Clarke, K. C. (1995) *Analytical and Computer Cartography*, Prentice Hall Series in Geographic Information Science, Englewood Cliffs, New Jersey.

Dubinsky, E. and G. Harel (1992) "The concept of function", *Mathematical Association of America.*

Feeman, T. G. (2002) "Portraits of the earth, a mathematician looks at maps", *The American Mathematical Society, Mathematical World,* Vol. 18, The American Mathematical Society.

Feynman, R. P. (1999) *The Pleasure of Finding Things Out*, Penguin Books, London.

Firby P. A. and C. F. Gardiner (1982) *Surface Topology,* Ellis Horwood Ltd., Chichester, England.

Harley, J. B. (1991) "Un cambio de perspective", *El correo de la UNESCO*, Mapas y cartógrafos.

Haseloff, O. W. (1970) "Cuestiones fundamentales de la cibernética", (Grundfragen der Kybernetic), *Editorial Tiempo Nuevo*, Caracas, Venezuela.

Jackendoff, R. (1994) "The architecture of the linguistic-spatial interface" in Bloom P., M. Peterson, L. Nadel and M. F. Garrett (eds.), *Language and Space, A Bradford Book*, The MIT Press, Cambridge Massachusetts, London, England, pp. 1–24.

Jordá, M. and C. Reyes (1993) "Petróleos Mexicanos corporate geographic information system (SICORI)", *GIS/LIS Proceedings*, Vol. 1, pp. 331–340.

Keates, J. S. (1996) *Understanding Maps*, Addison Wesley Longman Limited.

Kuipers, B. (1978) *Cognitive Modeling of the Map User*, MIT Artificial Intelligence Laboratory.

Lauriault, T. P. (2003) Cybercartography and the New Economy Final Report, Research Collaboration, Geomatics and Cartographic Research Center, Carleton University, Ottawa, Canada.

Linstone, H. A. and M. Turoff (eds.) (1975) "The Delphi Method: Techniques and Applications, Addison Wesley, Reading, Massachusett.

Lluis-Puebla, E. (2004) Cuadrivium: La Matemática en la Musicología, *Opción, Revista del alumnado del ITAM*, No. 124, año XXIII, marzo 2004, Instituto Tecnológico Autónomo de México.

Luscombe, W. (1986) The Strabo Technique, PhD dissertation, Simon Fraser University.

Mazzola, G. (2002) *The Topos of Music*, New Publications, Birkhauser.

Myers, M. D. (1997) *Qualitative Research in Information Systems*, www.auckland.ac.nz/msis/isworld/.

O'Connor, J. J. and E. F. Robertson (2002) *The MacTutor History of Mathematics Archive*, School of Mathematics and Statistics, University of St. Andrews, Scotland, accessed 14 February 2005 from http://www-history.mcs.* st-andrews. ac.uk/history/index.html

Patel, A. D. *et al.* (1998) *Processing Syntactic relations in language and music: An Event-Related Potential Study*, The Neurosciences Institute, La Jolla, accessed 7 February 2005 from http://neurocog.psy.tufts. edu/papers/pdfs/Patel1998.pdf

Peucker, T. (1972) *Computer Cartography, Commission on College Geography*, Resource paper No. 17, Association of American Geographers.

Reyes, A. (2005) *Algebra Superior*.

Reyes, C. (1986) Neighborhood Models: An Alternative for the Modeling of Spatial Structures. PhD Thesis, Simon Fraser University.

Reyes, C. (1997) A Qualitative Approach for the Design and Implementation of Community-Based Geographic Information Systems, World Bank, Unpublished.

Reyes, C., F. Taylor, E. Martinez, and F. Lopez, (2005) "Geo-cybernetics: A new avenue of research in Geomatics?," *Cartographica* (in revision).

Rodriguez Y. (2000) "Environmental Reporting by Government in Mexico", paper prepared for the Seminar on Public Access to Environmental Information, Working Party on Environmental Performance, OECD.

Russell, B. (1967) *The Impact of Science on Society*, George Allen and Unwin Ltd.

Russell, B. (1962) *The Scientific Outlook*, George Allen and Unwin Ltd.

Tall D. (1991) *Advanced Mathematical Thinking, Mathematics Education Library*, Kluwer Academic Publishers.

Taylor, D. R. F. (1997) *Maps and Mapping in the Information Era*, Keynote address to the 18th ICA Conference, Stockholm, Sweden, in Ottoson, L. (ed.), Proceedings, Vol. 1, Swedish Cartographic Society, Gavle, pp. 1–10.

Taylor, D. R. F., C. Reyes and M. L. Alviar (2001) "Capacity building for cybercartography: The cybercartography for the Americas project", Scientific and Technical Program Committee LOC for ICC 2001, Beijing, China, *Proceedings of the 20th International Cartographic Conference, Mapping the 21st Century*, Vol. 1, Chinese Society of Geodesy, Photogrammetry and Cartography, Beijing, pp. 69–75.

Taylor, D. R. F. (2003) "The concept of cybercartography", Chapter 26 in Peterson, M. P. (ed.) *Maps and the Internet*, Elsevier, Amsterdam, pp. 405–420.

Von Bertalanffy, L. (1950) "An outline of general systems theory", *British Journal of Philosophy of Science*, Vol.1, pp. 134–164.

Weinberger, N. M. (1994) *Musical Building Blocks in the Brain*, University of California, accessed 7 February 2005 from http://www.musica.uci.edu/mrn/V1I2F94.html#building

Wiener, N. (1950) *The Human Use of Human Beings*, Cybernetics and Society, Avon Books.

Wiener, N. (1967) Comienzo y progreso de la cibernética, in Haseloff O.W., Cuestiones fundamentales de la cibernética, (Grundfragen der Kybernetic), 1970, Editorial Tiempo Nuevo, Caracas, Venezuela.

Winograd, T. and F. Flores (1987) *Understanding Computers and Cognition, A New paradigm for Design*, Addison-Wesley Publishing Company Inc. Reading, Massachusett.

Wood, M. (2003) "Some personal reflections on change: The past and future of cartography", *The Cartographic Journal*, Vol. 40, No. 2, pp. 111–115, ICA Special Issue.

Zonneveld, I. S. (1989) "The land unit- A fundamental concept in landscape ecology, and its applications", *Landscape Ecology*, Vol. 3, No. 2, SPB Academic Publishing, The Hague, pp. 67–86.

CHAPTER 5

Cybercartography and Society

ELVIA MARTÍNEZ

Centro de Investigación en Geografía y Geomática J. L. Tamayo (CentroGeo), México, México

MARÍA DEL CARMEN REYES

Centro de Investigación en Geo grafía y Geomática J. L. Tamayo (CentroGeo), México, México

Abstract

During the past five years, CentroGeo has developed several cybercartographic atlases. This chapter explores the circular relationships of these prototypes with the social systems they interact with. This exploration is guided by the conceptual framework of second order cybernetics that has the potential to approach communication processes in social contexts, as well as to explain more specifically issues related to language, cognition, and meaning; issues of utter importance when exploring the interaction between cybercartography and society. The cybercartographic atlases of CentroGeo play a major role as empirical Cybercartographic prototypes. This is a field where cartography and cybernetics converge generating messages in a complex language, which is able to transmit the spatial dimension in a holistic and effective way. The execution of the cybernetic character of these cybercartographic atlases demands active participation of their users and involvement in social interaction processes. Second order cybernetic concepts are used to explain the circular relationships between research groups, relevant stakeholders, and geographic space, during the design and development stages of the atlases; as well as, the feedback loops and second order cycles present in the interactions between users and technological prototypes. Users and stakeholders act as observers of the geographic space and observe themselves as participant actors in that space. Also, in group communication processes, these atlases may play a relevant role in the construction of a spatial language and in the triggering of processes of decision making and planning of joint actions to improve situations prevalent in the geographic space. Cybercartographic atlases are described as social products, fed by society in their development, and once embodied in technological prototypes, as entities that may impact the social context in which they are inserted. Three possible cybercartographic atlas roles are explored: their role in the construction of a spatial language among the users; their role in the user's new perceptions of the geographic space including the awareness of his/her role in shaping such space; and their role in social steering processes. Some issues are advanced for further research in the field, such as the role of cybercartographic atlases in cognitive processes involving the spatial dimension, in the construction of social knowledge about the geographical space, in the use of this knowledge in collective action, and in the building of self-steering capacities among the intervening social groups and organizations. Also, work remains to respond to the user's needs, demands, and purposes; to facilitate the effective social insertion of the prototypes and to include the cultural dimension in the design process.

1. Introduction

Cybercartography and society is an interesting topic to explore from a cybernetic perspective. This chapter approaches the topic by analyzing the results of empirical work on cybercartography developed over the last five years within Mexican society. This empirical work produced five cybercartographic atlases described in Chapter 6. These were developed by an interdisciplinary research team at CentroGeo supported by advisors involved with RedGeo, the CentroGeo Virtual Research Network.

These cybercartographic atlases are examples of the communication paradigm that is emerging in cybercartography, and the exploration of the interaction process with society is necessary for further development of the paradigm of cybercartography. A cybernetic perspective is the approach used in this chapter.

Why such a perspective? First, because cybercartography, as argued by Reyes in this volume, includes a strong cybernetic element. Second, as will be discussed later, the metadiscipline of cybernetics offers a useful conceptual framework to explain the behavior of open, dynamic, and complex systems, including social systems. But most importantly, because the purpose of this chapter is to explore the non-linear relationships between cybercartographic atlases and the social systems they interact with in complex ways. These products offer spatial information and knowledge to society, while taking from society the relevant elements required in the construction of the atlases themselves. Cybercartographic atlases are social products built through interaction with society and evolving along with the social systems in which they are embedded.

Heylighen and Joslyn's summary on the origins, the actors, and the evolution of cybernetics, turns out to be most enlightening in explaining the meaning of the prefix cyber. Today, the prefix is commonly used to denote, in a very restrictive way, something that is related to computer technology. However, the prefix actually means to steer or to govern (Heylighen and Joslyn 2001, Reyes in this volume). The core of Wiener's definition of cybernetics "the science of control and communication in the animal and the machine" is the concept of control and the potential of cybernetics to consider social systems.

In its first stages of development cybernetics concentrated on studying the teleological behavior of servomechanisms and automata. The first cyberneticians viewed machines as if these were able to imitate human behavior. Nevertheless, their initial discoveries, such as those in the field of neurophysiology, pointed out the need to study purposeful systems in general. This was accomplished by focusing on circular mechanisms, such as negative and positive feedback, that allows self-steering systems (machines, organisms, human beings, and social systems) to control environmental disturbances and attain their goals or preferred states.

Since the dawn of cybernetics, the social character of communication has been recognized. Wiener, in his pioneer work, acknowledged the social character of communication. He approached society as an intercommunication system and noted that the individual constituent parts had permanent topological relationships "while the community consists of individuals with shifting relations in space and time, and no permanent, unbreakable physical connections" (Wiener 1948: 182). The relationship that gives unity to the community is the "intercommunication of its members" (Wiener 1948: 182); and the size of such systemic units is determined by effective transmission of information (Wiener1948: 186).

However, Wiener found obstacles to apply cybernetics to the social sciences. These include the contextual nature of social data and the fact that social scientists are biased rather than objective observers who influence their subject matter and may even change it (Geyer 1994).

Since 1970s, the field has witnessed the emergence of what some authors call second order cybernetics or sociocybernetics. Others prefer to view this as a new stage in the developmental process of the field "firmly ingrained in the foundations of cybernetics overall" with a "stronger focus on autonomy and the role of the observer rather than a clean break between generations or approaches" (Heylighen and Joslyn 2001: 4). It was von Foerster, who in 1970 and in 1974 made a distinction between first and second order cybernetics by creating new opportunities to study society from a cybernetic perspective (Geyer and van der Zouwen 1998: 1). Second order cybernetics, building on the concepts of classic cybernetics, has considerable potential to approach communication processes in social contexts, as well as to explain more specific issues related to language, cognition, and meaning. These issues become of paramount importance when exploring the interactions between cybercartography and society.

In the following sections some concepts of second order cybernetics will be introduced as basis for and exploration of the interactions mentioned above. Approaching the subject of cybercartography and society from a second order cybernetics perspective will allow an understanding of the processes observed during the creation of cybercartographic atlases within their social context. It will also identify avenues for further research.

A conceptual framework is required in order to derive hypotheses to guide future research in the open and circular relationship between cybercartography and society.

2. The Observer and the System

Von Foester made a distinction between classic cybernetics, as proposed by Wiener, Rosenblueth and Bigelow, and second order cybernetics. He defined classic cybernetics as "the cybernetics of observed systems," and second order cybernetics as "the cybernetics of observing systems" (Geyer 1994).

First order cybernetics follows an engineering approach, focused on the observed system, and steered by feedback processes that allow a controlled course toward desired goals or preferred stable states. This theoretical approach concentrates on observing, defining, or designing steering system states. The emphasis is on the distinction between the system and its environment; the management of that environment through corrective action; the relationships of the system's internal variables; the stability of the system; and the mechanisms that support its stable states (Geyer 1994).

Second order cybernetics builds on the scientific discoveries of 1970s. Most relevant to the field were the discoveries in neurophysiology by Maturana, whose research led to the formulation of a theory about the organization of living systems, language, and cognition (Winograd and Flores 1986: 38). Second order cybernetics focuses mainly on the interaction between the cybernetic system and the human observer, who is in himself/herself a cybernetic system. The main interest of this approach is in the interaction processes that include interaction with the observer who is considered a part of the system.

The shift of emphasis from the observed object (steering artifacts as servomechanisms or automata) to living systems with purposes of their own, including the system that studies them (the observer or researcher), makes possible the exploration of the relationships between cybercartography and society. Cybercartographic products are not merely engineered systems; they include software and hardware design, but they are complex open communication systems that emerge from the observation of the behavior of social systems in a given geographical space and evolve through continuous interactions with different social actors. Second order cybernetic approaches are therefore, required to study them.

It might be argued that as researchers play a detached observer role in modeling cybercartographic atlases they can be viewed from a classic cybernetic perspective. However, as argued by Reyes in Chapter 4, the production of cybercartographic atlases also includes processes. Researchers and users become a component of the unfolding helix described by Reyes in Chapter 4 that depicts the dynamic dimension of the atlases. The more observers perceive themselves as part of the process, the more cybercartographic atlases become interesting artifacts of second order cybernetics.

To explore the nature of the circular relationships between cybercartographic atlases and the social groups and organizations with which they interact, it can be observed that: there is an interaction between the research group creating the atlases and the relevant stakeholders in the geographic space being mapped; that users interact with a cybercartographic atlas in ways that may include feedback loops leading to incremental improvements of the prototype. This is mainly through the insertion of new information, but users also demand the development of new modules and better functionality; that second order cycles emerge when users act as observers of the geographical space and observe themselves as participant actors in that space. In cybercartographic atlases users can perceive the impact of their actions in the phenomena and situations that occur in the geographic space, as well as the way in which particular user actions relate to those of other social actors; and cybercartographic atlases play an important role in group communication processes including the construction of a spatial language. The decision making processes include the planning of joint actions to improve situations prevalent in the geographical space being mapped.

Cybercartographic atlases as a new form of communication may aid collective attempts for self organization.

Initially, the scientists involved in the pioneer work of cybernetics did not realize the potential of this approach when applied to the social sciences. It can be argued that this was due to the involvement of social scientists with the subject of their research. For example, Wiener in his early work argued that all the great success in precise science had been made in fields where there was a "certain high degree of isolation of the phenomenon from the observer." In the social sciences, "the coupling between the observed phenomenon and the observer is hardest to minimize" (Weiner 1948: 189–190). From this he draws attention to two shortcomings of the social sciences. The first deals with the considerable influence that the observer is able to exert on the phenomena that come to his/her attention. The second relates to the difficulty of making reliable predictions due to the limited availability of time series social data.

Despite Wiener's early opinions, many social scientists were very enthusiastic about the epistemological perspectives that cybernetics presented to study society. Heylighen and Joslyn recall that cybernetics grew out of the Macy conferences on cybernetics, a series of interdisciplinary meetings between 1944 and 1953, in which some social scientists, such as the anthropologist Margaret Mead, and the management specialist Stafford Beer participated. They also point out that the scope of this field quickly encompassed the study of minds and social systems (Heylighen and Joslyn 2001). In addition, despite his early views about social sciences, Wiener (1948: 183) recognized the role of the observer in the cybernetic process by observing that "social animals give meaning to signals that have not (linguistic) content by means of observation." Later, in his classic work on *Cybernetics and Society*, he stated "communication and control belong to the essence of man's inner life, even as they belong to his life in society" (Wiener 1956: 18).

In the process of development and use of cybercartographic atlases, the social dimension appears in the following: the social interactions that take place in the geographical space being mapped that are modeled as both information and knowledge in the messages contained in the atlases; the phenomenological domain in which the interaction between the cybercartographic atlases and the users of the information and knowledge contained in them takes place; and the consensual domain in which a spatial language is constructed among the relevant stakeholders through their joint interaction with the atlases.

The designers and the users of the cybercartographic atlases, as observers, may perceive the geographical space being mapped as a unified entity and see themselves as integral parts of this holistic geographical space. In this sense, they may become involved in second order cybernetic processes.

The question of how observers build models of the systems with which they interact is essential to second order cybernetics; the answer is found by focusing on the dynamics of the system. Although structure and function are inseparable qualities of systems, cybernetics is more interested in the functional aspects, while the emphasis of system theory is on their structure. Hence, in building a cybernetic model, issues concerning the way in which systems function, control their actions or communicate with other systems or with their own components are central issues.

One dimension of the analysis of cybercartography and society is the interaction between the researchers (designers and producers) and the socio-spatial system needed to model a cybercartographic atlas.

The circularity in the system–observer relationship allows for the modeling of purposeful behavior, that is, behavior oriented toward a pre-established goal. This differentiates cybernetic models from mechanistic models that are built upon cause and effect relationships.

Von Foerster, who was a major participant in the *Macy Conferences*, conceptualized the organization as a "constraint in a state space"; that is, as something whose order tends to increase. In order to characterize this organization as a system, an observer is needed, as it is the observer who continually revises his frame of reference (Scott 2003: 141). Thus, the definition of the system's boundaries depends upon the reference perspective of the observers, who have their own purposes and values, which influence their descriptions of the system.

Cybercartographic atlases are modeled by selecting the geospatial messages that are relevant for the observers (the research group). The information selected has to respond to the characteristics and needs of the social network of stakeholders and expected users and relate to their needs as well as to their culture and world view. "A second order cyberneticist... recognizes the system as an agent in its own right, interacting with another agent, the observer... and the result of observations will depend on their interaction. The observer too is a cybernetic system, trying to construct a model of another cybernetic system" (Heylighen and Joslyn 2001: 3–4). In the process of self referencing the modeling process evolves to include both the observed and the social interactions present in the process.

The observer's knowledge about the system is mediated by the schematic representations or models build to conceptualize that system. Such models do not necessarily take into account information that is not relevant for the purposes of the observer. In the process, the observer generates various potential models and the role of the external world is to reinforce some of these models and to eliminate the others. This selection serves the purposes of the control process. When control is pursued as part of a social process, selection is achieved through interaction (Heylighen and Joslyn 2001).

The social context poses the relevant problematic situations that researchers might tackle; the term is used here to mean a situation characterized by problems interlinked in complex and dynamic systems and builds on Ackoff's (1999: 117) concept of "Messes." The social context is the laboratory used by researchers to define the borders between the system they want to model and its environment. In this sense, society feeds the modeling activity.

The Cybercartographic atlases of CentroGeo are not mere depictions of the physical characteristics of the regions they encompass. For example, the *Chapala Atlas*, as a model of the geographic space, reflects the socioeconomic, political, and cultural dimensions that converge in that space, and its development began from a problem solving perspective to illustrate important interactions with socio-economic reality.

During the development of the cybercartographic atlases, researchers and stakeholders became part of the socio-spatial context. The messages exchanged in their dialogues were grounded in the social world and make explicit the spatial dimension of this world in the language of cybercartography.

As described in by Reyes in Chapter 4, while designing and developing the cybercartographic atlases, the research group applied an action research methodology (qualitative research) and became involved in the physical and social environment of the geographic space they were modeling.

Qualitative research as a methodology can be viewed from what Diesing (1971: 137) calls the holistic standpoint that "deals with the various ways of studying a whole human system in its natural setting." This methodology, applied in the modeling of cybercartographic atlases, combines action research with some elements of ethnography.

According to Rapoport (1970: 499) "action research aims to contribute both to the practical concerns of people in an immediate problematic situation and to the goals of social science by joint collaboration within a mutually acceptable ethical framework."

The use of ethnography by cultural anthropologists is reflected, for example, in the pioneer works of Malinowsky or Radcliffe Brown (Diesing 1971: 139). This method requires that the researchers become constituent parts of the social systems they investigate, in the belief that the social "reality" can be grasped only by focusing in the worlds that human communities construct for themselves. This notion is shared by Von Foerster, who considers the idea that the observer entered the domain of his own descriptions as an ethical principle (Scott 2003: 138).

The pursuit of a qualitative research methodology requires that the research group adopts a particular angle or perspective to approach the subject matter. This angle does not refer exclusively to the social relationships established with the different stakeholders involved in the process; it also refers to the capacity to understand the experiences of others – researchers or stakeholders. In this sense, the members of the research group become involved in the situation they try to approach, build processes of empathy with different social actors, and interpret the information through a shared experience with the stakeholders.

Cybercartographic atlases, as prototypes emerging from a qualitative research process, are social products. Their social character can be perceived in many ways. For example, they include the problematic situations that interest the relevant stakeholders – local and institutional – they contain information and knowledge of holistic situations rather than a collection of data; the form and content of their messages reflect elements of the local culture; but above all, they are systems open to interaction with different user groups: users can improve the atlases by providing better and up to date information or by building models to help guide the improvement of their geographic space. The purposeful character of the cybercartographic atlases becomes framed in the interaction with the users and the relevant stakeholders.

Cybercartographic atlases include information and knowledge related to a geographically limited space; the systemic organization of this information and knowledge derives not only from a holistic research methodology, but also from an interdisciplinary perspective that allows capturing of the complex situations occurring in that space. The integration of a multidisciplinary group for the development of the atlases allows the inclusion of a wide variety of perspectives and, above all, to focus on the blurry intersections of disciplinary borders in order to reflect the complexities of the phenomena and situations tackled. This is more fully explored in the chapter by Lauriault and Taylor in this volume. Such information and knowledge can be perceived from different points of view, depending on the cultural and historical background of the observer.

3. Cybercartography and Society

Cybercartographic atlases have been incorporated in different social settings. These include decision making processes involving actors from different governmental institutions at the local and federal levels, and local planning processes with the participation of organizations and actors of the private, social, and public sectors. They are also used in an educational setting. Reyes and Martinez give an account of such processes for the *Lake Chapala* cybercartographic atlas in this volume.

As discussed above, cybercartographic atlases are social products, fed by society in their development, but once embodied in technological prototypes they also have an impact on society. This impact was illustrated empirically in each particular setting and in some instances, described qualitatively as case studies (Rodríguez 2000, 2001). However, in order to understand this impact fully, much more research is required. In the following paragraphs some conceptual elements from second order cybernetics will be explored to suggest guidelines for research aiming to understand the ways in which cybercartography interacts with society. These elements cover three interlinked questions, which seem relevant to the exploration of this matter: What is the role of cybercartographic atlases in the construction of a spatial language among actors of the social systems that interact with them?; How do users build, through their interactions with cybercartographic atlases, (perhaps new) perceptions of the geographic space that include the awareness of their role in shaping such space?; and What is the role of cybercartographic atlases in social steering?

3.1. Language and Communication

The generation of knowledge in the field of communications has evolved following the expansionist paradigm of science. Ackoff provides a neat synthesis of this process. In 1940, Langer proposed the concept of symbol as an entity that provokes an external response independent of its physical properties. Two years later, Morris proposed language as the framework for the scientific study of symbols and proposed that symbols are to be approached as a part of a larger whole. Shanon, in pursuing the interest to optimize the amount of information transmitted through channels of communication, created a theory of information and proposed that the nature of reality could be found in communication, in which language is only one element, through which a message is transmitted from an emissary to a receptor. In 1947, Wiener postulated the concept of control and proposed communication as an element of it. Cybernetics was thus, born as the science of control through communication (Ackoff 1974: 12–13). Communication as a systemic and social activity deals with multiple languages constructed with different symbols and codes.

Taylor (2003) outlines the major elements of cybercartography picturing a highly interactive communication process that involves all the human senses, multimedia, and new telecommunications technologies, and involves an information/analytical package. Cybercartography can be thought of as a next step in the expansion process described above, this new step then may represent a framework for the scientific study of the effective communication of spatial information.

Experience in the development of cybercartographic atlases shows the emergence of an integral form of information and knowledge that links communication elements and media that collectively articulate the spatial dimension, but which lose their effectiveness in isolation.

Cybercartographic atlases create a new language, which is able to express the holistic nature of the geographical space. Their function is to guide social actors engaged in learning about the phenomena and situations that exist in that space. In terms of second order cybernetics, this language is not merely denotative, but connotative as it implies more than its literal semantic meaning.

Language exists in a consensual domain. Winograd and Flores argue that, in order to talk about a "reality," a language is needed and this language is a consensual system of interactions. They argue that "a language exists among a community of individuals, and is continually regenerated through their linguistic activity and the structural coupling generated by that activity" (Winograd and Flores 1986: 49).

Cybercartographic atlases are not mere computational systems. They are prototypes of a new paradigm of communication, and cybercartographic language involves new possibilities for the transmission and reception of messages about geographic space.

Some people lack the ability to think in spatial terms and are unable to conceptualize the geographical space in which they live. In that sense, the geographic space is not part of their reality. Cybercartographic atlases aim to involve people in this spatial "reality," help them to perceive it, talk about it, describe it, and eventually, act upon it. "Language is a form of human social action, directed towards the creation of what Maturana calls 'mutual orientation'. This orientation is not grounded in a correspondence between language and the world, but exists as a consensual domain – as interlinked patterns of activity" (Winograd and Flores 1986: 76).

Cybercartographic atlases articulate a communication process in a language with great potential to transmit spatial concepts and in a form that seems to link effectively with the users' processes of cognition. If the interaction between cybercartography and the user is effective, one would expect a relevant impact on the way in which the geographic space is conceptualized and acted upon.

In sum, the goal of cybercartographic atlases is to embed the spatial dimension in communication processes. This is reflected in a number of ways: in the way people talk about the environment in their daily lives and engage in actions to improve it; in the way decisions for collective action are taken; or in the way public policies that promote certain processes or activities and deter how others are built. Hence, there is a need to imprint the spatial dimension in people's cognitive processes, and to emphasize on its role in a variety of different situations that occur in particular geographic spaces including problem identification and solutions. If this dimension can be articulated in the language built in cybercartographic atlases, this new paradigm may play a major role in the actions required to improve the environment. As Toledo (1998) says, language is a useful instrument to describe facts; it can bring together the facts that can be seen and is intrinsically linked to individual and social perceptions of the world.

3.2. Perception of Geographic Space

The conceptual shift from first to second order cybernetics has epistemological implications. The observation of the system from a classical cybernetic perspective depends on an impartial observer and does not affect the dynamic or states of these systems. Second order cybernetics differs from this classical view; by including the observer as part and parcel of the system under observation. Second order cybernetics recognizes that the knowledge of the observed system is mediated by the simplified representations that the observer makes of that system.

From the traditional point of view of scientific rationalism, reality is established by facts that exist previously and is independent of the observer or the society and culture in which the system is found. Toledo summarizes research conducted following a rationalist epistemology: "facts constitute reality and the impartial observer has to collect them, classify them and arrive, through inductive reasoning, to generalizations that may lead to universal and necessary conclusions, avoiding prejudice and all forms of subjectivity" (Toledo 1988, author's translation from Spanish). This approach sees logic as the instrument that transforms empirical observation into laws and theories.

One of the most criticized aspects of the social sciences is that the investigator's observations are not objective, as they are filtered through the social position, values, and the self-interests of the researcher. This critique is based on the basic positivist ontological belief of the existence of an external reality, independent of the observer. According to this belief, objects pose intrinsic qualities; the task of the scientific endeavor is to discover these qualities in order to arrive at the "truth." By following this line of thought, scientific research builds upon the isolation of its object of study. Behind all of these, lies the main assumption of objectivity: "that the results of scientific processes are independent of the agents that generate them" (Jara 1998: author's translation).

The epistemological position of second order cybernetics rejects the claim of objectivity and embraces subjectivity in the construction of knowledge by including the interaction that takes place between the observer and the system being studied during observation. This position also rejects relativism by recognizing that knowledge is both built and validated socially. Toledo points out that the transmission of information, from the reception of a stimulus, to the formation of a sensation (through specific nervous channels), and then to the formation of a perception (for example by "seeing," or "hearing"), is influenced by previous learning. Formal and informal education influence what is seen and at the same time, censor those "visions," which do not correspond to the paradigm, by which the members of social groups regulate their behavior; as a result, individuals that share the same system of language-culture will tend to see things the same way, where others who do not will tend to see the same things differently (Toledo 1998).

Initial followers of cybernetics were more interested in the observation and design of systems within a framework in which they could be considered as impartial actors with no impact on the behavior of their subject matter. The systems they designed were controlled by mechanisms designed from the top-down. These efforts gave rise to the development of disciplines such as artificial intelligence and computer science.

The experiments of Maturana *et al.* had a considerable impact on the study of the relationships between perception and the perceived world. They observed for example, that different people perceive colors differently and not necessarily according to the scientific color wave length. They concluded that in the study of perception the nervous system should be considered as a generator of phenomena rather than a filter to map reality (Winograd and Flores 1986: 41–44). The results of Maturana's experiments influenced researchers working in cybernetics by drawing their attention to "the observer and the biological basis of perception and knowledge acquisition processes" (Geyer 1994). This resulted in review of fundamental cybernetic concepts and attempts to develop a new approach, which could supply insights for studying living organisms, humans, and social systems.

Deutsch argues: "men think in terms of models. Their sense organs abstract the events that touch them; their memories store traces of these events as coded symbols; and they may recall them according to patterns they learned earlier, or recombined them in patterns that are new." He goes on to say that when these patterns resemble a situation or class of situations in the outside world, they are understood with the aid of the model (Deutsch 1963: 19).

Heylighen and Joslyn (2001) note the interest of cybernetics in the properties of systems that "are independent of their concrete material components." These authors stress that to discover such properties, observers need to consider relationships or connections between the system's components. The observers themselves are a system that operates on, and is part of, how a system is described. The observer's model of the observed system is derived from experience but is not a literal description of that experience. The authors also remind us that cybernetics has always emphasized psychological, epistemological, and social issues, and that its founders always agreed upon the subjective character of modeling.

Geyer remarks that second order cybernetics agrees with the main stream of the Newtonian philosophy of science in the need to differentiate scientific from non-scientific knowledge. However, it rejects the idea that observations are not influenced by the characteristics of those doing the observing. Second order cybernetics argues that individuals construct their own reality according to their personal experience (Geyer 1994).

Constructivism is the epistemology of second order cybernetics. It denies the conception of knowledge as a passive reflection of reality. The observer and the observed situation cannot be separated; and the product of the observation will depend on this interaction. Heylighen (1993) argues that constructivist epistemologies "put the emphasis on the relativity or situation – dependence of knowledge, its continuous development or evolution, and its active interference with the world and its subjects and objects."

Jara recalls that Ludwick Fleck, in 1934, as a response to resist positivist orthodox scientists, postulated the social conditionality of all knowledge. Fleck considered that the scientific enterprise has historically had a collective character. Jara (1998) also points out Fleck's influence on Kuhn and other epistemologists that have built upon Fleck's ideas, which gave rise to new epistemologies of the observer.

Arnold argues that "scientific information cannot be supported in neutral observers of a transcendental ontology; they are relative to the point of view and possibilities of the observer, that is, to a context and to a background" (Arnold 1997; author's translation). Modern epistemologies, centered on the observer do not work upon objects but upon the observations that allow understanding of such objects. Scientific observers and the product of their observations contain each other. Any proposition is inextricably tied to a collective thought, and the knowledge is legitimated by different people sharing the same intellectual concepts (Jara 1988).

The arguments of Searle regarding the existence or nonexistence of a "reality" is important. Here, he argues that the fact that different people in different cultures perceive "reality" differently does not imply the nonexistence of a reality totally independent from the observer. He makes a distinction between features intrinsic to nature, whose existence is completely independent of any observer; and features whose existence depends on the intentionality of the observer (Searle 1995).

According to the constructivist point of view, subjects organize the world experience and actively construct knowledge. In this construction, a selection process is involved. The subject generates potential models of the external world, the role of the external world is limited to the reinforcement of some of these models, so that others are eliminated in the process of selection. The most important issue is the process of individual choice between different constructions in order to select the appropriate one. However, in order to avoid having a completely relative point of view, selection criteria, such as coherence and consensus, are needed. Coherence is the "agreement between the different cognitive patterns within an individual's brain, and consensus is the agreement between the different cognitive patterns of different individuals" (Heylighen 1997a). Selection criteria can be of different types, and their use in selecting knowledge depends upon the objectives and actions that are relevant to individuals or social groups (Heylighen 1993). Hence, objects and properties are relevant to the distinctions made by the observer. Although observers are placed in worlds they did not entirely construct, the perceptions they build emerge from their purposes and perspectives. As Deutsch (1963: 5) remarks: "To know thus always means to omit and to select. In this sense, no knowledge is completely objective."

Cybercartographic atlases contain information and knowledge about events happening in a territory. The research issue is whether individuals interacting with the atlases extract this information and knowledge and incorporate it in their individual or social cosmology.

Cosmology is a concept intrinsically related to the process of selection, as it defines a larger context of what people understand as "ordinary world" that grounds their sense of reality, their identity, and their codes of behavior (Abrams and Primack 2001). Cosmology is related to the way in which individuals and groups give meaning to daily events and make decisions based on this meaning.

Following the reasoning of Feyerabend, Toledo (1988) argues that the act of perceiving is subject to theoretical and cultural conditioning, and that the individuals, sharing the same language and culture will tend to perceive things similarly. According to Feyerabend, "meaning" is the product of a social construction and has both a cultural and theoretical base. This causes people to think differently, and to view (observe) the world differently. People's discourses emerge from the way they view the world (Toledo 1988). Because of its social character meaning cannot be reduced to the activity of isolated individuals. It is the product of social activity. Therefore, meaning is both contextual and historical.

Toledo, in revising Feyerabend's view, argues that science is a cognitive style or form to understand or express reality. This form coexists with other cognitive styles that can be equally successful, on their own terms. The selection of a cognitive style and a consequent "reality" is a social act that depends on the historical situation. In regard to this cognitive style Toledo recalls the notion of paradigm, proposed by Kuhn as "a cognitive gestalt of a group of individuals that infuses all aspects of their life generating tacit knowledge" (Toledo 1988, author's translation).

Toledo argues that cognitive style refers to the processing of information, while the paradigm deals with the product of that cognition. The paradigm is the knowledge sanctioned by the community. The process and the product are intrinsically related, thus, the knowledge of one leads to the understanding of the other and to the satisfactory explanation of the individual and collective behavior. Cognitive style and paradigm constitute the cognitive dimension of culture, although, as a way of life, culture is wider than the cognitive dimension, as it encompasses other dimensions, such as institutions, rites and norms of individual and collective behavior, interpersonal relationships, and the interchange of goods (Toledo 1988).

The dissemination of knowledge and information has an impact on society's cosmology, in particular when social groups within society legitimate such knowledge and make it useful in their daily lives. As Heylighen comments: "Most of the beliefs a subject has were not individually constructed, but taken over from others. This process of diffusion plays an essential part in the selection of ideas. Only ideas that are transmitted frequently are likely to be assimilated frequently. Each time an idea is communicated, it replicates, i.e. copied into another cognitive system" (Heylighen 1997b).

When cybercartographic atlases were imported into a social group dynamic, the resulting process embodied a new paradigm of communication of the geographic space that had an impact on the way these groups perceived the space and the phenomena occurring in it. It is worth mentioning that such groups shared a territory and a history and, consequently, a world vision, a cosmology. The challenge is to discover how these perceptions influence the way these groups give meaning to their reality; and whether this meaning leads them to generate and integrate an image of shared problems that need a consensual vision and collective action if acceptable solutions are to be found.

Incorporating cybercartographic atlases into communication processes about the geographic space, may have an impact on the way social actors perceive the situation and perceive themselves as part of the situation.

The interaction of these actors with the complex patterns of information and knowledge contained in the atlases takes place in a common cultural context but becomes filtered by the subjectivities of different social groups.

However, it is worth mentioning that the social actors, who interacted with cybercartographic atlases were members of different social, economic, political, religious, or ethnic groups, and had therefore, different and sometimes diverging or conflicting interests. Even though they share a cosmology, their social positions had an impact in the way they construct their perceptions of the territory they share – either physically or institutionally.

An action may have different meaning according to the values of each social group. For example, to cut down trees in a forest may mean, for families in poverty, the possibility to prepare their food, while, for a group of ecologists, it may mean threat to the environmental equilibrium. Instead of exploring ways to reconcile diverging interests, it would be worthwhile to find a way for the members of both groups to separate themselves from their current role in the system and to act as observers of their own behavior, the behavior of others and the impact of such behavior in the same geographical space.

Cybercartographic atlases offer the opportunity for the stakeholders involved to act as observers of their own system. By interacting with the atlases they may select information about the geographical space they share in order to construct knowledge about it and about their role in it, and to complement or change the views they hold about the nature and causes of the problematic situations and phenomena occurring in that space.

Cybercartographic atlases execute their true cybernetic character through the interactions with users and groups of users. They have been inserted in social contexts in which the interaction is not a one to one user-prototype relationship but rather one in which the atlas is just one more actor in a group-communication process, the role of which is to facilitate human communication. Users and atlas get engaged in a process of conversation and knowledge construction with the atlas. Social actors can make explicit their different perceptions, interests and preferences and can become involved in consensus building processes (Rodríguez 2001).

Pangaro (1996) comments: "cybernetics, which is itself a formal inquiry into what we can know and how we know it, is always concerned with conversations... As observing beings we learn what we learn by interacting with our environment: the spaces, objects, processes and others-who-are-also-observing all around us.... Metaphorically speaking, we "converse with everything in our environment."

3.3. Social Steering

Deutsch in his classical political science text discusses cybernetics and says that it is communication that holds together any kind of organization, from living cells in the human body or pieces of machinery in an electronic calculator to thinking human beings in social groups. He also considers that the understanding of problems of communication and control in organizations could be increased by the study of self-steering systems (Deutsch 1963: 77–78). In this regard of particular interest is Geyer's assertion on the value of the social sciences in delivering useful knowledge to improve the steering capacities of social systems (Geyer 1994).

Self-steering refers to "the notion that a given system actively controls its course of activity within some external environment or a general set of possible states" (Whitaker 1995). Steering is a process fed by cycles of communication. Improving the steering capacities of organizations or social groups means finding ways to guide these systems toward desired goals or preferred states. This issue has been the main problem of planning, which in itself is encompassed by second order cybernetics. Scott (2003: 144) recalls that Von Foerster noted that "at any moment we are free to act toward the future we desire."

Little (1994: 159) argues that: "a central issue of sociocybernetics is the question of societal steering," and paraphrasing Kickert, he recalls that the last half of the twentieth century made evident the limitations of our ability to control society. Deutsch (1963: IX) also comments that "it would be more profitable to look upon government somewhat as a problem of power and somewhat more as a problem of steering."

Beer (1994: 104) remarks that to exert control "is to facilitate the speaking of the language of a recognized structure... (and) any system can be made responsive if it's talked to in the right language." Cybercartographic atlases talk the language of geographical space and they unfold its cybernetic character through their interactions with social actors.

At a social level, cybercartographic atlases can be viewed as a model of communication to help steer social systems. They contain information and knowledge framed in a communication paradigm that involves the spatial dimension, and information and knowledge are key elements in decision making and social action. According to Deutsch (1963: 129), the steering of a social system requires information about the world outside, the past (with a wide range for recalling and recombining) and itself and its own parts. The atlases provide this kind of information linked to geographical space, and therefore, are useful instruments in supporting the steering of social systems in a territory.

When cybercartographic atlases were incorporated in planning efforts at local levels it was observed that they helped identify and diagnose problematic situations in geographic space, and allowed the exchange of ideas and suggestions about viable alternatives for collective action, that could help deal with environmental challenges (Rodríguez 2001). All of these issues are relevant to the development of action and planning strategies.

Strategy involves developing the means to achieve the control of interacting processes. A major problem is that, in complex situations, control is distributed as opposed to being concentrated in a central unit making it impossible to predict what will arise in the dynamics of a system. As a result, strategy cannot be built only by logical processes; it must include processes that build meaning and identity among relevant stakeholders. This is the challenge that planning has to face.

Multiple actors intervene in the building of strategy. As isolated individuals, they interpret the internal and external relationships of the system, construct knowledge about it, and act accordingly. But while building a strategy, they have to negotiate and adjust their mutual relationships in order to generate a process to move toward the desired direction. Cybercartographic atlases support the design of strategy through social interaction processes. The atlases provide information and knowledge that help reduce uncertainties, they support communication processes that help generate consensus and shared agreements.

Nevertheless, the effectiveness of these processes still has to be assessed, as the implementation of many local initiatives need more than the consensus of the actors involved. They need actual strategy and energy from their environment, mainly in the form of resources and institutional rules.

The issue about the energy needed from the environment leads us to a central question about the steering system: Should this be part in local processes or processes induced externally? Geyer asks whether the behavior of individuals should be planned from the top-down or by actors operating at every level with adequate competences to handle their environment more effectively. He concludes: "since human societies are not simply self-regulating systems, but self-steering systems aiming at an enlargement of their domain of self-steering, there is a possibility nowadays..., for a coexistence of societal governability with ever less control, centralized planning and concentration of power" (Geyer 1994).

Governance from a second order cybernetic perspective is a matter of autonomous self-control and if that is so, cybercartographic atlases have a relevant role to play in regional planning processes as communication tools that may support self-steering efforts of social groups. These atlases, if adequately inserted, may help steering efforts aimed to reproduce the social system, adapt it to the environment, and attain its accepted goals and cultural integration.

Van der Zouwen remarks that "social systems are more than boundary maintaining, goal seeking, input-output machines... social systems can observe themselves, may learn from their experiences, can organize and steer themselves, change their structure in order to better cope with challenges coming from their environment, and even reproduce themselves" (Van der Zouwen and Van Dijkum 1998: 224).

4. Conclusion

Kuhn considers that scientific knowledge is social. He uses the concept of paradigm to depict the network of conventions and compromises, created among scientific communities in order to produce and legitimate scientific knowledge (Kuhn 1970).

Taylor (2003) argues that cybercartography is a new paradigm for cartography. In order for a model to be followed by a group of scientists, the group has to share an epistemological view and generate a body of knowledge that can be validated from such a view. In building this paradigm, the cybercartographic atlases of CentroGeo play a major role as empirical prototypes inserted in different social settings.

Cybercartography establishes the field where cartography and cybernetics converge generating a message, which can transmit the spatial dimension in a holistic and effective way. The experience in the development and social insertion of the cybercartographic atlases of CentroGeo depicts a new form of articulating elements, developing languages, and improving the means of communication about and within geographic space.

The main hypothesis of this chapter has been that the cybernetic character of cybercartographic atlases demands the active participation of the users and their involvement in social interaction processes; in these processes the atlases represent extensions of the users' minds. Users interacting with the cybercartographic atlases do not get linear answers to particular inquiries but complex holistic answers framed in a spatial context, meaning of which depends, to an extent, on the users' worldview. Thus, these answers depict an action space in which different social actors intervene, rather than leading to a unique course of action. The result of these interactions cannot be translated in predictable tasks but in processes that discover unfolding holistic situations in the geographical space that in turn, may be used as guidance in social steering efforts.

One could pose the hypothesis that the impact of cybercartographic atlases in social settings lies in the facilitation of consensus building among social actors. Through their interaction with the atlases, social actors, observing the spatial dimension and their role within it, may assign a shared or convergent meaning to different problematic situations, and may decide to act collectively toward the improvement of their environment.

This chapter has discussed the circular relationship between cybercartography and society. Such a discussion does not lead to definitive conclusions so much as to the identification of topics for further research in the field. These include the impact of cybercartographic atlases in people's cognitive processes involving the spatial dimension. Several chapters in this book address this challenge.

The role of cybercartographic atlases is the construction of social knowledge about the interaction processes in geographical space. This includes the use of this knowledge in collective action and in the building of self-steering capacities among the intervening social groups and organizations. The improvement of the design of the atlases in order to respond to the user's needs and demands, and to allow for a better information and knowledge base about geographical space, and its relevance in meeting user's needs. Several chapters in this book deal with this important topic. The conceptualization of cybercartographic atlas users as social networks, as opposed to isolated individuals, may facilitate the effective social insertion of their prototypes. Finally, the consideration of the cultural dimension in the design of cybercartographic atlases, that allow local actors to observe the spatial dimension in contexts familiar to them. Also seeing themselves as an integral part of this context should encourage users to identify themselves as relevant actors in their geographic context.

Cybercartographic atlases are the result of local social interactions. Their replicability lies more in the process than in the product itself, and a generalization on their usefulness is yet to be investigated. However, in a globalized world with a population that believes more and more in uniformity, cybercartographic atlases become a form of expression of the plural coexistence that enriches humanity. The transmission of atlases through modern communication technologies may take advantage of this asset.

References

Abrams, N. E. and J. R. Primack (2001) "Cosmology and 21st-century culture", *Essays on Science and Society, Science Magazine*, Vol. 293, No. 5536, pp.1769–1770,7 September 2001 accessed February 2005 from http://www.sciencemag.org/cgi/content/full/293/5536/1769

Ackoff, R. L. (1974) *Redesigning the Future: A Systems Approach to Societal Problems*, Wiley-Interscience Pub., John Wiley and Sons, New York.

Ackoff, R. L. (1999) *Ackoff's Best: His Classic Writings on Management*, John Wiley and Sons Inc., New York.

Arnold C. M. (1997) "Introducción a las epistemologías sistémico/constructivistas". Cinta de Moebio, No. 2, December, *Revista Electrónica de Epistemología de Ciencias Sociales.* Universidad de Chile, accessed February 3, 2005 from http://www.moebio.uchile.cl/02/frames32.htm

Beer, S. (1994) *How many Grapes went into the Wine? On the Art and Science of Holistic Management*, John Wiley and Sons, Chichester.

Deutsch, K. W. (1963) *The Nerves of the Government: Models of Political Communication and Control*, The Free Press of Glencoe, Collier Macmillan, London.

Diesing, P. (1971) *Patterns of Discovery in the Social Sciences*, Aldine-Atherton, Chicago-New York.

Geyer, F. (1994) "The challenge of sociocybernetics", Paper prepared for Symposium VI Challenges to Sociological Knowledge 13th World Congress of Sociology, Bielefeld, 18–24 July, accessed February 2005 from http://www.unizar.es/sociocybernetics/chen/pfge2.html, Subsequently published in *Kybernetes*, Vol. 24, No. 4, pp. 6–32, 1995 Copyright MCB, University Press, 1995.

Geyer, F. and J. van der Zouwen (ed.) (1998) "Sociocybernetics: complexity, autopoiesis, observation of social systems", *Contributions in Sociology*, No. 132, Greenwood Press, Westport, Connecticut, London.

Heylighen, F. (1993) "Selection criteria for the evolution of knowledge", Paper for the Proceedings of the 13th International Congress on Cybernetics, Association International de Cybernetique, Namur, in Heylighen, F., C. Joslyn and V. Turchin. (eds.), *Principia Cybernetica Web* (Principia Cybernetica, Brussels), accessed 3 February, 2005 from http://pespmc1.vub.ac.be/KNOWSELC.html

Heylighen, F. (1997a) "Epistemological constructivism", in Heylighen F., C. Joslyn and V. Turchin (eds.), *Principia Cybernetica Web* (Principia Cybernetica, Brussels), accessed 3 February, 2005 from http://pespmc1.vub.ac.be/CONSTRUC.html

Heylighen, F. (1997b) "Objective, subjective and intersubjective selectors of knowledge", *Evolution and Cognition*, Vol. 3, Number 1, pp. 63–67. in: Heylighen F., C. Joslyn and V. Turchin (eds.): *Principia Cybernetica Web* (Principia Cybernetica, Brussels), accessed February 3, 2005 from http://pespmc1.vub.ac.be/papers/knowledgeselectors

Heylighen, F. and C. Joslyn (2001) "Cybernetics and second order cybernetics", *Encyclopedia of Physical Science & Technology* (3rd ed.), Academic Press, New York. in: Heylighen. F., C. Joslyn, and V. Turchin (eds.), *Principia Cybernetica Web* (Principia Cybernetica, Brussels), accessed February 3, 2005 from http://pespmc1.vub.ac.be/Papers/cybernetics-EPST.pdf

Jara Males, P. (1998) "Las revoluciones de la ciencia o una ciencia revolucionaria. Convergencias y contrapuntos antes y después de khun", cinta de moebio, No. 4, December, *Revista Electrónica de Epistemología de Ciencias Sociales*, Universidad de Chile, accessed 3 February, 2005 from http://www.moebio.uchile.cl/04/frames05.htm

Kuhn, T. (1970) *The Structure of Scientific Revolutions*, 2nd edition enlarged, University of Chicago Press.

Little, J. H. (1998) "Autopoiesis and governance: societal steering and control", in Geyer F. (ed.), *Democratic Societies*, *Op Cit.,* 1998.

Pangaro, P. (May 1996) "Cybernetics and conversation, or conversation theory in two pages", published in *Communication and Anticommunication, American Society for Cybernetics,* accessed 3 February, 2005 from http://www.pangaro.com/published/cyb-and-con.html

Rapoport, R. N. (1970) "Three dilemmas in action research", *Human Relations*, Vol. 23, No. 4, pp. 499–513.

Rodríguez, A. Y. (2000) *Environmental reporting by government in Mexico*, paper prepared for the Seminar on Public Access to Environmental Information, Working Party on Environmental Performance, OECD, Working paper for the workshop included in the OECD's site OLIS.net

Rodríguez, A. Y. (2001) "Los servicios de información para la gestión ambiental en México: información, conocimiento y comunicación", in Anaya D. R. (ed.), *Ecología de la Información: escenarios y actores para la participación ciudadana en asuntos ambientales*, Nueva Sociedad, México.

Scott, B. (2003) "Heinz von Foerster – an appreciation revisited" *Cybernetics and Human Knowing*, Vol. 10, No. 3–4.

Searle, J. R. (1995) *The Construction of Social Reality*, The Free Press, New York.

Taylor, D. R. F. (2003) "The concept of cybercartography", Chapter 26 in Peterson, M. P. (ed.), *Maps and the Internet*, Elsevier, Amsterdam.

Toledo, N. U. (1998) "La Epistemología según Feyerabend", Cinta de Moebio, No. 4, December, *Revista Electrónica de Epistemología de Ciencias Sociales*. Universidad de Chile, accessed 3 February, 2005 from http://www.moebio.uchile.cl/04/frame01.htm

Van der Zouwen, J. and C. Van Dijkum (1998) "Towards a methodology for the empirical testing of complex social cybernetic models" in Geyer F. and van der Zouwen (eds.), 1998

Wiener, N. (1948) *Cybernetics or control and communication in the animal and the machine*, The Technology Press, John Wiley and Sons, Inc., New York.

Wiener, N. (1956) *The Human Use of Human Beings: Cybernetics and Society*, Doubleday Anchor Books, New York.

Winograd, T. and F. Flores (1986) *Understanding Computers and Cognition: A new Foundation for Design*, Ablex Publishing Corporation, Norwood, New Jersey.

Whitaker, R. (1995) *Self-Organization, Autopoiesis and Enterprises*, accessed 3 February, 2005 from http://www.acm.org/sigs/sigois/auto/Main.html

CHAPTER 6

Technology and Culture in Cybercartography

MARÍA DEL CARMEN REYES

Centro de Investigación en Geografía y Geomática J. L. Tamayo (CentroGeo), México, México

ELVIA MARTÍNEZ

Centro de Investigación en Geografía y Geomática J. L. Tamayo (CentroGeo), México, México

Abstract

Cybercartography emerges at a time of revolutionary advances in computing sciences and communication technologies. Consideration of the cultural characteristics of a society is essential for developing the content of the geospatial messages embedded in the communication process of cybercartography. The circular relationship between culture and technological development is recognized throughout the process in cybercartography to generate, support, and communicate geospatial information and knowledge. In this chapter, the authors explore the relationship among technology, culture, and cybercartography; share their experiences in the design and production of cybercartographic atlases that were developed within the Latin-American context and present a description of the *Cybercartographic Atlas of Lake Chapala* utilizing the theoretical and methodological frameworks described in Chapters 4 and 5 of this volume. This exercise enables the authors to complete the first stage of the scientific cycle of performing empirical work; making explicit the knowledge acquired in the process; rediscovering the atlases using the proposed theoretical and methodological frameworks; and finally, to propose new avenues of research.

1. Introduction

Technology is a revolutionary force, which continues to change many aspects of life at an increasingly accelerated rate. Bertrand Russell's (1967) reflections about the role of science in society underline the way in which science has been a dominant factor among educated men and women for more than 300 years, while the impact of technology has been significant for less than two centuries.

Russell (1967: 23) comments "science, ever since the time of the Arabs, has had two functions: (1) to enable us to know things, and (2) to enable us to do things." Technology is, therefore, an intrinsic component of science and in the twentieth century, technology had a profound effect on many aspects of society. For example, the atomic bomb in the last century changed geopolitical world scenarios in a similar but more extensive way than the introduction of gunpowder in the middle ages. Throughout history, technology has been both a threat and a source of hope.

Alternatively, Medhurst *et al.* (1990: ix) argue that the technological age commenced not with the atomic bomb or with computers but with the invention of language and points to the strong relationship between symbols and technologies. Medhurst also quotes Galbraith's definition of technology as "the systematic application of scientific or other organized knowledge to practical tasks" (Medhurst *et al.* 1990: x).

The generations, who have lived the contemporary technological age can easily identify a large number of scientific developments that have affected their personal lives. Millions of people are still alive due to the technological benefits, such as penicillin, and people have entered the 21st century in the midst of exponential growth in the number of innovations that impinge upon the way in which people live their daily lives.

The concept of technology has often been misused and misunderstood, especially in how it relates to science. Technology is sometimes viewed as less intellectually important than "pure knowledge," and there can be insufficient awareness of the basic science behind a technological development.

The fascination of society with technology and its unquestionable glamor often lead scientists to concentrate their attention on the artifacts themselves, rather than on other essential elements. These include, the processes involved in the incorporation of technology into societal activities, the intrinsic cultural aspects of technological development and use, and the needs of users.

There has been a trend over the last 50 years, toward increased specialization in either very specific areas of scientific knowledge or in the development of highly specialized technologies. The result has been that public policy actors and scientific communities alike have given insufficient attention and importance to the feedback process between the "knowing" and the "doing," as conceived by Russell. It can be argued that this has been the case in some of the spatial sciences in the field of geomatics.

For example, in the area of Geographic Information Systems (GIS) it is not unusual to find cases where governmental offices have acquired hardware and software with totally inadequate conceptual support for their applications. As a consequence, the technology is often used only for map reproduction and often the projects as a whole do not fulfill the expectations. The lack of appropriate knowledge frameworks to take advantage of available technological resources, and a misconception of the role of technology in cognitive and social processes has limited the potential of this emerging discipline. On the other hand, there is insufficient awareness in the geomatics scientific community about the importance of the interaction with society through geospatial technology to identify interesting avenues of fundamental research.

In cybercartography, as argued in Chapter 4 of this volume, both knowledge and technology play a fundamental role. In the first part of this chapter the role of technology and culture in cybercartography is analyzed. In the final section the results of seven years of empirical work are outlined by describing the cybercartographic atlas of Lake Chapala, both as a technological artifact and as an interactive social artifact. This encompasses the lessons learned from the incorporation of the atlas into the societal processes of this region of Mexico, as well as from the interactions with different groups of users. By describing the *Atlas of Lake Chapala,* we provide an empirical example that illustrates the cybernetic cycle, which we argue is inherent in cybercartography.

2. Technology and Culture

Don Ihde (1990), in his book on the Philosophy of Technology, defines a "notion of technologies as those artifacts of material culture that we use in various ways within our environment." As such, technologies have a core role in our daily lives, in fact, the author claims that "our existence is technologically textured" (Ihde 1990: 1).

Idhe differentiates the human experiences that are technologically mediated from those that are not, such as sensorial experiences or direct bodily contact with an environment. He concludes that "virtually every area of praxis implicates a technology. From burial to birth to eating and working, the use of artifacts embedded in a patterned praxis demarcates the human within his or her world.... the technological form of life is part and parcel of culture, just as culture in the human sense inevitably implies technologies" (Ihde 1990: 20).

By focusing on the phenomenology of the human experience of technology, Ihde proposes a typology that characterizes the relationships between humans and technologies. The types are:

(1) *Embodied relations*: These encompass humans' bodily engagement with a technology, as in the case where eye glasses become part of the way users ordinarily experience their surroundings. In this case, there is a binomial human-technology relationship to the world.
(2) *Hermeneutic relations*: These are an act of interpretation that is both textual and contextual and involve the user's language abilities. The reading of a text is an instance of this relationship in which linguistic transparency presents the world of the text. The world is brought to human perception by the technology.
(3) *Alterity relations:* By these, humans relate to technologies-as-other. Although there may be a relationship between humans and the world, through the technology, they need not be related and the world may remain as context and background. An example of this type is provided by human–computer relationships.
(4) *Background relations:* These are represented by functioning technologies, which ordinarily occupy background or field positions such as lighting or heating systems.
(5) *Horizontal phenomena:* The term is derived from "horizon" and these relations occur in the border area between the humans and the artifact, which itself practically ceases to be technology. One example may be the contraceptive pill that once taken becomes part of the organism's processes.

Although each of these elements in the typology can be observed, with different degrees of emphasis and in variable combinations, across cultures, the cultural diversity of societies introduces another dimension in the analysis of the relationships between man and technologies and in fact, in those between social systems and technologies. According to Ihde, this can be observed in the process of technological transfer from one socio-cultural system to another. The recipient receives not only the artifact per se but also a set of cultural relationships (values and processes). When the new technology relates itself with practices already familiar in the culture, the adoption of the artifact within its immediate use-context will be simpler than in those instances when the recipient culture requires learning and interpretation of the practice itself. However, in both instances the cultural interface involves the complex level of higher level cultural values. Furthermore, as Ihde argues, "technologies may be variantly embedded; the 'same' technology in another cultural context becomes quite a 'different' technology" (Ihde 1990: 144).

Technology has been recognized as a leading force in the development of the new paradigm of cybercartography. As such, it is necessary to explore the cultural dimension of these technologies: How do cultural contexts feed the construction of cybercartographic artifacts and how can these artifacts be introduced successfully in different social contexts?

The experience and knowledge of Dr Taylor provided him with the vision of the concept of cybercartography. The demand of society was the driving force for the development of cybercartographic atlases. The construction of these cybercartographic artifacts was supported as a methodological process as well as by the intuition of the research group at CentroGeo, who built these atlases. The vision and the intuition of the form, color, media, and the general characteristics of the atlases could not have become a reality without technology. Technology was the basic tool by which the researchers from CentroGeo could communicate with society. It should be noted that in the atlas creation experience, differentiation is required between the technologies themselves and the artifact; in a similar manner as the technology used to produce this book differentiates itself from the contents of the book.

In cybercartography, the technologists have to either use existing tools available in the market (such as multimedia, WEB and computer graphics technologies) or develop "tailor made ones." In this sense, the knowledge of the availability of information technology (IT) and communication technology (CT) is essential as well as developing trends in both fields.

As in many other areas of geomatics or geographic information sciences, geo-specialists take advantage of existing developments in IT created for other disciplines or adapt existing technologies (e.g. data base and visualization technologies). The best scenario has been where the geo-specialists have been able to obtain a suitable solution and to create specific geo-technologies, such as the implementation of topological data structures or cartographic generalization algorithms.

The development strategy in geomatics, in the last 20 years has given rise to the growth of software firms, which supports geospatial applications. For the software and hardware developer, the specialist and the user are often seen in separate "silos"; while the consultants in geomatics will bring together and integrate a solution by recommending or developing technology for specific user applications.

It is common in analyzing the "human factors" in the use of geospatial technology to take an approach, which focuses solely on issues, such as the user interface and usability, which are essential characteristics of the technology as such (Frank 1993: 13). Some of these are described in several chapters in this book. It is argued that cybercartography differentiates itself from the one traditionally followed by industry and opts for a systemic and holistic view, where the main issues are aimed in satisfying the wider needs of the users. This view includes an overall modeling process that incorporates various aspects, such as content, the knowledge frameworks of the users; qualitative characteristics of the process by which the artifact is embedded (organizational, community, or research environments among others), adaptability; openness; and of course, usability.

In empirical work in creating cybercartographic atlases it is possible to distinguish the role of technology (as an essential part of the artifact); the content of the artifact, the process of embedding the artifact in the suprasystem and the role played by the cultural context in Mexico and Latin America, where the atlases were designed and produced (Reyes in Chapter 4 of this volume).

The cybercartographic atlases of CentroGeo are not limited to the depiction of the physical, economic, and environmental characteristics of the regions they describe; they also reflect the richness and specificities of the socio-cultural context of these regions.

As will be discussed later in this chapter, the design and production of CentroGeo cybercartographic atlases were supported by a qualitative research methodology. By following this process of inquiry, anthropologists and graphic designers played a significant role, as they were able to capture in the cybercartographic atlases the cultural particularities of the geographical regions and express them artistically. This approach is cumulative. The knowledge gained during the design and production of the *Cybercartographic Atlas of Lake Chapala,* which was the first atlas produced, could be built on during the development of the other cybercartographic atlases (*Lake Pátzcuaro, Lacandona Region, Fire Risks and Lake Chapala: children's vesion*). For example, issues of ethnographic research and their relation to design were tackled in a more explicit way in the *Atlas of Lacandona*, where the role of anthropological research extended to music, design, and images included in the atlas. Besides being aesthetically important these elements allowed local actors to perceive the atlases as referents, which reflect characteristics of the local culture.

It is not yet clear whether the capturing of cultural elements in the cybercartographic atlases helps local users to adopt them more easily. In local processes that include showing the atlases information and processing capacities, using them as instruments to aid decision making and planning processes, or building local capacities for their use. The research group has observed that local users show great interest in the artifacts and relate to them in an easy way. However, these are the only observations, and further research is required in order to understand the way in which the incorporation of cultural elements in the design of the atlases helps the adoption process. In addition, there is also a great need to focus research efforts on more effective ways to capture cultural particularities in atlas design.

The cybercartographic atlases are expressions of "modern" or "sophisticated" technology. However, they address regions, which face complex environmental, economic, social, political, and cultural conditions. In all the regions infrastructure is poorly developed and there is extensive environmental degradation. They all have ethnically diverse populations with large sectors living in extreme poverty. The dynamics of these phenomena vary among the regions according to their particular geopolitical situation and history.

The adoption of cybercartographic atlases in such contexts poses major research questions. The visualization of geographical space and its correlated problems holistically is not a common practice in those contexts; nor is it in governmental institutions that deal with the environmental problems of the regions involved.

Hence, the process of adoption has to be intentionally promoted, by means of building local capabilities to perceive and understand the geographic space and to act upon it in ways that allow its harmonic development. Such endeavor needs to take into account the specificities that arise from the local cultures.

CentroGeo's cybercartographic atlases are technological prototypes developed to integrate and communicate spatial information and knowledge about structures and processes that occur in specific geographical contexts and to be used in local contexts as instruments that support decision making, collective action, and education. Accordingly, it might be concluded that they are technologies useful only in the socio-cultural context they target. However, this is not so; any user in any part of the world can relate to these atlases and learn in a holistic way, issues related to their knowledge model. For example, people from any nation can appreciate the ecological problems related to water resources by using the cybercartographic atlases of Lake Chapala or of Lake Pátzcuaro. Furthermore, the cultural roots of the atlas's designs may promote a global vision of a culturally plural planet.

2.1. *The User*

As mentioned by Taylor (2003a), the importance of focusing on a user-centered approach to the development of technology is essential for the success of cybercartography. Authors, such as Johnson (1998), have stressed the importance of focusing on users in the analysis of the role and trends of technology. Johnson is particularly interested in technical documentation centered on the user rather than on the designer or developer of the technology. Several authors in this volume consider this topic.

Johnson (1998: 13) argues that the Greeks treated technology from the point of view of the "use of the product, not in the design or making of the product itself." He introduces the concept of the "knowledge of use" indicating that, particularly, in the case of computers, there is a tendency to think that the computer specialists are the knowledgeable actors and that the users who do not understand the artifacts are some kind of "dummy" (Johnson 1998: 13). In more recent developments this view is changing, for example, computer scientists, such as Draper and Norman, have focused more on people rather than on the artifacts themselves expressing that their "book is about the design of computers, but from a user's point of view" (Johnson 1998: 12).

Johnson (1998: 12) also explicitly acknowledges the importance of culture in the development of technology and quotes the historian John Staudenmaier who argued that "all artifacts are affected by the social sphere, the cultural ambiance, thus making technological artifacts and systems dependent upon, instead of autonomous of, human intervention."

As Medhurst *et al.* (1990: x) mention, besides the "purist" definition of technology, it is essential to focus on the culture of technology. The political motives to develop technology, its psychological impact and its relations with the "economy, ideology and power" and the manner in which it affects the communication process are also issues of prime interest.

Wood (2003: 113–114), in reviewing the relationship between the instinctive nature of mapping and the evolution of cartography over the last decade, argues that technological developments provide new "interactive exploratory facilities" that allow their use in a more holistic sense. This represents a shift from previous trends that reproduced the dichotomy between professional mapmaking and map-use: "the mapper as percipient merely studied the map with a view to expanding her/his knowledge. Now she/he can truly interact with the map (and with its data sources) and become, in the process, also a mapmaker."

2.2. Communication as Technology

There is a vast literature that considers the topic of the role of technology in communication including all forms of media from paper to telegraph, telephone, voice and e-mail. Some aspects of the literature considered relevant to the communication process in cybercartography will be explored briefly in this section.

Medhurst *et al.* (1990: xiv) in their book on Communication and the Culture of Technology consider both language and logic as technologies in spite of the fact that they are not physical artifacts, such as telephones, televisions, or computers. Similarly, Beach while studying processes of group communication stresses "how language is used as a technology for accomplishing a task" and conceives a conversation as a "technological resource, the organization of which is ultimately rooted in practical circumstances of everyday choice and action" (Medhurst *et al.* 1990: xiv). These concepts point to relevant issues regarding the role of cybercartography in social processes. As will be presented in Section 4, there is an empirical evidence that shows the power of cybercartographic atlases as discussion tools in community groups that have quickly resulted in public policy action.

Penland (1974) in Communication, Science, and Technology, explores important issues, on communication services, especially those related with information, library, and media specialists. More than 30 years later, many of the technological limitations of that time, have been overcome. However, the main cognitive guide includes some of the basic concepts adopted for the theoretical framework presented for Cybercartography in Chapter 4 of this volume. In this regard, Penland takes a social-science research approach and a behavioral psychology perspective focusing on human communication and assumes that human behavior is the product of "a cybernetic system composed of a perception-concept subsystem, a motive-selector subsystem and an effector-feedback subsystem" (Penland 1974: vii).

2.3. The Spatial Dimension in Organizational Communication Processes

There is a voluminous literature regarding the use and/or impact of communication technology on organizations. For example, Büchel (2001: 18) reports rankings of the use of communication media in organizations, where face-to-face meetings have the highest score and faxes the lowest. Similarly, in terms of feedback the face-to face communication process is immediate while electronic mail is moderate and formal letters are slow. Büchel outlines the manner in which workers in an organization adopt different media to take into consideration a variety of factors, such as urgency, ability to convey personal feelings or communicate multiple cues, and the speed of feedback, among others. Visual communication is, however, not considered, and the role of cybercartography in knowledge organizations would be an interesting line of research.

Minoli (1994), in reviewing current and future electronic imaging systems, identifies the main users as bankers, medical, publishing, and scientific communities. The major benefits in using this technology include reduced storage costs and reduced chance of document loss among others. He only tangentially, mentions engineering and geographic applications. Cartographic applications have many of the same characteristics, such as scan, store, display, print, and transmit. Minoli wrote a decade ago, and technology has developed rapidly since. The importance of geospatial technology has been increasingly recognized, together with nanotechnology and biotechnology, as one of the leading technologies of the information era (Gewin 2004).

In more recent developments, Vriens (2004) analyzed, from different perspectives, the role of information and communication technology for competitive intelligence. In an organizational context, competitive intelligence is described as "producing and processing information about the environment of an organization for strategic purposes" (Vriens 2004: vi). The main argument rests on the fact that the business have always acquired and analyzed information regarding their competitors, suppliers, and consumers. A number of factors have emerged requiring a more detailed consideration of the resources to support organizational tasks and strategic decision making. These include the rapid change in the availability of information technology, the globalization of the markets, and a general increase in the complexity of the business environment. Vriens reviews basic concepts (data, information, knowledge, and intelligence), looks at competitive intelligence, both as a product and as a process, and gives guidelines for the acquisition and use of available software products. The author outlines the different models used by Dutch organizations, including simulation, scenario analysis, and competitor profiling to support competitive intelligence activities and points to the Internet as the best resource for information collection (Vriens 2004: 17).

On the topic of competitive intelligence, Hendriks (2004) looks at Geographical Information Systems (GIS) as a resource to incorporate the spatial dimension into the process. The author argues that this resource has been underused and presents the key elements for linking GIS and competitive intelligence. He recognizes that in order to take advantage of GIS it should be coupled with competitive intelligence at a conceptual level, and analyzes "the spatial perspective" in business. By presenting different lines of thought in spatial analysis and connecting them with specific interests in business, such as territorial market definition, routing, and geodemography, Hendriks illustrates the importance of a geographical approach in competitive intelligence. A detailed linkage with several of the competitive intelligence processes and spatial analysis potentialities are presented, though the limitations of the current GIS commercial software are being recognized. Although many of the topics considered by

Hendriks have been explored in other publications, such as the journal of "Business Geographics," the value of Hendrik's argument is in the effort to make the important conceptual link between competitive intelligence and the spatial perspective.

Cybercartography has the potential to be embedded in an organizational context and can open up the areas of empirical and theoretical work in organizational communication and competitive intelligence. Organizations need to learn how to use geographical space in strategy building and decision making processes and how to derive a competitive advantage from it. As cybercartography uses a spatial language, transmits holistic messages, and utilizes interactive tools, it can help its interacting social actors in perceiving geographical space through a systemic view. This will allow them to gain knowledge about the complex dynamic processes that occur in such space and enter in processes of communication and conversation that support consensus building and collective action toward shared purposes.

2.4. Society and Communication Technology

There are critical and thoughtful voices in this era of high-speed communication. Authors, like Inayatullah and Leggett (2002), question the role of communication in societal development. They refer to the fact that not only the message, information, and communication are involved, but also the "social, gender and civilisational context embedded in the process." In their edited volume several scholars analyze the costs and benefits of the communication era and its new technologies. Cyberspace can contribute to the creation of a more egalitarian world and support collaborative efforts among groups. For example, from a gender perspective, women can appropriate communication technologies to strengthen their networks. Other authors explore possible scenarios that could lead either to human well-being or to an overload of cybermarket and cyberfantasy via video games. The need to analyze the role of technology from a societal perspective is clearly highlighted. This includes issues, such as the need to select the technology appropriate to different cultural environments, methods of "culturally appropriate communication", and the politics of empowerment.

In synthesis, as mentioned in the statement by Inayatullah and Leggett (2002: 7) "technologies are part of culture not outside it." The design and development of technology cannot be disconnected from the cultural context, where it takes place. As cybercartography is, undoubtedly, based on information and communication technology, issues regarding its role in social processes and the identification of cultural elements that have an influence in the development of the concept itself are of considerable importance.

In the literature on Geographic Information Systems (GIS) some authors from a critical perspective, have expressed their concerns on the possible dual role of GIS as the empowerment vis-a-vis the marginalization of societal groups (see Harris and Wiener 1998). Some of the issues discussed include the social and ethical implications of the use of GIS and public access to geospatial information. With the exponential growth in the development and use of geotechnologies, in the near future avenues of research that focus on geomatics, culture and society, and human factors will certainly multiply.

At a national scale, Gupta (1999) analyzes the role and impact of media communication on Indian society. He recognizes electronic and satellite communication as the driving forces in the growth of the media industry and tackles their impact on "human activity-life styles, entertainment, work culture, modes of communication and... social and cultural values in society" (Gupta 1999: 1). The information revolution in India he describes includes issues, such as the convergence of computing, telecommunication and broadcasting technologies, and the actions taken to establish policies for developing education, the empowerment of women, and sustainable development.

Gupta (1999: 92) presents in a very inspiring manner, the concept of communication: the word communication is derived from the Latin word "Communis," which means:

> To make common, to share, to impart, to transmit. Communication is the basic instinct of man and a social, economic, political and cultural need. Without communication no society can exist, much less develop and survive. For the existence as well as the organization of every society, communication is a fundamental and vital process. The word communication means not only transmission but also community participation. Communication integrates knowledge, organization and power, and runs as a thread linking the earliest memory of man to his noblest aspirations through constant striving for a better life.

Information technologies, such as multimedia and the WEB, can be resources for economic development besides being "information carriers." The Cybercartography and the New Economy Project views cybercartographic atlases as powerful tools for communication, with a clear vision of the potential role of this new concept in economic development (Taylor 2003b).

Ramussen (2000) draws our attention to the fact that until recently, in the era of mass communication, the focus was mainly on technical, economic and policy issues. The cultural and social impact of artifacts, like the telephone were probably ignored due to its "domestic" character. With the rapid advance of the societal adoption of new technologies, such as the Internet, cellular phones, and video games the impact of these new technologies on "human action in an everyday world" and on human behavior and culture, is a cause of increasing concern. There is a need to incorporate more sociology and social theory in the study of communication technologies (Ramussen 2000: 1). Ramussen (2000: 6) presents a theoretical framework to advance the social understanding of what he calls the "new media" and "communication technologies" that include multimedia, hypermedia, and interactive media (Ramussen 2000: 1).

Two issues related to cybercartography are of special interest: the role of the user-technological artifact relations including the communication feedback process; and the role of geospatial information in broadcasting media.

The new paradigm of cybercartography needs to respond to these challenges by developing knowledge and technology to improve the performance of human-artifact interaction and by developing prototypes for the broadcasting industry.

2.5. Technology and Geo-communication

Traditional geo-communication technology is the "hard-map." For centuries it has served as a valuable resource in the most relevant societal activities and in almost every culture around the world as Monmonier argues in this volume. Throughout this book, the importance of information and communications technologies for the development of cybercartography has been emphasized. In the following review of some of the main technological resources there is a special reference made to existing shortfalls and possible new areas of technological development that could be explored within cybercartography.

2.5.1. Multimedia, the WEB and Computer Graphics

There is a voluminous literature on multimedia, the WEB, and computer graphics. A number of these studies are of particular interest for the production of cybercartographic atlases. Mahbubur (2001) explores some key issues including: the standardization of terminology, its study from a semiotic perspective, the use of fuzzy query languages, spatiotemporal data modeling, and educational applications.

Ditsa (2003: Preface) analyses the development of multimedia to support systems for information management. He argues that multimedia has transformed the interaction between individuals and computers, since it has "made it possible for us to see, hear, read, feel and talk."

In the realm of the World Wide Web, in addition to more technical research, efforts are being made to understand the relationship between technology, identity, and culture (Wood and Smith 2001). Several topics are of special interest for further research in cybercartography including: the analysis of the WEB as a cybernetic organism; the construction of online identities; interpersonal communication; and conversation processes including virtual support systems and legal issues.

Computer graphics is an area of computer sciences, which has rapidly advanced in the last decades (Newman and Sproull 1981). The development and application of technology in the entertainment industry, medicine, and education, is amply demonstrated in the annual conferences of SIGGRAPH. Although some of these resources have been used for communication in cartography, such as the developments in visualization, there are many others that remain unexplored, such as interactive music, animation, human–computer interfaces, and mixed reality technology among others.

Cartography, geography, and GIS have responded to the rapid development of computer science and telecommunication technologies. As examples one can mention the efforts of Cartwright *et al.* (1999) in *Multimedia Cartography*, Dodge and Kitchin (2001) in *Mapping Cyberspace*, Kraak and Brown (2001) and Camara and Raper (1999) in *Spatial Multimedia and Virtual Reality,* and Peterson (2003) in *Maps and the Internet.*

Taylor's (1999: 315–321) chapter on *Future Directions for Multimedia Cartography* outlines the origins of major ideas behind the concept of cybercartography. Since multimedia cartography is one of the main resources for cybercartographic atlases, the topics discussed in the Cartwright and Peterson (1999) book are a point of departure for the exploration of new avenues for research and development in cybercartography, as are those in Gartner's (2001).

Kraak and Brown (2001: 1) argue for the need to examine the "new opportunities and challenges offered by the WWW" and the incorporation of multimedia cartography. In their book, the perspective of the cartographer is emphasized by various authors, who present new visualization processes, ideas on semiology, applications, and forms of cartographic publishing. Dodge and Kitchin (2000) pose questions on the meaning of space and time in an era of Information and Communication Technology (ICT) and cyberspace. Geographical ideas such as distance and friction of distance are re-evaluated, and concepts such as "spaceless" and "placeless" within globalization are discussed. Overall, the focus is on the interest of geographers in the social, cultural, and political implications of ITC and cyberspace. The concept of "cybergeography" is introduced as the geography of cyberspace and the definition of cyberspace as a "navigable, digital space of networked computers" (Dodge and Kitchin 2000: 1). This contrasts with the approach adopted in Cybercartography as presented by Reyes in Chapter 4 of this volume, where the initial building blocks for a theoretical framework are three areas of knowledge: cybernetics, modeling, and general systems theory. The "common place" of cyberspace as described by Dodge and Kitchin could however be considered as part of applied cybernetics.

2.6. *Technological Trends for Cybercartographic Modeling*

Two different levels of modeling can be identified in cybercartographic atlases. The first is a metamodel that comprises the main characteristics of a cybernetic artifact as was discussed in Chapter 4 of this volume. The second level finds more specific modeling efforts that cover a wide range of purposes. Some trends in computing science and representations of the geographical landscape that might be relevant for the modeling process in the atlases are considered here.

Since the developments in computer science have been a driving force in cartography for several decades, these are worthy of further examination. Wegner and Doyle (1996) outline several avenues of research in computing that are of interest for the development of cybercartography. These ideas originated in the MIT Laboratory for Computing Research and three main trends are identified: foundations, systems applications and infrastructure. In each trend there are topics of special interest to cybercartography. These include computational geometry, artificial intelligence, networks and telecommunications, object-oriented programming, and Human Computer Interaction (HCI). The HCI has been extensively studied and is of special interest for cybercartography, since a user-centered approach is one of its main characteristics (Myers *et al.* 1996). Several chapters of this volume explore this topic.

Developments in object-oriented modeling are found in Gartner *et al.* (2001). Major studies of visualization include the work of Dykes (1997), Evans (1997), Breuning (1999), Uhlenkuken *et al.* (2000), Dransch (2000), Brainerd and Pang (2001), Megrey *et al.* (2002), Ogao and Kraak (2002) and Takatsuka and Gahegan (2002), among others. From a more technical perspective, there are developments toward a more effective use of artifacts such as video devices (Peker and Divakaran 2003).

Also of special interest for cybercartography is the work of Harding *et al.* (2002: 260), where a multi-sensory system was developed for the investigation of scientific data. In the prototype, stereo vision, touch and sound are used simultaneously to explore "overlapping surface properties."

In terms of the specific topic of the content of a cybercartographic atlas, a wide variety of spatial modeling approaches can be incorporated. In the case of the atlases produced by CentroGeo, the main modeling approaches used were based on landscape ecology, territorial planning, and environmental science. The atlases do not incorporate the spatial analytical tools themselves, but rather conceptual representations or interpretations of results that are expressed through a variety of geospatial messages, such as geotext, virtual maps, and diagrams.

3. The Cybercartographic Atlas of Lake Chapala: An Empirical Case

In order to describe the empirical experience from which the *Cybercartographic Atlas of Lake Chapala* emerged, the first part of this section describes how the atlas was conceived by utilizing an intensive interaction process with Mexican society. The second part describes the artifact and the processes involved, adopting the theoretical and methodological frameworks presented in Chapter 4 and 5 of this volume. The authors are an active part of the cybercartographic process and the conceptualization of cybercartography has evolved through a cycle of scientific research that involves empirical work, creation of knowledge, and feedback between users and developers of atlas prototypes.

3.1. The Network of Actors

The main social actor involved in developing an instrument to help understand the interlinked problems of *Lake Chapala* was the *Comisión de la Cuenca Directa del Lago de Chapala*, a Non-Governmental Organization (NGO), with representatives from the main sectors interested in environmental problems of the largest lake in Mexico. Actors in the Commission include fisherman, businessman, peasants, local, state and federal government officials, and a large retirement community of Canadians and Americans. In 1999, there was a complete lack of agreement over the actions required to deal with the environmental problems facing the lake among different groups represented in the NGO.

The Mexican Ministry of the Environment (SEMARNAT) identified the Lake basin as one of its priority regions for environmental restoration. Inside the Ministry there was an informal network of public servants with a special interest in the region's ecology who supported the NGO's efforts. There were two core members of this network. One was in charge of the international agenda of the Ministry. He had a background in regional planning and suggested to NGO membership that they should utilize geographical information to support planning and decision making processes. The second individual was from the statistical and environmental information services division, who had a good sense of the information needs of the NGO.

The lead author of this chapter, as an external consultant, had participated in this informal network, since 1998, in the design of a National Environmental Geospatial System (SNIARN). Also, in the same year, she started working with Dr Fraser Taylor and highly specialized groups from Canada and Latin America on a project supported by the Inter-American Development Bank (IADB) and executed by the Pan American Institute of History and Geography (PAIGH). One of the main purposes of the project was capacity building in Latin America on topics such as the construction and use of electronic atlases. In the course of the project, the vision of Dr Taylor on cybercartography had a strong influence and the final results of the project, an electronic *Atlas of Latin America,* was a prototype of things to come (www.centrogeo.org.mx/atlaslatinoamerica).

CentroGeo was established in March 1999 as a public research center in geography and geomatics supported by the Mexican Council for Science and Technology (CONACYT). The lead author was designated as the Director of CentroGeo and the Ministry of the Environment, formally requested support for the Lake Chapala NGO in developing digital mapping.

Although the network and collaborative process was implicit, it was not formal. This process helped to create favorable conditions for the development of a cybercartographic proposal to satisfy the needs of the Commission. Moreover, the trust among the different members of the network that came from interpersonal relationships over many years was a key factor in the success of the project.

Between April and September 1999, a small group of researchers and specialists in theoretical and applied GIS and cartography, landscape ecology and territorial planning, computing science, graphic design, anthropology, and communication science of CentroGeo, designed and produced the *Cybercartographic Atlas of Lake Chapala*. Important actors from the community and government were involved in the production process of the atlas. In many cases the information and knowledge incorporated was provided by a number of people including graduate students, fieldworkers, and governmental officials.

In late 1999, CentroGeo presented the *Cybercartographic Atlas of Lake Chapala* at a formal meeting of the Commission that took place at the Lake. The unanimous applause after the presentation and the positive reaction of the audience indicated that, for the first time, the members of the NGO agreed on the main environmental problems. Each of the constituent groups recognized that they were both part of the problem and part of the solution. Until that time, action by the federal government had been perceived as the only answer. After the presentation groups realized that by working together meaningful action could be taken at the local level. From then on, the atlas was used as a point of departure in local and state meetings and was widely used by local media in articles and commentary on the environmental problems of the Lerma-Chapala basin of which Lake Chapala is a key part.

In January 2000, at a meeting at the Lake, with the participation of members of the NGO, the Ministry of the Environment and the Governor of the State of Jalisco and approximately five hundred interested people from public and private sectors, the *Cybercartographic Atlas of Lake Chapala* was presented. At that meeting all groups signed an agreement to provide resources for the first stage of restoration of the Lake.

The *Cybercartographic Atlas of Lake Chapala* by conveying the right messages to the relevant stakeholders, had accomplished one of its main purposes: being used as a political interaction and a discussion tool among the different actors. It should be noted that the members of the NGO had never agreed earlier on any issue due to their evident conflicts of interests.

The atlas was also presented to a gathering of several hundred retired Canadians by the lead author and Dr Taylor to garner their support for the endeavor.

3.2. Communication, Modeling and Knowledge in the Cybercartographic Atlas of Lake Chapala

As mentioned in Chapter 4 of this volume, the *Cybercartographic Atlas of Lake Chapala* can be modeled as a multi-dimensional system composed of three axes: communication, models, and knowledge. The methodology presented in Chapter 4 of this volume was used for the overall conceptual design of the atlas. The utilization of these three axes to create the atlas as an artifact is described below.

3.2.1. The Communication Axis

The potentiality of the atlas to communicate geospatial messages was focused on two main issues: the holistic character of environmental problems in the Lake Chapala Basin and the need to strengthen the awareness of the community of the nature and causes of the environmental problems of the Lake. To accomplish this, multimedia (maps, remote sensed imagery, videos, photographs, texts, diagrams, and three-dimensional "flights"), interactive mapping (zooms, overlays, queries to attributes, distances, drafting of new geometrical elements, and metadata among others), "user-friendly interfaces" (user-oriented language and menus), and explicit knowledge representation (via the structure of "navigation" features and information recovery) were used.

3.2.2. *The Modeling Axis*

As discussed in Chapter 4 of this volume, modeling is an essential element of the design process of a cybercartographic atlas. The structural methodology and the content framework as described in Chapter 4 of this volume, were applied and as a result territorial planning and landscape ecology approaches were adopted. Both were selected as a result of the User Needs Analysis (UNA) and qualitative research analysis. As a result, the six subsystems of the atlas respond to a planning framework (characterization, diagnostic, scenarios, and management). In both communication and knowledge axes the models play different roles. In some cases the models are used to communicate geospatial messages and in others they are the backbone for visualization processes. Some of the resources used include: virtual and digital maps of the region, data structures with the relevant environmental information, digital terrain models of the Lake Chapala Basin, dynamic three-dimensional satellite images showing the Lake and its surroundings, geo-texts, diagrams, videos, and photographs.

3.2.3. *The Knowledge Axis*

As discussed in Chapter 5 of this volume, the main content framework for the atlas is based on landscape ecology theory and territorial planning. This decision responds to the results of the qualitative research exercise undertaken as part of the structural methodological process and to the UNA. To have such a well-defined structure for the development of the atlas has certain advantages. The guidance provided by the conceptual framework adopted is of great help in the definition of the geospatial messages to be included and as a result the communication process is improved.

3.2.4. *The Resulting Cartographic Artifact*

In the design and development of the atlas, several elements have to converge in an "interaction space" as outlined in Chapter 4 of this volume. These include: the complementarities of the different components; the synergetic effect that has to be present; the appropriate integration of the different models and languages; the aesthetic component; and the critical cultural and societal factors.

The purpose was to create a cybernetic artifact with all of the characteristics required including openness, interactivity, and adaptability. It was also important to ensure that the geospatial messages were able to communicate the desired information. Both hearing and seeing are involved in the process of communication, but this involvement goes beyond the linguistic interpretation of messages by the receiver. In the interaction between the users and the atlases, the visualization of images plays a major role in the meaning derived from the complex messages involved.

3.2.5. Technology Used in Building the Atlas

As part of the overall scientific strategy of CentroGeo, technological development is conceived as essential to advance basic research in cybercartography. For that reason, a small group of technologists (one at the beginning) was involved in the design and production of the atlas. The experience has been highly enriching and has given strength to the institution and the possibility to satisfy users without having to rely always on expensive (for Mexican standards) commercial software. The main technologies used include: Map Objects (ESRI); visual basic programming language; and HTML schemas.

The technological functionality of the atlas is innovative in several respects, for example, the users are able to draft and save new geometrical figures to represent specific areas of study together with new photographs, videos, or text that are incorporated into the atlas.

CentroGeo approaches technological development as an opportunity for a number of reasons. There is a gap between the theoretical developments in spatial analysis and geography in general and the commercial software available. In Mexico and Latin America educated human resources need to be enhanced to adopt and create new geo-technology; the competitiveness of the geomatics industry must be increased by developing in-house software; and tailor-made technological solutions for cybercartography are required to allow further research.

3.3. The Sociocybernetic Character of the Atlas

In the unfolding virtual helix presented in Chapter 4 of this volume, the requirements of both the Commission and the Ministry of the Environment (SEMARNAT) were taken as the point of departure to describe the development cycle. According to the methodological framework proposed, a qualitative research approach was used in the design of the atlas. The design and production of the prototype took approximately six months (April–September, 1999). During the final stages, the atlas was presented to some of the main government stakeholders, to researchers knowledgeable about the environmental situation of the basin, to the regional community, and to other members of society. This was to ensure that their views were incorporated and to increase the acceptability of the atlas in political and social terms. This experience turned out to be very successful for the embedding of the atlas into environmental public policy processes. In other words, the selection of the geospatial messages and the manner in which they were incorporated into the atlas, the functionality of the prototype and the graphic design, among others, was conducted by interaction processes that included designers, producers, and users in what can be called a "qualitative dynamic modelling" cycle (See Chapter 4, Fig. 4.10).

Sociocybernetics as presented in Chapter 5 of this volume, provide a theoretical framework to explain some of the social phenomena that were observed during the development of the atlases, which marked the first stage in the development of cybercartography. As a result of research in the creation of theoretical and methodological frameworks for cybercartography that has taken place during June 2003–June 2004 the second order cycles and their role in the construction of a spatial language, as outlined in Chapter 5 of this volume have become more evident. For example, the perception and understanding of the designers and producers of the atlas has evolved and the feedback of more "mature" users reflects its impact on both their spatial understanding and their ability to plan and manage the environmental challenges of the Chapala region.

The *Cybercartographic Atlas of Lake Chapala*, as a social product in its initial stage, had a substantial impact on the interaction among the different actors and was a driving force in building environmental public policy. The community recognized the importance of the Atlas as a means for teaching children to take a holistic view of the region where they live and the importance of "taking care" of the Lake. In fact, CentroGeo produced an educational version of this atlas. Each of the actors in the region became aware of their responsibility regarding the environmental problems of the Lake and members and working groups of local, state, and federal governmental institutions have used the atlas in diagnostic and planning activities. CentroGeo has organized workshops in order to familiarize different groups of stakeholders in the use of the atlas. In such workshops, the facilitator is part of the cybercartographic process and communicates the conceptual basis of cybercartography.

Although the efforts to embed the, *Cybercartographic Atlas of Lake Chapala* into different societal processes (organizational, educational, community, or governmental) is far from complete, the initial results clearly point to the fact that the methodological and the theoretical approaches adopted are taking the overall research in the right direction.

In the use of the *Cybercartographic Atlas of Lake Chapala* by the "mundane," as described by Johnson (1998) several lessons were learned. These include the establishment, at the early stages of design and throughout the production process, of an adequate environment that promotes cordial personal relationships, effective conversations, trust, and collaboration. This required an explicit strategy of relationships and processes aimed at building a common language among stakeholders. It also involves the adoption of an adequate methodological and technological framework to allow the research group to focus on the appropriate ways to embed the artifacts in organizational and community processes. In the interaction with society the use of geospatial information and knowledge was a key factor to obtain consensus on environmental issues and other topics of public policy. In addition, the importance of the continuous feedback with society in building the atlas became evident as did the need to follow up the insertion of the atlas in the communities and/or organizations in order to complete the cybernetic cycle. The atlases were presented in different social contexts, involving academics, public servants, politicians, and community groups, and it is worth noting that community groups accepted these artifacts much more readily than some highly specialized traditional cartographers.

4. Conclusions

This chapter discussed the relevance of technology and culture for cybercartography. In describing the cybercartographic atlases produced by CentroGeo two main recurrent themes emerged: the influence of the social and cultural contexts in the construction of these technological artifacts and the feedback cycles created by their insertion in diverse organizational and community processes. However, much research needs to be done in order to learn about the different social, cultural, political, and technical factors that influence the development of cybercartographic technological artifacts that effectively respond to the social needs in different cultural contexts, and at the same time, that allow theoretical and methodological advancements in this discipline. Several key research questions remain:

- Does the incorporation of cultural elements in the design and construction of cybercartographic artifacts help the transfer and adoption of these technologies? And if that is so, which are the best ways to identify and incorporate such elements?

- How can the role of cybercartographic artifacts in the course of different social processes, such as decision making, awareness generation, consensus building, planning, public policy construction, or education be identified, and how relevant is this role in the steering of such processes?
- What is the role of cybercartographic artifacts in knowledge organizations and what is cybercartography's contribution in competitive intelligence?

Additional theoretical and methodological questions for cybercartography include:

- The possible role of cybercartography in the broadcasting media and in the communication-feedback processes
- The impact of multimedia advances on cybercartography
- The influence of the WEB on future trends in cybercartography

The development of geo-technology, may benefit from common areas of research with computing science, geospatial modeling, and computer graphics.

The development of new technological frameworks and artifacts will contribute to providing answers to these questions and will support theoretical advances in cybercartography. Similarly, as in the advancements on the construction of theoretical and methodological frameworks, these first efforts to explore the new concept of cybercartography, from a technological and cultural perspective, point to a wide range of possible avenues. Both, empirical and theoretical work is needed, and as mentioned in Chapters 4 and 5 of this volume, a multidisciplinary and collaborative approach should be strengthened and the interaction and feedback with society should remain as the driving force of research in cybercartography. The various chapters in this book make a major contribution to advance this research.

Acknowledgments

CentroGeo is a public research center supported by the Mexican Science and Technology Commission (CONACYT) MSC. Fernando López was the main programmer and designer of the technological solutions for the cybercartographic atlases. A team of researchers, programmers, and graphic designers from CentroGeo as well as members of the Ministry of the Environment (SEMARNAT) and of the Commission of the Direct Basin of Lake of Chapala among others were involved in the design, testing, and development of the *Cybercartographic Atlas of Lake Chapala.*

Jack Dangermond (president of ESRI) supported the project by providing Map Objects software. Fraser Taylor's advice was invaluable in the overall process.

References

Brainerd, J. and A. Pang (2001) "Interactive map projections and distortion", *Computers and Geosciences*, Vol. 27, No. 3, pp. 299–314.

Breunig, M. (1999) "An approach to the integration of spatial data and systems for a 3d geo-information system", *Computers and Geosciences,* Vol. 25, pp. 39–48.

Büchel, B. S. T. (2001) "Using communication technology", *Creating Knowledge Organizations*, Palgrave Publishers Ltd., Great Britain, pp. 18.

Camara, A. S. and J. Raper (eds.) (1999) *Spatial Multimedia and Virtual Reality,* Taylor and Francis, London.

Cartwright, W. and M. P. Peterson (1999) *Multimedia Cartography*, Springer, Germany.

Ditsa, G. (ed.) (2003) "Information management", *Support Systems and Multimedia Technology,* Idea Group Inc.

Dodge, M. and R. Kitchin (2001) *Mapping Cyberspace*, Routledge, Taylor and Francis Group, London.

Dransch, D. (2000) "The use of different media in visualizing data", *Computers and Geosciences,* Vol. 26, pp. 5–9.

Dykes, J. A. (1997) "Exploring spatial data representation with dynamic graphics", *Computers and Geosciences*, Vol. 23, No. 4, pp. 345–370.

Evans, B. J. (1997) "Dynamic display of spatial data- reliability: Does it benefit the map user", *Computers and Geosciences*, Vol. 23, No. 4, pp. 409–422.

Frank, A. U. (1993) "The use of geographical information system: The user interface is the system", in Scott and H. M. Hearshaw (eds.) (2003), *Human Factors in Geographical Information Systems*, Belhaven Press, London, pp. 3–14.

Gartner, H., A. Bergmann and J. Schmidt (2001) "Object-oriented modelling of data sources as a tool for the integration of heterogeneous geoscientific information", *Computers and Geosciences*, Vol. 27, pp. 975–985.

Gewin, V. (2004) "Mapping opportunities", *Nature*, Vol. 427, pp. 376–377, 22 January.

Gupta, V. S. (1999) *Communication Technology, Media Policy and National Development*, Concept Publishing Company, New Delhi.

Harding, C., I. A. Kakadiaris, J. F. Casey and R. B. Loftin (2002) "A multisensory system for the investigation of geoscientific data", *Computers and Graphics,* Vol. 26, pp. 259–269.

Hendriks, P. (2004) "Intelligence from space: Using geographical information systems for competitive intelligence", Chapter IX, in Vriens, D. (ed.), *Information and Communication Technology for Competitive Intelligence*, Idea Group Publishing, pp. 194–226.

Harris, T. and D. Weiner (1998) "Empowerment, marginalization and community -integrated GIS", *Cartography and Geographic Information Systems*, April 1998, Vol. 25 i2 67(1).

Ihde, Don (1990) *Technology and the Lifeworld. From Garden to Earth*, Indiana University Press, Bloomington and Indianapolis.

Inayatullah, S. and S. Leggett (2002) "Transforming communication; technology, sustainability and future generations", *Praeger Studies on the 21st Century*, Praeger, Connecticut, USA.

Johnson, R. R. (1998) *User-Centred Technology; A Rhetorical Theory for Computers and Other Mundane Artifacts*, State University of New York Press.

Kraak, M. and A. Brown (eds.) (2001) *Web Cartography*, Taylor and Francis, London.

Mahbubur-Rahman, S. (ed.) (2001) *Design and Management of Multimedia Information Systems: Opportunities and Challenges*, Idea Group Publishing.

Medhurst, M. J., A. Gonzalez and T. R. Peterson (1990) *Communication and the Culture of Technology*, Washington State University Press, Pullman, Washington.

Megrey, B. A., S. Hinckley and E. L. Dobbins (2002) "Using scientific visualization tools to facilitate analysis of multi-dimensional data from a spatially explicit, biophysical, individual-based model of marine fish early life history", *ICES Journal of Marine Science*, Vol. 59, pp. 203–215.

Minoli, D. (1994) *Imaging in Corporate Environments; Technology and Communication*, Bell Communications Research Inc. and New York University, McGraw-Hill, Inc.

Myers, B., J. Hollan and I. Cruz *et al.* (1996) "Strategic directions in human-computer interaction", *ACM Computing Surveys*, Vol. 28, No. 4, December.

Newman, W. M. and R. F. Sproull (1981) *Principles of on Interactive Computer Graphics*, McGraw-Hill, Inc., Japan Ltd.

Ogao, P. J. and M. J. Kraak (2002) "Defining visualization operations for temporal cartographic animation design", *International Journal of Applied Earth Observation and Geoinformation* Vol. 4, pp. 23–31.

Peker, K. A. and A. Divakaran (2003) "An extended framework for adaptive playback-based video", *Proceedings of SPIE Vol. 5242 Internet Multimedia Management Systems IV*.

Peterson, M. P. (ed.) (2003) *Maps and the Internet*, Elsevier, Amsterdam, pp. 405–420.

Penland, P. R. (1974) *Communication, Science and Technology*, University of Pittsburgh, Marcel Dekker, Inc., New York.

Rasmussen, T. (2000) *Social Theory and Communication Technology*, Ashgate Publishing Limited, England, Hampshire.

Russell, B. (1967) *The Impact of Science on Society*, George Allen and Unwin Ltd.

SIGGRAPH, www.siggraph.org

Takatsuka, M. and M. Gahegan (2002) "GeoVISTA studio: A codeless visual programming environment for geoscientific data analysis and visualization", *Computers and Geosciences*, Vol. 28, pp. 1131–1144.

Taylor, D. R. F. (1999) "Future directions for multimedia cartography", in Cartwright, W., M. P. Peterson and G. Gartner (eds.), *Multimedia Cartography*, pp. 315–321.

Taylor, D. R. F. (2003a) "The concept of cybercartography", Chapter 26, in Peterson, M. P. (ed.), *Maps and the Internet*, Elsevier, Amsterdam.

Taylor, D. R. F. (2003b) *Cybercartography and the New Economy, Project Proposal*, Carleton University, Ottawa, Canada (www.gcrc.carleton.ca).

Uhlenkuken, C., B. Schmidt and U. Streit (2000) "Visual exploration of high-dimensional spatial data: Requirements and deficits", *Computer and Geosciences,* Vol. 26, pp. 77–85.

Vriens, D. (2004) *Information and Communication Technology for Competitive Intelligence*, Idea Group Inc.

Wegner, P. and J. Doyle (1996) "Editorial: Strategic directions in computing research", *ACM Computing Surveys*, Vol.28, No. 4, December.

Wood, M. (2003) "Some personal reflections on change... the past and future of cartography", *The Cartographic Journal*, Vol. 40, No.2, pp. 112–115.

Wood, A. F. and M. Smith (2001), *Online Communication; Linking Technology, Identity and Culture*, Lawrence Erlbaum Associates, Publishers, London.

CHAPTER 7

The Cartographer as Mediator: Cartographic Representation from Shared Geographic Information*

PETER L. PULSIFER

Geomatics and Cartographic Research Centre (GCRC), Department of Geography and Environmental Studies, Carleton University, Ottawa, Ontario, Canada

D. R. FRASER TAYLOR

Geomatics and Cartographic Research Centre (GCRC) and Distinguished Research Professor in International Affairs, Geography and Environmental Studies, Carleton University, Ottawa, Ontario, Canada

Abstract

The advent of the Spatial Data Infrastructure (SDI) and distributed, agent-based computing present new challenges for the development of cartographic systems. The ability to effectively share meaning between system elements (semantic interoperability) remains an area of active research. The authors suggest a concept of geographic mediation (geo-mediation) that may be useful in framing current and emerging processes of integrating and representing geographic information in the context of distributed information. It is proposed that this concept can be used to guide practice and theory related to the design and evaluation of cartographic systems that can be seen as central to concepts

*This chapter is based on a PhD comprehensive paper presented as part of the lead author's PhD program at Carleton University, Ottawa, Ontario, Canada.

like Taylor's cybercartography. The chapter presents a high-level architecture, the "open cartographic framework", based on a mediator approach. Implications for using such an approach are discussed in this chapter. The chapter concludes by suggesting that an effective geomediation process necessitates consideration of both formal and negotiated processes for establishing its meaning.

1. Introduction

Historically, as people moved beyond local environments and geographic knowledge increased, so did their geographic information in the form of oral narratives, texts, artistic works, and maps. As geographically stable human settlements developed, early clearinghouses for geographic information emerged. The library at Alexandria during the centuries between 2300 and 2000 years before present is often sited as a repository containing geographic information from many parts of the then known world (Brown 1979: 12). Thus, the sharing of geographic information was facilitated by gathering information resources at a central place. Although millennia have passed since the existence of the library at Alexandria, the strategy of providing a single point of access to geographic information still continues. However, rather than a single building, that single point of access now exists at the interface of networked digital information devices. In this chapter, the authors suggest the concept of geographic mediation that may be useful in framing current and emerging processes of integrating and representing geographic information in the context of distributed information.

The two world wars in the first half of the twentieth century prompted increased interest and activity in terms of recording geographic data as maps and other forms of geographic information (e.g. aerial photographs, tabular records, etc.) (Brown 1979: 3). The development of new remote sensing and survey technologies in the latter half of the century coupled with concerns about environmental, resource and global security issues, has lead to the collection of unprecedented amount of spatially referenced geographic data. At the same time, there have been extensive developments in the disciplines of cartography and what can be referred to as Geographic Information Science (GIScience) (Goodchild 1992). With post WWII development of the digital computer came dramatic changes regarding how geographic data were stored, retrieved, manipulated, and represented.

The practice of cartography saw the emergence of forms of automated mapping (Tobler 1959), digital display, and later multimedia (Peterson 1995; Cartwright 1997; Van Dijk and Leusink 1983) and Internet cartography

(Peterson 2003). Theoretically, cartography moved from what Robinson (1953) saw as the current, essentially artistic approach to the creation of maps to what he saw as a more objective and functional approach to cartography based on psychophysics. Robinson's contributions led to the communications model of cartography prevalent throughout the 1970s and 1980s (Kolacny 1969). The 1990s saw the shift from a communication paradigm that saw the map-user as a reader simply reacting to the cartographer's message, to a visualization paradigm, where the map-user is an active participant in constructing geographic information through the use of maps and other visual forms (DiBiase 1990; MacEachren and Taylor 1994; MacEachren 1995).

Recent changes in the practical and theoretical directions of the discipline of cartography have been taking place in the context of broader developments in information technologies and the Internet in particular. Peterson (2003) identifies the emergence of the Internet as an important method for map distribution. He states that in 2001 a single map service was providing more than 20 million maps per day. High-level politicians in developed nations started to envision the possibilities for new representations of the Earth underpinned by Internet-based geographic information technologies (Gore 1998). Cartographers (Peterson 2003; Taylor 2003; Zaslavsky 2003) and non-cartographers (Goodchild 2000) in the same manner comment about the implications of Internet mapping for the practice and discipline of cartography. Taylor (2003) argues that cartography, now and for the foreseeable future, can be described by seven primary elements that together comprise the concept of cybercartography. Taylor's elements of cybercartography are as follows.

Cybercartography:

- is multisensory using vision, hearing, touch, and eventually smell and taste;
- uses multimedia formats and new telecommunications technologies such as the World Wide Web;
- is highly interactive and engages the user in new ways;
- is applied to a wide range of topics of interest to society, not only to location finding and the physical environment;
- is not a stand alone product like the traditional map but part of an information/analytical package;
- is compiled by teams of individuals from different disciplines; and
- involves new research partnerships among academia, government, civil society, and the private sector.

Thus, Taylor's concept suggests that emerging forms of cartography are interactive and multisensory representations of geography built upon technical and human networks. In his view, these systems can be used to create representations for a variety of social contexts.

Zaslavsky (2003) suggests that emerging cartographic systems will rely on the ability to share map data through interoperable distributed systems.

> As modern online cartography matures from displaying static map images to creating adaptable, dynamic, interactive designs, which incorporate multimedia and integrate map information from distributed sources, it becomes critical to enable standards-based interoperability of map data and cartographic services (Zaslavsky 2003: 172).

The term Spatial Data Infrastructure (SDI) is often used to denote the relevant base collection of technologies, policies, and institutional arrangements that facilitate availability of and access to spatial data. Proponents of SDI initiatives suggest that the resulting systems can support access, integration and use of spatial data from disparate sources at a variety of scales (Nebert 2004). Current SDI programs and systems can be seen as the human–technical network as identified by Taylor. The interoperability requirements identified by Zaslavsky are an integral element of these SDI programs. Interoperability is a term used to describe a situation where data and operations can easily be shared. In addition, interoperability implies that users do not require product specific expertise to use a given software system (Goodchild *et al.* 1999). The later part of this section discusses SDI programs and interoperability as background to a subsequent section describing the aforementioned concept of geographic mediation. It is proposed that this concept can be used to guide practice and theory related to the design and evaluation of cartographic systems that can be seen as central to concepts like Taylor's cybercartography. Section 3 provides an overview of the architecture of the aforementioned system. Section 4 provides a review of theoretical discussions that may be applied to the work presented.

1.2. Spatial Data Infrastructures

Increased connectivity resulting from the development of the Internet has resulted in the emergence of new political/social/technical structures designed to manage the distribution of digital geographic information. The SDI concept was popularized in the early 1990s and early adopters started to establish implementation strategies (Clinton 1994; Coleman and McLaughlin 1994; Morain 1997). SDI programs are typically led by public agencies interested in facilitating the cooperative production and use and sharing of geospatial information (Boxall 1998; Masser 1998; Lachman 2002).

These infrastructure programs encompass policies, standards, technologies and procedures, and often depend on collaboration between public, private and academic agencies. SDIs are being developed at a number of geographic scales including local (Harvey *et al.* 1999), national (McLaughlin and Nichols 1994; Rhind 1998), continental (Coleman and Nebert 1998), and more recently, global (Coleman and McLaughlin 1994; Lachman 2002; Nebert 2004). Global SDIs are typically used for applications such as evaluating global environmental change or disaster response.

Funds, primarily public, being spent on the development of spatial data infrastructures are significant, with countries like Canada directly spending $60 million ($CAD 1998) in establishing the Canadian Geospatial Data Infrastructure (CGDI 2004). Korea, in contrast, is spending in the order of hundreds of millions of dollars (Masser 1998). These investments are likely a fraction of those made to support major programs in countries like the United States and United Kingdom (Rhind 1998; FGDC 2002). While investments in SDIs are large, so too are the expectations from these programs. In launching the Canadian program, The Honorable Ralph Goodale, the then federal Minister of Natural Resources had these comments:

> We want to exploit the vast potential of integrating information from maps, satellite images, statistics and other sources into geospatial data infrastructures. These data infrastructures are opening up a whole new way of understanding our world (Goodale 1999).

As governments and institutions develop SDIs, it will be increasingly possible to obtain public and private geographic information using Internet-based delivery services based on SDIs. Some national mapping agencies have already moved to this approach for delivering geographic information (Rhind and Nanson 1998). The developing Canadian program aims to provide citizens a "common-window" access to geospatial services and information through the Internet by "harmonizing" the public and privately collected geospatial information into easily accessible and searchable databases (CGDI 2004). In the early stages of SDI development, the Internet was used for data discovery through online directory searches. Data were then provided on digital media such as CD-ROM disks and shipped rather than being visualized or accessed directly on the Internet – thus, data were not harmonized to any extent and the "common-window" was not a reality. New technologies and approaches are supporting the transmission of geographic information over the Internet in a way that the information can be visualized or accessed directly. A key element in these developments is the concept of interoperability.

1.3. Interoperability

A major impediment to the exchange of digital geographic information has been interoperability between systems. In the formative years of automated cartography and Geographic Information Systems (GIS), the technology development process resulted in systems with proprietary structures, analytical operations, and user interfaces. A lack of standards meant that data stored using different platforms or operating systems could not be shared. The result was that information and functionality could not easily be shared between systems produced by different vendors. Moves toward developing data interchange formats started decades ago. However, due to the cost and effort involved in translating data, these efforts have done little to alleviate interoperability problems. In the last decade there has been increased activity in addressing the problem as the geomatics industry has realized that industry growth has been hindered by the lack of interoperability between systems. As a result, organizations such as the Open Geospatial Consortium (OGC) have formed. The OGC is a member-driven, nonprofit international trade association that appears to be leading the development of geo-processing interoperability and computing standards. The OGC is working closely with the International Organization for Standardization's (ISO) Technical Committee 211 (TC211) to wards establishing standards for geographic information and the geomatics industry.

There are three primary areas of research and development related to the goal of achieving interoperability, namely syntax, schema, and semantics. The willingness of information gathering entities to share information is a fourth important dimension of interoperability, however, one that does not receive treatment in typical interoperability research. The organizational dimension of interoperability is discussed in Section 4 of this chapter. Out of the commonly recognized areas of interoperability research and development, syntax relates to the basic way that geographic entities are represented in terms of geometry or the representation model (object versus field representations). Schema refers to the strategies, rules, and relationships established to organize and structure spatial data within a computer database. To use a language metaphor, syntax can be seen as vocabulary with schema being grammar (Salgé 1995). Schemata are the formalizations of classification systems used to communicate meaning and are thus, closely associated with semantics. The bulk of the GIScience literature suggests that syntactic and schematic interoperability are being resolved by the private sector (Egenhofer 1999; INTEROP99 1999) and thus, are currently beyond academic treatment. Developments in syntactic and schematic interoperability are apparent in the present ability to readily exchange data using OGC/ISO standards. However, exchanging information is not equivalent to sharing information.

If semantic interoperability is not considered, then a situation can arise where data transfer is successful but information is not in fact shared – that is, meaning is not communicated.

In simple terms, geospatial data semantics encompass how – typically through the use of language – computer-based representations stored in information systems are related to entities and concepts in the real-world. Central to data semantics is the process of categorization of geographic phenomena. Semantic interoperability is fundamentally related to the overall concept of interoperability in that pure data transfer makes no sense if the members of the user community interpret the data differently (Bishr *et al.* 1999b). There are a number of reasons for semantic difference in geographic information including differences in the classification scheme, difference in the rules that define if an instance of a geographic phenomenon is included in a particular class definition and lastly, differences in geometric representation.

Landuse classification schemes are a good example of the potential for semantic difference or heterogeneity. Differences may exist either in the label used to describe a class (i.e. fallow land versus set aside land) or the label may be the same with the class inclusion rules being different. For example, in North America for the class of "wetland" – the inclusion rules are very different depending on the context and the individual stakeholders (Harvey and Chrisman 1998). While class definition and inclusion rules may be similar, the geometric representation may be different. The conceptualization of a road is an example, where a transportation engineer may "see" roads as line features (vectors that carry traffic) whereas a utilities engineer will see the same road as an area feature containing other features (i.e. linear features such as gas mains, and point features such as water valves that are spatially located under or near the road).

The concept of semantic interoperability is an extensive area of ongoing research and has been recognized by those driving the development of the Internet (Berners-Lee *et al.* 2001) and the OGC (OGC 1999). In general, the geomatics community is dealing with semantic heterogeneity in three ways: metadata, standardization, and semantic translation.

The first approach mentioned involves an attempt to understand the semantics of a given domain primarily through the production of metadata. In addition to the metadata approach, there is a movement by industry towards standardization within information communities. In general terms, an information community can be defined as "a group of spatial data producers and users who share an ontology of real-world phenomena" (Bishr *et al.* 1999a). Academia has been primarily concerned with semantic translation. That is, translating meaning within and between information communities. This approach begins by attempting to formally model the semantics of an information community. This model is then used to attempt translation between semantic domains using methods of formal logic. Semantic information collected as part of these processes is typically stored in a formal ontology, defined here as a formal conceptualization of specific domains that attempts to capture and constrain a set of conceptualizations (Guarino and Poli 1995) and henceforth is denoted by an upper case Ontology.

In cartography and GIScience, research that falls under the rubric of Ontology is emerging as a key area of study (Brodaric and Gahegan 2002; Maozhen *et al.* 2002; Worboys and Duckman 2002; Galton 2003). Alternatively, the term "knowledge base" can be used in place of Ontology (Lehto 2003) and has been a more common term in literature related to artificial intelligence and expert systems in cartography (Forrest 1999; Shea 1991). Within the Internet, distributed computing and intelligent systems communities, the use of Ontology is becoming commonplace (Berners-Lee *et al.* 2001; W3C 2004c).

In previously cited works, the importance of establishing effective human and technical networks has been identified as significant in supporting the development of new forms of cartography. Interoperability in general, and semantic interoperability in particular are key factors in developing effective networks for sharing information and associated meaning. SDI programs are driving interoperability research and development and, it seems, they will ultimately be the information delivery mechanisms that implement the results on a broad scale. The process of integrating the syntactic, schematic, and semantic elements of a variety of heterogeneous information sources is central to the practice and discipline of cartography. Thus, cartographers are well positioned to contribute to the development of the interoperable systems required to support information sharing.

2. Mediation of Geographic Information

This section proposes that the concept of geographic mediation may be useful in conceptualizing a system that supports the sharing of geographic information for the purposes of developing cartographic systems.

2.1. *Geographic Mediation*

The terms mediation and mediator have many definitions. A dictionary defines mediator as "one that mediates, especially one that reconciles differences between disputants." Currently, the term mediator is widely used within the disciplines of computer science, information engineering, interoperable GIS, and cartography (Wiederhold 1999; Fonseca *et al.* 2002; Zaslavsky 2003). In this context, the mediator accesses multiple heterogeneous information sources and services, retrieves different pieces of the result and seamlessly assembles these pieces into a composite response. These software components have embedded knowledge that is derived and maintained by experts. Wiederhold discusses mediation in terms of a model for interoperability. In Wiederhold's model, the mediator is seen as an intermediate element between digital sources of information and the end user interface in a (typically) networked infrastructure. Much of this work focuses on computational and resource efficiencies for this type of model.

Another definition of mediation sees the map as a form of "mediated seeing" (Fremlin and Robinson 1998: 30). In this model, mediated seeing permits observers to see in times and places in which they are not physically present, or perhaps cannot possibly be, and in projections other than that supplied by the lens of the eye. Mediated seeing can be applied beyond maps (i.e. television). In the case of a map, "lateral inputs", such as projection, scale, generalization and others are part of the process of producing a mediated representation on the retina, which in turn projects a representation onto user mentality. Although this work focused solely on the paper map, some interesting ideas are put forward that may be applied to digital cartography.

Gumpert and Carthcart (1990) examine mediation with respect to the role of technology in communication and human information processes – and the nature of the relationship between mass and personal communication processes. They posit that all forms of mediation produce unique forms of information which, in turn, have potent effects on producers, programmers, messages, receivers, and the social construction of reality. In more recent works, cartographers and GIS researchers have discussed the effects of mediation (in a media sense) on various forms of geographic information (Peterson 1995; Sui and Goodchild 2001).

2.2. *The Cartographer as Mediator*

Informed by the various definitions of mediation/mediator discussed, the present work proposes the abstract concept of geographic mediation (geo-mediation) and a discipline specific extension of cartographic mediation as useful for the development of cartographic practice and theory. The concept of geo-mediation proposes that a mediator is the nexus between observations of the real-world and representations based on those observations. In this definition, the mediator forms a representation of geographic phenomena based on processes of abstraction and integration applied to heterogeneous sources of (geographic) information. In this context information is defined as spatially referenced observations of the world accompanied by contextual data after Masser (1998). This can include: (i) the actual world; (ii) a way that the actual world might be arranged; (iii) an interpretation; and (iv) a possible world. The goal of geographic mediation is to produce a representation that is meaningful to an end user. The term is used in the broader sense in an attempt to integrate discourse in the domains of applied cartography, cartographic theory and critical social theory. The abstract term can be used in the multiple, but related contexts often experienced in cartography and GIScience. Figure 7.1 provides a generalized illustration of the geo-med iation process.

In many ways, the geo-mediation process describes an element of what cartographers have practiced and studied for some time. However, the end result of contemporary geographic information processing is not necessarily a map. For example, some forms of atmospheric modeling can be seen as geo-mediation with a strictly numerical output. Similarly, modern in-vehicle navigation systems use an audio (voice) representation to convey geographic information. Thus, a distinction is made between geographic and cartographic mediation. Cartographic mediation is a specific instance of geographic mediation whereby cartographic knowledge is applied to produce some form of graphic representation of geographic phenomena.

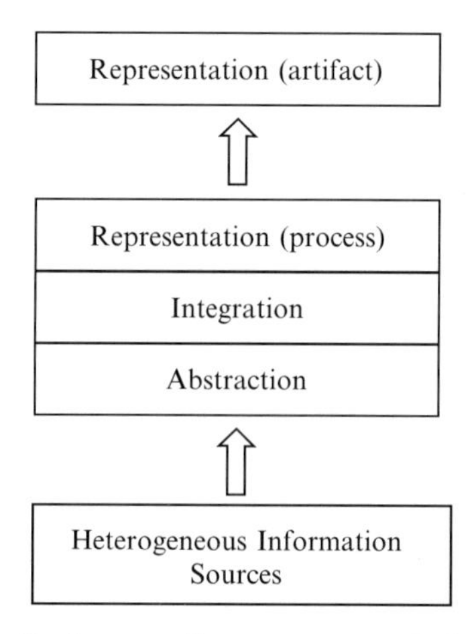

FIG. 7.1. Geographic mediation applies processes of abstraction, integration and representation to heterogeneous information sources with the end result being a representation of the real-world that is meaningful to an end user.

In the case where a cartographer manually produces a map, cartographic knowledge is applied directly to form a representation. Knowledge embedded in a machine-based mediator or mediators such as an expert system can be used by an entity other than the original cartographer to create representations. In the latter case, the cartographic knowledge will be reused as part of a process. For cartographers in a digital environment, knowledge is often embedded even if it is simply through the process of assigning a line weight in a desktop mapping environment. As long as the file exists, it can be reused. It should be noted, however, that in some cases (i.e. expert systems) cartographic mediators formalize knowledge so that it can be applied to more than one application or instance of cartographic mediation.

Although the idea of the cartographer as mediator is central to the present work, it is important to recognize the distinction and relationships between geographic mediation and cartographic mediation. Geographic mediation is a general term that may describe a variety of individual and collective mediation processes including the production of art (i.e. graphic and literary) and the construction of geographic narratives (Doubleday 1993). Cartographic mediation assumes the result to be some form of cartographic representation. Thus, cartographic mediation is a form of geographic mediation focused on a particular type of representation. To expand on some of the implications for developing systems based on the concept of cartographic mediation, the following section outlines the use of geographic and cartographic mediation in an applied setting.

3. A Mediator Based Development Model

This section presents a development model in which the concepts of geographic and cartographic mediations are used. Current and emerging development approaches inform the architectural design. The present work builds on previous work detailed elsewhere (Pulsifer 2003, Pulsifer, and Taylor 2003).

Conceptually, the model design uses a four level approach to describe elements of the system. Figure 7.2 illustrates the general relationships between the top levels of Infrastructure; Mediator, Interface, and User. In a contemporary applied sense, the Infrastructure level is modeled after SDI programs and thus comprises data, technology, standards, and management policies. Standards and policies are typically embedded within systems at the Infrastructure level. That is, technologies are based on standards and typically incorporate elements that manage policies (i.e. security, purchasing, etc.). Generally, the Mediator level is the level at

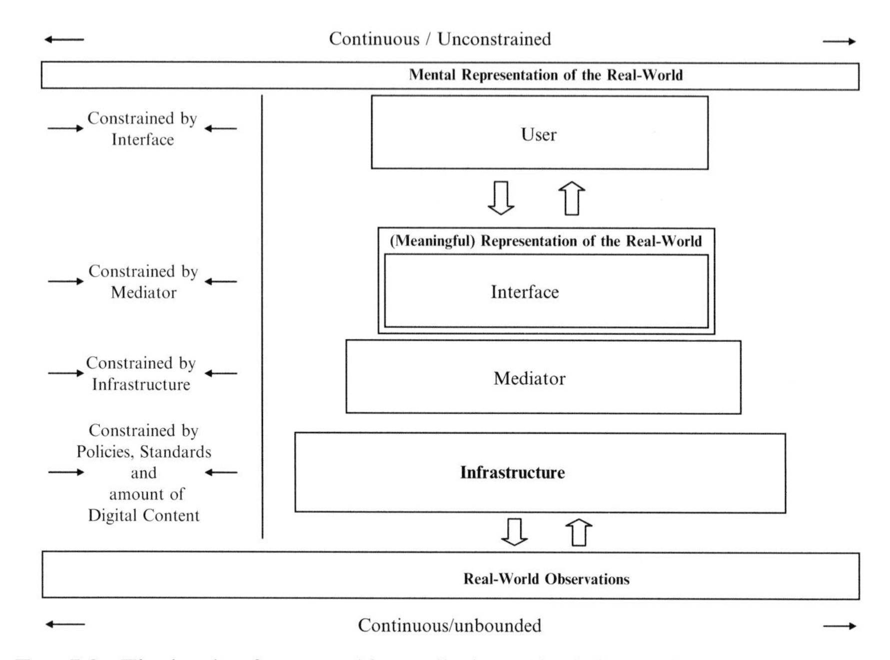

FIG. 7.2. The levels of geographic mediation. The information Infrastructure is built upon observations of the real-world and serves the Mediator level. At the mediator level, the process of geographic mediation produces a meaningful representation of the world that the user accesses via the Interface level. Once accessed, the representation is mediated by the User's mental process. The column on the left indicates how levels are constrained in relation to other levels.

which domain specific geographic and cartographic knowledge is applied to information acquired from the Infrastructure level and results in the formation of a representation of the real-world presented at the Interface level. An interface can be defined as a boundary across which two independent systems meet and act on, or communicate with each other. It is important to note that the proposed model considers two types of interfaces, user and software. The Interface level depicted in Fig.7.2 refers to the user interface, or the point at which a human user interacts with a given representation.

Software interfaces are also an important element of the model and are discussed in subsequent sections. The User level refers to human cognitive processes and is not discussed in detail in this chapter.

From subsequent discussion, we see that the information flow between Infrastructure, Mediator, Interface and User is a bidirectional and nested process that occurs at various levels of the system. The following sections provide a high-level description of a system architecture under development. The intent is to present a concrete example of how a system might incorporate the concepts of geographic and cartographic mediation.

3.1. *The Open Cartographic Framework*

A development system, labeled the Open Cartographic Framework (OCF) is being designed for use in the Cybercartography and the New Economy (CANE) project (Taylor 2003). This section provides a necessarily high-level discussion of the design and selected components of the OCF. The OCF is established on an Internet-based, distributed database and computing approach, and is enabled by emerging agent-based computing and Web service strategies.

3.1.1. *Agent-Based Computing*

The OCF uses a software agent architecture approach. A software agent is a software entity that functions continuously and autonomously in a particular environment. Each software agent has specific functions and responds to specific events, based on predefined knowledge rules, the collaboration of other agents, or users' instructions (Ming-Hsiang 2003). While a detailed discussion of agent-based computing is beyond the scope of this work, the idea of primary interest is that by using this method, tasks are performed by distributed entities in a collective process with a requirement that each agent present both data and semantic (contextual) information.

Agent-based computing is central to visions of the next generation World Wide Web currently termed the "Semantic Web" (Berners-Lee *et al.* 2001). Berners-Lee *et al.* (2001: 34) state that "to date, the Web has developed most rapidly as a medium of documents for people rather than for data and information that can be processed automatically." The Semantic Web aims to make up for this. Central to the Semantic Web strategy is the establishment of meaning through the use of formal ontologies and distributed software agents to negotiate between system elements to achieve a requested result (see Section 1.3) (W3C 2004b). There are strong relationships between developments related to the Semantic Web and the previously discussed work in the domain of semantic interoperability of digital geographic information. Thus, the Geo-Semantic Web, as it is known is emerging from within academia and industry. In recognition of the relationships between geomatics and mainstream information technology, developments in both domains are considered in the design of the system presented.

3.1.2. *Web Services*

To facilitate communication between software agents, a Web Services development approach is adopted. Generally, the term Web Service describes a standardized way of integrating Web-based applications over an Internet protocol backbone. Web services allow organizations to communicate data without users (human or software) needing detailed knowledge of how a given service is implemented. Standard interface protocols are established and responses to requests are delivered in a standard format. The interoperability specifications put forward by the OGC and ISO are based on a Web Services model. In the case of the OCF, services include the following examples (OGC 2003b):

- *Web Map Service* (*WMS*): Standardizes the way in which clients request and receive maps.
- *Web Feature Service* (*WFS*): Standardizes the retrieval and manipulation of geographic feature databases including the ability to INSERT, UPDATE, DELETE, QUERY, and DISCOVER geographic features.
- *Web Coverage Service* (*WCS*): Supports the networked interchange of geospatial data as "coverages" (imagery, rasters, DEMs, etc.) containing values or properties of geographic locations.
- *Gazetteer Service*: Retrieves the known geometries for one or more features, given their associated identifiers (i.e. place name).

The OGC specifications for geo-processing interfaces largely eliminate the need for costly batch data conversion. A query returns not a whole data file, but only the "result," or the answer to the query, and thus, the network model eliminates the need for users to keep (usually outdated) copies of whole datasets. These Web services can be seen as system agents.

An integral part of the Web Service model is the use of the eXtensible Markup Language (XML). XML is a simple and very flexible text tag based markup format (W3C 2004a). In contrast to the tags used by the well-known HyperText Markup Language (HTML) to define how a Web page is formatted and displayed, XML tags reflect semantic content. The intention is to separate the content from its presentation. XML can still be used to display information as is the case when XML is translated into HTML. As an example, in the case of HTML, a person's name might be marked up as follows:

<H2> Peter Pulsifer </H2>

(displayed in a Web browser as a second level heading)

Using XML, the same information might be marked up as follows:

<FullName>Peter Pulsifer</FullName>

In the first example, a browser can display the information to be read by a human, however an information processing system cannot readily process the information between the tags as it lacks semantic description. The XML example states that the information between the tags is included in the category "FullName." To format this information for display, a simple transformation rule can be established, like all <FullName> items are marked up with an <H2> tag.

The usefulness of XML is in its extensibility. The browser software does not define the tags that can be used, but rather this is done by users or communities and stored in schemas that contain namespaces. Namespaces define tag vocabularies that give precisely defined meanings to each element and attribute used in an XML document. A namespace is made available on the Internet using a Uniform Resource Identifier (URI) that resembles a common Web site address. The ability to include several namespaces in a single XML document while avoiding potential homonyms across domains is important in the context of geographic and cartographic mediation where integration of information from multiple domains is necessary (Zaslavsky 2003). Additionally, the use of XML as a means to encode information presents the possibility of transforming information into a variety of representation forms including Web pages, paper maps, digital documents, etc. (Lehto 2003). This capability is important in the context of cartographic mediation where the goal is to produce a representation that is meaningful to an end user. This goal may be better achieved if multiple forms of representation can be generated at the interface level. Many organizations have created domain specific XML profiles. Two of that are the Scalable Vector Graphics profile (W3C 2004a) and the Geography Markup Language (OGC 2003a), which are relevant to the discipline of cartography.

3.1.3. Open Cartographic Framework Architecture

Figure 7.3 provides a graphic depiction of the OCF architecture. As seen on the left of the diagram, the design incorporates the Infrastructure–Mediator–Interface chain, previously described (Fig. 7.2). The "cycling arrow" graphic on the left of the diagram denotes that the cycle is nested within the system – i.e. it occurs at more than one level. This is discussed in more detail in subsequent sections. The OCF is discussed in reference to Fig. 7.3 and the numeric labels to the left of the diagram denoted in a square box, and square brackets in related text (i.e. [1]).

In the interest of facilitating information sharing within the system, and the developments related to the CANE project, open (nonproprietary)

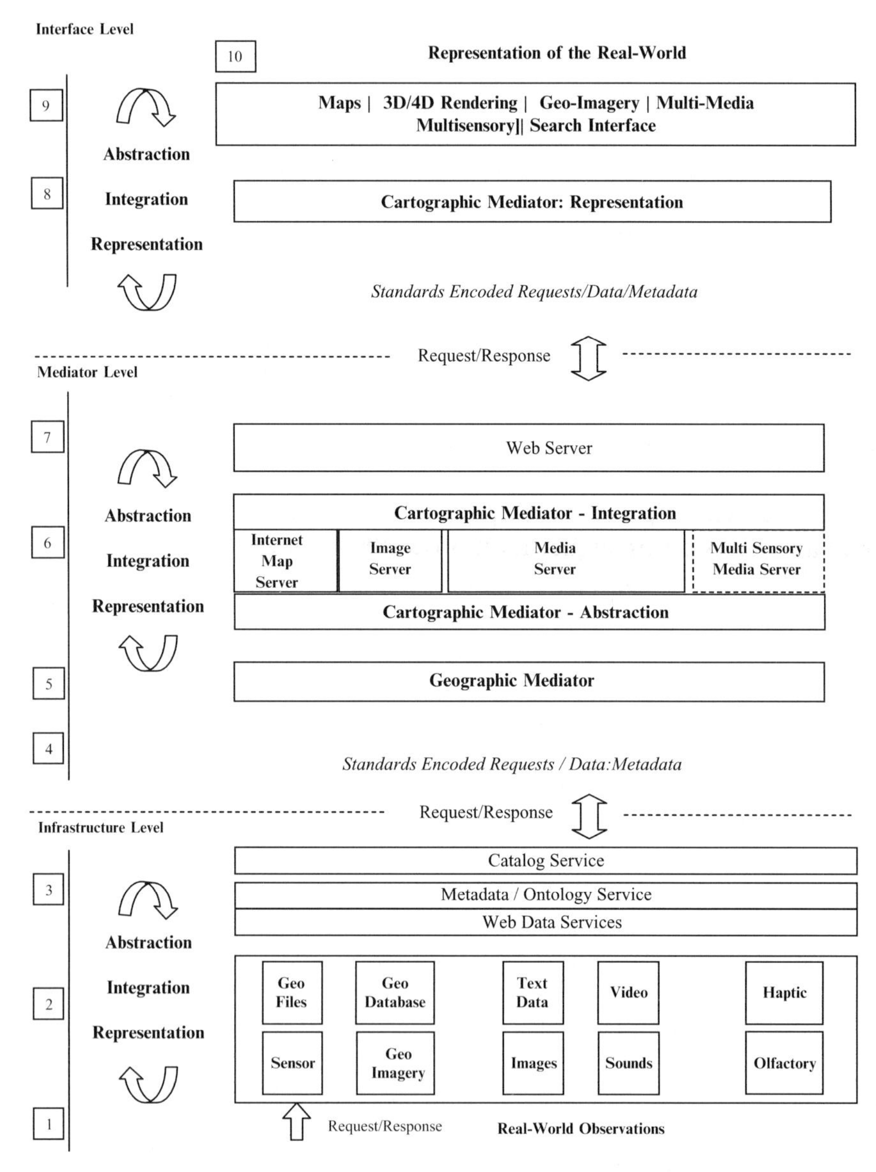

FIG. 7.3. The Open Cartographic Framework: An implementation of mediator based development model. Infrastructure elements are below mediator elements, which in turn are below interface elements. Mediation occurs at many levels of the system, however, the process may be predominant in middleware applications at the Mediator level.

standards such as those developed by the OGC and ISO and the World Web Consortium are being used wherever possible.

3.2. Infrastructure Level

Generally, the infrastructure level is where data and their descriptions of data are stored and managed. The data are distinct pieces of factual information derived from observations of the real-world [1]. At this level, contextual data is stored and associated with the data to which it refers. In practical terms, data [2], metadata and catalog services [3] exist in this layer. Although not yet well developed, inclusion of near real-time sensor data from in situ sensors existing in a Sensor Web is possible. Given the cybercartographic context in which the system is being developed, a number of data types not typically identified as spatial data are included at this level. Data types include sound, video, and olfactory.

Within the Infrastructure level, geographic data are stored in a variety of forms and formats [2]. Standard, file-based layers using a georelational data model (e.g. shp : ESRI shapefiles) are stored on a network accessible drive and accessed by higher level processes through Web services such as the OGC service models previously outlined. Alternatively, data are stored in a spatial relational database management system, such as ESRI's ArcSDE (ESRI 2004), or the open source PostGIS (Refractions 2004) referred in Fig. 7.3 as a geographic database. In the case of the OCF, the WMS capabilities of the University of Minnesota's open source Mapserver are used to provide necessary Web Services (UMN 2004).

Imagery data are delivered through a Web Coverage Service (WCS), again using the functionality of UMN Mapserver. For applications that require very fast, online rendering of large geospatial imagery data layers (e.g. full resolution LandSat scenes), two commercial products are being used. These are not yet compliant with Web Services standards, however discussions with the vendors suggest that this is a possibility in the near future.

Text and typical multimedia data (images, sound) are stored as files or in a database and are currently accessed through Internet HyperText Transfer Protocol (http) protocol requests. Current research is attempting to establish the best approach in developing a self describing, geospatially enabled media service. The strategy is based on the development of multi-namespace XML documents embedded within media files. The XML document includes metadata from various standards (i.e. Dublin Core, MPEG7, ISO19115) coupled with location data encoded in Geography Markup Language (GML). In this strategy, multimedia data are treated very much like a geographic feature rather than an aspatial data entity. Other types of data such as streaming media are delivered using appropriate server software (i.e. Apple's Quicktime or Microsoft's Windows Media Server). Again, these are not yet defined as spatially enabled Web Services.

At present, there is not a well developed plan for providing multisensory media services to deliver haptic or olfactory data that are reproduced by force feedback and scent producing devices, respectively. In terms of developing a Web Services approach, discussions with the OpenGIS Consortium are pending on this matter.

To facilitate data discovery and subsequent mediation, metadata, catalogue and possibly formal ontology services are required [3]. Metadata services manage detailed contextual information about individual datasets while catalog services record information about data collections and point to metadata services. In the case of geographic data, OGC Web Services include metadata services. For example, a request to a Web Feature Server will return metadata for the holdings in that service. Thus, if data services related to a collection are registered within a catalog server, detailed metadata can be retrieved by a user or agent as needed. Catalog services use a standard protocol called Z39.50. Developed by the American National Standards Institute's (ANSI), Z39.50 is a computer-to-computer communications protocol designed to support searching and retrieval of information, full-text documents, bibliographic data, images and multimedia in a distributed network environment. During the early years of SDI development, a profile specifically designed to deal with metadata for geographic data was developed (Z39.50 FGDC/GEO profile search protocol) (Nebert 2000). The Z39.50 protocol is recognized in the catalog interface standards of the OGC and is incorporated in the OCF design.

In examining the cycle represented by [1], [2], and [3] in Fig. 7.3, we see the emergence of a nested process. Real-world observations can be seen as a form of information infrastructure [1] (e.g. field observations with contextual data in the form of logs etc.). These observations are abstracted, integrated, and then represented as a file or in a database [2]. Thus, in this discussion, the process of creating data structures is a form of geo-mediation. The infrastructure-mediator-interface chain is completed through the establishment of (a software) interface in the form of Web Services [3].

The data, metadata, and catalogue services at the Infrastructure level can be seen as autonomous agents within the system. These agents can be used by human users or other agents through requests and resulting responses. Given the increasing number of Infrastructure services available, combined with the necessity to access them using standard request/response formats, these infrastructure agents are not typically useful to a human user. While the user can access the infrastructure agents directly using a Web browser or programming language, the resulting XML documents may not provide a representation that is meaningful to a human user. To create a meaningful representation, processes of abstraction, integration, and cartographic representation are performed by mediator agents at the mediator level of the system.

3.3. Mediator Level

In a system that supports new forms of dynamic cartography underpinned by various multimedia and map content from distributed sources, much of the processing required to generate these new forms of representation may be most efficiently executed at the Mediator level (Fig. 7.3). In the case of the OCF, the Mediator level handles encoded information from the Infrastructure level [4] using mediator agents for geographic [5] and cartographic mediation [6]. Geo-mediation does not consider cartographic representation, but rather is concerned with data modeling, manipulation of semantics, etc. These operations (e.g. semantic translation) can take place using information and a knowledge base related to the real-world phenomena. Operations need not consider elements related to cartographic representation. This separation is a logical construct, however, in a cartographic system, such as the OCF, the two types of mediator agents (geographic and cartographic) are tightly integrated.

Although Fig. 7.3 depicts the Mediator level as contained, agents may be distributed. For example, geo-mediation [5] may take the form of modeling and may take place on a server or peer separate from that which performs the cartographic mediation [6]. A geographic mediator [5] abstracts and integrates information from infrastructure level services to create a representation in the form of model output, for example. In response to a request, the model output is delivered over the Internet in standards encoded XML to a cartographic mediator agent, which in turn applies cartographic knowledge embedded within that particular agent. The cartographic mediator may be receiving input from a variety of agents (e.g. a multisensory media agent, a video agent, a legend agent, etc.) and thus will execute a process of abstraction and integration with the purpose of generating a meaningful representation for use in the Interface level of the system.

Mediation strategies can be implemented using a wide variety of tools. For cartographic mediation [6], the OCF employs a set of server-based technologies including a Web map server (UMN Mapserver), high-speed geo-imagery servers, and multimedia servers. Application logic and knowledge are embedded using the open source server side scripting language PHP and XML Stylesheet Language Transformations (XSLT) (Lehto 2003).

Geographic mediation carried out separately from the cartographic mediation, may be performed with a variety of tools including open source GIS packages, such as GRASS (http://grass.baylor.edu), commercial packages, such as ArcGIS, ENVI, PCI, etc. or possibly using high-level programming languages such as JAVA. For research related to semantic translation, standards, and technologies proposed by the Semantic Web initiative are being investigated. These standards can be implemented using a range of development platforms including the PHP scripting language mentioned. Results of the various mediation processes executed at the mediator level are delivered to the Interface level using the open source Apache Web server [7].

At the mediator level we again see a nested infrastructure-mediator-inte rface cycle with responses from the infrastructure level becoming the infrastructure layer of the mediator level [4]. Processes of geographic [5] and cartographic [6] mediation are applied and the resulting representation is passed to (a software) interface [7], which in turn responds to requests from the Interface level.

3.4. Interface Level

The Interface level is the level at which a human user interacts with the information system. The user requests information and responses are delivered at this level. Depending on the type of mediator strategy being used, the Interface level can simply render what is produced at the mediator level (i.e. HTML + multimedia) or provide interactive functions for the user. In the first case, much of the cartographic mediation will have been accomplished using limited HTML functionality such as hyperlinks. For more responsive, interactive applications dynamic server side scripts can be used. While this approach provides extended functionality, there are several potential limitations to maintaining a strictly server-side architecture. First, HTML provides very limited native support for interactive operations thus, reducing the potential for developing advanced cartographic applications. Second, processing application functions on the server side using scripting may result in unacceptable delays while results are returned from the server to the user's client application. Lastly, the use of server side scripting may be outside the skill set of the cartographer or may not be supported by the server infrastructure hosting an application.

Alternatively, mediation can occur outside the mediator level, nested within the Interface level. In this case, the cartographer incorporates elements of abstraction, integration and representation at the Interface level within the client environment (e.g. Web browser). The use of client side scripting (i.e. JavaScript) with SVG graphics is an example of this methodology. In this way, a local cartographer (one without access to mediator level processes) can implement cartographic mediation with the added performance efficiency eliminating unnecessary requests to the server. This approach also allows the cartographer to take advantage of powerful client side authoring tools such as Macromedia's Flash. In this case, the mediation functions are embedded within the Flash application file. A limit to this method is the inability to easily update information from the mediator or infrastructure levels. The UMN Mapserver being used in the mediator layer of the OCF can generate Flash vectors. When necessary, data within a Flash application can be updated through a request to the mediator layer. Further development of this strategy is being considered as part of the CNE development efforts.

Also available to support mediation at the Interface level is the use of emerging OGC specifications. Certain specifications allow the cartographer to modify cartographic elements of the representation at the Interface level. The Styled Layer Descriptor (SLD) specification is a means of controlling the portrayal of data that is rendered from OGC Web Services. In this way, a cartographer who may not have control over the styles defined at the infrastructure or mediator level can override with local settings. Similarly, OGC Web Map Context (WMC) documents can be used to provide interface level control of the user interface state and layout (map layers, position of legends, multimedia, etc.).

In summary, a nested infrastructure–mediator–interface chain at the Interface level is identified. Responses from the mediator level effectively become infrastructure within the interface level. At the interface level, in some cases, no mediation is required, or perhaps allowed. However, mediation at this level is possible [8]. This point of mediation may be one of the most important, as it provides the opportunity for the "local" cartographer to incorporate cartographic knowledge into the representation presented to the user. Depending on the system design, the end user can be included in the category of local cartographer. Increasingly, the end user has the opportunity, and in some cases is required to perform abstraction, integration and representation processes. To support this type of user mediation, some products now support online, multiauthor, collaborative map data publication and map design (Mapbuilder 2004).

3.5. Implications of a Mediator Based Development Approach

The review presented in Sections 3.1–3.4 reveals an architecture where information is exchanged and processed by a variety of agents at many levels. Within the parent levels of Infrastructure, Mediator and Interface are nested elements of the same chain (infrastructure–mediator–interface). In general, the expectation from those in government, industry and parts of the GI Science community is that such a system can provide single window access to and relatively easy integration of geographic information. The potential power to form new representations afforded by improved information access and integration is, however, coupled with increasing complexity and a lack of structure relative to previously existing forms of cartography. No standard methods exist to establish the allocation of various components and processes (i.e. mediation) within the system. In an agent-based approach, the designers of an application system may not have control of how or where particular service agents are implemented. Moreover, the potential for multiple authors (including end users) presents research, design, and implementation realities not present in previous generations of cartographic systems.

Distributed Web Services, such as those proposed in the architecture, can be chained to create what is known as cascading services. In the present scenario, data are delivered from one service agent to another. The information is in some way processed and then passed to another agent and so on. This service chaining presents powerful possibilities in terms of data access and integration. However, through this chaining process information can be remediated as it moves from agent to agent. In situations where multiple agent remediation takes place, establishing information lineage may be difficult or impossible. This can impede a user's ability to evaluate information quality and fitness for use. Ensuring that the Semantic Web previously discussed includes a collective process that manages information lineage is important for modern cartographic systems development.

A standard or typical usage scenario no longer exists (and thus, none was presented) for a system like the Open Cartographic Framework. Particular usage scenarios are defined by negotiations between stakeholders that may include field scientists, data collectors (human and otherwise), database designers, application developers, archivists, librarians, graphic designers, logicians, computer scientists, cartographers, human computer interaction specialists, and end user/producers. The last stakeholder in the list is of particular interest. Poore (2003) suggests that users of SDI, including people who build and integrate data should also be characterized as designers of the infrastructure. In her view, this "bottom up" approach contrasts the formalist, "top down" approach adopted by most SDI programs, and presents possibilities for more effective geographic information integration. In Poore's view, integration is a process of negotiation rather than technical translation. She recognizes that the negotiation process necessary to create effective systems is difficult, heterogeneous, diffuse, and unstable. If we accept Poore's account, then building the negotiation process into development will be a requirement if we are to design and implement effective cartographic information systems.

4. Praxis

Krygier (1999) asserts that the most vital issues in cartography are not technical issues. He suggests the need for "praxis" in cartography, an explicitly theorized practice. The work presented here is developed in this spirit. The proposed model can be evaluated solely as an implementation model, however, if considered from other perspectives it can open up new discussions in theoretical cartography, critical social theory, media studies, and cognitive science to name a few. Some interrelated questions and issues emerge and are presented briefly here.

4.1. *Theoretical Cartography*

Many approaches to interoperability are based on rational analysis and assume that geographic information and cartographic representations of that information can be rigorously defined and manipulated. Can this engineering – Formulating and applying precise rules for decision making – rules derived from a combination of the application of principles with iterative refinement through empirical testing (MacEachren 1995) – approach be reconciled with more artistic and holistic approaches to cartography as proposed by Medyckyj-Scott and Board (1991)? Can the two approaches be complimentary rather than mutually exclusive? Current agent-based approaches adopt a formalist, systems engineering approach to cartography. Does this model provide a solution that limits artistic approaches or does increased information availability provide new potential for artistic expression?

How do we analyze new maps in the tradition of J. Brian Harley, if in fact a given representation is the result of distributed cartographic mediation involving multiple stakeholders and authors?

As the practice of cartography moves forward, these theoretical questions and many others will need to be engaged.

4.2. *Critical Social Theory*

From a critical social theory perspective, we can examine the implications of developing systems based on standardized, government controlled SDIs.

In this context, questions about whose interests are being best served are compelling. For example, the Canadian SDI strategy sees a single, standards-based set of framework data:

> All framework datasets comply with standards for the content structure and semantics as well as for the metadata that describes it. Priority is given to CGDI endorsed standards. "Framework data for a specified resolution and geographic area will be unique." (CGDI 2004: 20)

The model outlined is based upon the assumed existence of a single "framework" geography. Recent works in the domain of alternative or "counter" mapping suggests that no single geography exists. Peluso (1995) reviews variations in forest classification results for a number of mapping projects in Indonesia. Discrepancies between forest classification maps created by different government programs are large. Some classes, such as "convertible forest" differ by a factor of five (Peluso 1995: 390). When comparing government planning maps with alternative maps generated by traditional land-users, dramatic differences in classification are observed. These differences are not attributed to methodology but rather epistemological differences between stakeholders. In many cases, areas traditionally used for production purposes by local people go unrecognized and are "considered available for development planning" by the state (Peluso 1995: 393). Similarly, Norheim (2001: published 2004) reveals that comparable forest mapping projects carried out by the US Forest Service and a conservation organization obtained acreages that differed by a factor of two. The differences are attributed to issues related to the organizational context of each project rather than methodological issues. Norheim's work identifies the importance of the institutional elements (e.g. protectionist data policies, funding, political agendas) involved in the collection and construction of geographic knowledge. According to some, the result of simply translating between this semantic heterogeneity without considering its root causes can be destructive. Rundstrom (1995) claims that to assimilate the geographic conceptualizations of one group into that of another may result in the destruction of cultural and epistemological diversity of the assimilated group.

Multiple maps of a single area challenge the concept of a unique, "standard" map for inclusion in framework SDI layers. This has implications for cartographic systems based on SDIs. Given the "unique" layer approach of some SDI proponents, cartographers must ask what this development approach means in terms of accommodating alternative perspectives? To achieve praxis in cartography, critical discourse of the type reviewed must be integrated into the research and development process.

4.3. *Origins of Meaning*

There is no agreement on whether spatial and geographic concepts are formed from innate, universal, mental spatial models that result from evolutionary psychology (Egenhofer and Mark 1995; Smith and Mark 2001) or social construction (Bowker 2000; Poore 2003). For example, Rundstrom (1995) notes conclusions drawn by scholars grounded in formalist GIS traditions that declare that the four cardinal directions (North, South, East, West) are axiomatic of spatial relations. Rundstrom identifies the failure of this work to recognize the seven dimensional spatial schematization of the Zuni nation of the Soutwestern United States (Frank and Mark 1991). Similarly, Hutchins (1995: sited in Poore 2003) maintains that cognition and establishment of meaning are active processes of interaction among people and their environments. What are the implications for SDI development and geographic and cartographic mediation? Will or should one perspective on the origin of meaning be dominant in terms of the knowledge embedded into SDI structures (i.e. formal Ontologies) and agent-based mediators? Given the diversity of perspectives on this issue, the perspective adopted during system design can have a significant impact on the nature of the resulting representations.

4.4. *Incommensurable Ontology*

In the domain of philosophy the study of ontology is concerned with the nature, constitution, and structure of reality. If one supposes a single reality, then it follows that there is only one ontology or world view that is representative of reality. Generally, current postmodern thought does not presuppose a single ontology but rather ontologies that are socially constituted, contingent, and partial. Discourse from this philosophical perspective suggests the concept of ontological incommensurability (Doubleday 1993; Rundstrom 1995; Bowker 2000). Contrary to the movement to capture and constrain meaning within machine-based formal ontologies, these works suggest that normalization of geographic conceptualizations is not possible. This difference in perspective presents a theoretical tension for those developing cartographic systems and representations. Using SDIs and agent-based development, approaches can effectively adopt a formalist perspective. Reflexive research and development, will be important for understanding resulting systems. The question of incommensurable ontologies is ultimately one that is not easily answered and depends upon the perspective of the researcher.

5. Conclusion

Cartographers are conceptualizing an emerging cartography that produces adaptable, dynamic, and interactive designs. This cartography incorporates multimedia and integrates map information from distributed sources using systems built by teams comprised of a variety of stakeholders and linked through new forms of partnership (Taylor 2003; Zaslavsky 2003).

This chapter provided a broad overview of a system architecture that may support these visions. Moreover, the paper identifies challenges and implications for developing such systems. Historically, a limit for cartographers in terms of producing meaningful representation of the world has been the limited access to geographic information. With the advent of SDIs, this reality is changing and a new challenge is emerging – how to mediate the information made available through SDIs. The need for abstraction and integration of heterogeneous sources of information will be increasingly relevant given the visions outlined above.

All of these new possibilities provide the cartographer not only with opportunities but also with new technical and theoretical challenges. In terms of creating products that are useful beyond the validation of technological methods, it will become increasingly important for cartographers to recognize the human context within which their work is situated. While this point has been well-established through academic discourse, emerging social and technical structures (i.e. SDIs) may have implications for the cartographer's ability to effectively evaluate these contexts.

Using the concept of cartographic mediation as a practical and conceptual framework may contribute to organizing and understanding emerging cartographic theory and practice. Understanding the mediation process, in addition to system components is critical both in terms of establishing data quality, fitness for use, etc., and understanding the human context(s) from which representations are constructed. In this chapter, the mediation process is one that is described as resulting in a "meaningful representation." Meaning may be defined in at least two ways:

Conventional: Common, or standard sense of an expression, construction, or sentence in a given language, or of a nonlinguistic signal or symbol.

Literal: Nonfigurative, strict meaning an expression or sentence has in a language by the virtue of the formal (dictionary) meaning of its words and the import of its syntactic constructions.

The term meaningful is deliberately left undefined in this work to represent the multiplicity of perspectives facing modern cartographers. Technical problems will be identified and addressed in turn. However, if contemporary cartographic systems are to address a variety of topics, beyond products used for location finding or describing the physical environment (Taylor 2003), then both stated definitions for meaning will need to be considered. If mediation processes within distributed systems are to provide meaningful representation, then the related questions and challenges that arise through the representation process will necessarily be heterogeneous and diffuse.

References

Berners-Lee, T., J. Hendler and O. Lassila (2001) "The semantic web", *Scientific American*, Vol. 284, No. 5, pp. 34–44.

Bishr, Y. A., H. Pundt, W. Kuhn and M. Radwan (1999a) "Probing the concept of information communities - a first step toward semantic interoperability", in Goodchild, M., M. Egenhofer, R. Fegeas and C. Kottman (eds.), *Interoperating Geographic Information Systems*, Kluwer Academic Publishers, Norwell, Massachusetts, pp. 55–70.

Bishr, Y. A., H. Pundt and C. Rüther (1999b) *Proceeding on the Road of Semantic Interoperability - Design of a Semantic Mapper Based on a Case Study from Transportation*, Paper presented at the Interoperating Geographic Information Systems, 10–12 March, Zurich, Switzerland.

Bowker, J. C. (2000) "Mapping biodiversity", *International Journal of Geographical Information Science*, Vol. 14, No. 8, pp. 739–754.

Boxall, J. (1998) "Spatial data infrastructures: Developments, trends and perspectives from converging viewpoints", *Cartography and Geographic Information Systems*, Vol. 25, No. 3, pp. 129–131.

Brodaric, B. and M. Gahegan (2002) *Distinguishing Instances and Evidence of Geographical Concepts for Geospatial Database Design*, Paper presented at the GIScience 2002, Boulder, CO, USA.

Brown, L. A. (1979) *The Story of Maps*, General Publishing Company, Toronto.

Cartwright, W. (1997) "New media and their application to the production of map products", *Computers and Geosciences*, Vol. 23, No. 4, pp. 447–456.

CGDI (2004) *Guide to The Canadian Geospatial Data Infrastructure*, Geoconnections, Ottawa.

Clinton, W. (1994) *Coordinating Geographic Data Acquisition and Access: The National Geospatial Data Infrastructure, Executive Order 12906,* Federal Register, Vol. 59, No. 71, pp. 17671–17674.

Coleman, D. J. and J. D. McLaughlin (1994) "Building a global spatial data infrastructure: usage paradigms and market influences", *Geomatica*, Vol. 48, No. 3, pp. 225–236.

Coleman, D. J. and D. D. Nebert (1998) "Building a North American spatial data infrastructure", *Cartography and Geographic Information Systems*, Vol. 25, No. 3, pp. 151–160.

DiBiase, D. (1990) "Visualization in earth sciences", *Earth and Mineral Sciences: Bulletin of the College of Earth and Mineral Science*, Pennsylvania State University, Vol. 59, No. 2, pp. 13–18.

Doubleday, N. (1993) "Finding common ground: Natural law and collective wisdom", in Inglis, J. T. (ed.), *Traditional Ecological Knowledge*, International Development Research Centre, Ottawa.

Egenhofer, M. J. (1999) "Introduction: Theory and concepts", in Goodchild, M., M. Egenhofer, R. Fegeas and C. Kottman (eds.), *Interoperating Geographic Information Systems*, Kluwer Academic Publishers, Norwell, Massachusetts, pp. 1–4.

Egenhofer, M. J. and D. M. Mark (1995) "Naive geography", in Frank, A. U. and W. Kuhn (eds.), *Lecture Notes in Computer Science 988*, Springer-Verlag, Berlin, pp. 1–15.

ESRI (2004) *ArcSDE: Advanced Spatial Data Server,* accessed 12 March 2004 from http://www.esri.com/software/arcgis/arcsde/index.html

FGDC (2002) *National Spatial Data Infrastructure*, accessed 22 January 2002 from www.fgdc.gov/nsdi/nsdi.html

Fonseca, F. T., M. J. Egenhofer and P. Agouris, P. (2002) "Using ontologies for integrated geographic information systems", *Transactions in GIS*, Vol. 6, No.3.

Forrest, D. (1999) "Developing rules for map design: A functional specification for a cartographic-design expert system", *Cartographica*, Vol. 36, No. 3, pp. 31–52.

Frank, A. U. and D. M. Mark (1991) "Language issues in gis", in MacGuire, D. J., M. F. Goodchild and D. Rhind (eds.), *Geographical Information Systems, Volume 1: Principles*, Longman, Harlow, Essex, United Kingdom, pp. 147–163.

Fremlin, G. and A. H. Robinson (1998) "Maps as mediated seeing", *Cartographica*, Vol. 35, No. 1/2, pp. i–xiii, 1–141.

Galton, A. (2003) *Desiderata for a Spatio-temporal Geo-ontology*, Paper presented at the COSIT 2003, Kartause Ittingen, Switzerland.

Goodale, R. (1999) *Notes for a Speech to the International Cartographic Association Conference*: 08-16-1999, accessed 05 March 2003 from http://www.nrcan.gc.ca

Goodchild, M. F. (1992) "Geographical information science", *International Journal of Geographical Information Systems*, Vol. 6, pp. 31–47.

Goodchild, M. F. (2000) "Cartographic futures on a digital earth", *Cartographic Perspectives*, Vol. 36, pp. 3–11.

Goodchild, M. F., M. J. Egenhofer, K. K. Kemp, D. M. Mark and E. Sheppard (1999) "Introduction to the varenius project", *International Journal of Geographical Information Science*, Vol. 13, No. 8, pp. 731–745.

Gore, A. (1998) *The Digital Earth: Understanding our planet in the 21st Century*, accessed 10 March 2004 from http://digitalearth.gsfc.nasa.gov/VP19980131.html

Guarino, N. and R. Poli (1995) "The role of formal ontology in the information technology", *International Journal of Human-Computer Studies*, Vol. 43, pp. 623–624.

Gumpert, G. and R. Carthcart (1990) "A theory of mediation", in Ruben, B. D. and L. A. Lievrouw (eds.), *Mediation, Information, and Communication*, Transaction Publishers, New Brunswick, New Jersey.

Harvey, F. and N. R. Chrisman (1998) "Boundary objects and the social construction of GIS technology", *Environment and Planning* A, Vol. 30, pp. 1683--1694.

Harvey, F., B. P. Buttenfield and S. C. Lambert (1999) "Integrating geodata infrastructures from the ground up", *Photogrammetric Engineering and Remote Sensing*, Vol. 65, No. 11, pp. 1287–1291.

Hutchins, E. (1995) *Cognition in the Wild*, MIT Press, Cambridge, MA.

INTEROP99 (1999) *Interoperating Geographic Information Systems*, Paper presented at the INTEROP'99, Zurich, Switzerland.

Kolacny, A. (1969) "Cartographic information - a fundamental concept and term in modern cartography", *Cartographic Journal*, Vol. 6, pp. 47–49.

Krygier, J. B. (1999) "Cartographic multimedia and praxis in human geography and the social sciences", in Cartwright, W., M. P. Peterson and G. Gartner (eds.), *Multimedia Cartography*, Springer-Verlag, New York.

Lachman, B. (ed.) (2002) *Lessons for the Global Spatial Data Infrastructure: International Case Study Analysis*, Santa Monica, California, Rand Corp.

Lehto, L. (2003) "A standards-based architecture for multi-purpose publishing of geodata on the web" in Peterson, M. (ed.), *Maps and the Internet*, Elsevier Science, Amsterdam, pp. 221–230.

MacEachren, A. M. (1995) *How Maps Work*, The Guilford Press, New York, NY, USA.

MacEachren, A. M. and D. R. F. Taylor (eds.) (1994) *Visualization in Modern Cartography*, Vol. 2, Elsevier Science, Oxford.

Maozhen, L., Z. Sheng and C. B. Jone (2002) *Multi-agent Systems for Web-Based Map Information Retrieval*, Paper presented at the GIScience 2002, Boulder, CO, USA.

Mapbuilder, (2004) *Project: Community Mapbuilder: Summary*, accessed 10 May 2004 from http://sourceforge.net/projects/mapbuilder

Masser, I. (1998) *Governments and Geographic Information*, Taylor and Francis, London.

McLaughlin, J. and S. Nichols (1994) "Developing a national spatial data infrastructure", *Journal of Surveying Engineering - ASCE*, Vol. 120, No. 2, pp. 62–76.

Medyckyj-Scott, D. and C. Board (1991) "Cognitive cartography: A new heart for a lost soul", in J. C. Muller (ed.), *Advances in Cartography*, Elsevier, London, pp. 201–230.

Ming-Hsiang, T. (2003) "An intelligent software agent architecture for distributed knowledge bases and internet mapping services", in Peterson, M. (ed.), *Maps and the Internet*, Elsevier Science, Amsterdam.

Morain, S. A. (1997) "A strategy for the national spatial data infrastructure federal geographic data committee", *Photogrammetric Engineering and Remote Sensing*, Vol. 63, No. 10, pp. 1159–1159.

Nebert, D. D. (2000) *Z39.50 Application Profile for Geospatial Metadata or "GEO"* 2000-05-27, accessed 20 March, 2004 from http://www.blueangeltech.com/standards/GeoProfile/geo22.htm

Nebert, D. D. (ed.) (2004) *Developing Spatial Data Infrastructures: The SDI Cookbook,* 2nd ed., Reston, VA, Global Spatial Data Infrastructure (GSDI).

Norheim, R. A. (2001) published (2004) "How institutional cultures affect results: comparing two old growth forest mapping projects", *Cartographica,* Vol. 38, No. 3 and 4, pp. 35–52.

OGC (1999) *The OpenGIS® Abstract Specification: Topic 14: Semantics and Information Communities,* 4th ed., Wayland Maryland, Open GIS Consortium.

OGC (2003a) *Geography Markup Language Specification,* accessed 01 March 2004 from http://portal.opengis.org/files/?artifact_id=4700

OGC (2003b) *OpenGIS, Reference Model,* 2004-03-04, accessed 01 May 2004 from http://www.opengis.org/docs/03-040.pdf

Peluso, N. L. (1995) "Whose woods are these? Counter-mapping forest territories in Kalimantan, Indonesia", *Antipode*, Vol. 27, No. 4, pp. 383–406.

Peterson, M. P. (1995) *Interactive and Animated* Cartography, Prentice Hall, Englewood Cliffs, NJ.

Peterson, M. P. (2003) "Maps and the internet: An introduction", in Peterson, M. P. (ed.), *Maps and the Internet,* Elsevier Science, Amsterdam, pp. 1–16.

Poore, B. S. (2003) "The open black box: The role of the end-user in GIS integration", *The Canadian Geographer,* Vol. 47, No. 1, pp. 62–74.

Pulsifer, P. L. (2003) *The Cybercartographic Atlas of Antarctica*, paper presented at the Cybercartography and the New Economy Project: 3 November, Collaborators Meeting, Carleton University, Ottawa, Canada.

Pulsifer, P. L. and D. R. F. Taylor (2003) *The Cybercartographic Atlas of Antarctica: Towards Implementation*, Paper presented at the 2nd International Antarctic GIS Conference, 9 April, IPG, Freiburg, Germany.

Refractions, (2004) *PostGIS, Geographic Objects for PostgreSQL*, accessed 12 March 2004 from http://postgis.refractions.net/

Rhind, D. (1998) *Establishing the UK National Geospatial Data Framework*, accessed 10 February 2003 from http://cgdi.gc.ca/sdi98/abstracts/2–20.html

Rhind, D. and B. Nanson (1998) *SDI National Overview*, Paper presented at the SDI '98, Vancouver, B.C.

Robinson, A. H. (1953) *The Look of Maps,* Unpublished PhD. Dissertation, University of Wisconsin at Madison, Madison.

Rundstrom, R. A. (1995) "GIS, indigenous peoples, and epistemological diversity", *Cartography and Geographic Information Systems*, Vol. 22, No. 1, pp. 45–57.

Salgé, F. (1995) "Semantic accuracy," in Guptill, S. C. and J. L. Morrison (eds.), *Elements of Spatial Data Quality*, Elsevier Science, Oxford, UK, (on behalf of the International Cartographic Association), pp. 139–152.

Shea, K. S. (1991) "Design considerations for an artificially intelligent system", in Buttenfield, B. P., and R. B. McMaster (eds.), *Map Generalization: Making*

Rules for Knowledge Representation, Longman Scientific and Technical, London, pp. 3–20.

Smith, B. and D. M. Mark (2001) "Geographical categories: An ontological investigation", *International Journal of Geographical Information Science*, Vol. 15, No. 7, pp. 591–612.

Sui, D. and M. F. Goodchild (2001) "GIS as media?", *International Journal of Geographical Information Science*, Vol. 15, No. 5, pp. 387–390.

Taylor, D. R. F. (2003) "The concept of cybercartography", in Peterson, M. P. (ed.), *Maps and the Internet*, Elsevier, Amsterdam.

Tobler, W. R. (1959) "Automation and cartography", *Geographical Review*, Vol. 49, No. 4, pp. 526–534.

UMN, (2004) *University of Minnesota Mapserver Homepage,* accessed 19 March 2004 from http://mapserver.gis.umn.edu

Van Dijk, E. and R. Leusink (1983) "Producing a videorecording 'basic audiovisual media and cartography' ", *Bulletin - Society of University Cartographers*, Vol. 16, No. 2, pp. 29–38.

W3C (2004a) *Extensible Markup Language (XML)*, accessed 27 January 2004 from http://www.w3.org/XML/

W3C (2004b) *Semantic Web*, accessed 15 March 2004 from http://www.w3c.org/2001/sw/

W3C (2004c) *Web-Ontology* (WebOnt) Working Group, accessed 27 January 2004 from http://www.w3.org/2001/sw/WebOnt/

Wiederhold, G. (1999) *Mediation to Deal with Heterogeneous Data Sources*, Paper presented at the Interoperating Geographic Information Systems, 10–12 March Zurich, Switzerland.

Worboys, M. and M. Duckham (2002) *Integrating Spatio-thematic Information*, Paper presented at the GIScience 2002, Boulder, CO, USA.

Zaslavsky, I. (2003) "Online cartography with xml", in Peterson, M. P. (ed.), *Maps and the Internet*, Elsevier Science, Amsterdam, pp. 171–196.

CHAPTER 8

Cybercartography and the New Economy: Collaborative Research in Action

TRACEY P. LAURIAULT

Geomatics and Cartographic Research Centre (GCRC), Department of Geography and Environmental Studies, Carleton University, Ottawa, Ontario, Canada

D. R. FRASER TAYLOR

Geomatics and Cartographic Research Centre (GCRC) and Distinguished Research Professor in International Affairs, Geography and Environmental Studies, Carleton University, Ottawa, Ontario, Canada

Abstract

As envisioned cybercartography is compiled by teams of individuals from different disciplines and involves new research partnerships among academia, government, civil society, and the private sector. Collaboration and integration of research require deliberate and directed effort to manage and coordinate to enable all partners, collaborators, students, and stakeholders to fully participate and contribute to the research process. This chapter explores collaboration with the use of concepts from organizational theory, transdisciplinary research, and methods to measure collaboration. The chapter describes the Cybercartography and the New Economy (CANE) project as an organization of researchers, developers, users, and stakeholders who are collectively engaged in creating cybercartographic products and theoretical constructs. The Cybercartography project is viewed as an organized process that is structured to integrate research from various academic disciplines.

1. Introduction

The history of technology shows that advances in specialized fields are often made by individuals from other disciplines (Epton *et al.* 1983: 15). Studies have also shown that heterogeneous teams are more creative than homogenous groups, particularly in scientific laboratories. Argote and Ophir (2002) argue that heterogeneous management and product development teams tend to be more innovative. Interdisciplinarity can also be "one of the most productive and inspiring of human pursuits – one that provides a format for conversations and connections that lead to new knowledge" (NAS 2004: 1). Cross-disciplinary collaborative research, however, does not come easily and it requires patience, flexibility, translators, and a time commitment at the beginning of the process to manage communication and understanding among individuals with diverse backgrounds (Epton *et al.* 1983; Levesque and Chopyak 2001; Argote and Ophir 2002; NAS 2004: 1).

Collaboration in geomatics is common practice but little research exists to explain this process. In geomatics collaboration has been considered from a number of perspectives: public administration mostly at the municipal level (e.g. Masser and Craglia 1997; Laurini 2001); geospatial data sharing processes have been studied (e.g. Onsrud and Rushton 1995; Pinto and Onsrud 1995; Wehn de Montalvo 2002) as have organizational and institutional issues related to geospatial data infrastructures (Coleman and McLaughlin 1997; Holland 1999; Lauriault 2003). The focus tends to be on technology and methods rather than on the complex organizational, institutional, research, and policy issues involved. In cartography, it has been argued that organizational and policy issues are of equal importance as technological issues but often ignored (Taylor 1999). Due to their complexity, the development of Internet cartographic products require teams of individuals for their creation. An examination of the cartographic literature confirms that technological processes are well described while collaborative processes are not.

The literature on collaboration in business is extensive and includes studies on learning organizations, knowledge management, and reasons why firms collaborate, how it is done, and evaluated. However, little research exists on how to organize or manage collaborative interdisciplinary research in a university setting, to measure its success or the complex social processes involved. Major research funding agencies in both the natural and social sciences are promoting interdisciplinary research collaborations but research on processes, management, and best practices are lacking. There has also been very little systematic study of "the people,

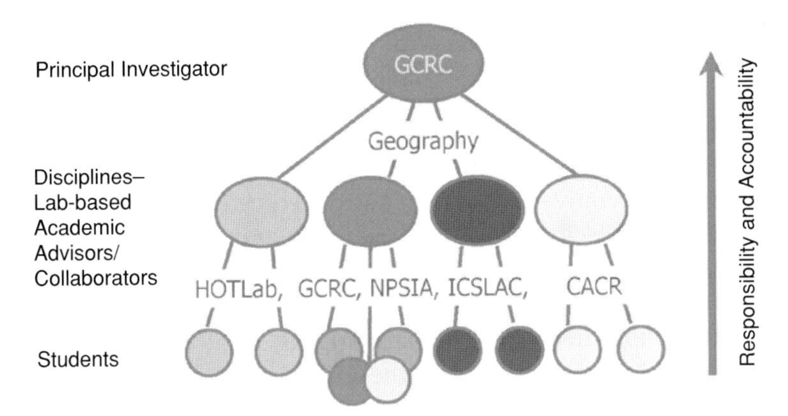

FIG. 8.1. Formal organizational structure of the project.

institutions, or funding organizations taking part in interdisciplinary activities" (NAS 2004: 166).

This chapter will consider collaborative research as an active process toward transdisciplinarity. The Cybercartography and the New Economy (CANE) Project at Carleton University is used as a case study. The CANE project is funded by the Canada Social Sciences Humanities and Research Council (SSHRC) Initiative on the New Economy (INE), a Federal Collaborative Research Initiative Grant described more fully by Taylor in Chapter 1. The CANE project involves teams from four separate research units: the Human Oriented Technology Laboratory (HOTLAB), the Institute for Comparative Studies in Literature, Art and Culture (ICSLAC), the Centre for Applied Cognitive Research (CACR), and the Geomatics and Cartographic Research Centre (GCRC) as seen in Fig. 8.1. More familiarly, the groups are the HCI (human–computer interaction) team or the psychologists, the English Literature team or the humanities group, the Cognitive Science Team, and the Geography team. These discipline-based teams are the formal research collaborators in the project.

This chapter will also analyze the CANE project as an organization of researchers, developers, users, and stakeholders who are collectively engaged in the process of creating cybercartographic products. It will therefore consider two of the seven elements of cybercartography as outlined by Taylor (2003): teams of individuals from different disciplines and involving new research partnerships. Concepts from organization theory, the small body of work related to doing transdisciplinary research and methods to measure collaboration. The objective is to understand how collaborative research fosters a culture of collaboration and how to structure work to enable transdisciplinarity.

2. Collaboration

Collaboration is defined as "to work jointly with others or together especially in an intellectual endeavor" (Merriam-Webster online Dictionary 2004). Cross-disciplinary research is a form of active collaboration commonplace in many universities. It has been argued that cross-disciplinary research is the most efficient way to investigate and apply research on social issues (Salter and Hearn 1996; NAS 2004). Cross-disciplinary research exists because some research problems cannot be satisfactorily resolved within a particular discipline. Others suggest that the drivers are: the complexity of nature and society, the exploration of problems that are not confined to a single discipline, the need to solve societal problems, or the power of new technologies (NAS 2004: 2).

The above reasons apply in the case of the CANE project which requires a collaborative approach to achieve its goals. Collaboration is however much easier said than done. Collaboration comes in many forms and the form it takes is contingent on the particular research problem, the resources available, and perhaps above all the individuals engaged in the process. In university-based research, collaboration is described as interdisciplinary, multidisciplinary, transdisciplinary, and cross-disciplinary.

Interdisciplinary research "refers to research teams in which the effort is integrated into a unified whole" (Birnbaum 1983: 3). It "involves the internal and substantive interlinking of various disciplinary analyses so that each considers the results of the others in its own development" (Rossini *et al.* in Epton *et al.* 1983: 4), or it "implies joint coordinated and continuously integrated research conducted by experts with different disciplinary backgrounds working together and producing joint reports and papers in which the specific contributions of each researcher tend to be obscured by the joint product" (Lindas in Epton *et al.* 1983: 4). According to the recent National Academy of Sciences report, it is a "mode of research by teams or individuals that integrates information, data, techniques, tools, perspectives, concepts, and/or theories from two or more disciplines or bodies of knowledge to advance fundamental understanding, or to solve problems whose solutions are beyond the scope of a single discipline or area of research or practice (2004: 2)." These definitions imply a joint effort by a group of individuals from different disciplines who produce joint products, and where research is influenced by a group that at the outset, aims toward a form of integration and where the whole may be greater than the sum of the parts.

Multidisciplinary research is defined as "research in which scholars from different disciplines work independently and are joined together externally through editorial linkages" (Birnbaum 1983: 3), and it is considered that "the result of the interrelation of disciplinary components occurs when they are linked externally only . . ." (Rossini *et al.* in Epton *et al.* 1983: 4). It is a problem-solving alliance where different disciplines come together to work collectively on a project. The terms multidisciplinary and interdisciplinary are often used interchangeably. The distinction, however, is the degree of integration for the latter and individual research autonomy for the former. Multisector research extends multidisciplinarity towards achieving a particular research goal or project by crossing institutional boundaries.

Transdisciplinarity (Nowotny 2003) is considered a new mode of intellectual activity not reducible to one discipline but involving those that emerge in the application of theory. Transdisciplinary and interdisciplinary are distinguished by the research outcomes, the former is likened to a hybrid discipline being developed out of the process while, interdisciplinary is an integration of the outcomes from the disciplines involved.

Cross-disciplinary research is defined as "tasks that require for their objective completion contributions from more than one discipline" (Epton *et al.* 1983: 6), since tasks often require "for effective completion contributions from more than one discipline" (Epton *et al.* 1983: 4). Different tasks may need to be carried out in different collaborative forms. In some cases a multidisciplinary form is required where "tasks are carried out by organizationally separate units each of which includes practitioners of only one discipline. The products of their activities are combined into a coherent whole by a task coordinator who bears ultimate responsibility for so doing" (Epton *et al.* 1983: 4). Tasks can also be carried out in an interdisciplinary form in which "the elements of the task are carried out within a single organizational unit consisting of the practitioners of the disciplines necessary for the completion of the task. The members of the unit share the responsibility for combining their individual products into a coherent whole" (Payne and Pearson, in Epton *et al.* 1983: 5).

Each collaborative form has its challenges. Interdisciplinarity requires time to resolve internal politics while the process is considered to yield better results and will identify more areas of innovation (Epton *et al.* 1983: 16). The coordinator/integrator is however tasked with the burden of making those linkages and making findings explicit. Multidisciplinarity requires significant integration and coordination among teams of researchers. The outcomes might be used by all, but not well integrated into the project as a whole. In transdisciplinarity, common problems are related to quality control particularly between and among disciplines, as what is quality or worthy of pursuit in one is not necessarily considered as such in another (Nowotny 2003).

Organizationally each has their particularities. Transdisciplinary research implies a heterogeneity of skills and expertise to problem solving, loose organizational structures, flat hierarchies, and open-ended chains of command (Nowotny 2003: 2). Research in multisector projects suggests a weak link with traditional research characteristics but with characteristics common in typical management strategies. Levesque and Chopyak (2001: 18) discovered that these managerial strategies included a heavy front-end process of negotiating acceptable parameters to decide how resources are allocated and the link to the social capital of that particular community. Multidisciplinarity implies a primary investigator (PI) or integrator to coordinate activities of each team while interdisciplinarity may require very close collaboration and frequent interaction possibly in a shared space.

The challenges of research integration arise from a number of factors, such as disciplinary differences, the magnitude of these difficulties, and academic tradition. In addition, the number, types, and degrees of difficulty play a role as does the depth of integration required (Gold and Gold 1983). It is suggested that when the burden becomes too great an individual with the primary work to integrate the components should be part of the process as a bridge scientist. Some of the obvious differences between disciplines are: specialized information, specialized language for expressing that information, and specialized methods for manipulating and adding to that information (Gold and Gold 1983). This is certainly the case in the CANE project where four discipline-based teams are involved, while two are in the same discipline (psychology). Different disciplines often attract while like disciplines may not, and at times student researchers get more connected to researchers outside their discipline (NAS 2004: 157). Mutual learning involves a commitment to sharing information, understanding another discipline's language, and using new applied and theoretical methods. Cross-sectoral work is cooperative in nature and requires communication which can be problematic as there are issues of not only physical separation between institutions, but also "there is a wider than average professional gap between the scientist and the technologists on the one hand and the politicians and administrators on the other" (Epton *et al.* 1983: 35).

The Cybercartography project from the outset was conceptualized as transdisciplinary in both form and process, to test and apply the theory of cybercartography and apply this theory in the development of cybercartographic products. The project's objectives are to ground the theory in tangible applications, outcomes, and impacts manifested particularly in two atlases but also including proof of concept products, reports, including dissemination activities to academia and the general public through the media. It was hypothesized that the creation of the Cybercartographic Atlases and related products would create a focus on integration in which all disciplines would contribute. A transdisciplinary approach would inform the overall collaborative approach, but elements of cross-, multi-, and interdisciplinarity would be used as appropriate to contribute to project goals, the context of the work at hand, the tasks to be completed, and how people like to work. A deliberate set of strategies and methods were used to facilitate collaboration, including the designation of one team member (Lauriault) with the specific responsibility for this task. Multiple approaches and tools were used and these changed over time as a result of consultation, observation, and adaptability. In their real application these theoretical concepts tend to blur and change over time.

Learning about each group's approaches was a long but essential time consuming process involving meetings, discussions, and presentations in different fora. A process of "cognitive tuning" occurred in which time was allotted to develop the knowledge base of some of the researchers on topics not related to their own research (Epton *et al.* 1983: 36). Time constraints proved problematic as some researchers could not foresee the benefits of early and intensive collaboration. The development and choice of appropriate tools and policies to create a work flow between the teams were time consuming and involved considerable experimentation, effort, and individual commitment to this process. Language also proved problematic. The same words have different meanings in different disciplines making communication and understanding difficult (Gold and Gold 1983: 95). For example, the term navigation in geomatics can mean moving through a virtual landscape, to human factors psychologists it means moving through the interface, and to the Cognitive Science Team (CST), it means traveling through hypertext, or navigating and memory in virtual spaces, while for cultural mediators it can mean traveling through ideological space.

Operationally, it was also necessary to develop methods to share documents and information, and a wide variety of tools were used for this purpose. The most useful of these was a Web-based knowledge forum with related sharing, referencing, storage, and descriptive functions. This provided a virtual space where researchers and developers could engage with their clusters and monitor activities. Typical information communication infrastructures, such as email, shared space on servers, backup systems, high speed connections, Web pages, and wikis, are also essential and not to be underestimated. Other record-keeping processes were also developed, such as file naming protocols, document metadata, data sharing policies, and bibliographic tools. Other managerial tools are milestone documents, cluster terms of references, research design statements from discipline-based teams, and production workflow diagrams.

The nature of the disciplines involved in a project as outlined earlier is important, and geography is the most interdisciplinary of the disciplines involved. The Geography Team led by the Primary Investigator (PI) has the responsibility for research integration, and was the most aware of transdisciplinary needs. Early processes involved seminars where individuals from each discipline explained their methods and approaches. Students were directed to explore the literature on topics identified as critical in the research proposal from their disciplinary perspectives. Integration meetings are held twice a year to identify research questions, evaluate team-based research design statements, collectively direct the way to work together, and to assess progress. Students presented and assessed their findings with each other on an ongoing basis at student meetings and continuously with their academic advisors through their own disciplinary teams. Concurrently, atlas content and infrastructure development was the responsibility of the Geography Team as discussed in several chapters of this volume.

A parallel meeting structure was also established. These included student only meetings, and meetings of the directors of the four groups with key players on the project management team. These bi-weekly sessions enabled learning at multiple levels, facilitated the flow of information and knowledge between groups, and provided opportunities and time for team members to get to know each other, to address administrative issues, to share their progress, and identify points of convergence and divergence.

After a year-and-a-half of collaborative efforts an assessment of the progress was made. It was concluded that the CANE project could be described as predominantly cross-disciplinary exhibiting some multidisciplinary and interdisciplinary elements. Existing processes, structures, and tools had failed to realize the transdisciplinarity which was desired (Taylor Chapter 1). The process to date was however effective at producing the necessary expertise to provide discipline-based direction, learning, and a collaborative culture. Change was required to promote transdisciplinarity and this was articulated by researchers at the May 2004 Integration meeting. Researchers frankly expressed their opinions regarding collaboration and the outcome was the consensus to develop a cluster-based matrix structure as seen in Fig. 8.2.

In the case of the CANE project, a research matrix is considered to be a flexible and dynamic organizational structure that adapts with the evolution of the research process, the task at hand, the context, the topic of research, and the researchers involved. Researchers move from one cluster to another based on where their expertise is most needed, where they can engage on a new topic of interest, or where they want to be creatively affiliated. Ideas and strategies are bridged as researchers engage in more than one cluster, all participate in identifying potential opportunities to collaborate and when ideas converge and interconnections are made. The collaborative culture, nurtured from the outset flourished in this new form.

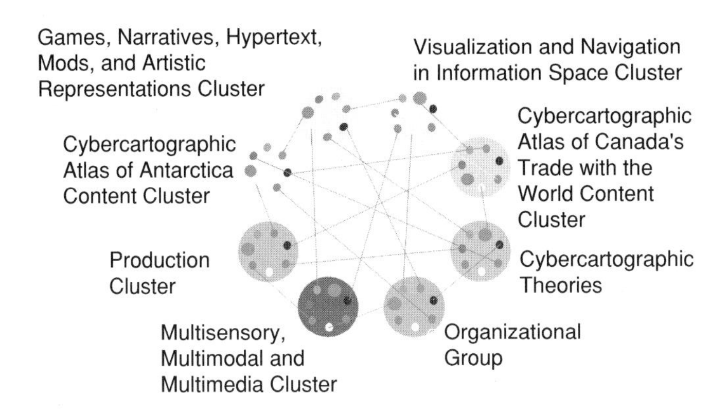

FIG. 8.2. Cybercartography research and practice clusters.

Membership in each cluster is based on research interests, preferences, expertise, and the intention to mix the disciplines. Researchers had already been exploring cluster topics within their discipline-based teams, the new groups simply facilitated the exchange of that expertise across disciplines. Cluster leaders were chosen based on expertise or previous responsibilities related to the cluster topic. Their roles are as coordinators or as facilitator of activities and not as managers. Leaders are research collaborators, PhD students, and postdoctoral fellows. There is already evidence that the new approach is leading towards a much stronger transdisciplinarity. Dormann and Lindgaard (2004) illustrate this with the concept of "creative ideation" emerging in the Games Cluster and the matrix approach is resulting in much more collaborative interaction among researchers from all four groups. An incremental learning approach within disciplines combined with a traditional hierarchical structure was transformed into clusters which shifted practice, theory, and leadership roles. Disciplines remain the home base of research.

Operationally, cluster leaders organize meetings and working groups as required with their members, and direction is provided through the development of a terms of reference document designed by each group. Cluster leaders and research collaborators meet to share results, express progress, discuss difficulties, and seek direction. Early observations suggest that clusters do particularly well when there is a strong presence of faculty, and where postdoctoral fellows can dedicate a significant portion of their time to cluster activities. All students make significant contributions, although as research assistants, time is limited to only 10 h per week during the school term and 20 h in the summer. It has been observed that since students enjoy working in the clusters, more time than expected of them is dedicated to cluster activities and it is argued that their contributions are the spirit of the collaborative effort. Further, students lead clusters or have their personal research tied to activities in these groups resulting in mutually beneficial research. Student research not only makes its way back to its discipline-based teams, but also throughout the home department.

The current clusters are: Antarctic Atlas Content; Trade Atlas Content; Cybercartographic Theory; Metadata, Navigation, and Visualization; Multimodality, Multisensory, and Multimedia, more familiarly known as MnMnMn, Production; and the Games, Narratives, Artistic Representation, Mods, and Hypertext (GNAMH) as outlined in Fig. 8.2. Each cluster operates differently but all include theory and practice that contribute to the Atlases or proof of concept products and publications. For example, the GNAMH cluster has two working groups: games and hypertext narrative; and each produces proof of concept products and then uses these objects to inform and illustrate theoretical concepts. The Theory Cluster has a formal component to produce papers, seminars, or promote transdisciplinarity, and a Cybercartographic Café where a research question is selected to be discussed and debated from myriad disciplinary and experiential backgrounds. The Trade Content cluster is more insular due to its intense data mining activities while the Antarctic Content cluster is more expansive with other researchers since it is in a production phase. The Visualization cluster meets less often, but is very active in small research intensive working groups, while MnMnMn is very interactive but behind in terms of production due to delays in technology acquisition.

3. Multisector Research: Partnerships

The multisectoral approach to research best exemplified in the CANE project's partnership arrangements. Partners are first and foremost stakeholders who share a common vision in the project. As part of the UNA, these stakeholders, at the outset, identified high level direction, determined the degree of commitment and generated ideas on the content and form of the atlases, and their target audiences. Some partners are members of the advisory board which assesses progress, question research direction, and ensure that the project meets its mandate with its funders (Taylor in Chapter 1). Partners are from a wide variety of organizations and institutions in the public, private, academic, and nongovernmental sectors. They are scientists, policy makers, researchers, teachers, explorers, expert advisors, data providers, technology specialists, association leaders, cartographers, or technology providers (see details at the GCRC Internet site at http://gcrc.carleton.ca).

Their contributions are invaluable in ensuring the work is relevant to external organizations, that the project positions its work to meet user needs and that GCRC meets its commitments to its funding organization. The relationships are mutually beneficial. Partners benefit by: contributing technology that can be tested; sharing data that can be represented in new and meaningful ways and beyond the scope or mandates of their own institutions; expanding the content of their atlases; having useful educational tools in their classrooms and accessing the expertise of leading researchers. GCRC benefits by having: access to technology; data content for the atlases and proof of concept products; access to users for testing; also by developing expertise; access to existing infrastructures; sharing human resources and exchanges with key policy makers, among many other benefits.

Operationally, a number of partnership arrangements exist in the CANE project including memorandums of understanding (MOU) between GCRC and its partners; deemed employee status arrangements to facilitate access to data; students pursuing relevant research with experts in external research institutions; membership in key scientific committees; installation of interoperable systems in international organizations to facilitate and test interoperability and inviting international researchers for extended research stays in Ottawa to work on the project. Partnerships have to be negotiated on an individual basis, need to be formal yet flexible, and require time. They are important aspects of collaborative research projects and the operational tools to facilitate the bridging of multiple sectors require careful attention at the outset, and continuing creativity and care throughout the duration of the project.

4. Some Perspectives on Collaborative Research

The Cybercartography project as outlined above was conceptualized as a transdisciplinary form of collaboration but involves multiple collaborative forms, activities and structures in order to meet different project needs. These utilize a number of research paradigms including: instrumental, conceptual, knowledge domains, and impact approaches.

The instrumental approach sees interdisciplinarity as an applied or problem centered activity which does not seek to challenge disciplinary boundaries or epistemological assumptions of particular disciplinary paradigms (Salter and Hearn 1996: 8). This approach can be likened to "methodological borrowing." Tools and methods from different disciplines are used but synthesis of knowledge is not necessarily required or produced. Interdisciplinarity is a "purely functional activity" which may lead to the establishment of new "hybrid disciplines" (Saltern and Hearn 1996: 9). The research matrix and the clusters of the CANE project reflect an instrumental approach.

A conceptual approach to interdisciplinarity emphasizes the synthesis of new knowledge. This is a "theoretical, primarily epistemological enterprise involving internal coherence, the development of new conceptual categories, methodological unification, and long-term research and exploration" (Salter and Hearn 1996: 9). The CANE project utilizes the Cybercartography Integrated Research Framework (CIRF) developed by Eddy (Eddy 2002) and its extension in the cybercartographic human interface model (CHI) as outlined by Eddy and Taylor in Chapter 3 of this volume, as the conceptual underpinning for the CANE project. As a result of the interdisciplinary and transdisciplinary processes, cybercartography may be emerging as a hybrid discipline in its own right as discussed in several chapters in this volume.

A knowledge domain approach looks at "forms of knowledge whose logic varies according to the institutions in which they are ascribed" (Pester 2003: 3). It is suggested that looking at disciplines alone negates the history of forms of techno-knowledge that vary according to where they are produced (e.g. academic world, the firm, start-ups, experts). Interdisciplinarity emerged when industrial laboratories (e.g. GE, ATT) were the norm as management tools, and it was a concerted and deliberate action (Pester 2003). These forms of research were designed to develop better and new products to overcome industrial bottleneck problems. This required a logical mix of professions, which led to a "recomposition of the disciplines" (Pester 2003: 4). The multisectoral research in Cybercartography is attuned to a knowledge domain approach. The knowledge domain approach is an incentive for government agencies, industry associations, and other partners to participate in CANE. The collaborative techno-knowledge approach used in CANE meets needs which other approaches do not.

A research impact approach occurs when "research makes a difference to the subsequent actions that people take" (Levin 2004: 3). Such research has a clear long-term agenda and third-party mechanisms are considered important (e.g. associations, popularizers, such as politicians, public officials, or lobbyists). Research dissemination is targeted and dependant on the type of research and can involve think tanks, foundations, professional organizations, and policy entrepreneurs (Levin 2004). Levin's research impact model has three elements: (1) *the context of research production* what gets done, who does it, how it is done, and related communication activities; (2) *the context of research use* – settings that have an interest in the application of research, government, educational organizations, and teachers, particularly, "the views, capacities and structure through which such organizations are able – or limited in their ability – to find, understand, and use research" (Levin 2004); and, finally, (3) *the connections and interactions between context and setting* involving direct and mediated interconnections. The CANE project has a strong application element and the research is designed to have the impacts which Levin's model suggests.

5. Organizations and University Research

Conducting collaborative research with many people requires an organization to coordinate and manage tasks, interactions and results. Organizations "are social structures created by individuals to support the collaborative pursuit of specified goals" and "they are deliberately structured activity systems with identifiable boundaries" (Withane, 1987: 4; Scott 1998: 10). Organizations vary in size, sector, structure, intensity – capital and human, historically evolve, and they pursue diverse interests. A detailed consideration of organizational theory related to collaboration is beyond the scope of this chapter, but can be found in the work of Scott (1998), Patriotta (2003), Agyris and Schön (1978), and Senge (1990), among others. Organizations are the places where work gets done, ideas are formed and shared, and where activities critical to human life are carried out. Organizations are complex entities and Leavitts's Model of an Organization provides a useful framework to understand them. The model divides organizations into five integrated interdependent elements: four intraorganizational elements, such as social structures, participants, technology, and goals with one extra-organizational element, the environment (Fig. 8.3.).

Applying Leavitts Diamond to cybercartography illustrates the utility of this model to visualize the organizational reality within which the theory and practice of cybercartography is being investigated. This also provides a way to visualize research as an organized process. The following is a simple illustration to demonstrate a few aspects of the Cybercartography project and captures only a part of the complexity of the project. Another important model to analyze cybercartographic processes is second-order cybernetics as discussed by Reyes and Martinez in this volume.

Organizations can be defined by a variety of characteristics, and form is a factor of group size, work complexity, and the degree of structuring of activities and tasks. Groups of less than 25 members are considered small and based on primary group affiliations, while larger groups require formal modes of engagement that are less reliant on implicit agreements among members. Institutional settings also produce different types of forms (e.g. hospitals, academia, etc.). The Cybercartography project is a mid-to-large

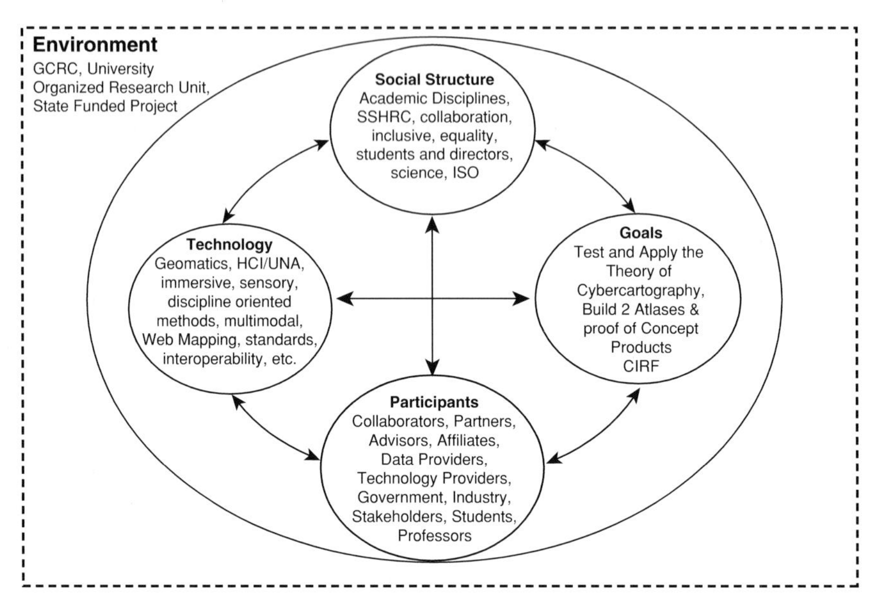

FIG. 8.3. Cybercartography and the new economy represented in Leavitts diamond.

sized organization with approximately 40 fully engaged researchers at any given time.

Epton *et al.* (1983) have studied university-based groups and provide some of the following descriptive characteristics. Small university-based groups are normally led by a university professor who has identified a problem that requires crossing disciplinary boundaries for resolution. Resources are normally acquired outside the university setting. Professors act as part-time leaders and may have the resources for full- and part-time researchers, technicians and for technical facilities that are normally acquired from the host institution. The difficulties and benefits of grant-based funding in these settings are outlined in Table 8.1.

Large university-based institutes described by Epton *et al.* are normally located in institutes or research centers, where budgets, expenditures, equipment acquisitions, and other administrative actions are centrally controlled. Organized research units (ORU) are a typical and some are large enough to become independent of the university. At times there are frictions between the university and the ORU for administrative and financial control. Often ORUs are application or problem oriented which is a reflection of the contractual or granting system. At times the university is a constraint for ORUs that are significant enough to exercise some influence. Larger ORUs can offer inducements to graduate students that increase their status in the university. Buffer institutions between the ORUs

TABLE 8.1
Descriptive Characteristics of University-based Research Groups

1. The reliance on short-term support from students thus causing turnover issues.
2. Students hired into these projects are given the "unfaculty" title as they do not adhere to a particular discipline and are of a different status than professors.
3. The limitations of funding often impedes travel and face-to-face interaction at conferences and thus time and resources allocated to networking is often scarce for smaller projects.
4. Balancing size and innovation; often the smaller the group the lesser the innovation; while a very large group requires too much structure.
5. Time pressures are significant since in academia the quantity of publications at times is considered more important than their quality.
6. Campus location and space is often an issue, in particular access to meeting space
7. Teaching tends to take priority over project demands
8. Researchers often do not like to disassociate themselves from their disciplinary institutions for fear of "out of mind out of sight" issues.
9. The advantage of a university setting however is access to a wide array of intellectual and infrastructure resources.
10. Organizational culture is a factor for these groups since, in a university setting, people choose to work on problems they like since there is no incentive to do otherwise, which increases the level of commitment
11. And often, but not always a university setting is generally considered egalitarian reducing the friction between the disciplines.

Adapted from Epton *et al.* (1983), *Managing Interdisciplinary Research.*

and the university are recommended in order to avoid power difficulties, while skilled leaders, who can develop good relations and open communication channels are nonetheless critical. People working at the interface of disciplines are often assigned boundary roles to bridge teams and ideas.

Adhocracies are "characterized by much horizontal specialization, high levels of professional training, little formalization of procedure, and organic rather than mechanistic relationships" (Epton *et al.* 1983: 29). These also tend to be young, highly technical with sophisticated technologies. They also exhibit much autonomy to lower levels of the organization, which is selectively allocated while coordination happens at the managerial level. The apex roles are to ensure liaison and negotiation with the external environment. Policy tends to be "what gets done is the policy" (Epton *et al.* 1983: 29). Innovation is the key in these types of organizations and a lack of standardization combined with decentralization enables an adhocracy to achieve this role.

The cybercartography project is based in an academic institution, Carleton University, and the project is based in the Geomatics and Cartographic Research Centre in an organized research unit (ORU) in the department of geography and environmental studies. There are 10 faculty research collaborators, 23 students, and 4.5 postdoctoral fellows associated to the CANE project. It is large enough to be innovative and small enough to maintain a loose structure. The project also includes advisory board members, research affiliates, and partners. The project is large, but it exhibits many of the characteristics of smaller university-based organization, where lead researchers also have teaching commitments beyond the project. There was no reluctance to cross disciplines, however, publishing in a researcher's own discipline has taken priority. Finally there has been an ongoing balancing act between setting rules and policies to streamline and standardize activities and enabling enough flexibility to nurture creativity and individuality. The project has also exhibited the common tensions between ORUs and Universities, particularly in relation to technology acquisition, the delivering of promised infrastructure in a timely fashion, and time delays on space allocation decisions. The status of the Principal Investigator (PI) in the institution, diplomacy and good technological skills sets within the project has enabled good communication between different branches of the university administration and solutions have been found. Persistence, patience, and pressure were however necessary. The project best fits the definition of an adhocracy as outlined above. The PI manages liaison roles, and the project manager allocates resources and human resources. Three PhD candidates and mature students who have collectively had NGO, private sector, and government experience lead atlas development and research coordination and integration, while policy is formulated as required. The organization is flat and students at all levels are actively engaged in directing the project along with collaborators and partners. It is flexibility with a roadmap, in this case a well-defined milestones document, production work flow, anchoring theory into applications along with an agile production process as introduced by the HOTLAB. This is however not for the faint at heart, since there is much ambiguity; roles are not explicit; tasks emerge; often tension arises on role overlap and divergent perspectives require mediation. This has however enabled participants to find their place in the process and contribute on equal footing resulting in creativity, innovation, and motivation.

6. Managing Collaboration: Practicalities from the CANE Project

As discussed, the Cybercartography project involves a number of individual researchers of all ranks, numerous disciplines, partners from multiple sectors, and technology specialists. Its organizational form is between that of a large ORU and a small research unit which very much resembles an Adhocracy. Research – depending on the context and timing – is multisectoral, multidisciplinary, cross-disciplinary, interdisciplinary, and is showing signs of transdisciplinarity. Further, the research approaches are instrumental, and conceptual with a knowledge domain approach and the intentions of a research impact approach. A collaborative culture was nourished and is flowering under the new cluster-based flexible matrix structure. Research, both applied and theoretical is organizationally structured toward transdisciplinarity.

On the practical side, university-based research projects need to recognize that innovators are not necessarily good accountants and it is critical that there be a good financial manager who is committed to interdisciplinary work. There are also some challenges that should be factored into the process. Groupthink can also be a problem if there is too much intimacy since collaborators can operate under their own self-deception and the "willingness to listen to the comments of outsiders and themselves is a prerequisite if this sort of folly is to be avoided" (Epton *et al.* 1983: 31). Managing interdependency is important as often there is a lack of clear defining lines between roles, or bureaucratic divisions, and line personnel often occupy the same status as scientists. Epton *et al.* (1983: 33) provide the following list of issues to be attuned to (Table 8.2.).

To overcome these common problems they recommend, at the outset of the research process, the creation of a shared appreciation or ideology toward interdependence. The key is to have an environment conducive to a shared ideology toward a belief structure of problem solving. If there is going to be equality between scientists, administrators, and students then there is a need for all to see themselves as part of the problem-solving community rather than hired hands, where there is more "pragmatic problem solving" than being "scientific," which may be necessary in cross-disciplinary work (Epton *et al.* 1983: 34). Further, an ideology that fosters commitment and sustains motivation is the key. The workloads are often

TABLE 8.2
Barriers to Managing Interdependency

1. Asymmetries in status and access to information exist
2. Different people and departments use different performance criteria and reward structures
3. Some people are focusing on short-term objectives and others on long term
4. Job dissatisfaction and task ambiguities stimulate unilateral action to change roles and to reduce ambiguities
5. Some shared resources are scarce
6. Different training and knowledge bases create semantic barriers and perceptual diversity

Adapted from Epton *et al.* (1983: 33), *Managing Interdisciplinary Research.*

uneven and members need to be selected to adapt to this type of environment and focus on work when and where they are needed most.

Motivation and commitment are issues as is research work that seemingly has few rewards. There are a number of other methods that are helpful for the management of research, such as the bridge scientist, cognitive tuning, and a collaboratory. Bridge scientists are individuals who may be considered as a "lesser mortal" since they acquire a familiarity with a number of disciplines or technologies, their basic paradigms, methodologies, and ways of working, and they must be sensitive to the motivations, aspirations and value orientations of individuals in several disciplines. This individual can help to facilitate communication among the teams, between specialists, can diagnose and point out differences or cross-purposes (Epton *et al.* 1983: 19). As mentioned earlier, people who work on the margins of disciplines are necessary in cross-disciplinary work. A collaboratory is a center without walls, in which researchers perform their research without "regard to geographic location – interacting with colleagues, accessing instrumentation, sharing data and computational resources, and accessing information in digital libraries" (NAS 2003).

Levin (2004) suggests that the context of research production should pay greater attention to impact, and he suggests the following methods: to include criteria to measure research impacts in proposals, provide resources to assist research impact, support knowledge mobilization in universities, strengthen the input of users, steps to extract maximum value from existing research, and build networks of researchers and include stakeholders. In the context of use he suggests: that more should be learnt about user organizations to understand their strengths and limitations; getting involved and engaged with user organizations; developing better means to exchange information; and placing researchers and students in user organizations. Linking production and use are also recommended, and he suggests employing people who are skilled at translating research results into plain language for non-specialist audiences; the use of media relation experts to build media connections, leverage existing venues of exchange, increase the connection between research and professional development organizations, learn more about Web interactions, reconsideration of electronic vehicles, expand literature reviews and draw on what works. However, he recognizes that research and development is a complex process and may require a series of complex tasks that may mix the forms and the modes of interaction which may vary along the research process of project initiation, project execution, and the results of implementations.

The integration of domains/disciplines into teams requires dedicated integrators who can identify the cross fertilization of ideas, and initiate exchanges (Pester 2003). The barriers to collaboration are often people's cognitive frameworks, differences in subject matter characteristics, normative, need for services from a disciplinary group, and an individual's identification with the disciplinary community (Gold and Gold 1983: 92). These are internalized and subconscious. Within one's community this is not an issue, however, in cross-disciplinary environments, it is important to bring the unconscious to the foreground in order to move beyond one's own discipline (Gold and Gold 1983: 95). As illustrated in Leavitts model of organizations, structures consist of complex interdependent elements that are not always obvious (Fig. 8.3.). Disciplines, like organizations, have their value structures and questions considered worthy of intellectual energy. Each discipline has its professional associations, but when outside that community, accepted norms no longer apply. Cross-disciplinary groups require a mechanism to exchange information, exchange recognition, provide tangible rewards, and find critical evaluation to share community services. Identification with the community occurs with training and socialization that connect a member of a discipline to an identified community. Loyalty to a particular community can cause professional chauvinism, providing severe barriers (Gold and Gold 1983).

The CANE project was aware of many of these issues at the outset, and when any one of these arose they were considered as part of the natural process in this type of collaborative endeavor. This flexible attitude and outlook, combined with an understanding of the collaborative process enables barriers and difficulties to easily be overcome.

7. Measuring Collaborative Transdisciplinary Research

To support and justify the need for transdisciplinary research, it becomes necessary to objectively evaluate its outcomes. Particularly since success may not be the same as what is expected in traditional academic settings. A recent report on interdisciplinarity by the National Academy of Sciences in the US confirmed a CANE project assumption: transdisciplinary and interdisciplinary research is hard to measure. The report indicated that it cannot be measured directly on a short term and on a regular basis because of its complexity and inherent unpredictability (NAS 2004: 149). It can be measured potentially on its quality in terms of research advancement, relevance in terms of its application, or in the development of new leadership (NAS 2004). These are however not always the measures used by funding institutes who still rely on traditional metrics of single authored peer reviewed papers. Also, within departments hiring and tenure track decisions are still gauged by the home discipline's standards. To measure the degree that a project can weave perspectives into a coherent whole, of influence of one field within another via solving of a common problem, of theoretical input into new technologies, is not an easy task (NAS 2004). A transdisciplinary project could be measured on how theory was applied, how the project advances understanding, inquiry, and problem solving. In the end "the contribution achieved by a research team may be more than the sum of the individual accomplishments" (NAS 2004: 153), which is the antithesis of current academic merit systems in some disciplines.

The collective ideals of transdisciplinary research are challenged on a daily basis and researchers who are engaged in the process have to weigh the cost of the collective good and their belief in the work against the measures in their departments, disciplines, and individual aspirations. The end result may be research with more depth and importance of achievement but less in the number of countable deliverables according to the academic gold standard. The CANE project will be producing atlases and proof of concept products, new and profound research relationships and potentially the creation of a hybrid discipline. These currently do not fall into academic metrics, take time, are notoriously difficult to measure and may not occur until the project has been completed for a number of years. The following are a few examples that demonstrate methods to evaluate or assess aspects of collaborative research. A full review of project or process evaluation is beyond the scope of this chapter, however, the following suggestions provide useful insights into the ways to assess transdisciplinarity.

Azad and Wiggins (1995) in their work on data sharing suggest that the degree of collaboration is a factor of relationship intensity. They empirically measure the intensity of interorganizational relationships defined as the magnitude of resource investment required by a relationship, where resources could include finances, personnel, clients, physical facilities, and authority. It also refers to the degree of commitment required by an organization where autonomy is considered a preference by most organizations and the "ability to make decisions independent of outside pressures varies with the type or contact it has with other units" (Rogers in Azad and Wiggins 1995: 32). Azad and Wiggins provide a five-step scale of relationship intensity as seen in Table 8.3. The NAS report also indicated that interdependency within a project as a whole and contributing to individual researchers, demonstrates that a culture of cross-disciplinarity is developing (NAS 2004: 155). This intensity scale could be used to assess the partnership arrangements in the multisector approach to research on the CANE project and could also be adapted to assess the intensity of the relationships between the discipline-based teams or clusters.

Collaboration varies, and in some instances it is the entire function of the organizations involved and in other cases it is a portion of an organization's work while the rest of its activities are separate from the shared function. Fishman *el al.* suggest that there are 10 collaborative functions as seen in Table 8.4. (adapted from Fishman *et al.* 2000)

The CANE project exhibits many of these functions, such as the joint consultation of stakeholders, coauthoring of papers, combining operations, co-training of students across disciplines, sharing human resources, and technologies, joint efforts involved in the Canada Foundation for Innovation grant to acquire core infrastructure technologies of benefit to all, and strategic planning between labs. The clusters exhibit less formal aspects outlined by Fishman *et al.*, as they do not have budgetary decision-making power. They however exhibit community leadership, information dissemination, and program direction.

Birnbaum (1983) studied the research performance of 59 cross-disciplinary research projects in Canada and the US. His methods involved questionnaires, interviews, and historical records. He suggests that there are two important measures: those of scientific products and those of scientific activity. Measures of scientific products require an assessment of the quality and quantities of publications, such as articles, books, technical reports, published papers, and patents. Measures of scientific activity involve measuring the process of interaction that take into account the characteristics of the scientists (e.g. eminence, recognition related to their publications, and quality and citations), their division of labor in a unit, the degree of integration, member turnover, conflict resolution methods, and relationships between leaders and members. Table 8.5 simplifies Birnbaum's (1983) measures.

Birnbaum's research concluded that scientific activity is product specific, that a clearly defined division of labor facilitates article quality and quantity while a loose division of labor produces more reports. Turnover tends to decrease article quality and increases patent applications while good leader member relations facilitate book production and inhibit patent activity and previous recognition facilitates patent activity. Member

TABLE 8.3
Five-step Relationship Intensity Scale

Least intensive	1	2	3	4	5	Most intensive
Most autonomous	Personal settings collaboration	Resource transfers	Board membership cooperation	Joint programs	Written contracts coordination	Least autonomous

Adapted from Azad, B. and L. L. Wiggins (1995) *Dynamics of Inter-Organizational Geographic Data Sharing: A Conceptual Framework for Research.*

TABLE 8.4
Ten Collaborative Functions

Strategic Planning	Research teams have their own set of goals, priorities, and strategic plans. Some work together; plans are shared there is a move and to increased levels of collaboration. Some choose to coordinate planning processes to develop mutual goals and initiatives, or develop a plan that incorporates their shared interests.
Resource Acquisition	Collaboration results in joint processes to acquire resources.
Resource Allocation	The allocation of resources or joint acquisitions to address common goals. Portions or entire budgets may be allocated to collaboration or decisions, or budgets are combined.
Policy and Program Direction	Collaborations may require an alignment of conflicting policies to better facilitate activities. This can entail comparing policy and making adjustments, or developing a shared policy framework that may affect the policies of the operations of all the organizations.
Staff Training	Organizations train their own staff while collaborations may require shared training initiatives. When strategic planning or policy programs have advanced to developing common goals and methods, organizations may opt to combine their efforts for mutual best interests.
Program Operations	Combining program operations may require a definition of which programs to combine, separate programs, or a new joint program, a new initiative, or to extend previous collaborations. The degree of collaboration may manifest in sharing a new location, interagency teams to align practices, and decision making, or the merging of programs.
Information Dissemination	As organizations collaborate, joint communication efforts may make sense while joint ventures are more conducive to joint reporting and communication.
Community Representation	Collaborations may lead to joint efforts to consult with stakeholders.
Community Leadership	Involves reaching out to build support for the collaboration's efforts and to raise awareness of the new initiative.
Program Evaluation	Collaborations may lead to either individual evaluation of the shared work, or a joint evaluation process by developing shared guidelines and conducting joint evaluations.

Adapted from Fishman *et al.* (2000) *Evaluating Community Collaborations: A Research Synthesis*.

perception or intermediate performance is a predictor of writing productivity. If a manager is interested in the production of articles and books it is advisable to have a clearly defined division of labor, but not a tight integrated effort, low turnover and to maintain good leader member relations. If reports are the priority then less well-defined divisions of labor,

TABLE 8.5
Measuring Team Collaboration

Horizontal Integration	Measured by counting the number of different functions performed within each group
Vertical Differentiation	Measured by the number of people involved in setting project policy
Integration	Assesses relationships, in this case the use of a language ladder (i.e. multiple choice questionnaire) of eight categories ranging from sound to no relationships at all
Turnover	Measures the ratio of the average number of personnel joining and leaving the project to the total group size
Leader-Member Relations	Measures the responses of leaders regarding the collaborative atmosphere using a series of bipolar scales such as pleasant–unpleasant
Open Discussion of Disagreements	Measures team member's disagreements to a six-point scaled statement of whether or not disagreements were discussed
Prior Scientific Recognition	Is measured as a factor score of the standardized average for each project of the number of books published, number of articles, national committee membership, and membership to a professional organization
In Process Performance Index	Is a factor scale derived from three indicators related to self-evaluated effectiveness, goal achievement, and the group process measure
Research Article Quality	Measured through the use of average project influence weight per year assigned to the journals in which articles appeared

Adapted from Birnbaum (1983) *Predictors of Long-Term Research Performance.*

encourage a high level of decision making, and encourage open discussions of disagreement. If patents are desired, then high scientific recognition is the key, and less discussion on agreements and a reduction in turnover are required (Birnbaum 1983). Some of Birnbaum's measurement ideas could be used to evaluate aspects of the CANE project.

Wasniowski (1983) researched the work of a Futures Research Group that included 40 scientists from both the natural and social sciences. He discovered a technique to assess the degree of influence that a particular field has on another in order to assess synergies and mutual interaction, and to develop interpretive structural models. Members were asked to indicate which disciplines influenced theirs and they assessed the degree of influence on a five-point scale ranging from no influence to a very strong influence. The results yielded a structure that enabled collaborations that influenced outcomes on a variety of interesting projects. The observations over time also demonstrated that different structures exist for different functions. A hierarchical structure emerged in some areas for some problems, such as policy, planning, technological, and scientific. This was useful for the implementation of tasks. However, a research structure also emerged which factored the research contributions of particular fields for particular problem which took considerable consideration. The transdisciplinary approach was directed at developing alternate goals for society and technology. This type of question could be used to inform social network mapping for the Cybercartography project.

Uta Wehn de Montalvo (2002: 186) explored the subject of geospatial data sharing from the perspective of intermediation, namely the "facilitation of coordination and interaction between the array of actors with overlapping roles that are involved in collecting and using spatial data." Wehn de Montalvo used a qualitative approach to survey actors involved in the development of the South African geospatial data infrastructure to assess the behavioural determinants enabling data sharing among and within organizations. He explored the incentives and disincentives to share data to provide "a basis for suggestions about how relationships between the actors contributing to the use of GIS might mean more effectively organized" (Wehn de Montalvo 2002: 187). He suggests that beyond organizational structures, the motivation to share is associated "with attitude and belief systems, which in turn, are influenced by a large number of cultural, social, political, and economic conditions" (Wehn de Montalvo 2002: 187). A qualitative approach to study collaboration on the CANE project is currently underway with members of the Eric Sprott School of Business at Carleton University and the School of Business at Ryerson University. It is expected that a study of the complex social structures will reveal clues to how transdisciplinary research actually works.

Wehn de Montalvo developed a model to assess the willingness to share geospatial data. His model includes three components to understand the intentions of participants as seen in Table 8.6.

TABLE 8.6
Wehn de Montalvo Model to Assess the Willingness to Share Geospatial Data

1. Attitude	The consequences and possible outcomes that organizations perceive as resulting from spatial data sharing;
2. Social Pressure	Insights about influential individuals or pressures within organizations and expectations;
3. Perceived Behavioural Control	Includes the skills, capabilities, and other factors that individuals perceive may be necessary to engage in sharing, or the extent to which they think they are in control or responsible for the activity.

Adapted from Wehn de Montalvo (2002).

The results indicated that beyond a supportive policy environment, standards and the implementation of technology, it is critical to understand the culture of data sharing and that intermediaries need to factor culture as part of a collaborative function. In the end, social and psychological determinants of data sharing needed to be taken into account to reduce the resistance to share and to increase a culture of data sharing munificence. Elements of this model could be adapted and used to assess and understand motivation as a critical factor of collaboration within and among clusters.

The work of Wehn de Montalvo and Wasniowski is helpful to understand the nuances of networks of practice – where "information does not always appear in immediate or traditional forms, such as publications in academic journals. Such networks may yield other important outputs, such as congressional testimony, public-policy initiatives, mass-media placements, alternative-journal publications, and long term product development." (NAS 2004: 156). The NAS report did not provide metrics or methods to measure interdisciplinarity but did suggest that positive responses to the following questions related to collaboration might be considered as indicators that it is occurring:

(1) Are students working and learning from students in other disciplines?
(2) Are researchers mastering more than one discipline?
(3) Do researchers get a sense of what it means to integrate more than one discipline in resolving problems?
(4) Are researchers learning instrumentation or techniques from other disciplines?
(5) Are researchers adding value to the project, and can they grasp aspect from another field?
(6) Can researchers interact with specialists in other fields?
(7) Is there a demonstrated learning of the language and content of another field culture, etc.?
(8) Is the interdisciplinarity of the project moving to conferences within fields?
(9) Are topics that could not otherwise be covered being studied?
(10) Is expertise extended in another direction?
(11) Is a new subfield being created?
(12) Is there an inclusion of other disciplines in one's own work?
(13) Is there a merging of disciplines?
(14) Are researchers giving lectures or presentation in other disciplines?
(15) Are the users of the research output involved in the process?

The methods discussed in this section provide ways to assess collaboration and how it could be structured. These did not provide a definitive method to assess the occurrence and success or failure of transdisciplinarity. Indicators at best are proxies. Transdisciplinary research is complex and while some of these measures may be of use, the underlying social processes are not addressed. As Klein, a leading humanities expert on the topic of interdisciplinarity indicates: "We need to do a better job of measuring results. . . There is not a lack of data. There is a profusion of invisible data that needs to be better collected and disseminated." (NAS 2004: 163). Transdisciplinarity is a knowledge-based approach to knowledge creation, and collaboration is a very deep personal and organizational process that is a part of the social structures of an organization. Developing a collaborative culture can increase knowledge creation activities (Sveiby and Simon 2003). It can be argued that the CANE project has created a collaborative culture, structured the work to facilitate transdisciplinarity, and in fact can provide positive responses to most of the NAS questions above. The CANE project cannot yet, however, explain the complex social process of how transdisciplinarity works.

8. Final Remarks and Next Steps

This chapter illustrated that research, as conducted by the Cybercartography and the New Economy (CANE) project, designed at the outset with transdisciplinarity in mind requires intention, structure, organization, and culture that fosters collaboration. The Cybercartography Project, as with other large collaborative projects, is expected to direct research, manage human resources and budgets, acquire equipment and services, disseminate results, develop groundbreaking products, produce theory, and are also asked to ensure that their work has impact. These and other project activities, to be successful, need to be carried out in an environment that is creative, conducive to knowledge creation, innovative and long lasting. Managing research projects, delivering on both the tangibles and intangibles, while structuring and organizing for transdisciplinarity are seemingly the most overlooked processes in the research literature and in practice.

The CANE project is ongoing, and early observations and a midterm review suggest that it will meet its commitments, and that it is doing so in a transdisciplinary fashion, where the integration of disciplines, clusters, and individual knowledge is conducted in an interesting and innovative fashion. Further, the students are actively engaged in leading the process.

The important lessons to date are: to allow time at the beginning for teams to get to know each other and their work – even though it is very enticing to gather data and begin programming. Transdisciplinary research requires more time, effort, and resources in the developmental phases of the research process than does conventional research. This suggests that both researchers and funders may need to allocate and negotiate for time and resources toward this activity in proposals and in evaluation criteria. Also, it is critical to accept that collaborative research is a difficult and complex process with numerous barriers, and the necessity to facilitate integration of the disciplines into task-based activities is essential as seen in the clusters. Further it is important to listen, consult, be patient, and flexible as the process is dynamic and unexpected directions and unplanned ways of working emerge. Having the creative space to do this is crucial while accepting that different management strategies are required at different phases for various activities, tasks, contexts, and most importantly to be styled to meet the interests and methods of the individuals involved. Finally, it is critical to dedicate time in workplans to negotiate with partners, advisors, collaborators, contributors and stakeholders, and also to dedicate human resources to the activity of setting the stage for these activities to occur. In short, heterogeneous teams are considered more innovative and creative, but there is a cost to manage these types of groups (Epton *et al.* 1983; Argote and Ophir 2002). The CANE project has at the outset dedicated resources to facilitate this process; researchers are committed to this process and the PI has much insight and experience to let the process emerge by overseeing all the parts on the one hand and paying attention to individual needs and particularities on the other. Nonetheless, it remains a learning by doing process with many challenges. This learning by doing is research in action that is not necessarily what is rewarded in academia since, as previously discussed, traditional measures of success remain refereed journal papers. It will take time to change traditional institutional and disciplinary reward structures.

The CANE project is doing transdisciplinary research but how and why it works is unknown. The NAS report also stated that "social-science research has not yet fully elucidated the complex social and intellectual processes that make for successful IDR (interdisciplinary research). A deeper understanding of these processes will further enhance the prospects for creation and management of successful IDR projects" (NAS 2004: 3). This chapter does not fully address this issue, however, the CANE project hopes to solicit support to design a research plan to better understand the complex social and intellectual processes involved with transdisciplinarity as an ongoing process, and to collect and analyze the rich and invisible process data that is being generated has enlisted on a daily basis.

Acknowledgments

The author would like to acknowledge Amos Hayes for designing the GCRC online Forum and his ongoing work related to the implementation of the project's information and communication infrastructure that is a key component of collaboration on the CANE Project. Also, Barbara George, the CANE project manager who ensures we have the resources required to do our work. To all those involved with the project, since without them there would be no need for a chapter on collaboration and transdisciplinary research. Finally, to Michael Peterson for his ongoing guidance.

References

Agyris, C. and D. A. Schön (1978) *Organizational Learning: A Theory of Action Perspective*, Addison Wesley Publishing Company, Reading.

Argote, L. and R. Ophir (2002) "Intraorganizational learning", in Baum, J. A. C. (ed.), *The Blackwell Companion to Organizations*, Blackwell Publishers, Malden, pp. 181–207.

Azad, B. and L. L. Wiggins (1995) "Dynamics of inter-organizational geographic data sharing: a conceptual framework for research", in Onsurd, H. J. and G. Rushton (eds.), *Sharing Geographic Information*, Vol. 2, Centre for Urban Policy Research, New Brunswick, pp. 22–43.

Baum, J. A. C. (ed.) (2002) *The Blackwell Companion to Organizations*, Blackwell Publishers, Malden, pp. 1–957.

Birnbaum, P. H. (1983) "Predictors of long-term research performance", in Epton, S. R., R. L. Payne and A. W. Pearson (eds.), *Managing Interdisciplinary Research*, Vol. 7, John Wiley & Sons, Toronto, pp. 47–59.

Coleman, D. and J. McLaughlin (1997) "Defining global geospatial data infrastructure (CGDI): components, stakeholders and interfaces", *Geomatica,* Vol. 52, No. 2, pp. 129–143.

Dormann, C. and G. Lindgaard (2004) "Developing innovative systems: creative ideation", *Interaction, Systems, Practice and Theory*, Sydney 16th–19th November 2004, pp. 451–471.

Eddy, B. (2002) "Image – cybercartographic integrated research framework", *Cybercartography and the New Economy Project Proposal.*

Epton, S. R., R. L. Payne and A. W. Pearson (eds.) (1983) *Managing Interdisciplinary Research*, John Wiley and Sons, Toronto.

Fishman, M., M. Farrell, A. Vincenca and E. Eisenman (2000) *Evaluating Community Collaborations: a Research Synthesis*, accessed January 2005 from http://www.lewin.com/NR/rdonlyres/eeqribl2odozkxlssh3ndiopok4nv4bpalk4yxcpm4c7r3622j4kru 7jdezoztbzjrx2qmuh6er247264tykdo6vefe/564.pdf

Geomatics and Cartographic Research Centre (GCRC), *Cybercartography and the New Economy Internet Site* at http://gcrc.carleton.ca

Gold, S. E. and H. J. Gold (1983) "Some elements of a model to improve productivity of interdisciplinary groups", in Epton, S. R., R. L. Payne and A. W. Pearson (eds.), *Managing Interdisciplinary Research*, John Wiley & Sons, Toronto, pp. 86–101.

Holland, P. (1999) *The Strategic Imperative of a Global Spatial Data Infrastructure*, accessed 5 January, 2005 from the GSDI Internet Site at http://www.gsdidocs.org/docs1999/ stratim.html

Lauriault, T. P. (2003) *A Geospatial Data Infrastructure is an Infrastructure for Sustainable Development in East Timor, Master Thesis*, Carleton University, Ottawa.

Laurini, R. (2001) *Information Systems for Urban Planning, A Hypermedia Co-Operative Approach*, Taylor & Francis, New York.

Levesque, P. and J. M. Chopyak (2001) "Managing multi-sector research projects, developing models for effective movement from problem identification to problem solving", *Public Management*, Fifth International Research Symposium on Public Management, Barcelona, Spain, pp. 2–48.

Levin, B. (2004) "Making research matter more in education policy," *Analysis Archives*, Vol.12, No. 56, accessed 2 December, 2004 from http://epaa.asu.edu/epaa/v12n56/

Masser, I. and M. Craglia (1997) "The diffusion of GIS in local government in Europe", in Couclelis, H. and M. Craglia (eds.), *Geographic Information Research Bridging the Atlantic*, Taylor & Francis, London, pp. 95–110.

Merriam-Webster Online Dictionary (2004), accessed 7 February, 2005 from http://www.m-w.com/

National Academy of Sciences (NAS) (2003) *Committee on Facilitating Interdisciplinary Research About Us Internet Page*. Committee on Facilitating Interdisciplinary, accessed 1 March, 2003.

National Academy of Sciences (NAS) (2004) *Facilitating Interdisciplinary Research. Report by the Committee on Facilitating Interdisciplinary Research, National Academy of Sciences, National Academy of Engineering, Institute of Medicine*. National Academies Press, Washington DC.

Nowotny, H. (2003) *The Potential of Transdisciplinarity. Rethinking Interdisciplinarity*, accessed 8 December, 2003 from the Rethinking Interdisciplinarity Internet Site.

Onsrud, H. J. and G. Rushton (1995) *Sharing Geographic Information*, Centre for Urban Project Research, New Brunswick.

Patriotta, G. (2003) *Organizational Knowledge in the Making. How Firms Create, Use, and Institutionalize Knowledge*, Oxford University Press, Oxford.

Pester, D. (2003) *The Evolution of Knowledge Domains, Interdisciplinarity and Core Knowledge*, accessed 8 December, 2003 from the Rethinking Interdisciplinarity Internet Site, http://www.interdisciplines.org/ interdisciplinarity

Pinto, J. and H. J. Onsrud (1995) "Sharing geographic information across organizational boundaries: a research framework", in *Sharing Geographic Information*, ed. Centre for Urban Policy Research, New Brunswick, pp. 44–65.

Salter, L. and A. Hearn (eds.) (1996) *Outside the Lines, Issues in Interdisciplinary Research,* McGill-Queen's University Press, Montreal, pp. 3–212.

Scott, R. W. (1998) *Organizations, Rational, Natural and Open Systems*, Prentice Hall, Saddle River.

Senge, P. (1990) *The Fifth Discipline: the Art and Practice of the Learning Organization*, Doubleday Currency, New York.

Sveiby, K. E. and R. Simons (2003) *Measuring Collaboration: an Empirical Study into Good and Bad Practice*, accessed 11 November, 2003 from http://www.sveiby.com/articles/ collaborations.html

Taylor, D. R. F. (1999) "Future directions of multimedia cartography", in Cartwright, W., M. Peterson and G. Gartner (eds.), *Multimedia Cartography*, 1, Chapter 29, Springer, Berlin, pp. 315–326.

Taylor, D. R. F. (2002) *Cybercartography and the New Economy Project Internet Site*, accessed 7 February, 2005 from http://www.carleton.ca/geography/geography/Taylor_research.html#Cyber

Taylor, D. R. F. (2003) "The concept of cybercartography", in Peterson, M. (ed.), *Maps and the Internet,* Elsevier Science Ltd.

Wasniowski, R. (1983) "Futures research as a framework for transdisciplinary research", in Epton S. R., R. L. Payne and A. W. Pearson (ed.), *Managing Interdisciplinary Research*, Vol. 33, John Wiley and Sons, Toronto, pp. 228–235.

Wehn de Montalvo, U. (2002) "The distribution of spatial data: data sharing and mediated cooperation", in Mansel, R. (ed.), *Inside the Communication Revolution. Evolving Patterns of Social and Technical Interaction*, Vol. 9, Oxford University Press, Oxford, pp. 186–205.

Withane, S. (1987) *Organization Design: theory and Applications*, Copley Littleton, Publishing Group.

CHAPTER 9

Interface Design Challenges in Virtual Space*

GITTE LINDGAARD
Professor, Natural Sciences and Engineering Research Council (NSERC) Industry Chair User-Centred Design, and Director, Human Oriented Technology Lab (HOTLab), Department of Psychology, Carleton University, Ottawa, Canada

ALLISON BROWN
Professor and Director, Teaching and Learning School of Electrical and Computer Engineering, Royal Melbourne Institute of Technology, University, Melbourne, Australia

ADAM BRONSTHER
HOTLab, Carleton University, Ottawa, Ontario, Canada

Abstract

There's no doubt about it: playing with multimedia is fun! But it takes more than an understanding of technology to design useful and usable multimedia interfaces. In this chapter we discuss several challenges that multimedia user interface designers face. In an effort to demonstrate why it is so tricky, we discuss a small selection of current cognitive models describing how sensory information may be perceived, interpreted, stored, and retrieved when we need it. Learning processes are then addressed followed by a brief discussion of mental models and the implications of this literature on the way static and animated graphics may be processed. In the context of the amount of information one should present on a page or screen we highlight ways that students'

*This Chapter is dedicated to the memory of our colleague and friend Dick Dillon who helped develop an early draft.

ability and prior knowledge of the topic portrayed in a multimedia application are likely to affect the way they deal with new information. Some frequently observed pitfalls in the way multimedia applications are tested and evaluated are discussed next. This leads into a discussion of the desirability for multimedia applications to invite immersion rather than merely being accessible or interactive. We end with a brief discussion on learning theory on user interface design.

1. Introduction

Multimedia is being used very effectively in film and computer games where the primary objective is to entertain and evoke certain positive or negative emotions in the user. On occasion, it is also employed successfully in educational applications where many issues must be considered to ensure that such applications achieve their educational goals. In this chapter we focus on some of the challenges facing designers of multimedia learning applications in which the user is expected to glean readily interpretable information or to learn something such as a particular principle, or a set of relationships or facts. We focus on multimedia involving text and still, as well as animated, graphics; it is impossible to do justice to the entire literature that includes sound, haptic, olfactory, and gustatory multimedia in a single chapter. However, additional contributions related to these are made in several other chapters in this book.

Multimedia is an important element in the creation of new cybercartographic products (Taylor 2003) and an increased understanding of the interface design challenges in virtual space is important if cybercartography is to advance. We present current cognitive models first to demonstrate how multimedia may facilitate learning. We then address learning processes and their implications for multimedia design. This leads into a discussion of mental models suggesting how textual, graphic, and animated information can affect the establishment of coherent mental models. Next, we discuss some of the pitfalls involved in testing multimedia applications, and finally the influence of learning theory on design, which leads to our conclusion.

2. Current Cognitive Models

The main objective of multimedia in education is to encourage and promote learning as discussed by Baulch *et al.* in this volume. To understand how and when learners may benefit from presentations of information in more than one modality that challenges without overwhelming them and encourages deeper processing, it is helpful to discuss some of the most relevant and current cognitive models. These models propose how we process, store, integrate, expand, and retrieve information.

We are constantly surrounded by far more sensory stimuli than we can possibly process. To make sense of the world around us, we must therefore select those we want to attend to at any moment in time. Although theories of cognition and information processing differ substantially at the detailed levels (see e.g. Miyake and Shah 1999 for a comprehensive review), most agree at a high level that incoming sensory information is integrated with knowledge that already exists in the learner's head – or other body parts – before a decision is made and an action taken regarding the input.

Incoming stimuli are processed in a "bottom-up" fashion, while information retrieved from long-term memory and employed to make sense of incoming stimuli is referred to as a "top-down" process. The integration of incoming and existing information is believed to be handled by a "central executive" (Baddeley 1986), usually referred as working memory. Working memory is generally described as a functional multiple-component of cognition that "allows humans to comprehend and mentally represent their immediate environment, to retain information about their immediate past experience, to support the acquisition of new knowledge, to solve problems, and to formulate, relate and act on current goals" (Baddeley and Logie 1999: 39). One important role of working memory is to allocate and coordinate cognitive resources, and attention to decide what information to hold in its temporary, limited-capacity store. As the central executive, working memory selectively controls the type and amount of information retrieved from its "slave-systems" (Baddeley 1986). Baddeley refers to two such slave-systems, the visuospatial sketchpad and the phonological loop, both of which comprise active and passive components. The passive components are considered as mere data stores, while the active components comprise the articulatory rehearsal process and the visuospatial process respectively. The visuospatial "inner scribe" maintains rehearsal of body movements and retains information of paths between objects and locations coded through vision, sound, or touch. Thus, this "inner scribe" deals with information from several sources including the auditory sensory system.

In Barnard's "Interacting Cognitive Subsystems" (ICS) model (1985), the "central executive" comprises four subsystems, which together, carry out similar activities to those described by Baddeley via a continued process of propositions and negotiations that resolve stimulus ambiguity and verify semantic interpretations of incoming information in the light of what is already known. Just like Baddeley describes a spatially-based "inner scribe," Barnard refers to an auditory-based component, the "inner voice" that we hear inside our heads that also processes information originating in the visual domain, for example, when we echo printed text inside the head while reading. Both theorists thus distinguish between visual and auditory processes at some level, but these seem to mix at higher levels of processing. Baddeley attributes the processing of sensory information to his "slave-systems" seen to be controlled by the "central executive," whereas Barnard incorporates this directly into the central (executive) subsystems that integrate and interpret information from different senses as well as initiating responses. Barnard further distinguishes between sensory registers, the input subsystems, and effectors, the output subsystems. Sensory registers encode visual, auditory or "body state" information; effectors control speech articulation and body/limb movements as well as information about body state, which includes taste, smell, temperature, and haptic information. Barnard's model comprises nine subsystems; Baddeley refers mainly to three, and Schneider (1999) proposes a set of eight such systems that include mood and what he calls "context." Schneider's explicit reference to "mood" is absent from both Barnard's and Baddeley's models, although Barnard's "body state" subsystem allows emotional and affective influences such as mood and motivation to attend to and absorb information. This absence of references to affective motivators is noted also in Paivio's (1990) "dual-code" theory, which concentrates exclusively on the processing of visual and verbal information. Paivio regards pictures and text as visual information but concedes that text is also translated into verbal information at some point in the process. This is problematic for his distinction between the two "channels," and so far, has not been resolved (Mayer *et al.* 2003).

Thus, theoretically it is not clear exactly how sensory information is processed, or where, in what form, it may reside at higher levels beyond the sensory receptor organs – eyes, ears, nose, mouth, and skin. In all fairness, none of these models purport to account for the entire repertoire of human information processing that includes affect-based characteristics. Nonetheless, it is important to note that mood and motivation underlie the decision to select aspects of the surrounding world, which, in turn, determine what may be learned when interacting with a multimedia application.

While some theorists propose that all components in their models share the same architecture and are regulated by similar processing properties, such as capacity and decay rate, others assume different domain-specific codes in working memory even if they do not postulate separate subsystems. However, despite major differences at the detailed levels of cognitive models, there is widespread agreement that working memory incorporates different codes and representations; various subsystems are, at least to some extent, independent, and they act in a cooperative rather than in a competitive fashion (Stanney *et al.* 2004).

At the sensory input level the nature of perceptible stimuli clearly differs: we pick up light waves with our eyes and sound waves with our ears. However, cognitive theories disagree with respect to the way information originating as sensory codes may be stored. Paivio (1990), for example, holds that information is stored in two codes even at higher levels: verbal and visual. Yet Pylyshyn (1981) and Johnson-Laird (1998) argue that different codes converge at a higher level of processing to a single amodal form of knowledge representation built from basic linguistic propositions. In their view, mental images are constructed from such propositional knowledge, not from analog representations as Paivio argues.

Barnard's framework allows for both analog and amodal knowledge representations. In every one of his nine cognitive subsystems, input information is copied and stored locally (analog) as well as being transformed into a code that is readable by the next subsystem in a potentially amodal form or expressed as, say, movement of a limb or an utterance at the effector level. Barnard's model is thus completely distributed and able to handle input from different sensory sources simultaneously and in parallel. This concurs with the assumption that multimedia presentations could be superior to presentations in a single medium.

If information at the highest level were stored only in an amodal form, then one would expect less localized functions in the human brain than is found to be the case. For example, damage to the visual (occipital) cortex has been found to disrupt both visual perception and visual imagery. Patients with brain damage who could not see objects present in the left side of space had similar problems when they imagined objects (Vekiri 2002). Likewise, the ability to maintain visuospatial information is affected by concurrent spatial tasks but not by concurrent verbal tasks and vice versa (Baddeley and Logie 1999). Further, the time required to generate, transform, and rotate mental images increases linearly with the amount of rotation required. This is true for the way we manipulate real images too (Kosslyn 1994), supporting the contention that visual perception and visual imagery are based on the same neurological mechanisms (Vekiri 2002). Converging evidence from cognitive psychology, neuroimagery, and neuropsychological research concur with the notion of separate multiple working memory systems (Stanney *et al.* 2004). In principle, then, the literature suggests that multimedia applications have the potential to enhance learning capacity as compared with information presented in a single medium. However, unless material presented in one medium also supports and supplements material presented in another, this advantage of multimedia would be lost because the organism would then be forced to divide attention between competing data streams. Indeed, under such circumstances learning may occasionally be inferior to situations in which material is presented in a single medium.

3. Learning Processes and some Implications for Multimedia Design

Learning may be described simply as a three-phased process: (1) the learner must select relevant material; (2) organize the selected material; and (3) integrate it with existing knowledge (Mayer 2002). Although one may expect educational multimedia products to support multiple learner goals, the material must be structured in a coherent fashion that guides learners towards their goals. Without a goal in mind, the learner will not be able to decide what is relevant, or to focus attention on the appropriate parts of the display. The responsibility for effective learning therefore, rests with both the learner and the developer of multimedia applications.

The fact that some learners will know more a priori about the subject matter than others complicates matters for the multimedia application designer. A knowledgeable learner who is under-stimulated by the apparent or actual complexity of the content is unlikely to be highly motivated to pursue demanding goals or even explore the application to the extent it might warrant. This lack of challenge is likely to result in shallow learning, or what Grace-Martin (2001) refers as "reflexive learning," which occurs in direct, quasi-automated response to external stimuli. At the same time, the unknowledgeable learner may be disheartened if he/she perceives or believes the material to be too complex and consequently fail to learn or even explore the material. The challenge of striking a balance between the level of complexity of material, invitation to explore it, and discoverability that appeals to learners along the entire knowledge-spectrum should not be underestimated.

Through empirical investigations of how users navigate applications, Lawless and Brown (1997) identified three types of learners that are relevant here. "Knowledge seekers" pursue information relevant to the context of the learning environment. They select a strategic, systematic navigation path that takes them in a logical sequence to screens containing material needed to enhance their comprehension of a particular domain. They impose what Grace-Martin (2001) refers as "intentional load" on themselves during learning. Thus, when a presentation fails to show why one step leads to another, knowledge seekers use prior knowledge to help in creating a "self-explanation" (Chi *et al.* 1989); they actively integrate existing knowledge with incoming material, which they then test to verify their understanding. These learners should be most stimulated by material providing incomplete information that invites discovery and inspires them to find out more for themselves.

"Feature explorers" (Lawless and Brown 1997) seem to spend a disproportionate amount of time interacting with the "bells and whistles" of the interface, investing more time in understanding how the application works and attending to interface features than gathering information from the content. Multilayered multimedia applications should allow the user to select one path among several that best corresponds to their learning needs, their previous and dynamically changing knowledge, and understanding as they proceed through the multimedia product (Grace-Martin 2001). It is a major challenge to provide materials that support both knowledge-seekers and feature explorers.

"Apathetic users" do not seem to care about the instructional environment or to explore the interface features. They take the most linear path through the application; they rarely deviate from a path once it has been selected, and they seem to have no goal and no notion of an educational outcome whatsoever. They appear too unmotivated to engage in more elaborate and meaningful exploration. Taking the three types of learner together, three major challenges arise for the interface designer. First, the display must address the learner's task and educational goals. Second, it must balance the application of novel, unusual and "interesting" interface features and avoid distracting the learner from the content. Third, the application must be flexible, providing multiple paths through the content and multiple layers of content complexity to satisfy learners of different ability levels while also recognizing that knowledge evolves dynamically in the interaction with the material. Thus, a learner who may be a natural "feature explorer" or even an "apathetic user" may turn into a highly motivated "knowledge seeker" if materials are presented in a manner that inspires curiosity, invites deeper interrogation, and awakens the desire to learn and know more.

Once the learner has selected relevant material, it must be organized to facilitate integration with existing knowledge. To achieve this, the material must be presented such that it is relatively easy for the learner to connect incoming material with knowledge already stored in memory and thus to construct new knowledge. We argued earlier that we are likely, at some level, to store material originating from different sensory sources in different "memories" working cooperatively, and that this arrangement provides an advantage of multimedia over a single medium. Good screen design principles prescribe physical proximity of items belonging together semantically, consistency in application of color schemes, scaling features, terminology, and so forth (e.g. Galitz 1993). Applying these principles will certainly enhance interface usability, and high levels of usability are always desirable. Indeed, the main role of the interface is to be transparent and ensure that it does not get in the learner's way: the learner should concentrate on the content, not on the interface. However, it is important to note that high levels of interface usability does not and should not imply that the learner's cognitive load inherent in the content should be minimized. Active engagement with the content is essential for learning to be effective, motivating, and enjoyable.

4. Mental Models and Information Presentation

Regardless of the particular code or codes in which existing knowledge is stored, it is unlikely that information is stored in a 1:1 direct mapping relationship with the external world (Schnotz *et al.* 1999). Therefore, some, but not all, attributes of presented information are probably represented in so-called "mental models" (e.g. Gentner and Stevens 1983). Mental models are believed to correspond with the outside world in analogous structural relations (Johnson-Laird and Byrne 1991). They can even represent subject matter that cannot be perceived by the sensory systems, for example, trade relations, such as those discussed in Chapter 22 by Eddy and Taylor.

When processing text, the reader is believed to construct a mental representation to generate propositional representation of the semantic content. Then this is applied to construct a mental model that describes the subject matter (van Dijk and Kintsch 1983). This would involve Barnard's (1985) central subsystems in bottom-up as well as top-down processes whereby task-relevant existing knowledge is activated, selected, and organized top-down, and incoming information is processed bottom-up. Once a mental model has been constructed, it can be read off and elaborated upon with new incoming information. Text thus provides an effective basis for building mental models.

5. Static and Animated Graphics

Research investigating how people perceive diagrams provides one line of evidence supporting this notion of mental model – inspection and elaboration (Schnotz *et al.* 1999). In her work Hegarty (1992), for example, used iconic diagrams and textual explanations to present components and configurations of mechanical devices. Eye-fixation measurements revealed that subjects systematically switched between text and diagram. After reading a unit of text describing particular relationships, subjects turned to the diagram to elaborate and clarify their understanding of the text sections in what Veriki (2002) calls "local" inspections. At the end of the text reading, subjects also made longer, more "global" diagram inspections in which they focused on more components. The visual display thus influenced cognitive activity by guiding, constraining, and facilitating cognitive behavior, thereby providing a "computational advantage" (Larkin and

Simon 1987). In that sense, visual displays, such as graphs, maps and pictures, can act as a kind of "external cognition" (Vekiri 2002) that allows subjects to make perceptual inferences. Thus effective visual displays can reduce the amount of information a person needs to maintain in working memory and the cognitive effort invested in interpreting it. The learner does not need to perform all the thinking processes but can think of a solution by manipulating parts of the visual image. Well-designed graphic representations are effective because they support thinking during problem solving and require fewer transformations than text processing for the learner to grasp the meaning, and are computationally efficient.

Maps tend to be better than text because when encoded as intact units, they preserve their visuospatial properties. That is, they contain both information about individual features (size, shape or color of discrete objects) and structural information about spatial relations among these features (distance, boundaries etc.) (Vekiri 2002). Schnotz and his colleagues (1999) discuss how realistic, meaningful pictures or drawings may be stored in a direct depictive form without much further processing, whereas diagrams are more abstract and require more cognitive processing for the student to extract meaning and establish a mental representation. When encoded as holistic units, students can thus generate and maintain mental images of maps without exceeding working memory capacity because features and structure are simultaneously available (Larkin and Simon 1987).

However, the mere presence of graphics does not ipso facto improve learning; to do so, these must support the text, they must not distract the learner, and the connection between graphic and text must be obvious (Najjar 1996). This connection is not always easy to make; on occasions, learners find it difficult to identify relevant information in static (Mayer and Gallini 1990) as well as in dynamic animated graphics (Lowe 2003).

Since a graph is an abstract representation of underlying data, the learner must identify and interpret critical information and relationships to provide meaning that corresponds to the accompanying text.

6. Animation

Animations are graphic structures that convey spatial and temporal pictures which change during presentation. Animated pictures can provide support for mental simulations because they present externally the corresponding model transformation and states of the model. The viewer can thus be a passive onlooker because relevant constraints are taken over by the medium, which reduces the learner's cognitive load. But a differentiated analysis of visuospatial structures is more difficult in this case because only a transient and fleeting source of information is available. Animations often provide more information than static pictures, but they also place greater cognitive demands on the learner because decisions must be made on how to use the various additional options available for exploring the subject matter. Cognitive resources may be taken up to handle the medium, leaving fewer resources for semantic processing of the learning content (Sweller and Chandler 1994). However, animated graphics are assumed to be superior to static graphics (Lewalter 2003), and static graphics presented together with text are presumed to be better than either of these alone (Mayer *et al.* 2003). Advantages of animated graphics are far from always supported by research (Lewlater 2003). The much-cited belief that "people generally remember 10% of what they read, 20% of what they hear, 30% of what they see, and 50% of what they hear and see" (Treichler 1967, cited in Najjar 1996) is completely unsupported by empirical evidence. Furthermore, it ignores that learning and understanding involve more than mere recall. Graphics may well be aesthetically pleasing, even humorous; they add color and attract attention, may keep the learner motivated (Najjar 1996), and may promote inference and discovery (Mayer 1989). The popularity of colorful graphics in interactive computing is therefore understandable, as is the tendency to provide increasingly complex and elaborate three-dimensional graphs with or without animation. Interestingly, simple graphics with less detail often convey the intended message more effectively, provided that they abstract the essential conceptual information (Tversky *et al.* 2002). While three-dimensional graphics are liked, it is not clear if they improve performance, speed of learning, accuracy or memory for the data (Carswell *et al.* 1991). Indeed, in some tasks, there is a disadvantage of three-dimensional graphics compared to two-dimensional; in others, there is little difference between these two types (Tversky *et al.* 2002).

According to Lewalter (2003), an animated illustration gives the learner a complete model for generating a mental representation of motion. This reduces the level of abstraction of concepts involving temporal changes and may enhance learning in conditions in which movement or change over time is conveyed. However, this is only true when the designer's model expressed in the application corresponds with, and is readily understood by, the learner's model (Norman 1986). Because static graphics portraying the same information force the learner to infer the temporal model, this should, by Lewalter's argument, lead to poorer learning outcomes. Since interactivity with the learning material generally improves learning (Tversky *et al.* 2002), it could equally be argued that static graphics would lead to better understanding precisely because it forces the learner to engage more deeply with the material. Dynamic graphics, then, may inspire learners to treat the material more superficially than the presented concepts warrant. For example, Lowe (1996) showed novices a series of animated weather maps that encouraged them in attending perceptually salient features, leaving them unable to discern the thematically important information from the display.

7. Problems in studies involving graphic information

7.1. Amount of Information Provided

Of course, there are situations in which graphics, whether static or dynamic, are better than text alone. For example, try verbally to explain how to mount and ride a bicycle or describe features of a map. For every study supporting the assumed superiority of graphics, animation, and sound (see e.g. Mayer 2002), there are as many that fail to provide such support. Careful inspection of these suggests a multitude of serious problems with experimental design, execution or interpretation of findings. For example, Large *et al.*'s (1996) study on the circulatory system included details about blood pathways in the animated version that were not provided in the static-graphic condition. In another study learners were asked to predict certain behaviors of a mechanical system after viewing an animation, a set of static graphics, or reading text, and then test their predictions (Mayer and Moreno 1998), but the animation provided information that was not available in other conditions thereby facilitating correct predictions.

At times, experimental procedures vary between conditions in ways that confound the findings, for example, by allowing interactivity with the learning material in the animated, but not in the static condition (Schnotz *et al.* 1999), or by providing a "prequestion" before the experiment to the animation-group only (Mayer *et al.* 2003). Telling students in advance that "this question will be in the exam" naturally motivates them to pay extra attention and try harder to process the learning material than students who do not have this information.

7.2. Students' Ability and Prior Knowledge

In addition to Lawless and Brown's (1997) profiles of knowledge seekers, feature explorers, and apathetic users discussed earlier, there is some controversy surrounding the benefits of multimedia on students differing in ability and/or prior knowledge. Some studies show that students with low ability and low prior knowledge benefit more than high-ability/high-knowledge students (e.g. Morrison and Tversky 2001); others show the reverse, namely that high-ability/high-knowledge students benefit more than low-ability/low-knowledge students (e.g. Hegarty and Just 1993). Hegarty and Just showed that high-ability/high-knowledge students were more strategic in their information search, better able to locate and select relevant information to form a mental representation of the material even when text did not provide all the information, and they spent less mental effort. However, the way these studies assess ability and knowledge differed, and these differences may account for the large variations in findings. Suffice it to say that as much care must be taken when designing tests and selecting participants or dividing them into different groups as when interpreting empirical findings.

Najjar (1996) discusses several pitfalls in his review of numerous studies claiming support for multimedia. First, the instructor who presents learning material in multimedia formats may organize and structure the materials better, which is likely to improve learning. Many successful applications of animation turn out to be consequences of superior visualization techniques in the animated than in the static condition. It is also possible that students have a better attitude towards learning in a multimedia environment and therefore they learn faster. Self-pacing and ability to rerun a session are likely to impact on learning, and a novelty effect cannot be excluded, especially since most studies focus on learning in a single session in which the application is assessed.

Indeed, most of the reported successes of multimedia advantages seem to be attributable to additional information conveyed, unequal procedures between conditions, differences in assessing students' prior knowledge, or to a failure to consider alternative, equally possible, interpretations of findings. In order to pinpoint the presumed facilitating effects of multimedia, experimental conditions must be so designed that in no way it favors one over another by way of procedure and they must be informationally equivalent (Tversky *et al.* 2002).

7.3. Experimental Effects

Users will not only learn what the application provider wants them to learn; indeed, they may miss the intended point of a lesson all together. "Incidental learning" refers to aspects of presented material that the learner was not intended to attend to, for example, irrelevant details of a picture supporting written or spoken material. Although some authors attempt to measure incidental learning (e.g. Rieber 1991), it is unclear that what is the advantage of such learning or whether/when it may present a barrier to learn the intended material. Attention to irrelevant details may cause the user to miss the point of the illustration or of the concept.

The notion of "transfer" of learning has two meanings, both are relevant here. One refers to the ability of learners to solve problems that rely on aspects of the concepts conveyed but are not directly explained in the lesson (Mayer 1989, Mayer, Dow and Mayer 2003), the other concerns potential experimental effects. There are at least two major problems with the first application: selection of suitable questions and specifying expectations. Depending on the complexity and the learner's prior knowledge of the material, it may be unreasonable to expect learners to comprehend, extrapolate some element of a message and translate it into another problem to find and articulate the correct solution. Thus, if learners perform poorly in transferring problems, it is difficult to discern whether they failed to learn the intended aspect of the material, whether they have difficulties in translating it, whether they are unable to articulate the concept, or whether the problem itself is unreasonable. Indeed, it is difficult to determine when a transfer question is reasonable and when it is not. Typically, domain experts have little insight into the extent of their own expertise. Given that relatively small samples of participants are usually tested in an experiment, and since individual differences in ability, imagination, and prior relevant knowledge also play a part, setting standard expectations for performance in transfer tests is problematic.

Great care must be taken to control the transfer effects when designing experiments unless the intention is to assess the carryover from one experimental trial or from one condition to the next. If participants are required to complete more than one task and performance is assessed in several conditions, the sequence in which these are presented should be varied systematically. A participant will learn something in each task that is carried over to the next. Therefore, performance on the second and subsequent tasks will differ from those on preceding tasks. This transfer effect will contaminate the data if all participants are given all the tasks in the same order.

8. Interaction or Immersion in the Material?

The difference between passive viewing/listening and active interaction with the material presented is recognized by some researchers to be significant for learning in a multimedia environment (Lawless and Brown 1997, Baulch *et al.* in Chapter 21). However, accumulating evidence suggests that the key to learning lies in immersion rather than in mere interaction with the lesson material. Relevant literature in this effect is reviewed in this section.

Observation of youth interacting with computer games gives many insights into this area. The multimedia effects of the game interface serve to stimulate and create interest and emotion, whereas the primary motivation of the game lies in the challenge it provides to the user. It is the challenge and the interaction required to take up the challenge that initially motivates the user to play the game and to keep on playing. In taking up the challenge, users don't work their way systematically through the "how to" files or the help files; they simply try things out. The genre of the game also provides many unspoken clues about how to interact and the nature of the challenge, and users transfer their learning from past experience with games in a similar genre to new games. The "first level" of the new game provides more contextual clues about the nature of the interactions required to meet the challenge, and users learn these interactions while still being stimulated by the challenge. Thus the teaching of the system is subtle and inbuilt and draws on the context, the user's prior knowledge, genre theory and activity theory, rather than requiring the user to go through a series of directive instructions before interacting at any level of the game. The help files to provide self-accessible further information on a need-to-know basis; when, how and how frequently they are accessed, are determined solely by the user.

9. The Influence of Learning Theory on Design

The key challenge for any designer of a multimedia educational product is the extent to which the product successfully enables the user to learn from as opposed to simply access the product. The designers of multimedia have a model of the user in mind when they are designing and that model also encompasses assumptions about how that user will interact with their design and assumptions about how they learn. Saljo's (1979) study of instructors and their conceptions of learning showed that these conceptions strongly influenced the design of their instruction. Similarly designers make assumptions about the tasks the multimedia system is being designed to accomplish and about how the users will interact with and perform these tasks.

Multimedia designs based on instructivist theories of learning typified by Gagne (1985) and Dick and Carey (1996) are strongly grounded in transmission models of education, based on passive exposure of the learner to a multimedia depiction of what is to be learned. They assume that there is only one message to be received, that this message will be received in the way the designer intended simply because it has been transmitted in the same way, and that there is no case for ambiguity. Transmission models tend to treat the learner as a passive receptacle for this message and assume that learning comes simply by exposing this message even if this exposure is designed in a way to cause the learner to interact with a series of steps of instruction and to perform certain tasks. This is a view of learning where meaning exists externally or independent of the individual learner that Hannafin (1997) argues is a widely accepted approach among instructivist's multimedia designs that isolate, simplify, and sequence concepts and skills according to identified task milestones, and learning hierarchies are consistent with this approach. Such approaches, as Hannafin (1997: 114) argues, define learning in terms of "attainment of well-defined enabling and terminal objectives; depth of understanding results from highly practiced, successful performance in carefully isolated, systematically sequenced, and externally engineered learning activities. The approaches explicitly assume that requisite skills and associated cognitive processes can be broken down and learned separately from holistic contexts." What is often missing from these approaches is a stimulus for learning; the highly controlled movement through such a learning environment can be very constricting and frustrating for learners, rather than engaging them in the process. Students in such an environment become "passive learners attempting to mimic what they see and hear from the expert teacher" (Berge and Collins 1995: 6) or system.

Carroll and Mack (1985) found that: "learners do not do what designers want them to do; instead they tend to get actively involved and to think and plan and solve problems." Similarly, Striebel (1991) argues that:

> the coherence of the learner's experience ... is not tied in essential ways to the instructional designer's intent (no matter how detailed or explicit these intentions are spelled out as instructional objectives) nor to the instructional plan built into the instructional systems. Rather, the coherence of the learner's instructional experience is tied to the sense that such a learner constructs out of the actual situation ... (123).

In other words, the learner extracts from a program what sense they make of it, not necessarily what use the designer may have intended.

However, interaction is more complex than simply following instructions to interact with the presented material, or being immersed in a designed environment. Constructivist theorists (Hedburg 1993; Jonassen 1994; Duffy and Cunningham 1996) and situated cognition theorists (Brown *et al.* 1989) argue that users are much more complex than that they will take an active and independent role in interacting with a designed system and will actively bring their previous experiences to bear in learning or understanding how to use the system. Rather than following the designed steps of interaction, active learners are more likely to explore and experiment with the system and to assign a meaning or interpretation to a wide variety of inputs, regardless of whether that meaning or interpretation was intended by the designers. They also detect ambiguities which might remain unnoticed by designers, and spend time in attending these rather than the designed task and the intended outcome. Ramsden (1992) argues that these unintended consequences of planned educational interventions can result in an increase in superficial learning rather than the depth of understanding aimed by the designer. Contructivism asserts that we learn through a continual process of constructing, interpreting, and modifying our own representations of reality based on our experiences with reality (Jonassen 1994). Thus, while instructivists typically rely on traditional hierarchical designs that concentrate on recall and application of knowledge, constructivists seek to create designs that engage learners in a greater range of learning outcomes, which is necessary for meaningful learning. Learners need to be empowered by the system to design their own representations of knowledge, rather than trying to "absorb" representations preconceived by others. As the learner is given more freedom to navigate, access, and manipulate the data and cognitive tools embedded in the design, the range and extent of user interaction with the system increases. Hannafin and Land (1997) argue that engaging systems elicit and sustain learner interest through "drivers" which establish the conditions, problems and scenarios that influence how the multimedia design is accessed and utilized. Engagement is increased through techniques that cultivate transformation. Such techniques include placing the learner in control of what information is accessed, when it is accessed, where it is obtained, why it is retrieved and how it is manipulated and customized for their needs. In this way the multimedia system is best used as a cognitive learning tool rather than as a surrogate teacher. In the design of the cybercartographic products described in this volume, especially those created with educational purposes in mind, the observations made in this chapter are of special importance. Cybercartographic atlases must be cognitive learning tools, and when multimedia functionalities are used these must be carefully introduced to support and supplement each other and to facilitate not detract from learning.

References

Baddeley, A. (1986) *Working Memory*, Oxford University Press, Oxford, UK.

Baddeley, A. and R. Logie, (1999) "Working memory: The multiple component model", in Miyake, A. and P. Shah (eds.), *Models of Working Memory: Mechanisms of Active Maintenance and Executive Control*, Cambridge University Press, New York.

Barnard, P. J. (1985) "Interacting cognitive subsystems: Modelling working memory, phenomena within architecture", in Miyake A. and P. Shah (eds.), *Models of Working Memory. Mechanisms of Active Maintenance and Executive Control*, Cambridge University Press, New York.

Berge, Z. L. and M. P. Collins (eds.) (1995) "Computer mediated communication and the online classroom", *Higher Education*, Vol. II, Hampton Press Inc., New Jersey.

Brown, J. S., Collins, A. Z. and Duguid P. (1989) "Situated Cognition and the Culture of Learning", *Educational Research*, 18(1), 32–42.

Carroll, J. M. and R. L. Mack (1985) "Metaphor, computing systems, and active learning", *International Journal of Man-Machine Studies,* pp. 39–57.

Carswell, C., S. Frankenberger and D. Bernhard (1991) "Graphing in depth: Perspectives on the use of three-dimensional graphs to represent lower-dimensional data", *Behaviour and Information Technology*, Vol. 10, pp. 459–474.

Chi, M. T. H., M. Bassok, M. W. Lewis, P. Reimann and R. Glaser (1989) "self-explanations: How students study and use examples in learning to solve problems", *Cognitive Science*, Vol. 13, pp. 145–182.

Dick, W. and L. Carey (1996) *The Systematic Design of Instruction*, 4th ed, Scott, Foresman and Co, Glenview, Illinois.

Duffy, T. M. and D. J. Cunningham (1996) "Constructivism: Implications for the design and delivery of instruction", in Jonassen, D. H. (ed.), *Handbook of Research for Educational Communications and Technology*, Macmillan Library Reference, New York.

Gagne, R. (1985) *The Conditions of Learning*, Holt, Rinehart and Winston, New York.

Galitz, W. O. (1993) *User-Interface Screen Design,* QED, Boston, Massachusetts.

Gentner, D. and A. L. Stevens (1983) *Mental Models*, Erlbaum, Hillsdale, New Jersey.

Grace-Martin, M. (2001) "How to design educational multimedia: A 'loaded' question", *Journal of Educational Multimedia and Hypermedia*, Vol. 10, No. 4, pp. 397–409.

Hannafin, M. (1997) "The case for grounded learning systems design: What the literature suggests about effective teaching, learning and technology", *Educational Technology Research and Development*, Vol. 45, No. 3, pp. 101–117.

Hannafin, M. and S. Land (1997) "The foundations and assumptions of technology-enhanced, student-centred learning environments", *Instructional Science*, Vol. 25, pp. 167–202.

Hedburg, J. G. (1993) "Design for interactive multimedia", *Audio-Visual International*, September, pp. 11–14.

Hegarty, M. (1992) "Mental animation: Inferring motion from static displays of mechanical systems", *Journal of Experimental Psychology: Learning, Memory, and Cognition*, Vol. 18, pp. 1084–1102.

Hegarty, M. and M. A. Just (1993) "Constructing mental models of machines from text and diagrams", *Journal of Memory and Language*, Vol. 32, pp. 717–742.

Jonassen, D. (1994) "Thinking technology: Toward a constructivist design model", *Educational Technology*, Vol. 34, No. 3, pp. 34–37.

Johnson-Laird, P. N. and R. M. J. Byrne (1991) *Deduction*, Erlbaum, Hillsdale, New Jersey.

Johnson-Laird, P. N. (1998) "Imagery, visualization, and thinking", in Hochberg, J. (ed.), *Perception and Cognition at Century's end. Handbook of Perception and Cognition*, Academic Press, New York.

Kosslyn, S. M. (1994) "Image and Brain", *The Resolution of the Imagery Debate*, MIT Press, Cambridge, Massachusetts.

Large, A., J. Beheshti, A. Breuleux and A. Renaud (1996) "Effect of animation in enhancing descriptive and procedural texts in a multimedia learning environment", *Journal of the American Society for Information Science*, Vol. 47, pp. 437–448.

Larkin, J. H. and H. A. Simon (1987) "Why a diagram is (sometimes) worth ten thousand words", *Cognitive Science*, Vol. 11, pp. 65–99.

Lawless, K. A. and S. W. Brown (1997) "Multimedia learning environments: Issues of learner control and navigation", *Instructional Science*, Vol. 25, pp. 117–131.

Lewalter, D. (2003) "Cognitive strategies for learning from static and dynamic visuals", *Learning and Instruction*, Vol. 13, pp. 177–189.

Lowe, R. K. (1996) "Selectivity in diagrams: Reading beyond the lines", *Educational Psychology*, Vol. 14, No. 4, pp. 467–491.

Lowe, R. K. (2003) "Animation and learning: Selective processing of information in dynamic graphics", *Learning and Instruction*, Vol. 13, pp. 157–176.

Mayer, R. E. (1989) "Systematic thinking fostered by illustrations in scientific text", *Journal of Educational Psychology*, Vol. 81, pp. 240–246.

Mayer, R. E. (2002) *Multimedia Learning*, Cambridge University Press, Cambridge, UK.

Mayer, R. E. and J. K. Gallini (1990) "When is an illustration worth ten thousand words?" *Journal of Educational Psychology*, Vol. 82, pp. 715–726.

Mayer, R. E. and R. Moreno (1998) "A split-attention effect in multimedia learning: Evidence for dual processing systems in working memory" *Journal of Educational Psychology*, Vol. 90, pp. 312–320.

Mayer, R. E., G. T. Dow and S. Mayer (2003) "Multimedia learning in an interactive self-explaining environment: What works in the design of agent-based microworlds?" *Journal of Educational Psychology*, Vol. 95, No. 4, pp. 806–813.

Miyake, A. and P. Shah (1999) "Toward unified theories of working memory: Emerging general consensus, unresolved theoretical issues, and future research directions", in Miyake, A. and P. Shah (eds.), *Models of Working Memory: Mechanisms of Active Maintenance and Executive Control*, Cambridge University Press, New York.

Morrison, J. B. and B. Tversky (2001) "The (In)effectiveness of animation in instruction", in Jacko, J. and A. Sears (eds.), *Extended Abstracts of the ACM Conference on Human Factors in Computing Systems*, ACM Press, New York, pp. 377–378.

Najjar, L. (1996) "Multimedia information and learning", *Journal of Educational Multimedia and Hypermedia*, Vol. 5, pp. 129–150.

Norman, D. A. (1986) "Cognitive engineering", in Norman, D. A. and S. W. Draper (eds.), *User-Centred System Design: New Perspectives on Human-Computer Interaction*, Erlbaum, Hillsdale, New Jeresy.

Paivio, A. (1990) *Mental Representations: A Dual Coding Approach*, Oxford University Press, New York.

Pylyshyn, Z. W. (1981) "The imagery debate: Analogue media versus tacit knowledge", *Psychological Review*, Vol. 87, pp. 16–45.

Ramsden, P. (1992) "Learning to teach", *Higher Education*, Routledge, London.

Rieber, L. P. (1991) "Animation, incidental-learning, and continuing motivation", *Journal of Educational Psychology*, Vol. 83, pp. 318–328.

Saljo, R. (1979) *Learning in the Learner's Perspective: Some Common-Sense Conceptions*, Internal Report, Department of Education, University of Göteborg, No. 76.

Schneider, W. (1999) "Working memory in a multilevel hybrid connectionist control architecture (CAP2)", in Miyake, A. and P. Shah (eds.), *Models of Working Memory: Mechanisms of Active Maintenance and Executive Control*, Cambridge University Press, New York.

Schnotz, W., J. Böckheler and H. Grzondziel (1999) "Individual and co-operative learning with interactive animated pictures," *European Journal of Education*, Vol. 14, No. 2, pp. 245–265.

Stanney, K., S. Samman, R. Reeves, K. Hale, W. Buff, C. Bowers, B. Goldiez, D. Nicholson and S. Lackey (2004) "A paradigm shift in interactive computing: Deriving multimodal design principles from behavioural and neurological foundations", *International Journal of Human-Computer Interaction*, Vol. 17, No. 2, pp. 229–257.

Striebel, M. J. (1991) "Instructional plans and situated learning", in Anglin, G. J. (ed.), *Instructional Technology Past, Present and Future*, Libraries Unlimited, Englewood, Co, pp. 117–132.

Sweller, J. and P. Chandler (1994) "Why is some material difficult to learn", *Cognitive Instruction*, Vol. 12, No. 3, pp. 185–233.

Taylor, D. R. F. (2003) "The concept of cybercartography", in Peterson, M. P. (ed.), *Maps and the Internet*, Elsevier Science Ltd., Amsterdam

Tversky, B., J. B. Morrison and M. Betrancourt (2002) "Animation: Can it facilitate?", *International Journal of Human-Computer Studies*, Vol. 57, pp. 247–262.

Van Dijk, T. A. and W. Kintsch (1983) *Strategies of Discourse Comprehension*, Academic Press, New York.

Vekiri, I. (2002) "What is the value of graphical displays in learning?", *Educational Psychological Review*, Vol. 14, No. 3, pp. 261–312.

CHAPTER 10

Cognitive Theories and Aids to Support Navigation of Multimedia Information Space

SHELLEY ROBERTS

Human Oriented Technology Lab (HOTLab), Department of Psychology Carleton University, Ottawa, Canada

AVI PARUSH

Professor of Psychology, Human Oriented Technology Lab (HOTLab), Department of Psychology, Carleton University, Ottawa, Canada

GITTE LINDGAARD

Professor, Natural Sciences and Engineering Research Council (NSERC), Industry Chair User-Centered Design, and Director, Human Oriented Technology Lab (HOTLab), Department of Psychology, Carleton University, Ottawa, Canada

Abstract

Navigation in information cyberspace presents difficulties for users leading to reduced satisfaction. How navigation can be facilitated is not well understood, particularly with complex systems involving spatial or geo-referenced information such as in cybercartographic products. There is extensive research on the way people navigate and orient themselves successfully in real and virtual environments, which could provide insight regarding ways to aid navigation and orientation in information cyberspace. This chapter discusses the similarities and differences of navigation and orientation between real/virtual spaces and cyberspace in terms of: (1) acquisition and representation of space; (2) tasks; and (3) navigation aids. The degree of generalization between these two spaces is discussed and serves as a basis for a conceptual framework, as well as guidelines, to support navigation in information cyberspaces.

1. Introduction

As technology advances, new opportunities arise for more innovative, useful, interactive, and customizable products. Cybercartographic atlases are one such product and these are described in several chapters of this volume. The cybercartographic environment is defined as having a large amount of geo-referenced information presented through multiple formats and mediums including text, video, three-dimensional visualizations, sound, and touch. Cybercartographic environments have the potential to engage the user in innovative ways. However, while the technology to provide users with novel and highly interactive ways of presenting information is available, the knowledge regarding the best ways to aid the user in navigating through such complex spaces is lacking. The primary task for users should not be navigation but rather be engaged with the content. Therefore designers must ensure that navigation does not impede the users primary task by making navigation as transparent as possible while promoting discovery and exploration of the content.

Navigation poses the most difficult task for users in electronic environments and cyberspace; disorientation, the tendency to lose one's sense of location and direction, is a well-known problem for users of cyberspace and can result from a variety of reasons (Edwards and Hardman 1988; Elvins 1997; Rumpradit and Donnell 1999). More specifically, Edwards and Hardman (1988) have categorized disorientation into three types of problems: the user does not know where to go next, the user knows where to go but not how to get there, and the user does not know where they are in relation to the overall structure.

In order to create products that are satisfying and easy to navigate, certain issues need to be explored. This chapter reviews and examines the spatial cognition literature in terms of how spatial information in real and virtual worlds is acquired and represented, and to what extent findings from this body of literature can be applied to the way cyberspace is acquired and represented. If navigation and orientation tasks are comparable between these environments, then aids that have proven effective in real and virtual environments should prove to benefit the user by decreasing or eliminating disorientation in cyberspace.

This chapter has three main objectives:

1. *Review and Analysis*: To inform the reader of the literature concerning user-related issues of navigation in real/virtual environments and

cyberspace, as well as analysis of the similarities and differences that exist between these two types of spaces.

2. *Analysis Implications*: To understand, based on the analysis, as to what extent those similarities and differences allow us to apply similar navigation and orientation aids in both types of spaces.
3. *Guideline Synthesis*: To provide a set of guidelines aiming to assist software developers to provide applications that will ease and facilitate navigation through electronic information cyberspace based on our findings.

1.1. Definition of Navigation, Orientation, and Information Cyberspace

Two types of abilities are involved in moving through real and virtual worlds; navigation and orientation. Navigation consists of two linked abilities; locomotion and wayfinding (Darken and Peterson 2001). Locomotion is the motor element of navigation and involves the ability of getting from "here" to "there". Wayfinding is the cognitive element of navigation and includes abilities, such as planning efficient routes, finding specific locations, and recognize destinations when reached (Elvins 1997). Orientation requires the knowledge of where one is with respect to objects and other cues (Satalich 1995). The ability to navigate and orient oneself within an environment requires the development and use of a cognitive map, also referred to as a mental map. Cognitive maps are mental representations of the layout of one's environments and will be discussed in more detail throughout this chapter.

Various types of environments will be discussed and definitions of each are essential. Throughout the first part of this chapter, research regarding navigation in real and virtual environments is discussed. The physical world is referred to as a "real" environment, whereas virtual environments encompass any type of environment that has a real world metaphor, including two- and three-dimensions, and with all levels of immersion. It has been discovered that differences do exist with regard to navigation between real and virtual environments, however, navigations are similar enough that no distinctions are made between them here, therefore they are combined and discussed as one type of environment.

Information space, also referred to as information cyberspace, is discussed in the second part of this chapter and includes any hypermedia system, which can consist of hypertext, sound, images, video etc, but is not limited to having a spatial metaphor. The main distinction between real/virtual environments and information cyberspace is that information cyberspace represents information, not real geographic space as its main focus.

1.2. Conceptual Framework for the Analysis

A conceptual framework is used as a guide for the analysis. This conceptual framework is based on what we know from studies on spatial cognition in real and virtual worlds, and may facilitate our understanding of how users acquire and represent information cyberspaces. The basic components are the three levels of spatial cognition: context, phases, and representation or knowledge. This framework as seen in Fig. 10.1 is used to look at similarities and differences across the two space types.

Context: The context includes the specific task the user is trying to accomplish, and the type of aid available to accomplish this task. For example the user may have a specific keyword that they would like to search for and the aid available is a search text box.

Phases: There are two basic phases, acquisition and usage. Acquisition includes learning the environment and leads to usage with time, while navigating and orienting throughout the environment. Acquisition usually occurs before navigation and orientation are carried out successfully.

Representation: There are three basic types of knowledge used to represent spatial information, landmarks, route, and survey knowledge. Landmarks refer to distinct features in the environment used as reference points to guide navigation. Route knowledge includes information relating to the distance and direction between locations or landmarks. Survey knowledge contains information about the topographic properties of the environment.

The three levels of this framework are related. The context (tasks and aids) is relevant to both phases (acquisition and navigation). Three types of

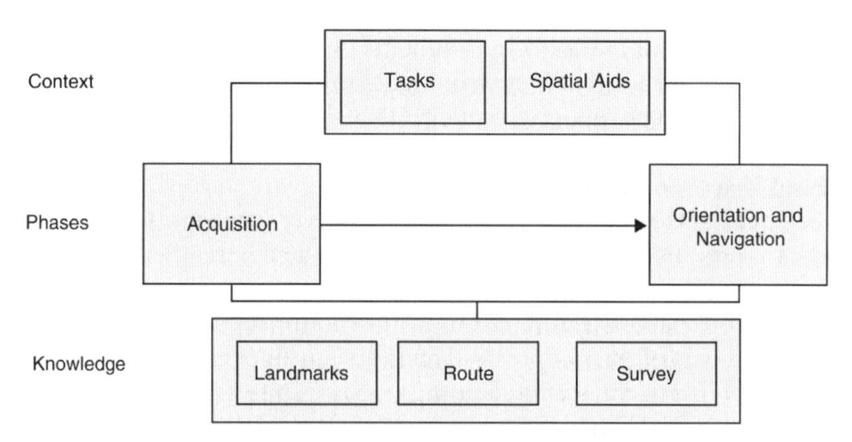

FIG. 10.1. A conceptual framework of three levels of spatial cognition; context, phrases, and representation.

representations are required for acquisition to take place, which also facilitates orientation and navigation.

The central question this chapter considers is, to what extent does this framework apply within information cyberspaces? Specifically, can we talk about the same kind of representation? Can we talk about the same kind of contextual impact? The analysis of similarities and differences is an attempt to answer these questions, and the guidelines are a product based on the answer to these questions.

2. Navigation and Orientation in Real/Virtual Environments

The aim of this section is to present the reader with an overview of the way in which people acquire and represent spatial information, the tasks performed, and aids that are utilized to assist navigation through real and virtual spaces. Real and virtual environments have been combined in this chapter because the ways people navigate through both are similar, although there are known differences between real and virtual spaces regarding navigation. The similarities include acquisition, representation, tasks, and aids (Arthur and Hancock 2001; Darken and Peterson 2001).

2.1. Acquisition and Representation of Real/Virtual Environments

The ability to form a cognitive representation of an environment in order to navigate has been the focus of research. Tolman (1948) observed the spatial behavior of rats in a maze and suggested that they were able to form a map-like representation of the maze. Tolman coined the term "cognitive map", also known as mental maps (Darken and Peterson 2001), referring to the ability to form an integrated layout of one's environment. There is also neuropsychological support for cognitive maps suggesting that cognitive maps are encoded in the hippocampus of animals (O'Keefe and Nadel 1978) and humans (Maguire *et al.* 1998).

Disciplines other than psychology have been aware of the way in which spatial knowledge of environments are acquired and represented. Lynch (1960), a geographer and urban planner, was the first researcher to consider how images of the environment might be formed in one's mind. Lynch's observations suggested that people recall and concentrate on certain features of a city more than others, for example, edges, paths, path junctions, and landmarks. The way these features are interpreted and organized affects what they see and how they interact with their environments. This work, along with Tolman's cognitive mapping theory, stimulated considerable research in human spatial cognition that currently suggests that people represent spaces mentally. How these spatial representations are formed and how cybercartography may benefit from these findings is discussed next.

Research on cognitive maps to date has shown that people learn the layout of an environment by acquiring three levels of knowledge; landmark, route, and survey knowledge (Siegel and White 1975; Thorndyke and Goldin 1983) and these appear to be formed progressively (Lynch 1960; Siegel and White 1975; Yeap and Jefferies 2001). However, research has recently supported the idea that aspects of landmark, route, and survey information may be learned simultaneously particularly with the use of aids (Montello 1998; Yeap and Jefferies 2001).

Landmarks refer to distinct features in the environment used as reference points to guide navigation. For a structure to be a landmark it must be an object that is imaginable, memorable, and distinguishable from other objects in the environment (Lynch 1960; Elvins *et al.* 1997). Landmark information has been shown to be very effective in supporting navigation and orientation in real (Lynch 1960; Elvins 1997) and virtual environments (Lynch 1960; Darken and Sibert 1993; Darken and Peterson, 2001; Parush and Berman, 2004). Route knowledge, information relating to the distance and direction between locations or landmarks, is primarily a verbal or graphical representation of instruction-like steps to aid movement from one place to another (Lovelace *et al.* 1999). Survey knowledge contains information about the topographic properties of the environment (Thorndyke and Hayes-Roth 1982) acquired by extensive direct experience, or indirect experience by way of aids such as maps. Survey information contains information about the topographic properties of the environment (Thorndyke and Hayes-Roth 1982), i.e. the knowledge of the location of objects relative to compass bearings, global shapes of the land and its features, and Euclidean distances between objects. Survey information is typically acquired through the use of maps, although it is possible to accumulate it through enough experience of multiple routes and their intersections. Thorndyke and Hayes-Roth (1982) distinguished between what they called "primary survey knowledge", which is acquired through direct interaction with the environment, and "secondary survey knowledge", acquired through indirect interaction, for example via a map. All levels of spatial knowledge seem to have varying importance depending on the type of navigation task and environment being navigated.

2.2. Tasks

People navigate for two reasons, to search or to explore (Darken and Cevik 1999; Tan *et al.* 2001). Searching can be broken down further into three search types, targeted, primed, and naïve (Darken and Cevik 1999). A targeted search is a task in which the target is in view. A primed search is a task in which the goal is known, but is out of view. A naïve search is one in which the location of the goal is unknown and there is no prior knowledge of the space. Navigation with the purpose of exploration is carried out when there is no specific goal or destination to reach.

2.3. *Navigation Aids*

As discussed in the previous section, not all navigation tasks are created equal. Research has suggested that the usefulness of the aid provided depends on the navigation task at hand. The type of navigation task being carried out is dependent on specific spatial knowledge and aids to support navigation (Hirtle and Sorrows 1998). Research in spatial cognition has revealed that the three types of knowledge, landmark, route, and survey, are relied on more heavily for different types of aids. Therefore, if designers are able to provide landmark, route, and survey information in the form of aids, navigation should be greatly enhanced.

2.3.1. Landmarks

For an aid to be recognized as a landmark it must be an object that is imaginable, memorable, and distinguishable from others in the environment (Lynch 1960; Elvins *et al.* 1997), for example, the big "M" that represents McDonald's is a well-known landmark. It is suggested that landmarks serve an important function in navigation because people can associate them with actions, such as when and where to make a change in direction (Vinson 1999), for example, "turn left after the McDonald's".

Landmarks, being the first environmental features to be encoded and recalled from memory, may be particularly important if users do not need to acquire a cognitive map of the environment, for example infrequent users who need to find information quickly. At any rate, the inclusion of landmarks within cybercartographic products will help structure the environment and provide location and directional cues to facilitate wayfinding and orientation, allowing people to navigate with confidence through the environments.

The benefit derived from a navigation aid will be dependent on the task. For example, a targeted task, navigating towards a specific target within one's view, would be affected primarily by landmark recognition. There are numerous studies supporting the use of landmarks, especially the need to rely on them while navigating through novel environments, given that they are the most salient cues in an environment (Tversky 1991; Tversky *et al.* 1994; Elvins 1997; Elvins *et al.* 1998; Tversky 2003). It has been found that users are less disoriented and find goals more efficiently within virtual environments when they are given the ability to "drop" their own landmarks, similar to adding a "pushpin" on a map to mark a location (Darken and Sibert 1993). In general, studies find degraded navigation and orientation performance in real and virtual worlds when there is a lack of landmarks (Darken and Sibert 1993; Vinson 1999; Parush and Berman 2004).

2.3.2. Routes

Route knowledge develops as landmarks are connected by paths. Using route knowledge consists of two abilities; traveling along a path until a landmark is reached and then changing direction to follow another path towards another landmark. The most important aspect of route information is to provide information for reorientation in order to switch from one path to another (Agrawala and Stolte 2000). Route information is sufficient for repeated journeys following the same path, but insufficient for planning new routes, such as shortcuts and detours. The amount of detail included in route information has been debated. Tversky (2001) has suggested that people schematize and distort the environment and remember only critical information. This would support the idea that simplified information presented in a way that emphasizes what the user needs and provides all other information in the background can support navigation. For example, a subway map typically consists of straight vertical and horizontal lines representing the tracks, intersections representing where subway lines meet at junctions, and nodes representing stops, and omits any detailed information.

Agrawala and Stolte (2000) have identified four design guidelines of efficient route directions based on Tversky's suggestions: readability, clarity, completeness, and convenience. Readability suggests that all important details be present, visible and identifiable, such as main roads. The route should be clearly marked and easily seen at a glance. The route must be complete by including only complete imperative information to navigate, and convenient by taking into account when, where and how the user will use the route information (Agrawala and Stolte 2000). These researchers analyzed different route depictions and concluded that hand-drawn styles of routes are most effective and contained all of the information in the principles. Using these principles along with the hand-drawn style, researchers produced usable maps across various situations (Agrawala and Stolte 2000), however, they suggested that they would like to include the placement of more important and salient landmarks in their principles.

Research on routes as aids is mixed. On one hand Thorndyke and Hayes Roth (1982) found that the route directions improved navigation when the environment was simple, whereas maps improved navigation of complex environments. On the other hand, a study of nurses' navigation through a spatially complex hospital demonstrated that they relied exclusively on route and landmark knowledge, even though overview maps of the hospital were placed throughout (Moeser 1988). It has also been found that drivers using in-vehicle navigation systems prefer verbal route descriptions while attempting to navigate novel environments (Streeter *et al.* 1985). Another study found that routes, as aids, provide quick navigation to goals; however they do not tend to lead to the acquisition of knowledge of the environment (Parush and Berman 2004). Results on route descriptions as aids suggest that they are beneficial, but are task dependent. Route descriptions have been found to be especially beneficial for primed searches, and tasks that are of high cognitive load, such as driving and navigating through a complex building. Overall, it seems that primed search tasks, locating a target when the location is not in view, would benefit from an aid such as route directions.

2.3.3. Survey

Survey information is generally portrayed in map form, providing the user with a "bird's eye view" of the area. Therefore, maps facilitate the effectiveness of navigation in real and virtual environments, particularly when the environment is novel. An overview of the space would greatly benefit someone performing a naïve search in which the location of the goal is unknown and there is no prior knowledge of the space, eliminating the need of an exhaustive search (Darken and Cevik 1999). Maps allow one to acquire quick survey knowledge without having to attain prior landmark or route knowledge. The downfall to learning an environment this way is that the type of survey knowledge gained is not as complete as it would be from direct experience. Direct experience provides people with a perspective-independent cognitive map of the environment whereas people who use a map to learn an environment acquire a perspective-dependent cognitive map (Thorndyke and Hayes-Roth 1982). Having a perspective-independent cognitive map has the benefit of being able to navigate an environment from any point of view. To further ease navigation it is suggested that when providing an overview of the environment the relative position of the user within the space should be presented (Levine 1982; Kim and Hirtle 1995). One type of aid that eases orientation is a "You-Are-Here" representation which contains a symbol marking one's current location and bearing within the environment (Levine 1982).

Tversky (Tversky 1991; Tversky *et al.* 1994; Tversky 2003) found that people typically sketch maps or verbally describe an environment from a "bird's eye view". She speculates that three-dimensional mental images are hard to create and even harder to comprehend and that a third dimension adds information that is irrelevant for recognizing specific landmarks. She argues that three-dimensional information included in spatial aids may impede performance and spatial relations among landmarks should suffice for effective navigation. Whether distinctive three-dimensional landmarks are pertinent for navigation and useful for aiding orientation under certain circumstances needs to be determined.

In order for a map to communicate effectively, research suggests that only pertinent information be displayed and the rest eliminated or hidden because the reduction of detail is one of the most useful aspects of a map by helping to control cognitive load (Tversky and Schiano 1989). Tversky also argues that maps should present environments with inconsistent scale and perspective. She uses the tourist map as an example, which presents an overview of the core of the area and then superimposes frontal views of important landmarks, allowing tourists to use these to navigate by recognition. There is some evidence that this type of map may provide navigational benefit (Tversky 1991; Tversky *et al.* 1994; Elvins *et al.* 1997; Elvins *et al.* 1998; Tversky 2003). Another important implication of Tversky's work is that maps as an external representation should be congruent to the users' internal representation of the area being portrayed. In sum, removing irrelevant information from maps facilitates the ability to navigate easily and quickly towards the goal in real and virtual environments and should be congruent with the user's internal representation.

2.3.4. Combination of Aids

Research suggests that mental representations of real/virtual space are subject to predictable, systematic distortions (Tversky 1981), that space is likely to be stored as schematized two dimensional images (Tversky 1981; Tversky and Schiano 1989; Tversky 1993; Tversky 2003), and that these simplifications reduce cognitive load. The usefulness of maps, it is argued, enables people to externalize information that would otherwise need to be stored in memory. Tversky (2003) holds that we learn about our environment through redundant mediums, which includes maps, verbal descriptions, direct experience, and a multitude of other aids. This multifaceted experience leads to a rich web of spatial information that resembles a "cognitive collage" more than a "cognitive map" and may be beneficial to the design of cybercartographic products.

Preliminary support has been found for this claim (Hirtle 1999). Hirtle (1999) argued that Tversky's cognitive collage framework might offer useful guidance when constructing navigation tools for systems that provide spatial information in particular. He argued that redundant information from a variety of media types might be most useful for navigation. Hirtle examined four systems that were used as tools to help guide real environment behavior and included, "you-are-here" maps, information kiosks, an information browser, and a landmark locator system (Hirtle 1999). Hirtle used these systems (Figs. 10.2 and 10.3) to find out whether or not the systems that use a variety of media in a redundant fashion would enhance navigation as Tversky (1993) suggested. He found that through multiple mediums, which offered landmark, route directions, and survey information simultaneously, people were able to find their goal effectively. This research suggests that the use of multiple modes and redundancy will aid in the formation of a cognitive map of the electronic environment, and the inclusion of Tversky's cognitive collage framework will aid in directing possible design considerations for navigational systems.

Landmarks, route directions, and maps are rarely utilized independently of one another in real and virtual environments. Parush and Berman (2004) compared the presence of two aids, route lists and maps in virtual

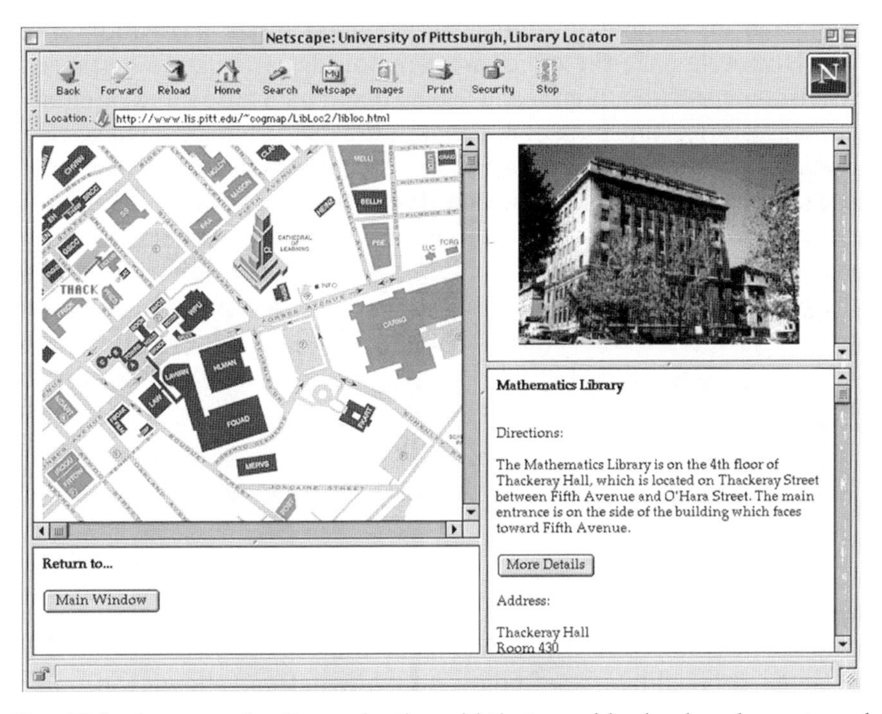

FIG. 10.2. An example of a navigation aid that provides landmark, route and survey information (From Hirtle and Sorrows 1998).

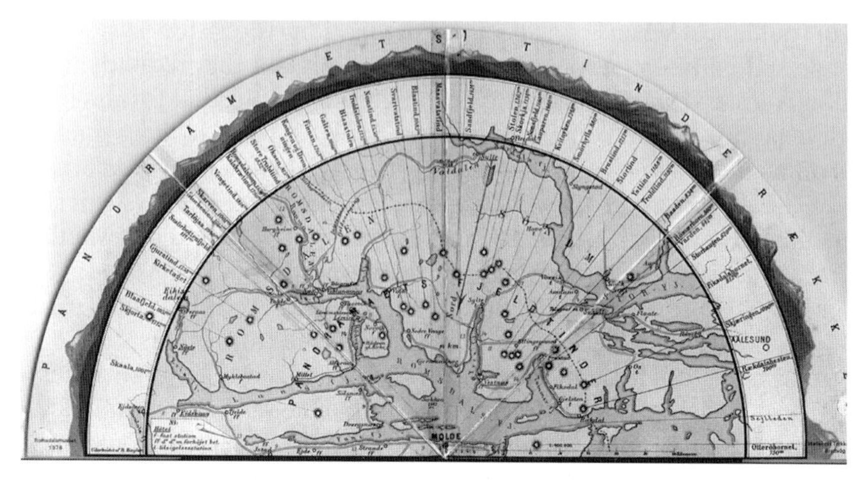

FIG. 10.3. 1891 map of Molde, Norway (From Hirtle 1999).

environments with and without landmarks, and how these conditions affected navigation and orientation tasks. Following a learning phase, aids were removed from the environment. It was found that navigating without aids was poorer in the environment without landmarks. While learning the environment, route lists provided faster navigation to the target. However, once these aids were removed, participants that learned the environment with a map made less errors in navigation and orientation tasks, suggesting acquisition of survey knowledge. Similar inferences on the importance of the map in this respect are reported by Gartner in this volume. Research has also indicated that the best way to represent space is by presenting the user with all three types of aids: landmark, route, and survey in a redundant fashion (Tversky and Schiano 1989; Hirtle and Sorrows 1998). It appears that aids, such as route descriptions, landmarks, and maps each provide additional, supplemental information for a single underlying spatial representation, and if all three are not provided, how these aids are combined should be considered.

2.4. Summary of Navigation and Orientation in Real/Virtual Environments

In summary, the literature on spatial knowledge acquisition suggests that landmarks help structure the environment and provide location and directional cues to facilitate wayfinding and orientation, allowing people to navigate with confidence through novel as well as familiar real/virtual environments. Route descriptions have been found to be specifically beneficial for primed searches, and tasks that have a high cognitive load, such as driving and navigating through a complex building. Overall, it seems that primed search tasks, locating a target when the location is not in view, would benefit from an aid such as route directions. Survey information can be acquired through direct experience or, more quickly, through the use of aids, such as maps, and includes the knowledge of the location of objects relative to compass bearings, global shapes of the land and its features, and Euclidian distances between objects. If users are to acquire survey knowledge of the space provided, topographic information in the form of a spatial overview of the area should be included as an aid.

There is strong evidence that the type of spatial knowledge acquired is dependent on the way the environment is presented. If distance information were more important for users to acquire, then a map would provide this. However, if route knowledge is to be acquired, a virtual tour is likely to be best. Lastly, the usefulness of navigational aid(s) will most likely depend on the task the user is trying to accomplish. Research has provided support for the use of different aids being dependent on the task at hand. An overview of the space would greatly benefit someone performing a naïve search in which the location of the goal is unknown and there is no prior knowledge of the space, eliminating the need of an exhaustive search. Overall targeted tasks would be affected primarily by landmark recognition. Primed tasks and tasks with a high cognitive load would benefit from route descriptions and naïve searching tasks would benefit from a survey map of the area.

Literature on spatial cognition has clarified how people represent space, how this representation is acquired, the tasks involved in navigating through space, and the aids that help. It seems that the type and cognitive demands of the task and the aid supporting navigation in real and virtual spaces should be complementary. Spatial cognition can provide methodologies and a theoretical framework for gaining insight into how we interact and move through cyberspace. An understanding of the similarities and differences between navigating in real/virtual spaces versus information cyberspace environments, will provide insight into how much generalization could be extended between the two realms. The way in which information is acquired, the types of tasks that are performed, and the aids that are used in cyberspace are discussed next.

3. Navigation and Orientation in Information Cyberspace

Cyberspace, or information space, consists of any hypermedia system, including hypertext, sound, images, and video as seen in Fig. 10.4. Disorientation within these spaces can be the result of various factors, such as

FIG. 10.4. An example of a three-dimensional information cyberspace used to browse a collection of online documents (From Chen and Carr 1999).

the cognitive load of the primary task or navigation difficulties. Understanding the cognitive aspects of how knowledge of cyberspace is acquired, represented, and navigated can promote the development of appropriate aids to improve performance in terms of navigation and information retrieval. This section reviews how knowledge of cyberspace is acquired and navigated, tasks that are carried out, and various aids that have been applied.

3.1. Acquisition and Representation of Information Cyberspace

The way cognitive or mental maps are formed is an important aspect of our understanding of how information cyberspace knowledge is acquired. The way information space knowledge is acquired is not well understood. However, if we are able to gain insight into how they are learned and developed over time, we can find ways to facilitate and reduce acquisition time and improve usage performance.

Foss (1989) has divided hypertext navigation problems into four causal groups: (1) disorientation; (2) the use of embedded links (hyperlinks that are located within the actual information on the site); (3) an overwhelming amount of information; and (4) the way in which the site itself is structured.

Although these problems were associated with hypertext, they are also associated with more interactive, multimedia, and multimodal products. Ways that these can be avoided and aids that can decrease these typical navigation problems are discussed. In order to navigate through a world with little or no prior knowledge or experience of its layout, each view the user encounters must present information such as signs, labels, and views of landmarks in the distance that will help users make their next navigational decision.

The way in which people navigate through virtual worlds is primarily through spatial means and metaphors such as flying, walking, and driving; however information cyberspace poses different navigation problems and strategies. Jul and Furnas (1997) suggested information spaces are structured in three levels, the inherent structure, the imposed structure, and the users own cognitive map of the structure (Jul and Furnas 1997). The inherent structure is the actual physical structure of files and networks, which is rarely seen by the user. The imposed structure contains the actual content of the site (or product) and how it is laid out. This structure is the one that is most often navigated. The third structure is the users own cognitive or mental map, which according to the authors, is based on the users' previous knowledge, experience, or views of the imposed structure. The authors suggest that the closer the imposed structure is to the users' cognitive map of the content, the more intuitive navigation will be. Very little is known about the way in which knowledge of cyberspace is acquired. The issue of cognitive or mental map acquisition raises questions regarding the ability to navigate through cyberspace. Analogous to the real/virtual world, do people acquire three comparable levels of spatial knowledge (landmarks, route, and survey) in cyberspace? Does the knowledge used depend on the type of task being carried out?

3.2. Tasks

Researchers have described numerous types of tasks cyberspace users perform (Benyon and Hook 1997; Jul and Furnas 1997; Boechler 2001; Pilgrim *et al.* 2004), however all can fit into one of two categories, open or closed. Open tasks include those that are exploratory in nature, where a clearly defined goal does not exist, or where the user wants to obtain a general overview of the space (Benyon and Hook 1997). Closed tasks are ones in which users have a clearly defined goal (Jul and Furnas 1997), whether they are searching for something that they have already found in the past, or searching for something specific and don't know where to find it (Boechler 2001). Given that open and closed tasks have different goals, the support provided should match the users' needs.

3.3. Navigation Aids

Various types of aids are currently being used throughout information spaces to navigate through the interface that provide information regarding where the users' are, where they have been, and how to move towards their goal. A short list includes landmarks, bookmarks, sitemaps or overviews, tutorials, links, backtracking, as well as various types of metaphors and ways to visualize information. This section provides the reader with an overview of important findings from various studies regarding general navigation aids.

3.3.1. Landmarks

Landmarks in information cyberspace are defined as nodes, points, or pages within the space that are connected to other nodes via links. Home pages or specific icons or hotspots are examples of cyberspace landmarks, acting as transition points where fundamental turns are taken (Chen and Czerwinski 1997; Westerink *et al.* 2000; Pilgrim *et al.* 2002). Pilgrim *et al.* (2004) found that landmarks are relied upon when performing closed tasks, whereas a browsing or exploratory strategy was used to perform open tasks. The exception was the novice user who relied on landmarks, specifically the home-page, regardless of the task type, presumably because of disorientation. One downfall of relying on a particular link or node as a landmark for locating information is that they can be removed, moved, or changed in appearance (Boechler 2001).

The importance of landmarks may be context dependent. Three-dimensional information visualizations in particular can be disorienting and landmarks are proving to be beneficial (Chen and Czerwinski 1997; Vinson 1999; Bederson 2001; Skopik and Gutwin 2003). It has been suggested that salient cues should be included in the design to reinforce users' cognitive map of the virtual space (Vinson 1999) and help them understand where they are within the space (Skopik and Gutwin 2003). For example, in a visualization of a large database of documents, an animation of how papers are organized would help users to understand the implicit organization (Chen and Czerwinski 1997). Having salient landmarks in the environment is likely to accelerate the development of such representations so that they can be more easily manipulated during navigational tasks.

3.3.2. Route

Route knowledge is represented in terms of intersections of landmarks used to make decisions regarding where to go next, or any cue for a specific action. Kim and Hirtle (1995: 241) claim that this type of navigation task is analogous to "planning and executing routes through the network" of hypertext. Route knowledge is not generally found in cyberspace for three possible reasons: (1) There is no way of accurately measuring distance between landmarks; (2) Hypertext is dynamic (Boechler 2001); distances between landmarks do not remain static in cyberspace because links and nodes are added intermittently; and (3) The ability to "teleport" between two spaces (Stoakley *et al.* 1995). Westerink *et al.* (2000) have found that people do establish measures of distance between landmarks in cyberspace corresponding to the number of steps required to get there. Given the above definition of route knowledge one could argue that there is a requirement for users to recall how to get from "point A" to "point B" with the use of landmarks. However, since cyberspace is quite dynamic, the need to support route knowledge is more problematic than in real/virtual worlds. Short walkthroughs of paths from one information node or landmark to another, particularly within three-dimensional information visualizations, may be one way to incorporate route information while facilitating navigation from one place to another.

3.3.3. Survey

Sitemaps provide the overall structure of the cyberspace in a textual or graphical format. It presents a "bird's eye view" of the space, similar to a map in real and virtual spaces, and has the potential to assist users in understanding the imposed structure (Jul and Furnas 1997), and form a cognitive map of the information architecture. Sitemaps are found to decrease the feeling of "lostness" by providing users' the ability to identify how much or how little of the space they have explored, where they are located currently, and where they have visited in the past (Bernard 1999). They can be used to bridge the gap between the sites' imposed structure and the users' cognitive map of the information (Dahlback 1998).

There are mixed findings as to how helpful sitemaps are as a navigational aid. Some have shown that users are faster and more efficient at finding information, dig deeper into the hierarchy, and are more satisfied (McDonald and Stevenson 1998; Bernard 1999; Danielson 2002). Other studies suggest that sitemaps do not provide any extra benefit for the user and may increase the cognitive workload (Nillson and Mayer 2001).

A study by Danielson (2002) investigated the use of sitemaps in five web sites, specifically how the constant presentation of sitemaps would affect navigational behavior and whether or not they provided a "bird's eye view" of the Web site. The sitemap was presented in a separate frame, and subjects were instructed to find specific facts during navigation. Danielson found that when sitemaps were constantly visible, subjects completed more tasks, were more confident information seekers, dug deeper into the site hierarchy, used the "Back" button less, and estimated more accurately how many separate pages they visited. These are certainly useful results, but users in this experiment were not "browsing" and their behavior was not exploratory in nature, they had specific information that they were required to find. Therefore one must be careful to transfer these findings to all task types. The real issue is what kind of tasks benefit from sitemaps and which ones, if any, are hindered. Pilgrim *et al.* (2002, 2004) found that participants relied on a search tool for closed tasks and relied on a sitemap for open tasks. These findings suggest that goal specificity influences the selection of different types of aids.

The use of overview plus detail interface is another way of providing survey information of cyberspaces. This type of aid is used to provide detail of where the user is within the space, while also providing them with an overview of the space in general, and thought to reduce the workload of remembering where one is within the overall context. There are numerous products containing overview plus detail interfaces and maps, while more novel information spaces are emerging using overview + detail in information spaces. When speed of finding information is of concern, using overviews should be considered carefully as it has been found in one study to reduce speed of navigation (Hornbaek, Benderson and Plaisant, 2001), however in the same study, although speed was reduced, user satisfaction was increased with the use of overview + detail interfaces (Fig. 10.5).

When deciding what type of aid to provide, designers must take into consideration the type of interface that users are interacting with. The use of spatial metaphors in cyberspace, in order to aid navigation, is reported in the literature (Kim and Hirtle 1995; Fuhrman and MacEachren 1999; Westerink *et al.* 2000) however, they may not be suitable for use in all information spaces. It is thought that the more similar the space is geographically to the real world, the more one can rely on tools that are relied on in the real world; however, navigating through hypertext space by providing a sitemap may not be so similar. The real world contains constant Euclidean distance and directions between elements, whereas in hypertext worlds, the relationship between elements is (can be) constantly changing (Dahlback 1998). The type of modes users are working in, needs to be taken into consideration. The more geographic the mode (i.e. virtual environments and maps), the more spatial tools and metaphors can be relied upon.

3.4. Summary of Navigation and Orientation of Information Cyberspace

In summary, it seems that there is information in cyberspace that corresponds to landmark, route, and survey knowledge. It is unclear as to whether there is an equivalent type of tiered or sequential acquisition of

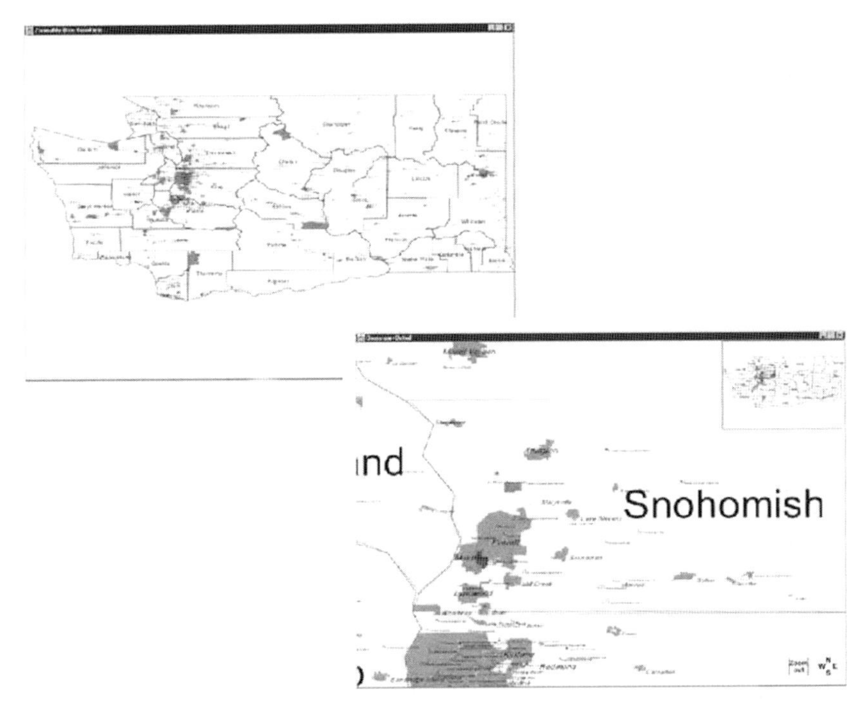

FIG. 10.5. The picture on the left displays an overview map, and an overview + detail map on the right (From Hornbaek *et al.* 2001).

this knowledge to form a mental map of the cyberspace. It may be that since the type of information that is relied upon is dependent on the users' experience and the task at hand that cyberspace may not be acquired in the same way.

4. Similarities and Differences between Real/Virtual and Information Cyberspaces

It is believed that if the domain is "spatial" in nature and contains a concrete common structure (e.g. a geographical site), designers will be able to rely on the "spatial metaphor" and aids relating to it (Dahlback 1998). For example cyberspaces involving information about geographic spaces as opposed to a domain that does not have a commonly agreed upon internal structure such as art or musical genres (Dahlback 1998). Dahlback suggests that perhaps we need to consider the limitations of the spatial metaphor if we want to develop systems that are easy to use for all types of users.

The literature review suggests that orientation and navigation activities are comparable across these environments, and that some aids, proven effective by decreasing or eliminating disorientation in real and virtual environments, have also proven efficient in cyberspace. The more "spatial" and concrete the domain in the cyberspace is, the more comparable these environments become. Regardless, tasks performed are found to be similar between the spaces, and the way in which they are supported in real and virtual spaces can provide, at a minimum, methodologies to further test the support needed for cyberspace tasks.

It is apparent that the aids available to users have strong influences on behavior (Darken and Sibert 1993). Therefore, the more designers are aware of the type of tasks users need to accomplish, the more they will be able to provide specific and multiple tools that will suit the users' needs. The overall goal of this chapter is to inform the design and development of cybercartographic products and to offer preliminary design guidelines aiming to assist software developers to provide applications that will facilitate navigation through multimedia/multimodal electronic environments.

Key similarities and differences of the acquisition, representation, tasks, and aids used in real/virtual in comparison to cyberspace are listed in Table 10.1.

5. Conclusions

To date, it is still not clear how and to what extent spatial cognition is involved in navigating through cyberspace. However, spatial cognition research can certainly provide guidelines as to how to provide navigation support in cyberspace. We believe that there are two main lessons to be learnt from spatial cognition of real and virtual worlds. The first is that while designing navigational aids for cyberspace, the tasks must be taken into consideration in order to choose the appropriate aid. Past research of navigating real and virtual worlds makes it clear that no single aid is good for all situations, major differences exist particularly between open and closed tasks.

The second lesson learnt is that methodologies from spatial cognition research can be used to gain insight into how cyberspaces are acquired, given that there is a lack of knowledge in this regard. For example, Taylor and Tversky (1992) asked people to describe real and virtual environments to others which revealed a great deal as to how people represent the environments. Perhaps if users were asked to either describe the cyberspace or explain how to get from one point in the space to another, insight may be gained into how people cognitively represent cyberspaces.

TABLE 10.1
Similarities and Differences between Real/Virtual and Information Cyberspaces

	Similarities	Differences
Acquisition	Landmark, route, and survey knowledge are all eventually acquired.	There may be differences in terms of the sequence by which landmark, route, and survey knowledge is acquired, if acquired at all.
Representation	Landmarks are represented in the same way by acting as transition points for action. Survey knowledge will only be gained by experience over an extended period of time if an overview of the space is not provided.	Cyberspace is dynamic and can constantly change. Cyberspace is lacking Euclidean and metric properties of real/virtual space and therefore the way the spaces are represented may be different. Mental rotation is not required in all types of cyberspaces and therefore one may not need to acquire perspective-independent survey knowledge.
Tasks	Goal oriented searching tasks or "closed" tasks are found in both environments and can be further categorized into targeted, primed and naïve searches. Exploratory or browsing tasks are also found in both environments.	
Aids	Landmarks are important, salient cues in the environments and aid in deciding where to go next. Closed tasks have been found to benefit most from landmarks as aids. Maps decrease acquisition time by allowing one to gain an understanding of how the space is organized. In both environments open tasks benefit most from overview maps.	Route information is generally not provided for cyberspace users, besides personally generated ones, such as bookmarks or histories, but is depended on in real/virtual worlds for simple primed searches and tasks with high cognitive workloads such as driving.
Performance		No significant correlation between navigation performance in real/virtual spaces and information cyberspace.

Overall, we believe that our conceptual framework of spatial cognition in real/virtual worlds can be applied to information cyberspaces. Certainly context in terms of tasks and aids is similar. In terms of the acquisition phases, there seem to be some difference between the two space types, particularly due to the differences in the representations. Yet, in spite of these differences, we can offer some general guidelines based on this analysis.

5.1. General Guidelines

Based on the comparison between real/virtual environments and information cyberspaces, we propose the following preliminary guidelines to support orientation and navigation in hypermedia cyberspace:

1. The more "spatial" and concrete the domain that is being represented in cyberspace, the more designers can utilize a "spatial metaphor".
2. Information spaces usually lack an inherent feature of geographic space, namely the explicit information of distance and direction. Users in cyberspace should be given "you-are-here" location information in sitemaps that tracks the users' movements through the application.
3. Given that landmarks are used to acquire the space as well as forming routes to wayfind, they should also be used throughout the information space. If the environment is spatial in nature, landmarks similar to ones found in real/virtual worlds should be applied. If the environment does not make use of a spatial metaphor, the landmarks should be of another nature. In either case, landmarks should be made salient, memorable, and distinguishable from other objects in the environment.
4. The nature and location of landmarks in hypermedia cyberspace should not be affected by the frequent changes that can occur in cyberspace.
5. Users should be allowed to control the placement of landmarks whenever and wherever possible.
6. Route information is required in real/virtual worlds for simple primed searches and tasks with high-cognitive workloads, such as driving. Route information is generally not provided for cyberspace users, besides personally generated ones relying on previously visited nodes and links, such as bookmarks or histories. Route information should be provided for users which should decrease the problem of primed searching, knowing what one is looking for but not where to find it.

7. When speed of finding information is of concern, consider using overviews carefully as it has been found in one study to reduce the speed of navigation.
8. If user satisfaction is of concern, the use of spatial overviews should definitely be considered.
9. Ensure effective, effortless translation between two-dimensional and three-dimensional views.
10. Enable presentation of displays to be schematized.
11. Ensure that route information is readable, clear, complete, and convenient.

Cybercartographic Atlases, as described in this volume deal with hypermedia cyberspace and these guidelines have been useful in the production of both the Cybercartographic Atlas of Antarctica (Pulsifer and Taylor in Chapter 20) and the Cybercartographic Atlas of Canada's Trade with the World (Eddy and Taylor in Chapter 22).

References

Agrawala, M. and C. Stolte (2000) "A design and implementation for effective computer-generated route maps", *Proceedings of Smart Graphics'00*, accessed 6 January 2005 from http://graphics.stanford. edu/papers/maps/

Arthur, E. J. and P. A. Hancock (2001) "Navigation training in virtual environments", *International Journal of Cognitive Ergonomics*, Vol. 5, No. 4, pp. 387–400.

Bederson, B. B. (2001) "Photomesa: a zoomable image browser using quantum treemaps and bubblemaps", *UIST 2001, ACM Symposium on User Interface Software and Technology, CHI Letters*, Vol. 3, No. 2, pp. 71–80.

Benyon, D. and K. Höök (1997) "Navigation in information spaces: supporting the individual", in S. Hammond, H. J. and G. Lindgaard. (eds.), *Human-Computer Interaction*: *INTERACT'97*, July, Chapman and Hall, pp. 39–46.

Bernard, M. (1999) "Using a sitemap as a navigational tool", in *Usability News*, 1, 1, accessed 6 January 2005 from http://psychology.wichita.edu/ surl/usability news/1w/Sitemaps.htm

Boechler, P. M. (2001) "How spatial is hyperspace? Interacting with hypertext documents: cognitive processes and concepts, *Cyberpsychology and Behavior*, Vol. 4, pp. 23–46.

Chen, C. and M. Czerwinski (1997) "spatial ability and visual navigation", *New Review of Hypermedia and Multimedia*, Vol. 3, pp. 67–89.

Dahlback, N. (1998) "On spaces and navigation in and out of the computer", in Dahlback, N. (ed), *Exploring Navigation: Towards a Framework for Design and Navigation in Electronic Spaces,* Technical Report T98:01, Swedish Institute of Computer Science, Kista, Sweden, pp. 15–29.

Danielson, D. R. (2002) "Web navigation and the behavioral effects of constantly visible site maps", *Interacting with Computers*, Vol. 14, pp. 601–618.

Darken, R. P. and J. L. Sibert (1993) "A toolset for navigation in virtual environments", *Proceedings of UIST* 93, pp. 157–165.

Darken, R. P. and H. Cevik (1999) "Map usage in virtual environments: orientation issues", *Proceedings of IEEE Virtual Reality*, pp. 133–140.

Darken, R. P. and B. Peterson (2001) "Spatial orientation, wayfinding, and representation", in Stanney, K. (ed.), *Handbook of Virtual Environments: Design, Implementation, and Applications*, Lawrence Erlbaum Associates, Inc, Mahwah, NJ, pp. 493–518.

Edwards D. M. and L. Hardman (1988) "Lost in hyperspace: cognitive mapping and navigation in a hypertext environment", *UK Hypertext*, pp. 105–125.

Elvins, T. T. (1997) "Virtually lost in virtual worlds: wayfinding without a cognitive map", *Computer Graphics*, August, pp. 15–17.

Elvins, T. T., D. R. Nadeau and D. Kirsh (1997) "Wordlets-3D thumbnails for wayfinding in virtual environments", *Proceedings of UIST* 97, Banff, AB, pp. 21–30.

Elvins, T. T., D. R. Nadeau, D. Kirsh and R. Schul (1998) "Wordlets: 3D thumbnails for 3D browsing", *Proceedings of CHI 98*, pp. 163–170.

Foss, C. L. (1989) "Tools for reading and browsing hypertext", *Information-Processing and Management*, Vol. 25, No. 4, pp. 407–418.

Fuhrmann, S. and A. M. MacEachren (1999) "Navigating desktop geovirtual environments", *Proceedings IEEE Information Visualization 99, Late Breaking Hot Topics Proceedings,* San Francisco, CA, pp. 23–28.

Hirtle, S. C. (1999) "Design of navigational systems", in Gero, J. S. and B. Tversky (eds.), *Visual and Spatial Reasoning in Design*, Key Centre of Design Computing and Cognition, University of Sydney, Sydney, Australia.

Hirtle, S. C. and M. E. Sorrows (1998) "Designing a multi-modal tool for locating buildings on a college campus", *Journal of Environmental Psychology*, Vol. 18, pp. 265–276.

Hornbaek, K., Benderson, B., and Plaisant, C. (2001). "Navigation Patterns and Usability of Overview + Detail and Zoomable User Interfaces for Maps", Technical Report ## 2001–11, HCIL, University of Maryland, 26 pages, accessed October 18, 2005 at http://www.diku.dk/~kash/

Hornbaek, K., B. B. Bederson, and C. Plaisant (2002) "Navigation patterns and usability of zoomable user interfaces with and without an overview", *ACM Transactions on Computer-Human Interaction*, Vol. 9, pp. 362–389.

Jul, S. and G. W. Furnas (1997) "Navigation in electronic worlds: a CHI 97 workshop report", *ACM SIGCHI bulletin*, Vol. 29, No. 4, pp. 44–49.

Kim, H. and S. C. Hirtle (1995) "Spatial metaphors and disorientation in hypertext browsing", *Behaviour and Information Technology*, Vol. 14, pp. 239–250.

Levine, M. (1982) "You-are-here maps: psychological considerations", *Environment and Behavior*, Vol. 14, pp. 221–237.

Lovelace, K. L., M. Hegarty and D. R. Montello (1999) "Elements of good route directions in familiar and unfamiliar environments", *Spatial Information Theory*, Vol. 1661, pp. 65–82.

Lynch, K. (1960) *The Image of a City*, MIT Press, Cambridge, MA.

Maguire, E. A., N. Burgess, J. G. Donnett, R. S. J. Frackowiak, C. D. Frith and J. O'Keefe (1998) "Knowing where and getting there: a human navigation network", *Science*, Vol. 280, pp. 921–924.

McDonald, S. and R. J. Stevenson (1998) "Navigation in hyperspace: an evaluation of the effects of navigational tools and subject matter expertise on browsing and information retrieval in hypertext", *Interacting with Computers*, Vol. 10, pp. 129–142.

Moeser, S. D. (1988) "Cognitive mapping in a complex building", *Environment and Behavior*, Vol. 20, pp. 21–49.

Montello, D. R. (1998) "A new framework for understanding the acquisition of spatial knowledge, in large-scale environments", in Egenhofer, M. J. and R. G. Golledge (eds.), *Spatial and Temporal Reasoning in Geographic Information Systems*, Oxford University Press, New York, pp. 143–154.

Nillson, R. M. and R. E. Mayer (2002) "The effects of graphic organizers giving cues to the structure of a hypertext document on users navigation strategies and performance", *International Journal of Human-Computer Studies*, Vol. 57, pp. 1–26.

O'Keefe, J. and L. Nadel (1978) *The Hippocampus as a Cognitive Map*, Clarendon Press, Oxford.

Parush, A. and D. Berman (2004) "Navigation and orientation in 3D user interfaces: the impact of navigation aids and landmarks", *International Journal of Human Computer Studies*, Vol. 61, No 3, pp. 375–395.

Pilgrim, C. J., Y. K. Leung and G. Lindgaard (2002) "An exploratory study of WWW browsing strategies", *Asia Pacific Computer Human Interaction Conference, APCHI 02*, Beijing, November 2002.

Pilgrim, C. J., G. Lindgaard and Y. K. Leung (2004) "An investigation into factors influencing users selection of WWW sitemaps", *Asia Pacific Computer Human Interaction Conference, APCHI 04*, New Zealand, June 2004.

Rumpradit, C. and M. L. Donnell (1999) "Navigational cues on interface design to produce better information seeking on the World Wide Web", *Proceedings of the 32nd Hawaii International Conference on System Sciences,* accessed January 5 2005 from http://csdl.computer.org/ comp/proceedings/hicss/ 1999/0001/05/00015028.PDF

Satalich, G. A. (1995) Navigation and wayfinding in virtual reality: Finding proper tools and cues to enhance navigational awareness, MS Thesis, University of Washington, accessed on 6 January 2005 from http://www.hitl.washington. edu/publications/satalich

Siegel, A. W. and S. H. White (1975) "The development of spatial representation of large-scale environments", in Reese, H. W. (ed.), *Advances in Child Development and Behavior*, pp. 9–55.

Skopik, A. and C. Gutwin (2003) "Finding things in fisheyes: Memorability in distorted spaces", *Proceedings of the 2003 Conference on Graphics Interface* (GI'03), Halifax, NS.

Streeter, L. A., D. Vitello and S. A. Wonsiewicz (1985) "How to tell people where to go: comparing navigational aids", *International Journal of Man/Machine Studies*, Vol. 22, pp. 549–562.

Stoakley, R., M. J. Conway and R. Pausch (1995) *Virtual Reality on a WIM: Interactive Worlds in Miniature, CHI 95* Mosaic of Creativity.

Tan, D. S., G. G. Robertson and M. Czerwinski (2001) "Exploring 3D navigation: Combining speed-coupled flying with orbiting", *Proceedings of CHI 2001, CHI Letters*, Vol. 2, No. 1, pp. 418–425.

Taylor, H. A. and B. Tversky (1992) "Spatial mental models derived from survey and route descriptions", *Journal of Memory and Language*, Vol. 31, pp. 261–292.

Thorndyke, P. W. and S. E. Goldin (1983) "Spatial learning and reasoning skill", in Pick, H. L. and L. P. Acredolo. (eds.), *Spatial Orientation: Theory, Research, and Application*, Plenum Press, New York, pp. 195–217.

Thorndyke, P. W. and B. Hayes-Roth (1982) "Differences in spatial knowledge acquired from maps and navigation", *Cognitive Psychology*, Vol. 280, pp. 560–589.

Tolman, E. C. (1948) "Cognitive maps in rats and men", *Psychological Review*, Vol. 55, 189–209, accessed 6 January, 2005 from http://psychclassics.yorku.ca/Tolman/Maps/maps.htm

Tversky, B. (2003) "Structures of mental spaces: How people think about apace", *Environment and Behavior*, Vol. 35, pp. 66–80.

Tversky, B. (2001) "Multiple mental spaces", in Gero, J. S., B. Tversky and T. Purcell (eds.), *Visual and Spatial Reasoning in Design*: Key Centre of Design Computing and Cognition, Sydney, Australia, pp. 3–13.

Tversky, B. (1993) "Cognitive maps, cognitive collages, and spatial mental models", *Proceedings of European Conference, COSIT 93: Spatial Information Theory – A Theoretical Basis for GIS* , Lecture Notes in Computer Science, Springer-Verlag, Marciana Marina, Elba Island, Italy, pp. 14–24.

Tversky, B. (1991) "Spatial mental models psychology of learning and motivation-advances", *Research and Theory*, Vol. 27, pp. 109–145.

Tversky, B. (1981) "Distortions in memory for maps", *Cognitive Psychology*, Vol. 13, No. 3, pp. 407–33.

Tversky, B., N. Franklin, H. A. Taylor, and D. J. Bryant (1994) "Spatial mental models from descriptions", *Journal of the American Society for Information Science*, Vol. 45, pp. 656–668.

Tversky, B. and D. Schiano (1989) "Perceptual and conceptual factors in distortions in memory for maps and graphs", *Journal of Experimental Psychology: General*, Vol. 118, pp. 387–398.

Vinson, N. G. (1999) "Design guidelines for landmarks to support navigation", *Virtual Environments*, CHI 99, pp. 278–285.

Westerink, J. H. D. M., B. G. M. M. Majoor and M. D. Rama (2000) "Interacting with infotainment applications: navigation patterns and mental models", *Behaviour and Information Technology*, Vol. 19, pp. 97–106.

Yeap, W. K. and M. E. Jefferies (2001) "On early cognitive mapping", *Spatial Cognition and Computation*, Vol. 2, pp. 85–116.

CHAPTER 11

Cybercartography: A Multimodal Approach

PATRICIA LEEAN TRBOVICH

Human Oriented Technology Lab, Carleton University, Ottawa, Ontario, Canada

GITTE LINDGAARD

Professor, Natural Sciences and Engineering Research Council (NSERC) Industry Chair User-Centred Design, and Director, Human Oriented Technology Lab (HOTLab), Department of Psychology, Carleton University, Ottawa, Canada

THE LATE DR RICHARD F. DILLON

Human Oriented Technology Lab (HOTLab), Department of Psychology, Carleton University, Ottawa, Canada

Abstract

Cartography is primarily a visual discipline. The concept of cybercartography, however, helps move cartography beyond the visual modality with the ultimate goal of using all sensory modalities. To this end, cybercartography permits us to migrate from a unimodal visual approach to a multimodal approach capitalizing upon combinations of sensory modalities, such as visual, auditory, and touch. The main reasons for developing multimodal interfaces are to help users achieve more efficient, natural, usable and easy to learn ways of interacting with computer applications, such as electronic maps. Furthermore, by offering users more extensive computing capabilities, multimodal interfaces permit more empowering applications relative to traditional interfaces. Such empowerment would be exemplified by a dynamic electronic map system, which could be interacted with using auditory, touch, and visual senses. Research is less advanced concerning the relationships between the map symbols and their meaning for nonvisual modalities (e.g. sound and haptic) than for the visual modality. This chapter examines the potential contribution of four nonvisual modalities to enhanced cartographic visualization. Specifically, these modalities are: speech, gesture, sound, and haptic (i.e. touch). Each of these four modalities is discussed in terms of their: (1) relative advantages and limits; (2) unimodal application; (3) multimodal application; and (4) cognitive workload implications.

1. Introduction

Traditional hard copy maps are not interactive and do not allow for user customization. Rather than using traditional hard copy maps, people are turning to cybercartography. That is, they are adopting new interactive and dynamic map systems that can be accessed from a variety of sources (e.g. online maps, in-vehicle route-navigation assistance devices) (Biocca *et al.* 2001; Clarke 2001; Gartner in this volume). These dynamic map systems permit the user to browse, navigate, and perceive information from displays that are data intensive. Furthermore, through the improvement of database search parameters, these map systems are allowing users to retrieve and filter information as they please, depending on the goal they are trying to meet (Egenhofer 1990; Williamson and Schneiderman 1992; Ahlberg and Schniederman 1994; Goldstein and Roth 1994; Fishkin and Stone 1995; Roth *et al.* 1997). Dynamic and interactive map systems are becoming increasingly popular and their functionality is expanding. The information provided by these systems may provide a new user experience. For example, the effect of wind direction and speed, currents, and the location of reefs may be provided to predict navigability of an ocean strait.

Cartography is primarily a visual discipline. The major goal when designing hard copy maps is to reduce the visual search time (Phillips 1979). Current dynamic map systems are reducing the design problems associated with visual search. The new generation of online maps reduces visual search time, by affording an active workspace that allows the user to directly manipulate the display by scrolling and zooming in and out (Kreitzberg 1991; Bederson and Hollan 1994). Further reduction of search time is achieved by using techniques such as transparency and blur. Using these latter techniques, dynamic maps can guide users' attention while maintaining the overall context of the maps intense information (Colby and School 1991; Lokuge and Ishizaki 1995). Interactive maps afford further potential for reduction of visual search times by permitting migration from a unimodal visual approach to a multimodal approach capitalizing upon combinations of sensory modalities, such as visual, auditory, and touch (Jacobson *et al.* 2002). That is, the concept of cybercartography helps move cartography beyond the visual modality with the ultimate goal of using all sensory modalities.

In the general sense, a multimodal interface supports communication with the user through various modalities. Literally "multi" refers to "more than one" and "modal" refers to the notion of "modality." In this chapter, modality is defined as the way in which communication is achieved (i.e. type of communication channel) to convey or acquire information. It is crucial to understand how to combine modalities in an interface to render the interaction "natural" and flowing without the user suffering information overload. This is especially true with the mobile devices of telecartography as described by Gartner in this volume.

The main reasons for developing multimodal interfaces are to help users achieve more efficient, natural, usable and easy to learn ways of interacting with computer applications, such as electronic maps (Benoit *et al.* 2000; Oviatt and Cohen 2000). Furthermore, by offering users more extensive computing capabilities, multimodal interfaces permit more empowering applications relative to traditional interfaces. Such empowerment could be achieved through the use of cybercatographic visualization techniques to create dynamic electronic map systems, which could be interacted with using auditory, touch, and visual senses (Hedley 2003).

Cybercartographic visualization is a term, which denotes the formation of mental models of geographic phenomena (e.g. the Great Lake system of Canada), which would otherwise be too large to observe. Individuals use maps to develop a mental model of geographic spatial relationships, such as distances between different cities, trade patterns, and movement of weather formations. Researchers suggest that visualization involves more than the development of visual graphics (Caivano, 1990; MacEachren 1992, 1995; Fisher, 1994; Flowers and Hauer 1995; Wheless *et al.* 1996; Flowers *et al.* 1997; Pereverzev *et al.* 1997; Cronly-Dillon *et al.* 1999; Griffin, 2001). The quality of visualization can be greatly enhanced through the integration of information communicated through different sensory modalities (e.g. vision, auditory, and touch). Thus, multimodality from a cartographic point of view is the capacity of a map to facilitate visualization of geographic information, using different types of communication channels.

Bertin (1983) enumerated the visual variables that cartographers should use when designing maps. These variables include: geographic position in the plane, size, value, texture, color, orientation, and shape. Bertin proposed rules for the use of these variables, based on whether the data was quantitative, nominal, or ordinal. His rules reflected his view that the perceptual characters of the variables and data should match. His visual variables syntax, however, was not empirically tested.

Subsequent researchers supplemented Bertin's syntax (e.g. Caivano, 1990; MacEachren, 1992, 1995). For example, in response to the impact of computer technological innovations, MacEachren (1995) added the following visual variables: hue, saturation, crispness, resolution, and transparency. This supplemental work has resulted in a better match between the objectives of the map designer and the map-user's perceptual capabilities. Thus, it is important to develop a commonly accepted visual variable syntax for cartography. Such a common base will promote consistent usage of map symbols amongst cartographers and interpretation amongst map-users.

Research is less advanced concerning the relationships between the map symbols and their meaning for nonvisual modalities (e.g. sound and haptic) than for the visual modality. This chapter examines the potential contribution of four nonvisual modalities to enhanced cartographic visualization. Specifically, these modalities are: speech, gesture, sound, and haptic (i.e. touch). Each of these four modalities is discussed in terms of their: (1) relative advantages and limits; (2) unimodal application; (3) multimodal application; and (4) cognitive workload implications.

2. Speech and Cartographic Visualization

Inclusion of speech in the design of online interactive maps has the potential to enhance visualization of cartographic information. Traditional maps are replete with visual and graphical information. Speech offers a promising way to interact with online dynamic maps, rather than solely relying on graphical and visual cues.

2.1. Advantages and Limitations

The speech modality affords rapid transmission of high bandwidth information and ease of use. Moreover, speech affords portability and allows the hands and eyes freedom to perform other tasks. This latter attribute facilitates performance of mobile and field work. Speech is preferable for description of: events, objects, hidden objects, conjoint information, and time status (Oviatt and Cohen 1991; Cohen and Oviatt 1995; Oviatt 2000). Likewise, it is efficient for issuance of action commands or iterative actions.

Speech should be assigned to tasks in a consistent manner. That is, certain task components should always be done via speech. Speech is not a good modality to use for describing the position or for manipulating objects. Speech, for example, is not a good input modality for commands such as "up," "down," "left," and "right." These commands need to be repeated iteratively to accomplish the desired action. The mouse, cursor key or other pointing devices are better suited to this type of task.

2.2. *Unimodal Applications*

There are a number of unique considerations with respect to systems that issue voice commands to users (Streerter *et al.* 1986; Freundschuh 1989; Kishi and Siguira 1993; Ross and Brade 1995; Takahashi and Takezawa 2002). Design of voice-guided interfaces (i.e. talking interfaces) should reflect appropriate diction and enunciations to ensure: (1) differentiation between similar sounding words; (2) clarity and coincidence as to the bearing and direction of both the user and the device (i.e. they both must have the same orientation perspective); and (3) a manageable cognitive information load. Kishi and Sugiura (1993) demonstrated these three considerations with respect to route-guidance systems that issue voice commands:

1. In Japanese, the terms for left turn (i.e. sasetsu) and right turn (i.e. usetsu) are difficult to discriminate;
2. The voice-guidance device should verbalize direction changes (e.g. "turn left"), relative to the driver's perspective of continuing straight along the route; and
3. Cognitive load can be minimized by restricting voice message duration to less than 7 seconds.

Kishi and Suguira (1993) conducted an experiment to compare user interaction with a route-guidance visual display system, and the same system supplemented by voice output. The average duration of visual fixation on the display with voice-guidance was significantly shorter, than without voice-guidance, $F(1, 8) = 26, 48$, $p < .01$. The average fixation frequency per minute with voice-guidance was also significantly lower than without voice-guidance, $F(1, 8) = 27, 76$, $p < .01$. These results are consistent with Streerter *et al.* (1986), who found that driver performance with voice-route-guidance was better than sole reliance on a map display. In sum, effectiveness of systems that issue voice commands is dependent upon: (1) provision of audible and comprehendible synthetic voice output; (2) appropriate timing at which the information is presented; and (3) appropriate amount of information delivered to the user.

Nass and Gong's (2000) paper provides experimental results and design implications for speech interfaces. They address the psychology underlying the way people react when confronted by a talking computer, and the design implications for such reactions. Participants in their study showed more socially acceptable and more cautious responses to a talking computer, when the input modality was speech versus mouse, keyboard, or pen. Results demonstrated that people "hyper-articulate" when they are talking but are not being understood. That is, participants spoke more slowly, louder, and paused for longer times between words. Oviatt *et al.* (1998) showed similar results. That is, their results showed that when the computer failed to recognize what was being said, participants elongated words and spoke with more precise diction. Thus, speech interface design should consider that users might interact with these systems by being more cautious in their answers and by hyper-articulating.

2.3. *Multimodal Applications*

Terken and Verhelst (2000) demonstrated that speech in combination with text resulted in superior communication relative to speech or text only. They observed that this result held despite the fact that the timing of speech and text presentation were asynchronous. They concluded that this result was attributable to constructive cross-modal compensation. That is, the limits to processing information via one modality are compensated by the information processing advantages of the other modality. Simply put, once the speech output is terminated, readers can still examine the text.

Although more than 20 years old, the original multimodal graphical interface demonstrated in Bolts (1980) "Put-That-There" work, is still an excellent example of an effective speech and gesture multimodal interface (see Fig. 11.1). The interface comprised a big room, with a back projection

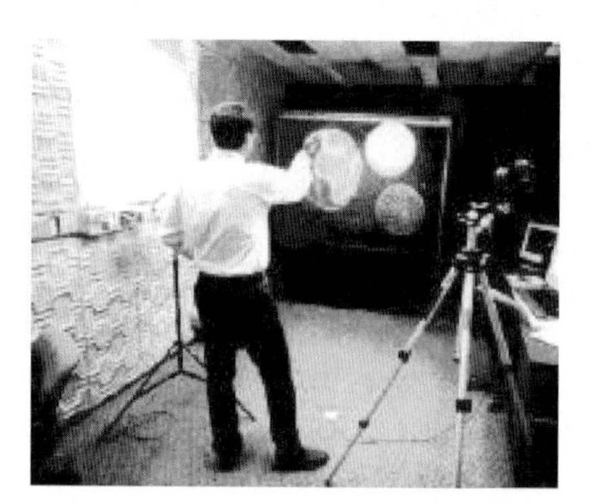

FIG. 11.1. An example of user interacting with the modern equivalent of "Put that there". This is from the Human Interface Technology Laboratory Web Site, http://www.siggraph.org/publications/newsletter/v32n4/contributions/billinghurst.html.

panel on a wall. Users were seated in the center of the room. To measure hand position, a magnetic position-sensing device was worn on users' wrists. Users were asked to move, add, or delete graphic objects on the wall projection panel. The computer responded to the users' commands by using both speech and gesture recognition, and by considering the current context. Such current context, or implicit situational information, could be communicated to the computer by pointing to the left of the projection panel, or verbally indicating the left of the panel. Therefore, neither the speech nor the gesture recognition had to be perfect, as long as they converged on the users' intended meaning. This early work demonstrated the importance of using contextual knowledge to get around the problem of imperfect speech or gesture recognition.

Another study that showed the importance of integrating speech and gesture with contextual understanding was Salisbury *et al.*'s (1990) "Talk and Draw: bundling Speech and Graphics" multimodal workstation for military operations. In this study, voice commands were integrated with direct manipulation of graphical objects using a mouse. Where speech recognition was incomplete, knowledge of the graphical context was used to prompt the user for relevant input. Thus, these studies underline the importance of contextual understanding in the design of multimodal interfaces.

Improving the computer's access to context, to enrich the communication in human–computer interaction, is especially important in multimodal interactive applications, such as handheld or ubiquitous computing or telecartography (Dey and Abowd 2000; Gartner in Chapter 16). Humans are good at using implicit situational information, or context, when talking with other humans. However, this ability does not transfer well when humans interact with computers. Therefore, effort must be made to enable computers to fully profit from the context underpinning human-computer dialogue. Oviatt (1997) illustrates the complementarity of speech and pen in ensuring accurate communication of context, when working with interactive maps.

Oviatt's (1997) research on interactive maps revealed advantages of multimodal (i.e. speech and pen) interfaces relative to speech only interaction. Under both conditions, participants undertook realistic tasks using a simulated map system. Results demonstrated in the multimodal approach exhibited the following advantages: (1) 10% more rapid task completion; (2) 36% less task-critical content errors; (3) briefer and simpler sentence structure, especially less complex spatial descriptions; (4) 50% reduction in fluency impediment; and (5) over 95% user satisfaction/preference.

Moreover, Oviatt's work demonstrated that performance of map tasks imposes a major demand upon cognitive spatial resources. Higher performance error rates in the speech condition versus the multimodal condition, were attributed to participants' difficulty in articulating spatially oriented descriptions. Oviatt concludes that optimal design of mobile interactive electronic map systems should adopt pen-voice input to afford users the flexibility to balance accuracy, efficiency, and cognitive load.

2.4. Cognitive Workload Implications

Several psychological factors must be considered when integrating speech recognition into graphics applications. The ability to perform certain tasks can be affected by the modalities in which the tasks are being performed. Consideration must be given to the cognitive interference that can be caused by performing tasks that require the same processing resources. For example, working memory can be conceptualized as a limited resource system (Baddeley and Hitch 1974). Baddeley (1981) proposed that working memory consists of a central executive, a phonological loop and a Visual-Spatial Scratch Pad (VSSP). The central executive is in charge of coordinating the actions of the phonological loop and the VSSP. The phonological loop is responsible for mental rehearsal, and is involved in temporary storage of verbal information (e.g. Baddeley and Logie 1992). The VSSP acts as a mental blackboard and serves as a temporary storage for visual and spatial material (Logie 1986).

Given that working memory can be conceptualized as a limited resource system, it is possible that two tasks being performed immediately after one another, or concurrently, require the same processing resources. This competitive demand for cognitive resources can cause decrements in tasks performance. Multimodal interfaces might help alleviate the problems caused by tasks that require the same processing resources, by permitting use of different modalities to perform various tasks (Hedley 2003). A multimodal interface may permit simultaneous performance of visual and verbal tasks, with little cognitive interference between the two modalities. For example, Martin (1989) showed that users performance improved by 108%, when using a Computer Assisted Design (CAD) program supplemented by a speech input device versus a traditional (i.e. strictly visual) CAD interface. Addition of the voice input permitted users to focus visually on the screen, while outputting speech commands.

3. Gesture and Cartographic Visualization

Gesture interaction with a computer may be defined as human interaction with a computer in which body movements, usually hand motions used to indicate direction and location, are recognized by the computer. Often, gesture techniques employ the use of an electronic pen or stylus, in addition to pointing or touching with the finger (see Fig. 11.2). Use of gesture has

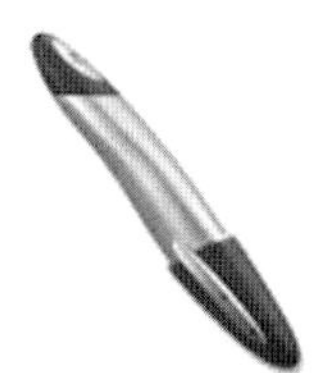

FIG. 11.2. An example of a digital pen. This is from the Techadvice Web Site, http://www.techadvice.com/info/item.asp?cid=413&pid=844&iid=16108

the potential to facilitate interaction with online dynamic maps. In combination with other modalities, gesture has the ability to accelerate visualization of geographic information.

3.1. Advantages and Limitations

Computer recognition of gesture input can make computers more accessible for the physically impaired, and render human–computer interactions more intuitive. It is important, however, to carefully choose the tasks that gesture input will be used to perform. Although gesture input is natural for certain navigation and direct manipulation tasks, it is inappropriate for tasks that require precision (Billinghurst 1998).

Gesture/pen interactions afford greater privacy in public situations, and offer an effective alternative, where voice communication is impaired due to noise. Pen input is highly practical in mobile and field operations, and therefore, affords portability. Pen permits writing or drawing of symbols, signs, digits, abbreviations, gestures, and graphics. Furthermore, pen may be used to point to select visible objects like the mouse does in direct manipulation interfaces. Pen input allows users to draw simple gestures such as scribbling over a word to delete it. Therefore, gestures can integrate the selection of the object to be deleted and the "delete" command. Another example of how gesture integrates two different commands is the moving of an object by circling it and dragging in to a new location (Hinckley *et al.* 2002).

Presently, pen technology is used on handheld, mobile phones, and PDAs for handwriting and gesture input. Pen-based computers support visual-spatial applications, such as: map-based interaction, sketching, flow chart design, user interface design, and circuit design. For map and other graphic displays, pen input is particularly useful for imparting spatial relations concerning precise points, lines, and areas.

For map displays, the fact that a pen input can deliver drawn graphics is beneficial because it allows for increased functionality of the map.

Tele-strators employed by TV sports commentators, such as in American football, exemplify this type of application. For example, the TV replay screen will be frozen as a still image. In turn, players will be highlighted with circles, and their ensuing movements indicated by pen drawn arrows. This example illustrates the power of the pen input interface in communicating spatial information. In sum, the pen input modality allows users to easily and efficiently communicate spatial information about points, lines and regions of interest on a map display.

3.2. Unimodal Applications

Wolf (1992) explored the use of pen-based interface techniques in conducting tasks on a spreadsheet. Three conditions were compared. In the first condition, a conventional keyboard was used to perform spreadsheet (Lotus 1-2-3) tasks. The second and third conditions used pen-based techniques. In the second condition, participants marked directly on the screen (i.e. with pen/stylus). In the third condition, an indirect pen-based technique was used. Specifically, participants used a stylus that was operated on an opaque tablet, which was physically separate from the display. Results showed that participants were 76% faster in completing the spreadsheet tasks when they used the pen-based techniques than when they used the keyboard. There was no significant difference, however, between the direct and the indirect pen-based techniques. Caution must be used when interpreting these results, however, because a smaller display was used in the keyboard condition than in the other two conditions. Therefore, it is possible that these results reflect a manipulation of readability of displays (i.e. small displays are harder to read than larger displays) more than the use of pen-based versus keyboard entry techniques.

3.3. Multimodal Applications

Researchers (Cohen 1992; McNeill 1992; Oviatt *et al.* 1997; Billinghurst 1998) have examined why we should think of using speech and gesture interfaces, as opposed to common desktop interfaces such as keyboard and mouse. As an ensemble, speech and gesture permit alternative ways of describing events, spatial layouts, objects, and their relationship to one another. The capabilities of speech and gesture are complementary. For example, examination of the relative roles of speech and gesture, during human communication, reveals that speech is more frequently used to convey the subject, verb, and object. Conversely, gesture is used to describe the relative location of things (McNeill 1992; Oviatt *et al.* 1997).

Hauptmann and MacAvinney (1993) evaluated a speech and gesture recognition system, by having users interact with typical graphics. They found that users preferred the combined speech and gesture interface (71%), relative to either modality alone (i.e. speech only: 13% and gesture only: 16%). They attributed this result to the greater expressiveness afforded by the combination of modalities. Hauptmann and MacAvinney (1993) conclude that the more spatial the task, the more users prefer a multimodal input system. When users were confronted with a verbal task, 56% preferred a combination of pen and voice interface over either modality alone. When asked to perform a numerical task, 89% of users preferred the pen–voice interface over either modality alone. Similarly, all users (100%) preferred the pen–voice interface to either modality alone, when performing a spatial map task. Thus, these results suggest that users prefer multimodal input devices to voice or gesture alone, when performing common tasks in graphic applications.

There are various advantages in combining speech and gesture modalities to graphical applications. First, this combination of modalities improves recognition accuracy and the time taken to input speech (Billinghurst 1998). For example, Oviatt and VanGent (1996) found that a combined speech and gesture input produced 10% faster task completion times than a speech only input device. Second, combined speech and gesture inputs reduce the number of errors associated with performing tasks. Oviatt and VanGent (1996) showed that a multimodal (voice and gesture) input produced 36% reduction in task errors. This is consistent with the earlier findings of Bolt in 1987. Specifically, Bolt (1987) found that using a combination of speech and gesture for input resulted in one modality compensating for the weakness in the other modality. Thus, a combination of speech and gesture interface allows users to complete their tasks faster and with fewer errors than a speech only interface.

3.4. Cognitive Workload Implications

A design challenge when using gesture input techniques is the number of gestures to include (Zeleznik *et al.* 1996). As the number of gestures increases, it becomes increasingly difficult for users to remember, learn, and correctly perform the gestures. Thus, although gesture/pen technology affords portability and flexibility, it is limited in the number of commands it can support. To overcome this challenge, multimodal interfaces often combine speech input with pen gestures. Combination of pen and voice input can reduce user's cognitive load by affording users the flexibility to use either modality (Oviatt 1997).

4. Sound and Cartographic Visualization

The use of sound in the design of online interactive maps offers potential for visualization of cartographic information, as it takes advantage of nonvisual human perceptual and cognitive resources. Researchers have developed applications that use sonic variables to represent geographic data (Fisher 1994; Flowers and Hauer 1995; Wheless *et al.* 1996; Flowers *et al.* 1997; Pereverzev *et al.* 1997; Cronly-Dillon *et al.* 1999; Holland *et al.* 2002). The auditory modality affords a third dimension for communication and visualization of data, traditionally displayed on two-dimensional maps.

Many researchers have suggested that abstract sounds (i.e. nonspeech sounds) can be used to improve performance and usability of human–computer interfaces (e.g. Gaver *et al.* 1991; Blattener *et al.* 1992; Brewster 1994; Brewster *et al.* 1996; Barras and Kramer 1999; Holland *et al.* 2002). The abstract sounds are structured audio messages called earcons (Brewster 1994; see Fig. 11.3). Earcons are abstract synthetic tones that can be used in structured combinations to create sound messages to represent parts of an interface. Such combinations employ the following basic elements of sound: duration, location, loudness, pitch, timbre, and rhythm. By varying the complexity of combinations of sound elements, hierarchical earcons can

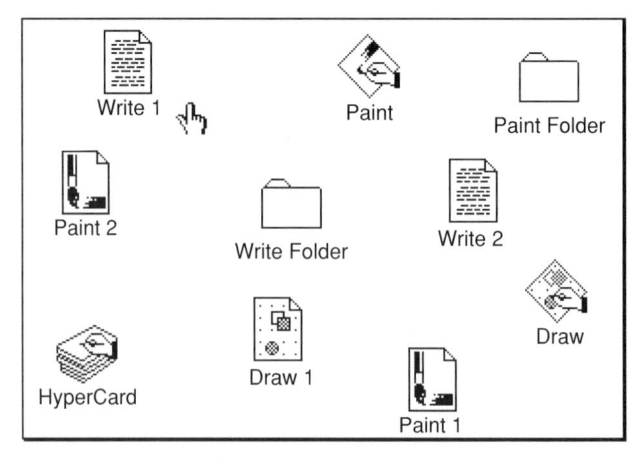

FIG. 11.3. Examples of Earcons by "Glasgow Interactive Systems Group". To listen to these earcons, go to the following Web site, http://www.dcs.gla.ac.uk/~stephen/earconexperiment1/earcon_expts_1.shtml and click on these images.

be created representing hierarchical map data structures (Blattner *et al.* 1992; Brewster *et al.* 1996).

Abstract sound elements may be referred to as sound variables. Just as Vasconcellos (1991) translated Bertin's (1983) visual variables into haptic variables, Bertin's variables may be translated into sound variables. This "sound variables" approach provides a useful heuristic for integrating sound into the design of cartographic visualization interfaces (Krygier 1994).

4.1. Advantages and Limitations

A benefit of incorporating sound in the design of human–computer interfaces is the use of sound to displace text on small screens such as PDAs, thereby avoiding visual clutter (Kristoffersen and Ljungberg 1999a; Pascoe *et al.* 2000). This is a particular challenge for telecartography, as discussed by Gartner in this volume. Using sound instead of text to convey information can also help with issues related to mobile devices where users cannot always look at the screen such as reading a map while driving. Information played in sound would be delivered to the user regardless of where the user is looking (Kristoffersen and Ljungberg 1999b; Pascoe *et al.* 1999). Moreover, sound is attention grabbing. Whereas, users can choose not to look at something, it is more difficult to avoid hearing something. Therefore, sound should be used to convey important information (Barrass and Kramer 1999).

Consideration should also be given to the limitations associated with sound. Kramer (1994) suggested some difficulties associated with using sound in human–computer interface design:

- Vision has a much higher resolution than sound. For example, only a few different values can be used to represent volume unambiguously, whereas vision affords fine gradation of values.
- It is difficult to present absolute data when using sound. Unless users have perfect pitch, it is difficult to present absolute data. Conversely, users can look at a value on a graph to get an absolute value.
- Changing one attribute of sound can have an effect on other attributes. For example, a change in pitch can affect loudness.
- Sound disappears and therefore, users must remember the information that sound contained. Visual information, on the other hand, allows users to look back at the display.
- Sound can potentially annoy users.

A difficulty posed by use of earcons is "learnability" (Tkaczevski 1996; Loomis *et al.* 2001). First time users of earcons are able to master between 4 and 6 symbolic sounds in a matter of minutes. However, mastery of up to 10 earcons can require many hours. Competency in the use of more than 10 earcons is problematic for most users, and often impossible for others. Tkaczevski (1996) illustrated this limitation of earcons in his research on the use of sonification, in the design of a nuclear power plant control system. Results demonstrated that use of earcons improved reaction times, diagnostic accuracy, and corrective response to alarms. However, operators were reluctant to further explore the diagnostic capabilities of sound, due to difficulties in learning the meanings of multiple earcons.

4.2. Unimodal Applications

Flower and Hauer (1995) assessed the efficacy of visual and auditory trend plots in communicating statistical characteristics and relationships (e.g. shape, magnitude, and slope). Participants demonstrated no significant differences in their ability to extract information from either the auditory or visual trend plots. Thus, their study suggests that in some circumstances, sound can be as useful as vision to represent data. Moreover, in instances where there is a visual cognitive overload, the sound modality may prove an equally viable alternative. More empirical research is required, however, to further substantiate this premise.

Holland *et al.* (2002) developed and evaluated a prototype spatial audio navigation system equipped with Global Positioning System (GPS). The system employed earcons to assist users in navigation tasks, where their attention, eyes and hands would be otherwise occupied. Use of the system was evaluated for participants proceeding on foot and by automobile. Using the prototype, navigational performance was evaluated in terms of determination of: the distance to destination, and the direction of the destination relative to the participants' current direction (i.e. relative bearing).

Results showed that the audio representations of distance and direction were clearly successful for distance but limited with respect to change in direction. The relative distance or proximity to destination was effectively communicated through a Geiger counter effect (i.e. the beep signal became increasingly frequent as the distance diminished). The system accurately denoted relative direction or bearing when participants were proceeding along a straight line. However, the system's responsiveness to changes in direction was subject to latencies of 10–15 s. Such responsiveness was barely acceptable on foot, and clearly unacceptable in a vehicle. Further development of the prototype seeks to compensate for this shortcoming by inclusion of a supplemental electronic compass.

A variety of authors have suggested that the use of sound is an effective means for communication of navigational information (e.g. Gaver 1989; Pocock 1989; Ackerman 1990; Krygier 1994). The Holland *et al.* (2002) experiment demonstrates the effective manipulation of sound elements to assist in navigation. Participants wore headphones to detect sounds. The relative bearing to the destination was denoted by panning a sound source (i.e. a briefly repeated tone). With simple panning, participants were easily able to identify sound sources originating from the far left, far right, and from intermediary points. Some difficulty was incurred in differentiating sound sources placed in front and behind the user. Variation in timber was used to compensate for this effect. That is, a harpsichord sound (i.e. sharp timber) was used to denote sound sources in front of participants, whereas a trombone sound (i.e. muffled timber) was used to denote sound sources behind the participant. Deviation from a straight line to the destination was denoted by a change in pitch (i.e. frequency of sound). These manipulations proved to be successful navigational aids.

4.3. Multimodal Applications

Wheless *et al.* (1996) illustrate an optimal combination of visual and sound modalities. They factored sound into the design of their virtual reality representation of the Chesapeake Bay ecosystem. Salinity data was represented by sound. Specifically, change in salinity value was denoted by change in pitch. When navigating through the interface, users could listen to changes in salinity while observing the Bay's topography. Thus, users were able to understand the linkage between the Bay's biological and physical systems.

A further illustration of optimal combination of sound and other modalities was provided by Pirhonen *et al.* (2002). They conducted a study comparing two different mobile music player interface types. One music player had a gesture and nonspeech audio interface, whereas the other music player had a standard visual interface. The premise for their study was that the design of mobile computers is problematic, because these devices have small screens on which large amounts of information are displayed. Users often need to keep an eye on their environment rather than on the devices they are using, which makes input into these interfaces difficult, for example, driving a vehicle while using a route-guidance system.

In the gesture and nonspeech audio interface, feedback was given via nonspeech sounds (e.g. beeps). The main functions of Windows Media

Player (i.e. play/stop, next/previous track, volume up/down) were offered on this device. Users interacted with the system by gesturing on the screen. For example, a sweep across the screen from the left side to the right side indicated that the user wanted to move to the next track, a single tap indicated start and stop, a sweep from bottom to top indicated that the user wanted to increase the volume. Earcons were used to provide feedback to the users. For each of the devices in turn, participants were required to proceed along a course and perform a series of prescribed tasks. Results showed that subjective workload was reduced when using the gesture/audio player (mean = 25.4) versus the standard player (mean = 43.8), ($p < .001$). Time taken to complete the experiment was significantly lower in the gesture/audio condition (mean = 143.3 s) than in the standard condition (mean = 164.7 s), ($p < .0001$). No differences were found in the number of errors made with input in the two interfaces. This study showed that it is possible to create an effective interface without a visual interface. In this study, participants had only a small number of gestures to remember. More research would be needed to see whether these types of interfaces could be used with numerous functions. If extensive functionality was added to a device, the amount of cognitive workload needed to remember the different gestures could potentially be higher than the mental workload needed for a visual interface.

4.4. *Cognitive Workload Implications*

Using sound instead of text to navigate an interface can reduce user's cognitive load on the visual system. Graphic interfaces use the human visual system extensively. Therefore, users can miss important information because the visual system is overloaded. Thus, to reduce this cognitive overload, information can be displayed in sound, or shared between senses.

5. Haptic and Cartographic Visualization

The haptic modality involves the sense of touch. Although there has been some work using nonvisual modalities to represent data for blind and visually impaired map-users, there has been little such research for fully sighted map-users (Griffin 2001). Vasconcellos (1991), and in her chapter with Tsuji in this volume, however, demonstrated the utility of haptic variables to represent map data for both fully sighted and blind users. She formulated a set of haptic variables by translating Bertin's (1983) visual variables into tactile variables. However, she did not develop any tactile variables that did not have a direct visual counterpart. Specifically, she only included variables that were perceptible by both touch and vision (e.g. shape and orientation of symbols). By adhering to a strict translation of Bertin's visual variables, Vasconcellos neglected to include variables that can only be perceived by touch (e.g. temperature, pressure, and kinesthetic resistance).

There has been little cartographic research till date on the use of haptics to communicate spatial relations (Griffin 2001). Virtual reality offers a promising medium to reflect spatial relations using haptic representations, since it permits the use of multimodal interfaces on an interactive basis. Devices used to navigate through or interact with haptic interfaces include: data gloves (see Fig. 11.4), data body-suits, computer mice, and joysticks (see Fig. 11.5). For example, pressure-feedback joysticks or computer mice are used to convey kinesthetic sensation (e.g. resistance and friction) and/or texture sensation (Minsky and Ouh-young 1990; Shi and Pai 1997); and data glove use fiber optic sensors to determine hand position, and sensory exoskeletons offer pressure-feedback (Luecke and Chai 1997).

Sokoler *et al.* (2002) developed a mobile navigation device called "Tact-Guide." The "TactGuide" is analogous to a mouse capable of tracking lines on an electronic map. The underside of the "TactGuide" exhibits a

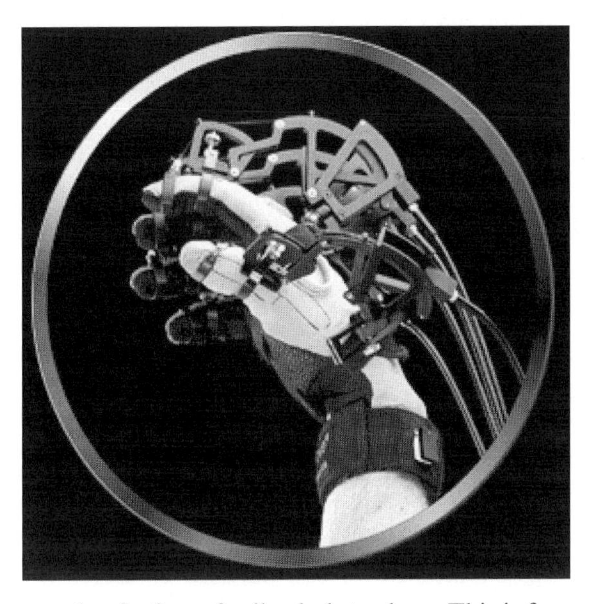

FIG. 11.4. An example of a force feedback data glove. This is from the Immersion Web Site, http://www.immersion.com/3d/products/cyber_grasp.php.

FIG. 11.5. An example of a pressure-feedback joystick. This is from the amazon.com Web site, http://www.amazon.com/exec/obidos/tg/detail/-/B00005NIMB/ref=e_de_a_smp/102-3989222-4456933?v=glance&s=electronics&vi=pictures&img=14#more-pictures.

central sensory aperture, surrounded by four smaller apertures, which indicate shift in direction, placed at 12 o'clock (i.e. forward), 3 o'clock (i.e. right), 6 o'clock (i.e. reverse) and 9 o'clock (i.e. left) respectively. A shift in direction triggers a solenoid, which drives a peg upward through the corresponding peripheral aperture. Placement of the user's thumb on the TactGuide display affords two sensory inputs to the user's thumb, one from the main central aperture and another from one of the four directional apertures. The position of the raised peg relative to the central aperture provides a tactile vector in one of the four directions. Thus, the TactGuide provides a physical tactile vector representation (i.e. center aperture to peg vector) analogous to the spatial physical relationship (i.e. direction in physical space relative to bodily orientation) depicted on the map.

Griffin (2001) lists the following potential examples of haptic representations of real cartographic events in a virtual reality environment: the vibration variable could be used to simulate the intensity of an earthquake in a seismic model map; the flutter variable could be used in a map model simulating the level of precipitation over time; and the temperature variable could be used in a map model to simulate cold water currents over time.

5.1. Advantages and Limitations

Traditionally, maps rely on visual graphics and text. In addition to alleviating the problem of potential visual clutter, haptic representations afford more intuitive interpretation of map information (Challis and Edwards 2000). For example, temperature can be used to differentiate ocean currents. With respect to limits of haptic perception, research suggests that there are physiological limits to the amount of haptic stimuli that humans can absorb and interpret in a given time frame (Von Uexkull 1957). Humans cannot differentiate more than 1 haptic sensation every 55 ms. Thus, if a person receives two haptic stimuli less than 55 ms apart, he will perceive these as a single stimulus. There is some dissonance among researchers as to whether this time interval of 55 ms is equivalent for all modalities (Von Uexkull 1957; White 1963; Eriksen and Collins 1968; Dufft and Ulrich 1999).

5.2. Unimodal Applications

Unger *et al.* (1997) conducted a study to evaluate tactile map-learning strategies. They compared a strategy employing overall tactile exploration of maps with sequential tactile route-based exploration of maps. They empirically demonstrated that an overall map exploration is superior to a sequential route exploration, for purposes of learning spatial relationships on a tactile map. Thus, if the user's goal is to acquire a better knowledge of the spatial relations on a map, the overall exploration strategy is superior to the sequential route approach.

Thompson (1983) conducted a study to explore how readers interpret different stimuli on a map. Specifically, he assessed the efficacy of various haptic symbols for representing different types of geographical data. He demonstrated that users could not estimate distances (e.g. mileage) from haptic symbols on the map unless they were able to compare all such symbols to symbol values contained in a legend. Similar results were found using graduated symbols on visual maps (Cox 1976). Thus, interaction of humans with haptic, and indeed visual maps is dependent upon the provision of adequate numbers of legend anchors necessary for magnitude estimations.

Bosman *et al.* (2003) developed and evaluated a wrist-mounted navigational aid called "GentleGuide." The prototype is worn on both wrists and is programmed to provide impulses indicating "turn left," "turn right," "stop," and "advance." This prototype was used by fully sighted participants to locate a room on the main floor of an office building. The "GentleGuide" prototype emitted increasingly frequent vibrations as the participant approached the target room. Conversely, frequency of the vibration diminished as participants proceeded away from the target location. Participants were asked to locate target rooms in two conditions: (1) using signage available in the building and (2) using the GentleGuide prototype in the absence of signage. All participants located the target rooms under both conditions. However, participants made significantly more errors when relying upon signage relative to "GentleGuide." Route completion was significantly faster using the GentleGuide prototype than in the signage condition.

5.3. Multimodal Applications

Combining vision and haptic modalities in virtual reality maps can prove beneficial. Researchers have demonstrated that both vision and touch can discriminate between textures (Lederman and Abbott 1981; Jones and O'Neill 1985; Lederman *et al.* 1986). For example, Lederman *et al.* (1986) illustrated that vision is used to segment spatial surfaces into different classes of texture, whereas touch is used to assess surface properties (e.g. sharp, abrasive, and smooth). Further, Heller (1982) showed that a combination of vision and touch enhanced people's ability to evaluate texture. Heller concluded that this was attributable to visual guidance of the hand during the evaluation process, not the added visual information related to the texture.

Oakley *et al.* (2000) used a haptic interface device (i.e. the PHANToM) to supplement traditional visual graphic interfaces. Their research demonstrated that task performance times remain constant, with and without the haptic device. However, results demonstrated reduction in error rates and subjective cognitive workload, when using the combined haptic–visual interface versus the traditional visual interface. Thus, this research demonstrates that haptic devices can provide significant performance benefits in a multimodal application. De Almeida Vasconcellos and Tsuji in this volume consider the combination of haptic and sound modalities in mapping for the visually impaired and the blind.

5.4. Cognitive Workload Implications

Although there has been considerable research on the mental processing implications of visual and phonological stimuli, to date there has been a little such research regarding haptic stimuli (Klatzky and Lederman 2003). Research on the cognitive processing of haptic stimuli is potentially complicated due to cross-modal considerations. Specifically, representations emanating from the haptic modality may be mediated by visual and phonological representations. Given the limited capacity of working memory, there is a limit to the number of haptic representations that can be retained. Researchers (Watkins and Watkins 1974; Millar 1975, 1999) demonstrated that a maximum of two to three haptic representations can be retained simultaneously in working memory.

When designing haptic interfaces for geospatial applications, it is important to safeguard against cognitive overload associated with the generation of too much information for the user to process simultaneously (Griffin 2001). Cartographers are already familiar with this issue when dealing with visual interfaces. The cognitive impact of excessive display of visual information can be somewhat mitigated by the fact that users do not have to retain the visual image in working memory because of the quasi-permanence nature of the visual display. This permanence allows the user to gradually assimilate the visual information, or refer back to it as required. Conversely, auditory and haptic information must be retained in working memory since they are not permanently displayed on the interface. Thus, when using these latter two modalities, interface designers should display one variable at a time (Griffin 2001).

Cognitive overload may also happen between modalities. For example, the Oak Ridge National Laboratory demonstrated the competitive overload impact, resulting from driving while having interface with a cell phone, a collision-avoidance system, a route-navigation system and an in-vehicle Internet-equipped device (ITS America 2001). Results showed that drivers coped best with this multitask load when they were able to complete one task before commencing the next. Thus, when designing interfaces, it is important to permit the user to control their cognitive load, by allowing the user to regulate the number and frequency of multisensory cues they must deal with. Conversely, simultaneous multisensory perceptual cues may prove beneficial under certain circumstances (Oakley *et al.* 2000). For example, in an intensive care unit, nurses may rely on an auditory warning as the priming cue to focus their attention on a patient monitor visual display read-out. Thus, further research is required to determine where the haptic modality would prove beneficial to cartographic design.

6. Conclusions

Given the pervasive technological innovation throughout all aspects of modern society, there is an increasing requirement for more efficient and effective modes of human–computer interaction. Accordingly, human–computer interface design is migrating from traditional graphic interfaces to more dynamic multimodal interfaces (e.g. speech, gesture, sound, and haptic). As systems become more complex and multifunctional, unimodal interfaces do not allow users to interact effectively across all tasks and environments (Oviatt *et al.* 2002). Furthermore, large individual differences exist in the preference and ability to use different communication modalities. Multimodal interfaces allow users to select a modality to suit their preference and situation, while balancing their cognitive workload (Karshmer and Blattner 1998). Finally, a multimodal approach to cybercartographic design affords a broad spectrum of choice for communication and visualization of geographic information.

References

Ackerman, D. (1990) "Hearing", in Ackerman, D. A. (ed.), *Natural History of the Senses*, Random House, New York, pp. 173–226.

Ahlberg, C. and B. Schneiderman (1994) "Visual information seeking: Tight coupling of dynamic query filters with starfield displays", *Proceedings of the CHI'94 Conference on Human Factors in Computing*, ACM, New York, pp. 313–317.

Baddeley, A. D. (1981) "The concept of working memory: A view of its current state and probable future development", *Cognition*, Vol. 10, pp. 17–23.

Baddeley, A. D. and G. J. Hitch (1974) "Working memory", in G. Bower (ed.), in *The Psychology of Learning and Motivation: Advances in Research and Theory*, Vol. 8, Academic Press, New York.

Baddeley, A. D. and R. H. Logie (1992) "Auditory imagery and working memory", in Reisberg, D. (ed.), *Auditory Imagery*, Lawrence Erlbaum Associates Inc., Hillsdale, New Jersey, pp. 179–197.

Barras, S. and G. Kramer (1999) "Using sonification", in *Multimedia Systems*, Vol. 7, pp. 23–31.

Bederson, B. B. and J. D. Hollan (1994) "PAD++: A zooming graphical interface for exploring alternate interface physics", *Proceedings of the UIST'94 Symposium on User Interface Software and Technology*, ACM, New York, pp. 17–27.

Benoit, C., J. C. Martin, C. Pelachaud, L. Schomaker and B. Suhm (2000) "Audio-visual and multimodal speech-based systems", in Gibbon, D., I. Mertins and R. Moore (eds.), *Handbook of Multimodal and Spoken Dialogue Systems: Resources, Terminology and Product Evaluation*, Kluwer Academic Publishers, Dordrecht, pp. 102–203.

Bertin, J. (1983) *Semiology of Graphics: Diagrams, Networks, Maps,* Madison, Wisconsin, University of Wisconsin Press, Translated by William Berg from Semiologies Graphiques, 1967, Edition Gauthier-Villars, Paris.

Billinghurst, M. (1998) "Put that where? Voice and gesture at the graphics interface", *Computer Graphics*, pp. 60–63.

Biocca, F., J. Kim and Y. Choi (2001) "Visual touch in virtual environments: An exploratory study of presence, multimodal interfaces, and cross-modal sensory illusions", *Presence*, Vol. 10, No. 3, pp. 247–265.

Blattner, M., A. Papp and E. Glinert (1992) "Sonic enhancements of two-dimensional graphic displays", in Kramer G. (ed.), *Auditory Display, Sonification, Audification and Auditory Interfaces, The Proceedings of the First International*

Conference on Auditory Display, Santa Fé Institute, Santa Fé, Addison-Wesley, pp. 447–470.

Bolt, R. A. (1980) "Put-that-there: Voice and gesture at the graphics interface", *Computer Graphics*, Vol. 14, No. 3, pp. 262–270.

Bolt, R. A. (1987) "Conversing with computers", in Baecker, R. and W. Buxton (eds.), *Readings in Human Computer Interaction: A Multidisciplinary Approach*, Morgan-Kaufmann, California.

Bosman, S., B. Groenendaal, J. W. Findlater, T. Visser, M. de Graaf and P. Markopoulos (2003) "Gentle guide: An exploration of haptic output for indoors pedestrian guidance", *Mobile HCI*, pp. 358–362.

Brewster, S. A. (1994) *Providing a Structured Method for Integrating Non-Speech Audio Into Human-Computer Interfaces*, PhD Thesis, University of York, UK.

Brewster, S. A., V.-P. Raty and A. Kortekangas (1996) "Earcons as a method of providing navigational cues in a menu hierarchy", *Proceeding of British Computer Society Conference on Human Computer Interaction, HCI '96, London, UK*, British Computer Society, Swinton, UK, pp. 169–183.

Challis, B. P. and A. D. N. Edwards (2001) "Design principles for tactile interaction", in Brewster, S. and R. Murray-Smith (eds.), *Haptic Human-Computer Interaction*, Lecture Notes in Computer Science, LNCS 2058 Springer-Verlag.

Clarke, K. C. (2001) "Cartography in the mobile internet age", *Proceedings of the 20th International Cartographic Conference*, ICC, Beijing, China, Vol. 3, pp. 1481–1488.

Cohen, P. (1992) "The role of natural language in a multimodal interface", *Proceedings of the UIST'92 Conference*, pp. 143–149.

Cohen, P. R. and S. L. Oviatt (1995) "The role of voice input for human-machine communication", *Proceedings of the National Academy of Sciences*, Vol. 92, pp. 9921–9927.

Colby, G. and L. Scholl (1991) "Transparency and blur as selective cue for complex visual information", *International Society for Optical Engineering Proceedings*, Vol. 1460, pp. 114–125.

Cox, C. W. (1976) "Anchor effects and the estimation of graduated circles and squares", *American Cartographer*, Vol. 3, pp. 65–74.

Cronly-Dillon, J. K. and R. P. F. Persaud (1999) "The perception of visual images encoded in musical form: A study in cross-modality information transfer", *Proceedings of the Royal Society of London*, Vol. 266, pp. 2427–2433.

Dey, A. and G. D. Abowd (2000) "Towards a better understanding of context and context-awareness", *CHI 2000, Workshop on the What, Who, Where and How of Context Awareness*.

Dufft, C. C. and R. Ulrich (1999) "Intersensory facilitation: Visual accessory signals can also shorten reaction time", *Zeitschrift fur experimentelle Psychologie*, Vol. 46, pp. 16–27.

Egenhofer, M. (1990) "Manipulating the graphical representation of query results in geographic information systems", *Proceedings of the IEEE Workshop on Visual Languages*, IEEE Computer Society Press, Los Alamitos, California, pp. 119–24.

Eriksen, C. W. and J. F. Collins (1968) "Sensory traces versus psychological movement in the temporal organization of form", *Journal of Experimental Psychology*, Vol. 77, pp. 376–382.

Fisher, P. (1994) "Hearing the reliability in classified remotely sensed images," *Cartography and GIS*, Vol. 21, pp. 31–36.

Fishkin, K. and M. C. Stone (1995) "Enhanced dynamic queries via movable filters", *Proceedings of the CHI'95 Conference on Human Factors in Computing*, ACM, New York, pp. 415–420.

Flowers, J. H. and T. A. Hauer (1995) "Musical versus visual graphs: Cross-modal equivalence in perception of time series data", *Human Factors*, Vol. 37, pp. 553–569.

Flowers, J. H., D. C. Buhman and K. D. Turnage (1997) "Cross-modal equivalence of visual and auditory scatterplots for exploring bivariate data samples", *Human Factors*, Vol. 39, No. 3, pp. 341–351.

Freundschuh, S. M. (1989) "Does 'anybody' really want (or need) vehicle navigation aids?", *IEEE Vehicle and Navigation Conference*, Conference Record, pp. 11–13, 439–442.

Gaver, W. (1989) "The sonic finder: An interface that uses auditory icons", *Human-Computer Interaction*, Vol. 4, No. 4, pp. 67–94.

Gaver, W., R. Smith and T. O'Shea (1991) "Effective sounds in complex systems: The arkola simulation", in Robertson, S., G. Olson and J. Olson (eds.), *Proceedings of CHI'91*, ACM Press, Addison-Wesley, New Orleans, pp. 85–90.

Goldstein, J. and S. Roth, (1994) "Using aggregation and dynamic queries for exploring large data sets", *Proceedings of the CHI'94 Conference on Human Factors in Computing*, ACM, New York, pp. 23–29.

Griffin, A. (2001) "Feeling it out: The use of haptic visualization for exploratory geographic analysis", *Journal of the North American cartographic information Society*, Vol. 39, pp. 12–29.

Hauptmann, A. G. and P. McAvinney (1993) "Gestures with speech for graphics manipulation", *International Journal of Man-Machine Studies*, Vol. 38, pp. 231–249.

Hedley, N. R. (2003) "Empirical evidence of advanced geographic visualization interface use", *Proceedings of the 21st International Cartographic Conference, ICC*, pp. 383–392.

Heller, M. A. (1982) "Visual and tactual texture perception: Intersensory cooperation", *Perception and Psychophysics*, Vol. 31, pp. 339–344.

Hinckley, K., E. Curtrell, S. Bathiche and T. Muss (2002) "Quantitative analysis of scrolling techniques", *Proceedings of Computer-Human Interaction 2002: Association for Computing Systems*, Association for Computing Machinery, New York, pp. 65–72.

Holland, S., D. R. Morse and H. Gedenryd (2002) "AudioGPS: Spatial audio navigation with a minimal attention interface", *Personal and Ubiquitous Computing*, Vol. 6, pp. 253–259.

ITS America (2001) *High-Tech Cars could Bring Cognitive Overload*, accessed April 2004 from http://www.itsa.org/ITSNEWS.NSF

Jacobson, R. D., R. M. Kitchin and R. G. Goolledge (2002) "Multi-modal virtual reality for presenting geographic information", in Fisher, P. and D. Unwin (eds.), *Virtual Reality in Geography*, Taylor and Francis, New York and London, pp. 382–400.

Jones, B. and O'Neil (1985) "Combining vision and touch in texture perception", *Perception and Psychophysics*, Vol. 37, pp. 66–72.

Karshmer, A. I. and M. Blattner (1998) *Proceedings of the 3rd International ACM Proceedings of the Conference on Assistive Technologies* (ASSETS '98), Marina del Rey, California, accessed April 2004 from http://www.acm.org/sigcaph/assets/assets98/assets98index.html.

Kishi, H. and S. Sigiura (1993) *Human Factors Considerations for Voice Route Guidance*, SAE technical paper 930553.

Klatzky, R. L. and S. J. Lederman (2003) "Representing spatial location and layout from sparse kinaesthetic contacts", *Journal of Experimental Psychology: Human Perception and Performance*, Vol. 29, No. 2, pp. 310–325.

Kramer, G. (1994) "An introduction to auditory display", in Kramer, G. (ed.), *Auditory Display*, Addison-Wesley, Reading, Massachusetts, pp. 1–77.

Kreitzberg, C. B. (1991) "Details on demand: Hypertext models for Coping with information overload", in Dillon, M. (ed.), *Interfaces for Information Retrieval and On-Line Systems,* Greenwood, New York, pp. 169–176.

Kristoffersen, S. and F. Ljungberg, (1999a) "Making place to make IT work: Empirical explorations of HCI for mobile CSCW", *Proceedings International ACM SIGGROUP Conference on Supporting Group Work (GROUP'99)*, Phoenix, AZ.

Kristoffersen, S. and F. Ljungberg (1999b) "Designing interacting styles for mobile use context", in Gellersen H-W (ed.), *First International Symposium on Handheld and Ubiquitous Computing (HUC 99): Lecture notes in Computer Science 1707,* Springer, Berlin, Heidelberg, New York, pp. 208–221.

Krygier, J. (1994) "Sound and geographic visualization", in MacEachren, A.M. and D. R. F. Taylor (eds.), *Visualization in Modern Cartography*, Elsevier, Oxford, UK.

Lederman, S. J. and Abbott (1981) "Texture perception: Studies of intersensory organization using a discrepancy paradigm and visual vs. tactual psychophysics", *Journal of Experimental Psychology: Human Perception and Performance*, Vol. 47, pp. 54–64.

Lederman, S. J., G. Thorne and B. Jones (1986) "Perception of texture by vision and touch: Multidimensionality and intersensory integration", *Journal of Experimental Psychology: Human Perception and Performance*, Vol. 12, pp. 169–180.

Logie, R. H. (1986) "Visuo-spatial processing in working memory", *The Quarterly Journal of Experimental Psychology*, Vol. 38A, pp. 229–247.

Lokuge, I. and S. Ishizaki (1995) "Geospace: An interactive visualization system for exploring complex information spaces", *Proceedings of the CHI'95 Conference on Human Factors in Computing*, ACM, New York, pp. 409–414.

Loomis, J. M., R. G. Golledge, R. I. Klatzky (2001) "GPS-based navigation systems for the visually impaired", in Barfield, W. and T. Caudell (eds.),

Fundamentals of Wearable Computers and Augmented Reality, Lawrence Erlbaum, Mahwah, New Jersey, 2001, 429–446.

Luecke, G. R. and Y. –H. Chai (1997) "Contact sensation in the synthetic environment using the ISU force reflecting Wxoskeleton", *IEEE Annual Virtual Reality International Symposium,* IEEE Computer Society Press, Albuquerque, New Mexico, pp. 192–198.

MacEachren, A. M. (1992) "Visualizing uncertain information", *Cartographic Perspectives*, Vol. 13, pp. 10–19.

MacEachren, A. M. (1995) *How Maps Work*, Guilford Press, New York.

Martin, G. L. (1989) "The utility of speech input", *User-Computing Interfaces, International Journal of Man-Machine Studies*, Vol. 30, pp. 355–375.

McNeill, D. (1992) *Hand and Mind: What Gestures Reveal About Thought*, University of Chicago Press, Chicago, Illinois.

Millar, S. (1975) "Spatial memory by blind and sighted children", *British Journal of Psychology,* Vol. 66, No. 4, pp. 449–459.

Millar, S. (1999) "Memory in touch", *Psicothema*, Vol. 11, pp. 474–767.

Minsky, M. and M. Ouh-young (1990) "Feeling and seeing: Issues in force display", *Computer Graphics*, Vol. 24, pp. 235–243.

Nass, C. and L. Gong (2000) *Maximized Modality or Constrained Consistence?* Paper presented at audio-visual speech processing conference.

Oakley, I., M. R. McGee, S. Brewster and P. Gray (2000) "Putting the feel in 'look and feel' ", *Proceedings of ACM CHI 2000*, pp. 415–422.

Oviatt, S. L. (1997) "Multimodal interactive maps: Designing for human performance", *Human–Computer Interaction* (special issue on Multimodal Interfaces), Vol. 12, pp. 93–129.

Oviatt, S. L. (2000) "Taming recognition errors within a multimodal interface", *Communications of the ACM*, Vol. 43, No. 9, pp. 45–51.

Oviatt, S. L. and P. R. Cohen (1991) "Discourse structure and performance efficiency in interactive and noninteractive spoken modalities", *Computer Speech and Language*, Vol. 5, No. 4, pp. 297–326.

Oviatt, S. L. and R. VanGent (1996) "Error resolution during multimodal human-computer interaction", *Proceedings of the International Conference on Spoken Language Processing*, Vol. 2, University of Delaware Press, pp. 204–207.

Oviatt, S. L. and P. R. Cohen (2000) "Perceptual user interfaces: Multimodal interfaces that process what comes naturally", *Communications of the ACM*, Vol. 43, No. 3, pp. 45–53.

Oviatt, S. L., A. DeAngeli and K. Kuhn (1997) "Integration and synchronization of input modes during multimodal human-computer interaction", *Proceedings of Conference on Human Factors in Computing Systems (CHI'97)*, ACM Press, New York, pp. 415–422.

Oviatt, S. L., M. MacEachern and G. Levow (1998) "Predicting hyperarticulate speech during human–computer error-resolution", *Speech Communication*, Vol. 24, pp. 87–110.

Oviatt, S. L., P. R. Cohen, L. Wu, J. Vergo, L. Duncan, B. Suhm, J. Bers,T. Holzman, T. Winograd, J. Landay, J. Larson and D. Ferro (2002) "Designing the user interface for multimodal speech and pen-based gestures applications: State-of-the-art systems and future research directions", in Carroll, J. M. (ed.), *Human-Computer Interaction: In the New Millennium*, Addison-Wesley, New York.

Pascoe, J., N. Ryan, and D. Morse (1999) "Issues in developing context-aware computing", in Gellersen, H-W (ed.), *First International Symposium on Hand-held and Ubiquitous Computing (HUC 99), Lecture notes in Computer Science* 1707, Springer, Berlin, Heidelberg, New York, pp. 208–221.

Pascoe, J., N. Ryan and D. Morse (2000) "Using while moving, HCI issues in fieldwork environments", in *ACM Transactions Computer Human Interaction*, Vol. 7, pp. 417–437.

Pereverzev, S. V., A. Loshak, S. Backhaus, J. C. Davis and R. E. Packard (1997) "Quantum oscillations between two weakly coupled reservoirs of superfluid He-3", *Nature*, Vol. 388, pp. 449–451.

Philips, R. J. (1979) "Making maps easy to read, a summary of research", in Koelers, P.A., M. E. Wrolstad and H. Bouma (eds.), *Processing of Visible Language*: Vol. 1, Plenum, New York, pp. 165–174.

Pirhonen, A., S. Brewster and C. Holguin (2002) "Gesural and audio metaphors as a means of control for mobile devices", *Proceedings of the Conference on Human Factors in Computing Systems (CHI'02)*, Vol. 4, pp. 291–298.

Pocock, D. (1989) "Sound and geographer", *Geography*, Vol. 74, No. 3, pp. 193–200.

Ross, T. and S. Brade (1995) *An Empirical Study to Determine Guidelines for Optimumtiming of Route Guidance Instructions, Colloquium on "Design of the driver interface"* IEEE Professional Group C12, London.

Roth, S. F., M. C. Chuah, S. Kerpedjiev, S., Kolojejchick and P. Lucas (1997) "Toward an information visualisation workspace, combining multiple means of expression", *Human-Computer Interaction*, Vol. 12, pp. 131–185.

Salisbury, M., T. Henderson, C. Lammers, Fu, and S. Moody (1990) "Talk and draw, bundling speech and graphics", *IEEE Computer*, Vol. 23, No. 8, pp. 59–65.

Shi, Y. and D. K. Pai (1997) "Haptic display of visual images", *IEEE Annual Virtual Reality International Symposium*, IEEE Computer Society Press, Alburquerque, New Mexico, pp. 188–191.

Sokoler, T., L. Nelson and E. R. Pedersen (2002) "Low-Resolution supplementary tactile cues for navigational assistance", *Mobile HCI*, pp. 369–372.

Streeter, L. A., D. Vitello S. A. Wonsiewicz (1985) "How to tell people where to go, comparing navigational aids", *International Journal of Man-Machine Studies*, Vol. 22, pp. 549–562.

Takahashi, K. and T. Takezawa (2002) "An interaction mechanism of multimodal dialogue systems", *Systems and Computers in Japan*, Vol. 33, No. 11, pp. 70–79.

Terken, J. and L. Verhelst (2000) *Multimodal Interfaces for Wearable and Mobile Computing: Information Presentation for a Wearable Messenger Device*, accessed April 2004 from http://www.ipo.tue.nl/ipo.

Thompson, N. R. (1983) "Tactual perception of quantitative point symbols in thematic maps for the blind", *Proceedings of the First International Symposium on Maps and Graphics for the Visually Handicapped,* Association of American Geographers, Washington, DC, pp. 103–113.

Tkaczevski, A. (1996) "Auditory interface problems and solutions in commercial multimedia products", in Frysinger, S. and G. Kramer (eds.), *Proceedings of the Third International Conference on Auditory Display ICAD'96, Palo Alto*, California, Web-proceedings, accessed April 2004 from http://www.santa fe.edu/~icad

Unger, S. M., M. Blades and C. Spencer (1997) "Strategies for knowledge acquisition from cartographic maps by blind and visually impaired adults", *The Cartographic Journal*, Vol. 34, p. 93–110.

Vasconcellos, R. (1991) "Knowing the amazon through tactual graphics", *Proceedings of the International Cartographic Association, International Cartographic Association*, Bournemouth, pp. 206–210.

Von Uexkull, J. (1957) "A stroll through the worlds of animals and men, a picture book of invisible worlds", in Schiller, C. H. (ed.), *Instinctive Behavior,* Madison, International Universities Press, Connecticut.

Watkins, M. J. and O. C. Watkins (1974) "A tactile suffix effect", *Memory and Cognition*, Vol. 5, pp. 529–534.

Wheless, G. H., C. M. Lascara, A. Valle-Levinson, D. P. Brutzman, W. Sherman, W. Hibbard, and B. E. Paul (1996) "Virtual chesapeake bay, interacting with a coupled physical/biological model", *IEEE Computer Graphics and Applications*, Vol. 16, pp. 52–57.

White, C. (1963) "Temporal numerosity and the psychological unit of duration", *Psychological Monograph*, Vol. 77, pp. 1–37.

Williamson, C. and B. Schneiderman (1992) "The dynamic homefinder, evaluating dynamic queries in real estate information exploration system", *SIGIR Conference Proceedings*, ACM New York pp. 339–346.

Wolf, C. G. (1992) "A comparative study of gestural, keyboard, and mouse interfaces", *Behavior and Information Technology*, Vol. 11, No. 1, pp. 13–23.

Zeleznik, R., K. Herndon and J. Hughes (1996) "SKETCH, an interface for sketching 3d scenes", *Proceedings of Association for Computing Machinery's Special Interest Group on Computer Graphics*, Association for Computing Machinery, New York, pp. 163–170.

CHAPTER 12

Art, Maps and Cybercartography: Stimulating Reflexivity among Map-Users

SÉBASTIEN CAQUARD

Geomatics and Cartographic Research Centre (GCRC),
Department of Geography and Environmental Studies,
Carleton University, Ottawa, Ontario, Canada

D. R. FRASER TAYLOR

Geomatics and Cartographic Research Centre (GCRC)
and Distinguished Research Professor in International Affairs,
Geography and Environmental Studies,
Carleton University, Ottawa, Ontario, Canada

Abstract

In this chapter the case is made for encouraging reflexivity by making map-users aware that any map is a "construction of the image of space" and is thus inherently subjective. Informing map-users of the constructed dimension of maps is particularly vital given the exponential production of maps via the Internet. This growth expands the presence of potentially dogmatic and misleading messages. To address these issues, the interrelation between aesthetics, science and technology, and its impact on the perception of maps by most users is discussed. On the one hand cartographers have widely used the aesthetic dimension of art to increase the impression of objectivity associated with maps and on the other hand, developments in the humanities provide grounds to challenge this concept of map objectivity. In this context the potential of cybercartography to combine multiple media, art, technologies, and perspectives into maps is highlighted. This chapter concludes by arguing for stronger presence of artistic creativity and social criticism in maps on the Internet to stimulate map reflexivity.

1. Introduction

While critical cartographic studies have brought issues of power and representation to the fore, mapmakers rarely acknowledge this dimension in the maps they design. As a result maps are often rendered as a dogmatic medium: they are representations that seem objective of information that in fact, is not. This impression of objectivity becomes particularly problematic with the exponential growth of maps on the Internet. Over 200 million maps are distributed daily through the Internet (Peterson 2003) and this number is increasing rapidly. These maps represent a substantial market and attract many new map-users. These novice users make all kinds of decisions based on maps (e.g. which route to take) without necessarily, being aware of the possible implication embedded in the decision they make. For instance, maps with commercial sponsors may encourage users to drive through particular shopping areas. In light of this situation the purpose of this chapter is to explore how to help map-users in generating greater and critical map reflexivity.

The first section of this chapter examines how art, science and technology have been woven together in a particular manner to strengthen the objective appearance of maps. In a second section, examples of maps designed by contemporary artists are revisited in order to highlight the ways these artists challenge the objectivity of maps. While substantial research has been done to study maps and art from a critical and historical perspective (Woodward 1987; Jacob 1992; Cosgrove 1999; Peterson 2003;), relatively few have focused on trying to integrate artistic perspectives into maps (Taylor 1987; Corner 1999). This integration requires a paradigm shift that could be provided by cybercartography. The potential of cybercartography in this context will be discussed in the third and last section.

2. Aesthetics, Science, Technology and Cartographic Objectivity

As Taylor (1991: 2) pointed out: "Art seems to have been driven out of contemporary cartography by the 'scientific objectivity' of modern production techniques. It can be argued, however, that if modern cartography is to progress then the need is not only for more science but also for more art." The question then becomes: what kind of art?

Maps and art share a common agenda, which is to imitate nature. Despite such affinity and the multifaceted nature of artistic endeavors, cartography has drawn primarily on the aesthetic dimension of art in order to make maps more visually appealing. This dimension is vital for the pleasure of making as well as using maps. It provides a "tonic for imagination" as Tuan (1999) points out. More generally, according to Girod *et al.* (2003), an aesthetic experience draws people into the world of science. Aesthetics provide a powerful and often unacknowledged link between art and science. Moreover, an aesthetic experience is emotional and dramatic. Emotion can augment engagement and enhance the pleasure of learning. It has been argued that the chief impediment and motivator of learning is not cognition but emotion (Pedretti 2004). If we want our maps to be seen, used and understood then they need to be aesthetically appealing.

In cartography the aesthetic dimension is emphasized because it inspires confidence. A beautiful map looks more accurate than an ugly one. This relation between attractiveness and impression of accuracy is widely accepted in conventional cartography (Lussault 1995: 1760) as well as in digital maps (Harrower *et al.* 1997: 33). But this dimension has also been recognized as being problematic. As Wright (1942: 528) pointed out more than 60 years ago the "clean-cut appearance of a map" exaggerates the "impression of accuracy" and characterizes the "indifference to the truth" by cartographers. It has been known for many years that the aesthetic dimension of maps is useful not only to provide refined, sensuous pleasure but also to disguise the limitations of a particular map. The well-known quotation by Swift on maps of Africa springs to mind "so geographers in Afric-maps, with Savage-Pictures fill their gaps; and o'er uninhabitable Downs, Place Elephants for want of Towns" (in Horrell 1958: 746). Nevertheless this aesthetic dimension is still required for a good map. In other words a good map is a map that hides its limitations and its subjectivity well; it's a map that appears to be objective.

Critical cartographic studies have developed theoretical tools to highlight the power of maps in relation to the image of objectivity they convey (Pickles 2004). These works were largely influenced by postmodern theorists such as Barthes (Wood and Fels 1986) or Derrida and Foucault (Harley 1989). Based on these studies most cartographers now agree that the correspondence theory – that maps portray an objective and neutral representation of the reality (Crampton 2003: 30) – is no longer valid among some GIS specialists. However the belief in an "objective" cartography still persists. Nevertheless mapmakers rarely openly acknowledge this dimension in the maps they design. They are still more interested in improving the apparent objectivity than in acknowledging the fact that maps are a "construction of the image of space" to restate Jacob's formula (Jacob 1992: 136) and are therefore inherently subjective.

The fascination for objectivity and accurate measurement in cartography emerged in the late 18th and early 19th centuries (Cosgrove 2003: 133) and increased during the 20th century. It became dominant with technological developments such as automated cartography, GIS, and remotely sensed satellite imagery, and apparently unlimited analytical potential they promise. Technology has increased map credibility as well as the faith put in it from the perspective of both mapmaker and map-user (Wood and Fels 1986: 99).

This strong technical bias in cartography was not only a question of improved map quality but also was a strategic issue. It helped cartography to acquire the status of a subscience (Harley 1990: 9). For many cartographers, recognition of cartography as a science was a vital goal throughout the 20th century. An example was the major effort by the International Cartographic Association (ICA) to obtain membership of the International Council of Scientific Unions (ICSU), which was achieved in 1996. A consequence of the scientific quest in cartography has been to fix the idea in public mind (as well as in the mind of some mapmakers) that maps are scientific, accurate and objective.

This confidence in map objectivity has been discussed by different authors such as Woods and Fells. For them "the most fundamental cartographic claim is to be a system of facts, and its history has most often been written as the story of its ability to present those facts with ever increasing accuracy" (Wood and Fels 1986: 3). Wood (1992) also points out that the connotation underlying objectivity of reference mapping is so strong that it becomes axiomatic, and is then widely accepted. This confidence does not affect only reference maps. As Edney (1996) points out, our culture has surrounded the map with objectivity, leading us to see them as a mirror of reality. This confidence in the information mapped, has also been highlighted in previous work done with water management stakeholders and multimedia maps (Caquard 2003). In this research, stakeholders concurred that the information mapped was objective. When the question "do you think that the information presented on these maps is objective?" was posed, all stakeholders answered "yes" (Caquard 2003).

This perception does not seem to have changed with the Internet. In theory, on the web any map can be accessed from anywhere on earth by any user. This exponential increase in map accessibility makes cartographic information more democratic and more ethical (Peterson 1999). It generates new map-users as well as new map producers. These new map-users are not necessarily aware of the power of map rhetoric. They often don't even challenge the neutrality and the objectivity of the map. This lack of critical perspective is reinforced by the technological dimension of most of the maps on the Internet, which involve interactivity, multimedia, or accurate remote sensing images. Maps on the Internet tend to be more technology oriented and look more scientific. In other words the Internet supports the exponential expansion, within the general public of the idea that maps represent an objective view of the world. One consequence is that it reinforces the conflation between map and reality. As Peuquet and Kraak (2002) point out the boundaries between reality and the image created in map form become even blurrier with the Internet.

This faith people put in maps is a recurrent source of humor. A scene in the film *Doctor Strangelove* (Director Stanley Kubrick 1964) offers a caustic example. In this movie, the anticommunist US general Turdgison uses the map – the "big board" – as an irrefutable proof of the fact that the USSR president is lying when he says that an American B-52s is still flying toward its Russian target. Turdgison argues that it is impossible that this B-52s is still flying when the "big board" shows the retreat of all the planes. In the reality – of the movie – this B-52s is not only flying but also it will lead to a mutual nuclear destruction. Another caustic and humoristic example of the excessive confidence people have in maps is provided by *Calvin and Hobbes* cartoon as seen in Fig. 12.1. Here the map obviously "precedes the territory" in the sense that the territory has to be modified to conform to the map (Fig. 12.1).

This confusion between map and reality has been discussed in the pre-Internet era by Baudrillard (1981). Baudrillard reversed Borges' famous tale about a map of the size of an Empire to argue that the territory does not precede the map anymore; now "the map precedes the territory." As Corner (1999: 222) points out "Baudrillard is going one step further here, claiming that late twentieth century communication and information

FIG. 12.1. The excessive confidence Calvin and Hobbes have in maps (Cartoon from Bill Watterson). Here the map obviously *precedes the territory* in the sense that the territory has to be modified to conform to the map. [CALVIN AND HOBBES © 1995 Watterson. Dist. By UNIVERSAL PRESS SYNDICATE Reprinted with permission. All rights reserved].

technologies have produced such a blurring of what is real and what is a representation that the two can no longer be distinguished." Corner uses this argument as well as the work of child psychologists such as Winnicott and philosophers such as Cassirer to highlight the complexity of the relations between maps and reality and to critique the lack of techniques in cartography for materializing a culturally constructed reality.

This leaves a paradox: on one hand cartographers are widely conscious that their maps cannot be true "representations" of reality, on the other hand, they have not developed techniques to inform the user of this lack of objectivity. The consequence is that map-users still appear to trust the map more than their own experience. The implications of this paradox can be insignificant such as digging a big hole to make the yard conform to the map, or catastrophic such as the potential destruction of humanity. Highlighting the false objectivity of maps could in fact help to make better decisions. For instance, as Corner (1999: 216) points out, planners usually take maps for granted as institutional conventions and they don't raise any kind of skepticism. This "indifference towards mapping is particularly puzzling when one considers that the very basis upon which projects are imagined and realized derives precisely from how maps are made" (Corner 1999: 216). To reduce the impact of poor decisions made on the basis of maps as well as to educate map-users, it becomes vital to stimulate reflexivity vis-à-vis maps. As Pedretti (2004) points out, reflexivity can create powerful learning opportunities. A fundamental change is required in the way cartographers perceive and construct maps. Artistic elements can be used to initiate this change.

3. Art for Stimulating Reflexivity

3.1. Humanities as a Source of Inspiration

Over the last few years several books have been published on maps in relation to the arts. For instance the *Atlas of Experience* (Van Swaaij and Klare 2000) represents an imaginary world of concepts and feelings. It is definitely in the lineage of the *Carte du pays de tendre* – Map of Tenderness – from Madeleine de Scudéry (1654). This Map of Tenderness similarly structures Bruno's very personal journey exploring the world of maps and geography from a cinematic viewpoint, at the intersection of art, architecture and (pre)cinema (Bruno 2002). The eclectic collection of maps published by Harmon (2004) highlights the richness and the diversity of unconventional maps. Rogoff (2000) utilizes contemporary art in relation to maps and geography, such as the work of Joshua Neustein or Mona

Hatoum, to explore the change in the contemporary world. Several exhibitions have also recently presented contemporary artwork related to map and space (Curnow 1999) and have provided the opportunity for collaboration among artists, curators and scholars from different disciplines. For instance *World Views: Maps and Art* was presented in 1999 at the Frederick R. Weisman Art Museum, University of Minnesota (Silberman 1999) and *Global Navigation System* was presented in 2003 at the Palais de Tokyo, Paris (GNS, 2003). The symposium *Mapping in the Age of Digital Media* organized at Yale in 2002 (Silver and Balmori 2003), describes some examples of the use of new technology by different disciplines (architecture, science, art) to design digital maps.

These examples show the real interest and fascination generated by maps outside the world of conventional cartography; conventional cartography means here cartography primarily focused on the accuracy of the phenomena mapped and/or on the efficiency of their communication. The existence of a parallel cartographic world dedicated to capture the complexity of the contemporary world with maps is highlighted. It offers an extensive source of inspiration. Maps are viewed as instruments of imagination, criticism, poetry, of emotion as well as an expression of *Magic Realism* (Ríos 2003) (http://www.public.asu.edu/~aarios/resourcebank/maps/index.html). This fascination does not seem to have yet been reflected in the practices or tenets of conventional cartography. Most of the maps produced in this context do not integrate the creativity of this parallel world.

This situation holds interesting parallels with cartography in the 19th century. At that time thematic cartography was emerging in Europe (Palsky 1996) and was mainly developed by other disciplines such as engineering (Robic 1997). Geographers and cartographers, who were more interested by the accuracy of topographic maps, largely ignored this development. For instance, Minard who designed the famous 1861 *Carte figurative des pertes successives en hommes de l'Armée Francaise dans la campagne de Russie de 1812–1813* – Loss of the French Army in the Russian Campaign 1812–1813 – was an engineer. It took several decades for geographers and cartographers to recognize the value and potential of thematic maps (Palsky 1996). It seems that at the outset of the 21st century, cartographers were mainly interested in the technological potentialities while artists were the ones bringing more challenging perspectives on society through maps. As Bourriaud (2003: 21) points out, "in this world, which has been deterritorialized and entirely remodelled by technology, geography is no longer simply an affair of hard science; it is equally a matter for artists, who come to the discipline from a perspective that is both poetic and critical." It may be vital for cartographers to better understand the ideas and trends emerging in the development of maps in the humanities.

Understanding and incorporating these might help cartographers to better portray the complexity of the contemporary world. In the context of this chapter it could also be inspiring to stimulate reflexivity.

3.2. Art, Maps and Subversions

Artists have been subverting maps and challenging their power for a long time. A famous example is provided by Artist-cartoonist Elkanah Tisdale in 1812 who used ridicule to denounce the partisan redistricting of the Essex County, Massachusetts (Monmonier 2001). During the 20th century, several artistic movements diverted maps for some political and intellectual purposes. Among these movements Dadaism and Surrealism had a wide influence on contemporary art. These two movements, emerging in the early 20th century, opposed Realism and Naturalism as "incapable of acting – or even commenting – upon a world sorely in need of change" (Rubin 1968: 14). These movements, like other avant-garde movements, have developed some techniques in many different fields such as painting, theater, cinematography, photography, poetry, music, literature as well as in mapmaking.

The famous map *Le monde au temps des surréalistes* is a fascinating example showing the way surrealists perceived the world in 1929 (Fig. 12.2). Obviously this map contains a strong political dimension (Ginioux 2004) that generated all kinds of controversial interpretations (Clair 2001; Debray 2003). But what is more important in the context of this chapter is that this map is a graphic demonstration of the power of maps, and of their inherent and indissociable subjectivity. It uses cartographic convention such as realistic shapes and toponyms to produce an unconventional perspective of the world. By doing so it puts into perspective the "official" western representation of the world including all of the connotations contained within the Eurocentric Mercator projection. In some ways it "precedes" the Peters projection by explicitly using the map for politics. While Peters projection uses a coordinate system to highlight the idea that any map conveys and disseminates a specific view of the world, this map developed the same argument using an artistic perspective. The notion of the map as a mirror of reality and its associated power is then challenged graphically by this map (Fig. 12.2).

Contemporary artists challenge map objectivity in a similar way. For instance Lordy Rodriguez's maps typically consist of misplaced cities and fictional states alongside recognizable landmarks and familiar sites. In reading these images, the viewer must suspend the popular notion of a

FIG. 12.2. Le *monde au temps des surréalistes* – 1929 – Surrealist Map of the World. The notion of the map as a mirror of reality and its associated power is challenged graphically by this map.

map as an accurate and factual document (Roland 2002) (http://www.art junction.org/articles/mapartists.html).

Surrealists and Dadaists were also interested in maps as political icons. For instance, maps appear regularly in collages and photomontages such as the map of Europe in *Cut with the Kitchen Knife* (Hannah Höch 1919). More than the function of the map in the photomontage what is interesting here is the technique of photomontage. Photomontage is a polyphonic mix of material and sign that takes away the sacred aura of photograph and bring it back to its status as an object (Dubois and Van Cauwenberge 1990: 236). Based on this interpretation it could then be possible to apply this technique to cartography, to develop some kinds of "cartomontages" able to take away "sacred aura" of the map and bring it back to status as an object. Contemporary artist Uta Riccius uses this kind of concept in *Surface Tension 2* (Fig. 12.3). To create this piece, she cut a world map into countries, plastic-coated them and put them in a sphere with water and oil. The idea was to relate global concern by transforming a two-dimensional well structured map into a three-dimensional chaotic and changing representation of the world (personal communication). This piece fits in the bigger category of contemporary art described by Curnow (1999: 261) that expresses urgent fears for the environmental, economic and political state of the globe with ironic and sometimes fiercely satirical works (Fig. 12.3). In the technological era artists combine not only multiple

FIG. 12.3. Uta Riccius, *Surface Tension 2*, 2004 (Map of the world, glass, plastic, oil, water) – diameter: 20 cm (Personal collection). The idea of the artist is to relate global concern by transforming a two-dimensional well structured map into a three-dimensional chaotic and changing representation of the world (personal communication).

materials but also multiple media to explore all kinds of unexplored worlds. Among these worlds the ones related to the body are recurrent in contemporary art (Rogoff 2000; GNS 2003; Silver and Balmori 2003). For instance, in her video installation *Corps étranger* artist Mona Hatoum projects video of her body onto a circular screen from a tiny endoscopic camera which she had inserted (Silver and Balmori 2003). As Imperiale (2003: 36) points out, "what was once interior and invisible became external and spatial." Making visible what was once invisible, is one of the major function of maps. This point is usually interpreted literally, such as mapping phenomena that are not visible in the reality or phenomena that have never been mapped previously. But it can also be interpreted as making visible/concrete what was once invisible/abstract. An illustration of this idea is provided by Andrews (2001: 24) who quotes Harley: "by dehumanizing a landscape, maps make it easier for military commanders or civil administrators to escape the pangs of conscience when inflicting cruelty or hardship on the inhabitants of the mapped area." ... a colored rectangle on a battle plan cannot easily be made to resemble a group of flesh-and-blood soldiers." But it could also be argued that a photo or a video of "flesh-and-blood soldiers" can do it. Making visible what was once invisible does not necessarily mean making it visible in the conventional cartographic language. It could also mean making it visible in many different forms more or less realistic. Contemporary art provides us with a multiplicity of forms of expression. Why should cartography restrict itself to traditional conventional symbols inherited from paper maps?

Perspective switch is another way used to challenge map conventions. The inverted map of South America illustrates this idea (Fig. 12.4). This map synthesizes Torres-Garcia's idea that art should not be created from top to bottom and his claim for unifying the future art of the Americas. Again it subverts cartographic conventions (e.g. accurate coordinates) to denunciate the oppression of the conventional cartographic representation. It marked the start of a movement by Latin American Artists to claim a prominent place for their art in a Northern dominated world. This was also an inspiration for several contemporary artistic creations (e.g. Milhé 2001 in GNS 2003; ODT 2002 in Harmon 2004), with this obvious notion of challenging conventional ideas and representations.

In his installation *America.Invention* artist Lothar Baumgarten (in curnow 1993) decorated the interior of the main rotunda of the Guggenheim Museum in New York with the name of the native societies of the Americas. Curnow (1999: 266) describes it in these terms:

> standing on the ground floor and gazing around and up at the red and grey names, it was clear that north was up and south was down (...) [But] the southern hemisphere inverted inside the northern. Moreover, to ascend the ramp was to head south to where at the top, under the dome, Tierra del Fuego met up with Alaska. What this folding, this implication of one hemisphere in another, served to disturb was the identification of polarity and orientation with hierarchy.

All the notions of (cartographic) convention were then challenged by the artist as well as by the personal experience of the public. The map becomes interactive in the sense that it's meaning evolves with the itinerary of the viewer. Each viewer experiences multiple perspectives of the same map.

3.3. Art, Maps and Personal Experiences

Multiplying the perspectives on maps and territories could help to generate some reflexivity vis-à-vis the information mapped. As the publisher of *What's Up? South!* reminds us: "it takes many points of view to see the

FIG. 12.4. *Inverted Map of South America* (Joaquin Torres-García 1936). This map introduced Torres-García's fundamental concept that "our North is the South," redirecting Uruguayan artists away from Europe toward their own American traditions. Cecillia de Torres-http://www.cecilliadetorres.com/jt/jt.html

truth" (Harmon 2004). Among these perspectives, personal experiences with the traditional representation might help the user to dissociate map construction from the reality. The famous series of maps The *Naked City* (1957) developed by the situationist Guy Debord illustrates this idea. This series presents the situationist concept of the *dérive* "the dream-like drift through the city, mapping alternative itineraries and subverting dominant readings and authoritarian regimes" (Corner 1999: 231). This series also call for an alternative approach to cartography based more on subjective perceptions of the space (Picon 2003: 63). These maps challenge conventional cartographic practices (Pinder 1996: 421) at a large scale by limiting the space represented on the map to the space experienced in the city. The psychogeography represented by the situationists is not geographically continued. It is formed by disparate pieces connected together (Fig. 12.5). This series of maps provide a visualization of the way we experience space without filling the blank with any kind of artifact. Our experience of space is indeed not continuous, it is an accumulation of places related or not. Beyond the political implications of this work, this series acknowledges the fact that the map is not a mirror of reality but a mask of this reality. This mask allows the user to see only some pieces of reality; pieces of territory or pieces of information. Through this mask, the map depiction of the world becomes more opaque than transparent. The notion of map as a transparent representation of the reality is challenged. One more time, by subverting

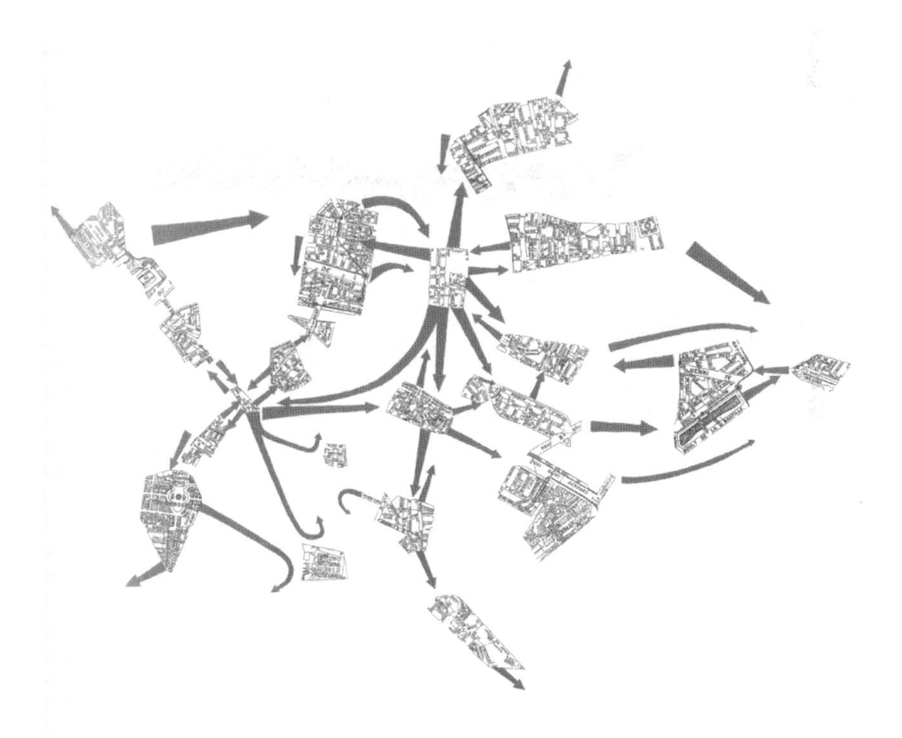

FIG. 12.5. Guy Debord, *The Naked City*, illustration *de l'hypothèse des plaques tournantes en psychogéographie, 1957* (printed material). This series acknowledges the fact that the map is not a mirrior of reality but a mask of this reality. This mask allows the user to see only some pieces of reality; pieces of territory or pieces of information. Through this mask, the map depiction of the world becomes more opaque than transparent.

cartographic conventions such as a base map, the author challenges the notion of map objectivity (Fig. 12.5).

The famous example the *Carte du pays de tendre* – Map of Tenderness – from Madeleine de Scudéry (1654), provides another way to map feelings and emotions. In this allegorical map, she replaced toponyms by distraction and pitfalls that lovers encounter. This map inspired many other representations (Van Swaaij and Klare 2000; Harmon 2004). Contemporary artists map their personal feelings and experiences in different ways. Some draw their personal map for example, *Mon plan de metro de Paris, Air de Paris* (in GNS 2003) or the *Fetish Map of London*, by Chris Kenny (in Harmon 2004) while others describe feelings and experiences on maps. In *Spatial Poem No. 5* (1972 in GNS 2003: 38) artist Mieko Shiomi asked people all over the world to describe what happened when they opened something that was closed. Then she located these comments on a world map (Fig. 12.6). As

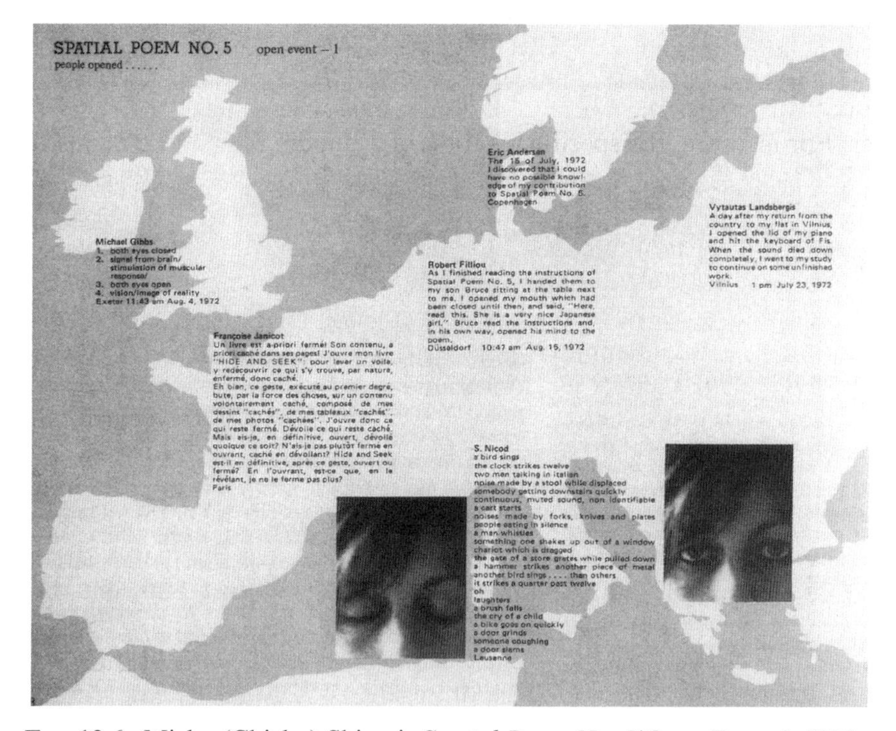

FIG. 12.6. Mieko (Chieko) Shiomi, *Spatial Poem No. 5/Open Event 1, 1972* – Off-set on paper – Walker-Special Purchase Fund, 1989 – In this work artist Mieko Shiomi asked people all over the world to describe what happened when they opened something that was closed. In the context of this chapter, these personal and subjective experiences can be seen as a way to oppose the apparent neutrality and objectivity of conventional maps.

Daniels (1994) points out, her work "combines approaches in conceptual art and *Mail Art* to form a new, global composition which anticipates aspects of telecommunication such as simultaneity and ubiquity." More recently, artist Lori Napoleon has created a project based on personal maps. In *Mapsproject* Napoleon had collected personal maps associated with personal documents (e.g. pictures). The "original goal of the project was to view each map as a portrait of the individual who created it" and to see how each person sees his/her place in the world (Napoleon 2004, http://www.subk.net/mapsabout.html). In the context of this chapter, these examples show how artists integrate personal experiences into maps. These personal and subjective experiences can be seen as a way to oppose the apparent neutrality and objectivity of conventional maps.

The integration of these experiences can now be extended with recent technological developments. The democratization of GPS and of mobile telecommunications – outlined by Gartner in Chapter 16 – opens the map to all sorts of information and perceptions provided by all sorts of people. Maps do not have to convey the perspective of a fixed group of people anymore. They can carry many perspectives at the same time and these perspectives can change at anytime. Anyone can become a content developer. For instance, the specific location of a building of particular interest can be found through "geocatching" using mobile location devices. Once the building is found the person can write her/his impressions/sensations/ideas on a collective weblog. It is then easy to integrate these weblogs into the map to make the map-user aware of the different feelings or opinions concerning the space represented. Other projects such as *MapHub* (2004) (http://hactivist.com/maphub/overview.html) or *Geograffiti* (Tuters 2004) (http://www.gpster.net/geograffiti.html) are designed to leave "location-dependent messages or media to be accessed by others not by a URL, but by the real location itself" (Tuters 2004). Map perspectives become diversified and the conventional information – the base map – is then associated with subjective personal experiences that are constantly evolving.

The different examples of maps diverted from their original "objective form" by artists that have been revisited in this section, highlight many strategies developed to challenge the objectivity of maps. This also emphasizes on the importance of modern artists in this matter as well as their influence in more contemporary works. The question that remains at the end of this section concerns the integration of these ideas and methods into conventional maps: How can these artistic representations could be effectively combined with conventional maps to stimulate reflexivity among map-users?

This section has highlighted multiple examples of the ways in which artists transform maps in a manner that challenges their presumed objectivity.

In so doing, it draws attention to the importance of work by modern and contemporary artists in interpreting the meaning and significance of mapping. Such work invites, furthermore, reflexivity from its audiences. The question that remains to be answered is whether such artistic techniques could be integrated into conventional mapping, for the particular purpose of stimulating reflexivity among map-users. The next section addresses this issue.

4. Cybercartography and Reflexivity: Combining Maps with "ANTI-MAPS"

The exploration of the relationship between art and maps pursued in this chapter so far, leads to three main strategies that can be used for integrating artistic and personal experiences into conventional maps. The first consists of linking existing artistic creations with conventional maps. For example, the surrealist map of the world could be combined with a conventional political map of the world. A second strategy would be integrating the methods and techniques developed by artists into conventional maps. The technique of perspective switch (e.g. Upside-down Map) could be applied to strongly institutionalized maps, such as country maps. A third strategy could be for contemporary artists to develop their own vision in relation to a conventional map, thereby making the artist part of the cartographic project. None of these strategies is exclusive of the other. They could be applied separately or together depending on their relevance and on the specificities of the project. Whatever the strategy chosen, the integration of unconventional maps into conventional ones requires more than just techniques and methods. It requires a new paradigm, which is how Taylor (2003) envisions cybercartography. In fact cybercartography provides not only the technological tools but also a unique environment for creativity as illustrated in several chapters of this book. This creative environment could be useful for combining conventional representations with less conventional ones.

As these strategies indicate, the notion of combination is central in the perspective of this chapter. Because existing map forms are so strongly connotated with objectivity, proposing new perspectives on maps can only be achieved using new forms of map construction. Producing new forms of maps is a challenge in terms of creativity. Moreover, new maps which completely ignore previous images are liable to be misunderstood and rejected by the users (Jacob 1992: 138). This point is emphasized by Peuquet and Kraak (2002: 89) for whom "no matter how gifted and imaginative the individual, he or she will not be able to grasp a concept that is too far unrelated to what they have already known and experienced." This point has been illustrated in a previous paper (Caquard 2003) where certain graphical elements were integrated to challenge map objectivity. The reactions of these map-users were negative such as "it's quite complex" or "it's better without it." In other words new representations should be developed while "traditional" representations should be kept. This dilemma is recurrent in contemporary cartography. Should we develop new maps based on our existing knowledge or should we create a new taxonomy based on the specificities of the cyber world? Monmonier addresses this question in this volume.

In the context of this chapter, the option proposed is to combine conventional representations with unconventional ones. The aim of this combination is to keep the attention of the user with the conventional map and to stimulate his/her reflexivity with the unconventional ones. Any map would have a double or – to reuse the metaphor "image/anti-image" introduced by Ruiz (1995: 60) for the cinematographer – an "anti-map." As the anti-image is the subconscious of the image and vice-versa, the anti-map could become the subconscious of the maps. The combination map/anti-map could provide an opportunity to open the world of maps to artists and their critical and poetic visions of the world. As Tuan (1999: 23) points out "good artists – serious artists – even in their most utopian aspirations in matter of design and color, can incorporate, without tortuous artifice, the negativities of life and the world."

From a technical perspective this combination could be easily done by using interactivity or animation. For instance, any map could have an associated anti-map that could be activated at any time by the user. The multimedia and multisensory dimensions of cybercartography also expand the field of possible representations. For example, anti-maps would not have to be graphic. They could be audio as well as tactile. This expands the field of artistic expression by using appropriate materials and techniques. This also opens the possibilities of using media that are not necessarily connotated with objectivity in cartography. For example, the tactile dimension of a map is a new dimension that does not have yet strong connotations. Such media diversity is also appropriate for facilitating multiple kinds of expression beyond graphic ones. It opens the world of maps to nongraphic artists and people in general. Furthermore, it takes the map outside of its usual framework, deepening the challenge of objectivity. As Curnow (1999: 264) points out "the question some conceptual art has raised about neutrality, transparency and objectivity of the map have also been directed at the museum, and its white cube interior decoration." If the map might keep a conventional structure, the anti-map definitely needs some new media to convey well its different perspective.

The Internet is another important strength of cybercartography. Theoretically it allows anybody to access information in real time. This means that maps and anti-maps could be accessed by anybody. This also means that anybody could develop or update anti-maps at any time. Any user having access to the technology could add multiple kinds of perception, feelings, information, and critiques remotely. While the map would remain under the control of its producer, the anti-maps would be shaped by diverse annotations, discussions, and controversies. In this sense, the map/anti-map combination opens up questions and discussions, rather than shutting them down. The distinction between user and creator would also be blurred and users would contribute to the construction of the anti-map as a living entity by integrating her/his perspective regarding the phenomena mapped.

Imperiale (2003: 39) suggests considering mapping "like writing a text that negotiates, travels and narrates without seeking a fixed conclusion." These notions of "non fixity" and "non answer" synthesize some of the ideas developed in this section. In order to challenge map objectivity and to stimulate critical map reflexivity, perspectives should be diverse and changing. New perspectives could be integrated at any time to highlight different points. This might also be interpreted as a return to a more narrative itinerary in maps, to restate De Certeau (1984). For him maps represent place while the narrative itinerary represents space. Place is static and passive while space is a practiced place. De Certeau notes that the map has slowly disengaged itself from itineraries and become place bound. By integrating multiple practices, experiences and perceptions of space, we might be able to reverse this tendency.

Finally, the integration of these multiple perspectives could be facilitated by the interdisciplinary dimension of cybercartography (Lauriault and Taylor in Chapter 8). This dimension gives a unique opportunity for cartographers to interact with people and perspectives from other disciplines as several chapters in this book illustrate. In this environment cybermaps become a combination of many different points of view. The diversity of these perspectives is definitely a part of the unique environment created by cybercartography. This kind of environment is vital in the context of this research for rethinking deeply the way we design maps, the reasons for their creations and the consequences thereof.

Beyond these strong assets, the use of cybercartography to stimulate reflexivity faces several challenges. First at a practical level, any user of cybermaps can potentially construct any kind of image using remote databases and online cartographic tools. These are appearing in increasing numbers. In these circumstances it is impossible to associate systematically each map with its anti-map(s). Conceptually it is also inappropriate to develop systematic anti-maps. Even if any map is a constructed image of space, the acknowledgement of this dimension is not systematically relevant. It might be more relevant to focus on some specific kinds of maps, such as political maps to reduce the risk of systematic rejection by the general public. This raises a second challenge which is to make anti-map(s) attractive to map-users. As described previously, users are often uncomfortable with critical perspective on maps. They want some "objective" and clear information. Engaging them in the use and production of anti-maps would then require the development of specific techniques such as humor. Finally a major challenge could be to convince cartographers of the vital interest of challenging map objectivity in their production. In a pre-cybercartographic world cartographers did their best to construct images as objectively as possible. In the approach suggested here, cartographers are required to go beyond the "most objective possible" concept and to accept the social construction of knowledge. Such a position poses a challenge to the ways in which cartography is currently institutionalized within scientific communities.

5. Conclusions

Stimulating reflexivity is a new challenge for cartography. Its fundamental function is to educate map-users and mapmaker in the power of the maps they use and design. As the last section of this chapter indicates, cybercartography provides a conceptual and technological framework vital for drawing a transitional path for cartography. Following this path will not be easy and will raise some fundamental questions. Is it the role of the cartographer to produce views challenging her/his own work? Yet, some artists do this as a matter of course as highlighted by Curnow (1999: 268):

> They are the artists who have most tellingly challenged the transparency of the map, its claim to semiotic and social innocence and objectivity; they are the artists who have most thoroughly questioned the hegemony of the visual implicit in the inscribed map, and have sought to restore mapping the task of articulating the kinds of spaces it saw fit to silence. And it is they who have looked for an ethics of mapping through their adoption, in the arts of performance and installation, of exemplary manners of use.

In this context, can new relationships be developed between art and cartography? In the past, artistic movements challenged the "objectivity" of realism in art leading to increased social criticism within art. Could new cybermaps create a similar movement in cartography and in the process of challenge claims to cartographic objectivity? If this is so, a major task is now to design cybermaps that would do so effectively.

Acknowledgments

We would like to thank Patricia M. Martin, and Tracey P. Lauriault for constructive critique, insightful comments, and considerable editorial assistance. William Cartwright and Georg Gartner also offered challenging and useful perspectives on an earlier version of this Chapter.

References

Andrews, J. H. (2001) "Meaning, knowledge, and power in the map philosophy of Harley", in Harley, J. B. (ed.), *The New Nature of Maps: Essays in the History of Cartography*, Johns Hopkins University Press, Baltimore, pp. 1–32.

Baudrillard, J. (1981) *Simulacres et simulation*, Editions Galilée, Paris.

Bourriaud, N. (2003) "Topocritique: l'art contemporain et l'investigation géographique", *Global Navigation System, Cercle d'art, Palais de Tokyo, site de création contemporaine*, pp. 9–37.

Bruno, G. (2002) *Atlas of Emotion – Journeys in Art, Architecture and Film*, Verso, New York.

Caquard, S. (2003) "Evolution cartographique et participation publique – un prototype de cartes multimédias pour la gestion de l'eau", *Revue Internationale de Géomatique, vol. 13, Numéro spécial Cartographie Animée et Interactive*, pp. 15–27.

Certeau, M. (De). (1984) *The Practice of Everyday Life*, University of California Press, Berkeley.

Clair, J. (2001) "Le surréalisme et la démoralisation de l'Occident", *Le Monde* 11-21-2001.

Corner, J. (1999) "The agency of mapping: speculation, critique and invention", in Cosgrove, D. (ed.), *Mappings*, Reaktion Books, London, pp. 213–252.

Cosgrove, D. (ed.) (1999) *Mappings*, London, Reaktion Book.

Cosgrove, D. (2003) "Historical perspectives on representing and transferring spatial knowledge", in Silver, M. and D. Balmori (eds.), *Mapping in the Age of Digital Media – The Yale Symposium*, Wiley-Academy, pp. 128–137.

Crampton, J. W. (2003) *The Political Mapping of Cyberspace*, The University of Chicago Press, Chicago.

Curnow, W. (1999) "Mapping and the expanded field of contemporary art", in Cosgrove, D. (ed.), *Mappings*, Reaktion Books, London, pp. 253–268.

Daniels, D. (1994) "The art of communication: from mail art to the e-mail", *Neue Bildende Kunst*, Vol. 5, pp. 14–18, accessed in December 2004 from http://www.medienkunstnetz.de/source-text/73/

Debray, R. (2003) "L'honneur des funambules", *Le Monde Diplomatique*, 9-1-2003.

Dubois, P. and G. Van Cauwenberge (1990) "De la verisimilitude a l'index – petite rétrospective historique sur la question du réalisme en photographie", in Dubois, P. (ed.), *L'acte Photographique*, Nathan, pp. 17–53.

Edney, M. H. (1996) "Theory and the history of cartography", *Imago Mundi*, No. 48, pp. 185–191.

Ginioux, P. (2004) "1929: L'Océanie au centre du monde?", *Séminaire Art d'Océanie, Musée du quai Branly, IUFM de l'académie de Créteil*, p. 11.

Girod, M., C. Rau and A. Schepige (2003) "Appreciating the beauty of science ideas: teaching for aesthetic understanding", *Science Education*, Vol. 4, No. 87, pp. 574–587.

GNS (2003) "Global navigation system", *Cercle d'Art, Palais de Tokyo, Site de Création Contemporaine*.

Harley, J. B. (1989) "Deconstructing the map", *Cartographica*, Vol. 26, No. 2, pp. 1–20.

Harley, J. B. (1990) "Cartography, ethics and social theory", *Cartographica*, Vol. 27, No. 2, pp. 1–23.

Harmon, K. (2004) *You are here – Personal Geographies and Other Maps of the Imagination*, Princeton Architectural Press, New York.

Harrower, M., C. P. Keller and D. Hocking (1997) "Cartography on the internet: Thoughts and a Preliminary User Survey", *Cartographic Perpectives*, Vol. 27, pp. 27–37.

Höch, H. (1919) "Cut with the Kitchen Knife, collage of pasted".

Horrell, J. (1958) "On poetry: a rhapsody", *Collected Poems of Jonathan Swift*, Vol. 2, Cambridge, Massachusetts, Harvard.

Imperiale, A. (2003) "Refiguring the figure – imaging the body in contemporary art and digital media", in Silver, M. and D. Balmori (eds.), *Mapping in the Age of digital Media – The Yale Symposium*, Wiley-Academy, pp. 35–39.

Jacob, C. (1992) *L'empire des cartes – Approche théorique de la cartographie à travers l'histoire*, Bibliothèque Albin Michel Histoire, Albin Michel, Paris.

Lussault, M. (1995) "La ville clarifiée, essai d'analyse de quelques usages carto – et iconographiques en oeuvre dans le projet urbain", in Maximy, R. (De) and L. Cambrézy (eds.), *La cartographie en débat – Représenter ou convaincre*, Karthala-Orstom, pp. 157–193.

MapHub, Accessed in December 2004, from http://hactivist.com/maphub/overview.html

Monmonier, M. (2001) *Bushmanders and Bullwinkles: How Politicians Manipulate Electronic Maps and Census Data to Win Elections*, University of Chicago Press.

Napoleon, L. (2004) Personal Web Site, accessed in December 2004, http://www.subk.net/mapsabout.html

Palsky, G. (1996) *Des chiffres et des cartes. La cartographie quantitative au XIXe siècle*, Ministère de l'Enseignement supérieur et de la recherche, Comité des travaux historiques et scientifiques, Paris.

Pedretti, E. G. (2004) "Perspectives on learning through research on critical issues-based science centre exhibitions", *Science Education*, Vol. 88, Special Issue in

Principle, in Practice: Perspectives on a Decade of Museum Learning Research (1994–2004), pp. 34–47.
Peterson, M. P. (1999) "Maps on stone: the web and ethics in cartography", *Cartographic Perspectives*, Vol. 34, Fall 1999, pp. 5–8.
Peterson, M. P. (2003) "Maps and the internet: an introduction", in Peterson, M. P. (ed.), *Maps and the Internet*, Elsevier, pp. 1–16.
Peuquet, D. J. and M.-J. Kraak (2002) "Geobrowsing: creative thinking and knowledge discovery using geographic visualization", *Information Visualization*, Vol. 1, No. 1, pp. 80–91.
Pickles, J. (2004) "A History of Spaces: Cartographic reason, mapping and the geo-coded world", Routledge, London and New York.
Picon, A. (2003) "Représenter la ville territoire : entre écrans de contrôle et dérives digitales", *Global Navigation System, Cercle d'art, Palais de Tokyo, site de création contemporaine*, pp. 54–67.
Pinder, D. (1996) "Subverting cartography: The Situationists and Maps of the City", *Environment and Planning A*, Vol. 28, pp. 405–427.
Ríos, A. (2003) *A Maps, Mapas, Cartes personal* web site, accessed December 2004 from http://www.public.asu.edu/~aarios/resourcebank/maps/index.html
Robic, M.-C. (1997) "Une école pour des universitaires placés aux marges de l'expertise : les années trente et la cartographie géographique", *Communication au Colloque 30 ans de sémiologie graphique*, Paris, pp. 25.
Rogoff, I. (2000) *terra infirma geography's visual culture*, Routledge, London.
Roland, C. (2002) *ArtJunction*, accessed December 2004 from http://www.artjunction.org/articles/mapartists.html
Rubin, W. S. (1968) *Dada Surrealism and Their Heritage*, Museum of Modern Art, New York.
Ruiz, R. (1995) "Poetique du cinema", *Dis voir*, Paris.
Silberman, R. (ed.) (1999) *World Views: Maps and Art*, Frederick R. Weisman Art Museum, University of Minnesota.
Silver, M. and D. Balmori (eds.) (2003) *Mapping in the Age of Digital Media – The Yale Symposium*, Wiley-Academy.
Swift, J. (1958) "Ou poetry a rhapsody", in Horrell, J. (ed.), *Collected Poems of Jonathan Swift*, Harvard University Press, Cambridge.
Taylor, D. R. F. (1987) "The art and science of cartography: the development of cartography and cartography for the development", *The Canadian Surveyor*, Vol. 41, No. 3, pp. 359–372.
Taylor, D. R. F. (1991) 'A conceptual basis for cartography/new directions for the information era", *Cartographica*, Vol. 28, No. 4, pp. 1–8.
Taylor, D. R. F. (2003) "The concept of cybercartography", in Peterson, M. P. (ed.), *Maps and the Internet*, Elsevier, pp. 405–420.
Torres, Cecilia de. (2004) *Joaquin Torres-Garcia*, accessed December 2004 from http://www.ceciliadetorres.com/jt/jt.html
Tuan, Y.-F. (1999) "Maps and art: identity and utopia", in Silberman, R. (ed.), *World Views: Maps and Art*, Frederick R. Weisman Art Museum, University of Minnesota, pp. 11–25.

Tuters, M. (2004) *Geograffiti*, accessed December 2004 from http://www.gpster.net/geograffiti.html

Van Swaaij, L. and J. Klare (2000) *The Atlas of Experience*, Bloomsbury, New York.

Wood, D. (1992) *The Power of Maps*, Guilford, New York.

Wood, D. and J. Fels (1986) "Designs on signs/myth and meaning in maps", *Cartographica*, Vol. 23, No. 3, pp. 54–103.

Woodward, D. (ed.) (1987) *Art and Cartography, Six Historical Essays*, The University of Chicago Press, Chicago.

Wright, J. K. (1942) "Map makers are human: Comments on the Subjective in Maps", *The Geographical Review*, Vol. 32, No. 4, pp. 527–544.

CHAPTER 13

Mapping Play: What Cybercartographers can Learn from Popular Culture

BRIAN GREENSPAN

Institute for Comparative Studies in Literature, Art and Culture (ICSLAC) and the Department of English Language and Literature Carleton University, Ottawa, Ontario, Canada

Abstract

Although cybercartography is a new field of study, its arrival has long been rehearsed in popular culture and entertainment, including novels, films, Web sites and digital games. The widespread representation of cartographic, navigational and location technologies in popular texts have created cultural expectations that far exceed the current technical limits of mapping tools. At the same time, these popular narratives enable a growing familiarity with the concept of cybercartography that cannot but shape the way users approach the cybermaps of the future. This chapter will explore cybermaps as a popular concept, examining some of the cybercartographic interfaces (both real and fictional) that have been imagined and implemented in printed fiction, hyperfiction, Web sites, films and digital games. Through various forms of cultural analysis, I will show how the stories we tell about cybermaps today can inform future cybercartographic theory and practice.

Note: View the SIM City Screen Capture Images on the Accompanying CD-Rom.

1. Introduction

Although cybercartography is a new field of study, its arrival has long been anticipated and rehearsed in numerous forms of popular culture and entertainment, including science-fiction narratives, cinematic technothrillers and videogames. Like virtual reality, cybermaps represent at one extreme a form of "vapourware", conceptual tools with developed cultural expectations that far exceed their current technical limits. At the same time, these inflated expectations are grounded in our collective experience of fictional representations of cybermaps in literature and film, as well as our interactions with actual, working cybermaps in various interfaces, including Internet mapping sites, consumer-grade GPS devices, and cartographically oriented digital games. Millions of people are already using new cartographic interfaces to compensate for the failure of traditional cognitive strategies, as they navigate both the material and immaterial landscapes of contemporary space. As Taylor (2003: 404) has argued, "[t]he development of computer hardware and software is driving research and production rather than an analysis of the uses to which these technologies should be applied. A new user-centered model is needed, in which the demand for cartographic applications has a larger role". Yet, while user-needs analyses and controlled usability testing are essential elements of cartographic design, as outlined by Lindgaard and Parush in this volume, this chapter will suggest that researchers must also pay attention to the cybermaps which people use every day to navigate their increasingly elaborate digital environments.

Those researching cybercartography have identified numerous geovisualization, cognitive and usability issues for further investigation, including the impact of spatial perception, cognitive overhead, and metaphors for nonspatial data on the use of digital mapping interfaces. There exist numerous studies of the design of navigation tools to assist wayfinding strategies in virtual environments, most of which involve user-testing and user-needs analyses grounded in well-established psychological theories of cognitive mapping. Following on the now classic studies by Lynch (1960) and Seigel and White (1975), psychologists have developed numerous tests to discover the strategies individuals develop for the cognitive mapping of digital spaces (Edwards and Hardman 1989; Frank 1995; Dieberger 1996; Boechler 2001). And yet, while the concept of cognitive mapping has had the greatest impact in the field of cognitive psychology, it is also associated with a rich tradition of materialist social and cultural theory, thanks largely to the influential Marxist theorist Fredric Jameson. When Jameson originally wrote on the topic in 1983, he was able to declare, "I am addressing a subject about which I know nothing whatsoever, except for the fact that it does not exist" (Jameson 1988: 347). Of course, cognitive mapping does exist, both as everyday practice and as a rich field comprising research in, for example, geography, urban planning and cognitive psychology. Much more than the "low-level subdiscipline" which Jameson claimed as a theoretical tool, cognitive mapping has come to represent a powerful interdisciplinary metaphor.

Not surprisingly, Jameson's particular use of the cognitive mapping metaphor has had its greatest impact on cultural critics. He describes cognitive mapping as an unrealized aesthetic so that it might stand for a new form of symbolic resistance to the forces of transnational capitalism, as a figuration of "the new kinds of representational processes demanded" by postmodern social spaces (Anders and Jameson 1989: 21). As one commentator puts it, for Jameson, "cognitive mapping is simply shorthand for transnational class consciousness" (Cazdyn 1995: 139), a strategy for resisting postindustrial assaults on spatial and social relations. In the wake of recent critiques of globalization, this view seems only to be gaining in influence. In his recent study *Cognitive Fictions* (2003), Joseph Tabbi describes cognitive mapping as an aesthetic through which "the 'individual subject' can be kept intact and... [postmodernity's] repressed material culture – Jameson's 'political unconscious' – can be brought to awareness much as the unconscious is brought into consciousness during psychoanalysis" (Tabbi 2002: 44–45) – a sort of "walking cure" for the schizophrenic postmodern subject.

Needless to say, this Lacanian subjectivity hardly bears upon the cognitive mapping models which are elaborated through empirical testing. To psychologists, Jameson's sense of cognitive mapping as a correlative of postmodern cityscapes will never appear more than, as Paul Patton puts it, "a madeup image of reality, the invention of an overheated theoretical imagination" (Patton 1995: 114). And yet, this very imaginary quality perhaps explains why Jameson's theory has been linked repeatedly to recent attempts to model future geographies in science fiction literature and film. As Farnell (1998: 460) writes in his study of William Gibson's speculative narrative landscapes, "a 'cognitive mapping' of our postmodern present-as-simulacrum is actually a prerequisite step to any speculative contemplation of our possible futures". Similarly, Kneale and Kitchin (2002: 9) describe science fiction as "a useful cognitive space, opening up sites from which to contemplate material and discursive geographies and the production of geographical knowledges and imaginations." Though fictional, such representations nevertheless play a role in determining how individuals visualize information spaces. In his study, *Cognitive Maps and the Conceptualization of Cyberspace*, Frank (1995) reports that "[t]he Gibsonian image of cyberspace as a neon grid stretching to an endless horizon seems to hold sway over many imaginations".

Of course, Gibsonian cyberspace is also a metaphor for the flow of data within today's networks of transnational capital: "familiar geographies of the real world persist in these narratives of virtual spaces". (McCallum 2000: 352). This critical function explains why the futuristic landscapes of popular literature and film have, by and large, emphasized the dystopian effects of our collective loss of cognitive maps (Sponsler 1993). By contrast, urban and architectural theorists working with digital media have emphasized the utopian potential of new cartographic practices to reclaim postmodern spaces. Declaring traditional geographies "useless" for representing contemporary spaces, the Italian urbanist Stephano Boeri has called for the production of "eclectic atlases":

> [t]hese are heterogeneous texts (reports, photographic surveys, geographic and literary descriptions, classifications, research reports, qualitative investigations, essays and articles, anthologies and monographs, collections of plans or projects...) but similar to the visual approach. They tend to be 'atlases' because they seek new logical relationships between special elements, the words we use to identify them, and the mental images we project upon them. They tend to be 'eclectic' because these correspondences are based on criteria that are often multidimensional, spurious and experimental (2004: 119).

Though enticing, Boeri's rhetoric of innovation ignores the likelihood that cybermaps will continue to depend on traditional geographies for some time. At the same time, Boeri underestimates the changes which new media will bring to mapping practices. Even his eclectic textual atlases are trapped within a visual paradigm, as examples of "plural visual thinking" (Boeri 2004: 120), and propose nothing like the audible, tactile and olfactory maps which cybercartographers are proposing (Taylor 2003: 405). Boeri does not acknowledge the effects which new media have on the storage, transmission and display of data. Nor does he consider the potential for cognitive overload, disorientation or other destabilizing effects which users might experience when encountering large-scale digital environments through radically pluralized cartographic perspectives (see Garling *et al.*1984).

Rather, Boeri reconceptualizes the map as artwork, framing it in an avant-garde aesthetic that disrupts the representational space of the map, alienating the user in order to enable new perceptions of the ideological power of cartography. Rogoff (2000: 83) describes such an alienating cartographic aesthetic in her fascinating survey of geography's visual culture. Rogoff (2000: 83) examines numerous international art installations in which "the map, the tool of location and emplacement, of clarity and illumination, is transformed into a hidden, veiled, somewhat indecipherable and evidently secret... object". Rogoff (2000: 97–98) notes that artist's maps "can be read as the formulation of a theory of cognition", but one "imbued with doubt concerning... the empirical basis... on which any theory of cognition must be founded . . . [F]or all of their cartographic detail these maps set up a strategic resistance to being read empirically". Rogoff's insights and examples are fascinating, but cybercartographers are not likely to take them as models for design and implementation, as such artistic obfuscations of the map threaten to further mystify the digital interface and disorient the cartographic subject. This aesthetic of dissonance and alienation remains at odds with both Jameson's project of raising transnational consciousness, and the broader cybercartographic goal of improving usability by naturalizing cybermaps for a wide user base.

2. "The Map is Clunky": Cybermaps and Their Users

Between better usability and Rogoff's sense of maps as "strategic resistance" there remains a third space, a space filled with the countless representations and implementations of cybermaps that already exist within literature, film, advertisements, games, Web sites, and other popular texts. Mark McGuire (2003) notes how Sony's use of the phrase "Third Space" to describe its PlayStation 2 gaming platform contradicts the term's familiar association with public sites of leisure distinct from both the private home and the public workplace. I am reclaiming the term somewhat differently, to describe a popular space of digital cartographic representations that act as interface between empirical and avant-garde understandings of cognitive maps.

In this chapter, I explore the popular dimension of cartography as a way of bridging Jameson's macropolitical analysis on the one hand, and usability studies of navigational interfaces based in cognitivist principles. Specifically, I argue that cybercartographers can learn valuable insights into the effective and appealing design of mapping interfaces from users of digital games. It is no secret that topographic and geospatial maps have emerged as the dominant metaphor of digital games, due in part to the evolution of game design, as well as the overwhelming success of titles like *Caesar, Age of Empires, Final Fantasy, The Getaway* and *SimCity*. Although cybercartography involves highly technical and complex issues of data management and display, its ultimate goal is to mediate and interpret this complexity for a broad user-base. As the most popular of digital applications, games would seem to present a natural source for design inspiration; cartographic researchers, by and large, have yet to acknowledge the prominence of mapping interfaces within digital games. The designers of the *Skylink Virtual Frankfurt Project*, for instance, seem entirely unaware of the state of digital games:

> it is quite interesting, what kind of images of a "city" are used by digital cities. Basically they are shaped after a mediterranean small town. . . . Superhighways, shopping malls

> and the sprawl of the suburban fabric, which are our urban reality of today, will not be found. A formulation of a new typology for the digital world is hardly ever found. It seems that the city-metaphors, which are employed by most digital-cities, seem not quite suitable for neither [sic] the structure of the modern city nor that of the digital world (Franken 1996).

Even for its time, the project's application of a "hypertecture" interface to visualize the economic dynamics of a metropolis seems quaint in comparison to the synoptic urban maps of *SimCity*, a best-selling digital game since 1989.

William Cartwright (2001b) is almost alone in arguing that the lessons of popular geographic interfaces, including electronic books, interactive television and games like *SimCity*, "could be applied to contemporary mapping package interfaces." As a genuine interactive urban simulation environment, *SimCity* is first and foremost a mapping game which simulates many advanced cybercartographic features. Although its represented territories are fictional, the game arguably provides the most thoroughly tested and usable cartographic interface available:

> the influence of *SimCity* is greater than that of academic geography, both in terms of the numbers of users, and in terms of the initial appeal of its 'message'...[T]he popularity of the program indicates that the more geographers can emulate the attractive and entertaining features of *SimCity* in their urban simulation models, the more likely they are to promote alternative understandings, either in the classroom or beyond (Adams 1997: 391–392).

Some usability researchers have turned to digital games for insight into the improvement of application design for both expert and novice users. The first-person shooter *Doom* has been used as an interface for UNIX systems process managers (Chao 2001); while Grice and Strianese (2000) argue that applying the adaptive learning curves of gaming interfaces to business and educational applications could help novices to master their use: "[gamers] develop strategies for improving their performance and become increasingly expert at achieving success through use – something that is not usually seen with other types of software "(2000: 536). For the most part, however, game authors and interface designers have remained, in Chuck Clanton's phrase, "two isolated design communities": "few software applications show any awareness of techniques of game design that could make them easier and more fun to learn and use" (2000: 301). Recently, Swartout and van Lent (2003) identified numerous features of digital games that could help to increase the appeal of applications generally, including self-scaling difficulty levels, multileveled goals, and a balance between automation and user-control. They further suggest that games provide ready-made virtual environments in which prototypes can be tested, as an improvement on more familiar methods of "paper prototyping".

Some researchers have pointed out that user satisfaction arises not only from ease of usability, but also from the effort required to overcome hurdles built into the interface. Digital games tend to multiply these hurdles in the form of puzzles, textual ambiguities, unreliable narration, red herrings and other forms of cognitive dissonance:

> In order to work, novelty and surprise must impair at least the external consistency of the software, a core principle of usability. In usability design, this might conflict with the task-related efficiency and effectiveness of the software system. This raises the question, whether it is appropriate to call for enjoyment and fun when it comes to work-related software systems (Hassenzahl *et al.* 2000: 202).

Hassenzahl *et al.* (2000) found that while such "hedonic" elements interfered with the ergonomics of an interface, they compensated by contributing to its overall appeal. Other research supports the view that discovering how users solve the deliberate problems posed by interfaces provides insight into how they confront inadvertent problems arising from design flaws: "game designers know that frustration is not an enemy but a friend.... The pleasure of mastery only occurs by overcoming obstacles whose level of frustration has been carefully paced and tuned...." (Clanton 2000: 301).

Champion (2003) notes the potential of user-testing with game-like environments to gather data on the navigation of nongaming applications, such as virtual heritage environments. His test case, a virtual Mayan city, is of particular relevance to cybercartography for its reliance on maps both as user navigational aids, and heuristically to monitor each user's progress:

> Engagement will be assessed by physiological measuring of users, questionnaires, and indirect monitoring, such as mapping of user paths, points scored in solving tasks, social roles obtained, and the inventories of artifacts collected. Users will be able to check navigation via maps in flyover mode, scaled down walk through mode, or axonometric mode. They can also choose to monitor their progress via point scores, maps of their path, or by inventories.

These studies emphasize the use of stand-alone game environments for gathering user data for new applications, but tapping into multiplayer online games presents a special challenge: "Testing Internet games is a major problem for games developers. It's impossible to fully predict how the servers and the game networking will work with really large numbers of players and a wide variety of connection speeds, connection qualities, and ISPs" (PCTest 2004). Although the usability company PCTest pioneered game-testing with such classic titles as *SimCity* and *Myst*, they note that analyzing Internet gaming is challenging because of the compatibility issues that arise when players compete using different browsers and platforms. Despite these challenges, the networked component of games offers a particularly valuable resource for cybercartographic design. Usability experts typically perform user needs analyses through direct observation of subjects in the lab and posttest questionnaires, but will also occasionally turn to "canned" feedback provided from sources such as Usenet discussions (Divitini *et al.* 2000). Hilbert and Redmiles (1998) have even argued that small-scale user testing in the usability lab limits the amount and kind of data collected. As an alternative, they advocate the use of intelligent agents to mine the Internet for usability data: "the Internet can be used as a medium for collecting 'direct' user feedback in the form of subjective user reports, as well as 'indirect' feedback in the form of automatically-captured data about application and user behaviour" (Hilbert and Redmiles 1999: 35).

For the kinds of user-data required by cybercartography, *SimCity* Usenet lists present a particularly rich source of canned, direct feedback. The game by its nature encourages reflection on and tinkering with digital maps. Unlike *Doom, Unreal* and other real-time first-person shooters, *SimCity* is relatively slow and static: gameplay comprises the space-time of an entire city as it grows over decades or centuries. With the Terrain Modification tools and "urban renewal" kit (or SCURK) released with *SimCity* 2000, the game has even allowed users to customize the appearance of their cities, and produce urban plans modeled after their favorite real or imaginary urban centers. Moreover, the *SimCity* fan base provides a particularly savvy community of users with clear directions for improving the game's mapping interface. *SimCity* gamers comprise a sophisticated user group, many of whom regularly build mods and fixes for design flaws they perceive in the original product. In addition, the designers of *SimCity* have been particularly attentive to their user community. Maxis not only consults its user base, but maintains newsgroups through which users can provide feedback on the game, give notice of any software glitches, or upload their own game modes. The company even presents the Golden Llama award to the most user-friendly game improvements submitted by users. To use this information as a source of design improvements for nonentertainment mapping applications, however, involves overcoming some negative preconceptions about digital games and their users: "Researchers have found that peer-to-peer teaching reinforces mastery; why, then, do we dismiss such information exchange in the context of gameplay (a Web site devoted to strategies for a particular game, or picking apart the rules of a simulation to ensure maximum efficiency) as somehow intellectually illegitimate"? (Holland *et al.* 2003: 29). *SimCity* users provide a source of direct feedback that can help in giving direction to more controlled studies of the usability of cybermaps, and suggest future design improvements.

Of course, not all user feedback or suggested game improvements will necessarily advance the project of improving cybercartographic interfaces in general. But *SimCity* users prove particularly adept at intuiting the primary usability issues of digital maps. Researchers have identified a number of primary goals for cybercartography, including: finding ways to avoid disorientation while navigating geospatial virtual environments; designing interfaces for enabling collaborative geovisualization; determining the appropriate balance of realism and abstraction; measuring the effectiveness of animations; determining the best mix of geographic and other statistical data; assessing different metaphors for geovisualization; evaluating the pedagogical benefits of cybermaps for different educational fields and levels; and developing methods for training people to use digital maps (Slocum *et al.* 2001: 69–71). Just as *SimCity* maps exhibit many of the dynamic and interactive features called for by cybercartographic research, so feedback from its users offers insight into all of these research challenges. For instance, one user feedback posted a series of questions about the design of *SimCity* maps that speak to multiple issues of cartographic representation, including scale, color legends, and even considerations of the political division of simulated regions:

- How can you make a region smaller or larger than the default (has to do with the config.bmp)?
- How can you control the division of a region into the different cities...?
- What is the size of one tile (approx. 16 m, I think I heard, but I'm not sure)?
- What is the maximum height above sea level (in m, too)?...
- What color values correspond to what height? What's sea-level at? (Goerz 2003).

Such feedback demonstrates the degree to which distinct aspects of cartographic design can become confounded for nonspecialists; however, it also suggests the need for more research to determine how users discern meaningful cartographic symbols and signs from those dictated by interface conventions. At times these distinctions collapse, leaving users uncertain as to whether particular game elements function aesthetically or as navigational aids. One *SimCity* user writes that architectural landmarks like the Eiffel Tower or Taj Mahal "are basically there to enhance the visual look of the game and allow the creation of more realistic, less-generic looking cities" (SimCity User, 1998). This user misses the possibility that gamers find these landmarks appealing precisely *because* they assist navigation through the virtual city. Since user comments cannot always be taken at face value, more research is needed to determine under what conditions the repurposing of a familiar real-world landmark either assists or interferes with navigation in virtual environments.

Many *SimCity* users express a desire to push the limits of the game's maps, suggesting new areas of research and development for cybercartographers in general. Ease of navigation is a major implementation challenge for the field, as it cuts across all aspects of cartographic representation:

> navigation implies that the user is able to both navigate the visualization itself and navigate the synthetic geography that it represents. Navigation can refer to navigation between 'map' representations; navigation within 'map' representations (e.g. pan, zoom, scale, generalization); navigation between and within datasets; navigation between spatial objects and related temporal and thematic attributes and related information; and navigation between spatial objects and metadata (Cartwright *et al.* 2001a: 47).

SimCity players understand these navigational challenges on a pragmatic level, since mastering the game involves learning how to move across the simulated territory by alternating between several varieties of navigation. Many gamers register their frustration with challenges in map navigation which interfere with the gameplay rather than augmenting it. For example, the game allows the map "edge scrolling" feature to be toggled on or off, but at least one user calls for finer control over the rate at which game maps scroll (Eep2 2003). Another admits to confusion over how the game interface labels its various map control features: "I had, of course, seen that Edge Scrolling option several times but subconsciously confused it with Auto Edge Reconciliation" (Hodson 2003). Such divergent feedback indicates the necessity to determine the appropriate balance between simplicity of use of a mapping interface, and the degree of control offered to users.

Successful gameplay also depends upon navigation between different map views, including those presenting nonspatial information, such as economic data or crime statistics. Similarly, "[a] major challenge [facing cybercartography] is understanding how different types of information should be structured and presented to enable users to navigate through complex databases" (Taylor 2003: 408). The best method for accessing different data views from the same map is a point of contention among gamers. In responding to improvements introduced in version 4 of *SimCity*, user Welch (2003) writes, "I miss the selective views of previous versions. I would like to be able to just see the transportation system, for example, for placing subway stations and bus stops". He also calls for clickable maps that provide information directly: "You cannot click on a road with the query tool and have a summary of how much traffic is passing on the road, you have to go to the overlays" for situation updates (Welch 2003). By contrast, other users would prefer more overlays, including "population density and job density overlays" (Little 2003). While such disparate user feedback alone cannot indicate whether clickable maps are preferable to overlays, they at least suggest an area for further investigation, or what

Holland *et al.* (2003) call an *illuminative tension*. As they argue in their study of massive multiplayer game design, "[m]apping these tensions can enable researchers to identify the core activities and predict sources of change within a social system", (2003: 33) such as the social system of user communities.

Another major area of targeted cybercartographic research involves the question of what metaphors can best represent a given geographic space. Fairbairn *et al.* (2001: 17–19) have documented the role of metaphors in cybercartographic research and development, including the map metaphor itself. One of the greatest challenges for cybercartography is the selection of adequate spatial metaphors to represent nonvisible data, such as mortality rates or temperature (Cartwright *et al.* 2001a: 46), a process known as *spatialization* (Slocum *et al.* 2001: 65). For instance, the closeness with which the interface metaphor corresponds to the information domain it represents is just one factor that has been identified as a predictor of the usability of digital maps: "If a metaphor represents a mapping of two closely-related conceptual domains, then the correspondence of properties is high. The greater the cognitive distance, the fewer... the number of properties that can be mapped, but the greater the novelty (through novel correspondences) and the greater the potential for new insights" (Slocum *et al.* 2001: 89). However, it is not safe to assume that the popular appeal of a particular cartographic interface metaphor necessarily indicates a design which can be successfully transposed to other systems or contexts. It is also likely that different metaphors might be used to represent the same datasets in different media or information environments (Slocum *et al.* 2001: 65–66). Moreover, it is possible that no single mapping metaphor will prove best for all cases. Since cybermaps comprise heterogeneous data sets, discourses and semiotic systems, their interfaces require users to negotiate and merge multiple metaphors (see Cartwright *et al.* 2001a: 54).

Feedback from *SimCity* users offers insight into many of these representational issues. Until the release of *SimCity* 4, the game provided no distinction between navigating the visualization interface and navigating the "synthetic geography" it represents. Cybermaps generally heighten our awareness that the prioricity traditionally assigned to the cartographic referent collapses into the representational metaphors of the interface itself. Koskimaa (2000) has even argued that mapping metaphors exist solely to enable this collapse of distinctions between representation and reality, at least in the context of entertainment interfaces such as hypertext fiction: "In fact the map offers only very limited possibilities for actual navigating.... The use of a representational map is, of course, one way of making the interface 'fuse' to the represented fictional world" (*Map as Metaphor: Victory Garden* 2003:3). Games like *SimCity* provide a measure of how users accept, negotiate and rationalize different levels of representational abstraction. Though often characterized as unable to distinguish games from reality, gamers are in fact poignantly aware of the trade-offs between usability and realism in computer representation. *SimCity* is often lauded for the "realism" of its visualizations, but gamers regularly debate the most appropriate balance of realism and abstraction in *SimCity* maps. Some gamers retreat from user-generated mods because they provide a level of control over the simulation that ruins its verisimilitude. As one user warns, "if you are a realism freak like me do NOT use these modes" (Karybdis 2004).

Although the majority of *SimCity* users complain of having to sacrifice game graphics for improved processor performance, many users state a preference for usability over realism, and engage in nuanced discussions about the appropriate balance between graphic realism and the presentation, management and migration of map data: "I would rather have lower texture resolution (which would still rival most, if not all, FPSes and all other three-dimensional games) and FULLY (smoothly, not in steps) rotatable/zoomable camera views". (Eep2 2003). In response to this posting, another user reveals a nuanced awareness of the technical challenges of representing realistic landscapes:

> The reason why they have the set zoom levels and camera angles is the illusion provided by the textures of the higher poly building map breaks down at certain angles (the perspective of the detail starts looking wrong).... Apparently, Maxis wanted to maintain the richness of the buildings and the 'depth' provided by the current level of detail... (Wynands 2003).

Similarly, a user calling for "a feature that allows you to rotate the map three-dimensional style with the mouse so you can get a perfect top view" (Slavin) is told that "a zone map with all public/transportation buildings should not be to much of a problem.... [However,] IMHO a full three-dimensional rendering engine is not important for a game like [*SimCity* 4]" (Møller 2003). That users distinguish (and even toggle) between various levels of realism indicates their developing awareness of the contingency of cartographic representation. Feedback from *SimCity* users thus reveals the degree to which digital mapping interfaces are already altering user perceptions of the representational function of mapping generally.

Many *SimCity* users exhibit a more nuanced appreciation of the aesthetic and representational functions of digital maps than game designers themselves. The researcher and game designer Stephen Clarke-Willson (1998), for instance, sees maps as a "handy" addition to a game interface, but one that is secondary to the perspectival "first-person presentation" of the realistic virtual environment. While computer game design generally tends toward increased realism, *SimCity* gamers seem particularly suspicious of what this user labels the "GUI addiction":

> The makers of the Sim series seem not to believe in that old standby of managers, the spreadsheet. I'd love to see a tabular report listing all of my facilities of a given type, their locations, and various statistics... I've never understood what the designers have against reports. They seem to be obsessed with using graphical interfaces, even when those interfaces are not the best way to present the information.
>
> There's two features they could add to the report, to satisfy their GUI addiction: clicking on name of a facility could move you to the right location on the map, and there could be pop-up help for each column on the report. I'd also like the option of saving a report to a file, in a format that can be easily imported into a spreadsheet program (Kuyper 2000).

The practicality of migrating data between different representations in this manner still needs to be tested, but this kind of feedback at least indicates a preference among some users for maps that draw upon familiar business applications, in addition to realistic graphic user interfaces.

Cybercartographers have suggested that digital maps need to provide users with an overall perspective through which they might integrate the various data maps associated with a given territory, "an overview interaction 'map' with a fully synchronized field of view that would enable navigation and 'placefinding'... the use of fisheye techniques, and the application of advanced labelling algorithms" (Cartwright *et al.* 2001a: 47). *SimCity* has always provided a topographic or "axonometric" overview of the entire visualization (Adams 1997: 386), and recently has added additional views, such as the subjective camera perspective offered by the *SimCopter* game module and the *MySim* mode of *SimCity* 4. With these upgrades, users now have the ability to view their cities from the street-level perspective of a citizen, a view that adds *immersion* to the game's interactivity, two features called for by cybercartographers (Slocum *et al.* 2001: 63). A change in perspective can bring a complete change in affect, another important dimension of the user experience into which gamers provide a window:

> I bought it [*SimCopter*] because I have a mess o' SimCity maps I've created over the past 2 years and I wanted to get a perspective of them I've only been able to imagine. You can land your chopper and walk around the streets of your ENTIRE map/city and view it from street level. Cities look A WHOLE LOT cheerier in SimCity than they do when you see them in *SimCopter* (Stolen Child 1996; emphasis in original).

While some users remain unimpressed by improvements to the zoom and navigation controls of game maps, many have reacted positively to this innovation. However, the *MySim* mode also presents additional navigational challenges. The ability to zoom to street level afforded by the *MySims* view introduces numerous usability challenges associated with the navigation of three-dimensional virtual realities. Cartwright *et al.* (2001a: 51) describe some of these difficulties as typical of cybermaps generally, including the inadequacy of the mouse as an input device; the disorientation caused by navigation over distances and at speeds alien to everyday human experience; and the lack of fit between the locomotion metaphor (for example, walking, driving or flying) and the virtual space being explored. Moreover, they warn of the disorientation that can result when users toggle between the interactive (synoptic) and the immersive perspectives. Even before the *MySim* view was introduced, the disorientation resulting from view changing was already perceived as a serious user interface problem:

> Changing between views on the map is clunky. Often times it will take several seconds to go between the subterranean view and the City view. Furthermore, navigation around the map is difficult. Sim City 3000 would have benefited greatly had they taken a cue from real time strategy games such as *StarCraft* or *Myth* with their mouse scrolling of maps (Brinkman 1999).

Like many gamers, this user draws comparisons between the mapping interface of *SimCity* and those provided by various other games to illustrate how a simple technique such as real-time scrolling can best exploit the dynamic nature of cybermaps.

SimCity maps have already altered the understanding of cartographic representation in the minds of millions of users worldwide, and enabled new and playful strategies for visualizing information. As simulations, *SimCity* maps represent data relationships, not actual territories, thereby collapsing the distinction between spatial representations of realistic data (e.g. geographic terrains and cityscapes) and abstract data (e.g. economic trends, weather patterns or social unrest). In other words, it merges the *database* with the *navspace*, the two basic categories of digital media (Manovich 2001: 218). Far from a limitation, however, this aspect of the game anticipates the effects of cybercartography on the geospatial maps of actual territories: "geographers are linking disparate datasets together across place, scale, time, theme, and discipline", in order to model relationships between data, patterns and processes (Gahegan *et al.* 2001: 29–30). The mapping metaphor "visually superimposes non-geographic features or elements onto a map that still depicts geographic space". The map, especially in its digital form, is no longer seen as a complete and closed representation, but rather as "an exploratory tool for gaining new information and insight from large, digital databases" (Peuquet and Kraak 2002: 81). Peuquet and Kraak (2001: 82) acknowledge that this mapping of different data types involves an element of play: "We are already instinctively using maps and other forms of graphics to interactively explore geographic databases and solve problems. We pan the map, zoom in and zoom out, and change colors. All of these involve 'playing' with the map to allow latent relations to emerge". Clearly, digital games are especially good at enabling the serious play necessary for navigating visualizations of abstract databases.

Of course, the simulated maps of *SimCity*'s interface differ from cybermaps of actual terrain like Canada or the Antarctic. But the distinction is not absolute, for gamers have used *SimCity*'s interface to design their own maps of actual cities, including Toronto, New York and Sao Paolo, often in tandem with other design software packages. The customizable user interfaces called for by Cartwright *et al.* (Divitini *et al.* 2001a: 57) are already part of the *SimCity* interface design, which includes the Terrain Modification tool and Urban Renewal Kit. Some sophisticated users have even managed to integrate the game with geographic databases of physical data, by writing shareware code "in QBASIC that could read the elevation data... [creating] a DXF file loadable by AutoCAD to generate a mesh of the landform" (Thompson 1999). One can anticipate future gamers who achieve the cybercartographic ideal of integrating the game interface with dynamic, real-time databases of real cities. "Imagine if *SimCity* wasn't just a Game", writes Steven Johnson (2003) enthusiastically of the *mirror world*, "a virtual reconstruction of a space that would showcase everything going on in reality". With *SimCity*, the movement from gamespace to realspace has already begun.

3. "So You Can Get a Perfect Top View": Toward a Critical Cybercartography

In his semiotic analysis of the game, Ted Friedman (1999) suggests that *SimCity* might well achieve the aesthetic of cognitive mapping called for by Jameson. User feedback suggests that it does provide users with an urban metaphor, if not a full mental model, for the negotiation of contemporary social and virtual spaces. Whether or not the game also enables the development of cognitive maps, it does represent an articulation between the different *concepts* of cognitive mapping as developed in usability studies and cultural studies respectively. An analysis of user feedback indicates that the game's users have sophisticated understandings of the advanced features of cybermaps. These findings, though preliminary, foreground some of the illuminative tensions which structure user interactions with cybercartographic systems. While not all aspects of games will be relevant to other cartographic applications, interface designers do need to pay closer attention to the developments within digital games.

This method of interface assessment has its limitations. As Divitini *et al.* (2000: 31) point out, gaming communities are qualitatively different from other user bases: though vast, they tend to be fairly well circumscribed by the goals of the particular game at hand: "developers can in most cases only guess which user community and what types of users will be using their products". It is difficult to anticipate the full range of uses for digital maps of actual territories. For this very reason, these researchers argue for the use of groupware like discussion lists to solicit direct user feedback in order to improve product design, rather than relying on "laboratory experiments with 'typical users' " (Divitini *et al.* 2000: 31–32), even though they also report problems, such as "the informal structure of mailing lists". While data mining does not replace empirical usability testing under controlled conditions, the vast community of *SimCity* gamers provides a source of ready feedback for evaluating the success of many cybercartographic elements already in wide use, as well as a source of modes that improve the usability of the basic software.

This method of gathering data could be broadened to include other games, many of which include digital maps as virtual environments and navigational aids, as well as other kinds of applications, such as online mapping services and consumer-grade GPS devices. Cybercartographic research needs to recognize the ubiquity of mapping interfaces in high art and popular culture alike, and needs to survey extant mapping practices, their assumptions, and their cultural effects. But along with new tools and interfaces for mapping, cybercartographers must also develop critical explanations of the new reading practices these tools demand. Disorientation from sudden changes of scale and perspective is just one of the many research challenges that arise from the specific characteristics of cybermaps as distinct from those of paper maps. Far from just another game feature, these perspectival shifts between different map views herald important ideological effects. Ronald J. Deibert explains how the uniform spatial layout and typography of print technology changed the nature of perceptual space, privileging visuality over orality, and enabling the shift from medieval manuscript culture to the early modern worldview. Similarly, the synoptic overview map, combined with the invention of perspective in Renaissance painting, fostered new realistic modes of representation, which in turn enabled emergent political realms of property, surveillance and nationhood: "[t]he rediscovery of Ptolemaic cartography, which imagined how the globe would appear from a vantage outside and looking down on it, coincided with both a commercial and a security interest in the surveillance of territorial space" (Deibert 1997: 102).

The shift to digital maps brings new and equally powerful political effects. Digital maps challenge the assumptions of cartographic completeness, correspondence and fixity associated with the traditional printed maps. Mapping games reproduce many of the conventions of printed maps, demonstrating the degree to which print technology continues to shape cognitive mapping strategies in the digital age. The question to answer is which features of printed maps ought to be retained in the interests of usability and better cognitive mapping, and whether these features can be implemented *critically*, without reproducing the broader ideological effects of earlier, print-based cartographies. It is likely that the most usable mapping interfaces have yet to be designed, and will resemble today's cybermaps even less closely than printed books resemble medieval manuscripts. Cultural analyses that pay attention to the ideological aspects of new cartographic media are essential if we are to ensure that future cybermaps represent our culturally richest, most inclusive and most widely accessible practices.

Acknowledgments

I wish to acknowledge the Social Sciences and Humanities Research Council of Canada for their generous support of this research, and to thank Chris Eaket, Megan Graham, Kris Moran and Esther Post for their invaluable assistance in the research and preparation of this chapter.

References

Adams, P. C. (1997) "Simcity (software review)", *Cities,* Vol. 14.6, pp. 383–392.

Anders, S. and F. Jameson (1989) "Regarding postmodernism, a conversation with Fredric Jameson", *Social Text,* Vol. 21, pp. 3–30.

Boechler, M. (2001) "How spatial is hyperspace? Interacting with hypertext documents: cognitive processes and concepts", *Cyberpsychology and Behavior*, Vol. 4.1, pp. 23.

Boeri, S. (2004) "Eclectic atlases", in Graham, S. (ed.), *The Cybercities Reader*, Routledge, London, New York, pp. 117–122.

Brinkman, M., 02 12 (1999) *Sim City 3000, insanely-great.com*, accessed 25 November (2004) from http://www.insanely-great.com/ reviews/sc3000.html

Cartwright, W., J. Crampton, G. Gartner, S. Miller, K. Mitchell, E. Siekierska and J. Wood (2001a) "Geospatial information visualization user interface issues", *Cartography and Geographic Information Science*, Vol. 28, No. 1, accessed 7 December (2004) from www.geovista.psu.edu/sites/icavis/agenda/PDF/Cartwright.pdf

Cartwright, W. (2001b) "Metaphor and gaming and access to spatial information", Paper for 4th *E-MAIL SEMINAR ON CARTOGRAPHY 2001*

Cartographic Education, accessed 26 August (2004) from www.uacg.acad.bg/UACEG_site/sem_geo/ William%20Cartwright.doc

Cazdyn, E. (1995) "Uses and abuses of the nation: toward a theory of the transnational cultural exchange industry", *Social Text,* Vol. 44, pp. 135–159.

Champion, E. (2003), *Applying Game Design Theory to Virtual Heritage Environments,* presented at Graphite (2003), Annual Conference, Melbourne Australia, accessed 7 February (2005) from http://www.arbld.unimelb.edu.au/~erikc/papers/papers.html

Chao, D. (2001) "Doom as an interface for process management", *CHI* 2001, Vol. 3.1, p. 152.

Clanton, C. (2000) "Lessons from game design", in Eric Bergman (ed.), *Information Appliances and Beyond,* San Francisco, Morgan Kaufmann, pp. 299–333.

Clarke-Willson, S. (1998) "Applying game design to virtual environments", in Dodsworth, C., Jr. (ed.), *Digital Illusion: Entertaining the Future with High Technology,* ACM Press, Redding, Massachusets, pp. 229–239.

Deibert, R. J. (1997) *Parchment, Printing, and Hypermedi: a Communication in World Order Transformation,* Columbia University Press, New York.

Dieberger, A. (1996) "Browsing the WWW by interacting with a textual virtual environment - a framework for experimenting with navigational metaphors" *Proceedings of ACM Hypertext '96,* Washington D.C. pp. 170–179.

Divitini, M., B. A. Farschchian and T. Tuikka (2000) "Internet-based groupware for user participation in product development: A CSCW 98/PDC 98 workshop", *SIGCHI Bulletin,* Vol. 32.1, pp. 31–35.

Edwards, D. M. and L. Hardman (1989) "Lost in hyperspace: cognitive mapping and navigation in a hypertext environment", in McAllese, R. (ed.), *Hypertext: Theory Into Practice,* Intellect Books, Oxford, pp. 91–105.

Eep2. 04 03 (2003) Re: *Poor use of 3D? alt.games.simcity* [online], accessed 25 November (2004) from http://groups.google.com/groups? group=alt.games.simcity

Fairbairn, D., G. Andrienko, N. Andrienko, Buziek, G. and Dykes, J. (2001), "Representation and its relationship with cartographic visualization", *Cartography and Geographic Information Science,* Vol. 28.1, pp. 13–28.

Farnell, R. (1998) "Posthuman topologies: William Gibson's 'architexture' ", *Virtual Light and Idoru, Science Fiction Studies,* Vol. 25.3, pp. 459–480.

Frank, S. (1995) *Cognitive Maps and the Conceptualization of Cyberspace* [online], accessed 22 February (2004) from http://www-scf.usc.edu/~sfrank/hypercm.html

Franken, B. (1996) from Architecture to Hypertecture, accessed 7 December (2004) from http://www.inm. de/people/bernhard/lect_v2.html

Friedman, T. (1999) *Semiotics of SimCity, First Monday 4.4,* accessed 1 October (2003) from http://www.firstmonday.dk/issues/ issue4_4/friedman/

Gahegan, M. et al. (2001) "The integration of geographic visualization with knowledge discovery in databases and geocomputation", *Cartography and Geographic Information Science,* Vol. 28, No. 1, pp. 29–44.

Garling, T., A. Book and E. Lindberg (1984). "Cognitive mapping of large-scale environments: the interrelationship of action plans, acquisition, and orientation", *Environment and Behavior*, Vol. 16.1, p. 3.

Goerz, M. (2003) *Terrain Editing with Image Maps, alt.games.simcity* accessed 25 November (2004) from http://groups.google.com/groups?group=alt.games.simcity

Grice, G. and L. Strianese (2000) "Learning and building strategies with computer games", *Proceedings of 2000 Joint IEEE International and 18th Annual Conference on Computer Documentation (IPCC/SIGDOC 2000)*, Cambridge, MA, pp. 535–540.

Hassenzahl, M., A. Platz, M. Burmester and K. Lehner (2000) "Hedonic and ergonomic quality aspects determine a software's appeal", *CHI 2000*, Vol. 1–6 April, pp. 201–208.

Hilbert, D. M. and D. F. Redmiles (1998) "Why let perfectly good usability data go to waste?" *Boaster paper at the Human-Computer Interaction Consortium Meeting (HCIC '98)*, Technical Report UCI-ICS-98-12, Department of Information and Computer Science, University of California, Irvine.

Hilbert, D. M. and D. F. Redmiles, (1999) "Separating the wheat from the chaff in internet-mediated user feedback", *Proceedings of the Workshop on Internet-based Groupware for User Participation in Product Development held at CSCW '98, Reprinted in SIGGROUP Bulletin*, Vol. 20, No. 1, pp. 35–40.

Hodson, J. (2003) *Re: Auto Scroll. alt.games.simcity*, accessed 25 November (2004) from http://groups.google.com/ groups?group=alt.games.simcity

Holland, W., H. Jenkins, and K. Squire (2003) "Theory by design", in Wolf, M. J. and B. Perron (eds.), *The Video Game Theory Reader*, Routledge, New York, pp. 25–46.

Jameson, F. (1988) "Cognitive mapping", in Nelson, C. and L. Grossberg (eds.), *Marxism and the Interpretation of Culture*, Urbana, U of Illinois P.

Johnson, S. (2003), *Imagine if SimCity wasn't just a game, Discover 24:5* [online], accessed 25 October (2003) from http://www.discover.com/issues/may-03/departments/feattech/

Karybdis (2004) Cheat/Radical Mods Set 1, accessed 25 November (2004) from http://www.simtropolis.com/modfiles/ content.cfm?section=MODS&detail=15

Kneale, J. and R. Kitchin (2002) "Lost in space [Introduction]", in Kitchin, R. and J. Kneale, (eds.), *Lost in Space*: *Geographies of Science Fiction, Continuum,* London, New York, pp. 1–16.

Koskimaa, R. (1998) "Visual structuring of hyperfiction narratives", Electronic Book Review, Vol.6 (winter 1997/98) from http://www.altoc.com.com/ebr/ebr6

Kuyper, J. (2000) *Re: Automatic Power and Water.alt.games.simcity*.3000, accessed 2 December (2004) from http://groups.google.com/groups?group=alt.games.simcity.3000

Møller, M. (2003) *Re: Building transparency/hide. alt.games.simcity*, accessed 2 December (2004) from http://groups.google.com/groups? group=alt.games.simcity

Little, C. (2003) *Re: Bus systems, ARGH!. alt.games.simcity* [online], accessed 25 November (2004) from http://groups.google.com/groups? group=alt.games.simcity

Lynch, K. (1960) *The Image of the City*, MIT Press, Cambridge, Massachusetts.

Manovich, L. (2001) *The Language of New Media*, MIT Press, Cambridge, Massachusetts.

McCallum, E. L. (2000) "Mapping the real in cyberfiction", *Poetics Today,* Vol. 21.2, pp. 349–377.

McGuire, M. (2003) *PlayStation 2: Selling the Third Place,* accessed 15 July (2004) from http://hypertext.rmit.edu.au/dac/ papers/ McGuire.pdf

Patton, P. (1995) "Imaginary cities: images of postmodernity", in Watson, S. and K. Gibson (eds.), *Post Modern Cities and Spaces*, Blackwell, Oxford, pp. 112–121.

Pausch, R., Gold, Skelly, R., T. and Thiel, D. (1994) "What HCI designers can learn from video game designers", *CHI '94 Conference Companion*, pp. 177–193.

PCTest, accessed 7 December (2004) from http://www.pctest.com/services/game.html

Peuquet, D. J. and M. J. Kraak (2002), "Geobrowsing: creative thinking and knowledge discovery using geographic visualization", *Information Visualization,* Vol. 1.1, pp. 80–91.

Rogoff, I. (2000) *Terra Infirma: Geography's Visual Culture,* Routledge, London, New York.

Seigel, A. W. and S. H. White (1975) "The development of spatial representations of large-scale environments", in Reese, H. W. (ed.), *Advances in Child Development and Behavior*, Academic Press, New York, pp. 9–55.

SimCity Use, (1998) Sim City 3000 FAQ, alt.games.simcity2000.SymPosiuM, accessed 8 February (2005) from http://planetsim.tripod.com/simpage2. html

Slocum, T. C. B. Blok, Jiang, A. Koussoulakou, D. R. Montello, S. Fuhrmann and N. R. Hedley (2001) "Cognitive and usability issues in geovisualization", *Cartography and Geographic Information Science*, Vol. 28.1, pp. 61–75.

Sponsler, C. (1993) "Beyond the ruins: the geopolitics of urban decay and cybernetic play", *Science Fiction Studies,* Vol 20.2, pp. 251–265.

Stolen Child (1996) *Re: SimCopter, rec.aviation.simulators*, accessed 7 December (2004) from http://groups. google.ca/groups?group =rec.aviation.simulators

Swartout, W. and M. van Lent (2003) "Making a game of system design", *Communications of the ACM*, Vol. 46, No. 7, pp. 32–39.

Tabbi, J. (2002) *Cognitive Fictions,* University of Minnesota Press, Minneapolis.

Taylor, F. (2003) "The concept of cybercartography", in Peterson, M. P. (ed.), *Maps and the Internet*, Elsevier, Amsterdam.

Thompson, A. (1999) *Re*: *SimCity 2000 File format?, alt.games.simcity*, accessed 7 December (2004) from http://groups.google.com/groups? group=alt.games.simcity
Welch, D. (2003) *Thoughts on Sim City 4, alt.games.simcity* [online], accessed 25 November (2004) from http://groups.google.com/groups? group=alt.games.simcity
Wynands, M. (2003) *Re*: *Poor use of 3D?, alt.games.simcity*, accessed 2 December (2004) from http://groups.google.com/groups? group=alt.games.simcity

CHAPTER 14

Linking Geographical Facts with Cartographic Artifacts

WILLIAM E. CARTWRIGHT

School of Mathematical and Geospatial Sciences, Royal Melbourne Institute of Technology (RMIT) University, Melbourne, Victoria, Australia

Abstract

The design of a paper atlas took into account how the various elements – maps, diagrams, photographs, etc. – could be "fused" to provide a harmonious and complete product. Paper atlases allowed users to explore geography through the use of cartographic artifacts alongside presentations of geographical facts. Now, we are presented with a myriad of Rich Media alternatives and interactive and online enhancements with atlases delivered through the Internet, and particularly the World Wide Web (WWW). It could be argued that cartographic publishers of online products may be guilty of planning to assault their potential users with a bombardment of geographical resources and tools.

The combination of an efficient delivery mechanism, the Web; engaging information displays, New Media products; powerful tools for exploring geography, maps; online access to current geographical information, data resources available via Web-enabled repositories; and good interface and interaction practice, innovative maps and map-related products; can provide a wonderful mechanism for data prospecting and information mining. Consequently, there is now much interest in ensuring that such packages are designed and built in a way that best exploits existing and evolving media, both delivery medium and access/usage medium. This chapter explores the elements of the design of contemporary New Media atlas products, and how designers need to focus on how best to deliver both geographical facts and cartographic artifacts so as to ensure the provision of an "exploitable" information resource.

1. Introduction

The basic premise of using maps and other (geo)visualization artifacts is to build mental models of reality. This evokes the image of "Being There," without (actually) being there (a concept developed in the late Peter Sellers's film *Being There*). Historically, maps have been used to provide information to users about places recently discovered or voyages completed to unknown worlds or hitherto seemingly impossible journeys. Ptolemy believed maps to be the means to "exhibit to human understanding... the earth through a portrait" (Crane 2003: 33).

But, what do users now want? And, how should cartography provide information to users (globally through communication systems like the Web)? And, how should cybercartography deliver product that is useful, timely, and usable? How can we build on what the Web offers to provide products that can be exploited for (geo)information access and understanding?

Today has been dubbed "The Age of Access", where connectivity drives towards the access of everyone to everyone, everything to everything, and everything to everyone. Knowledge has been promoted as one of the benefits of new technology and mapping. Cartographic products and systems need to be knowledge based. Dr Samuel Johnson (London 1775 in Boswell 2004) said about knowledge and information that "Knowledge is of two kinds. We know a subject ourselves, or we know where we can find information."

Contemporary cartographic products can provide information that enables expert users to enhance their knowledge of a particular subject. But, in the hands of an inexpert or novice user, such systems may only provide a "basket" of data and information, with no real way for understanding of what is contained within that basket or its relevance. T. S. Eliot has said that knowledge should not be confused with information – there is a need for New Media-enhanced cartographic products to provide the means of acquiring knowledge and not just voluminous amounts of information. These products are useful and accessible to expert and novice alike.

This chapter addresses how New Media atlas products might be designed and built to provide geographical information in ways that exploit the delivery and presentation modes that constitute New Media. It first describes how cartography has used New Media (pre- and post-digital) to facilitate the publishing and dissemination of maps and map-related objects. Then it looks at the media that has been used to deliver products almost immediately – the Web. Next, the chapter focuses on the design of contemporary New Media products that take advantage of the plethora of tools that now allow almost anything that is designed to be produced. This follows with a discussion about how best to deliver both geographical facts and cartographic artifacts so as to ensure the provision of an "exploitable" information resource. The method proposed to facilitate this is the application of geographical storytelling utilizing the methods of interactive multimedia storytelling developed in the 1990s when the use of hypertext optical storage made this possible and deliverable. Then a proposal for using tactile, discrete, and distributed media resources to build installations that provide geographical facts plus cartographic artifacts is made (Cartwright and Hunter 1999). This is supported by the idea of using nontraditional approaches to geographical information delivery – gaming strategies and game software development tools. Also, as the Web now make cartographic products more readily available there are now more inexpert map-users gaining access to cartographic products. This is addressed in a section that discusses the problems that inexpert users of New Media Cartography might have. Finally, the chapter looks forward to how immediate, innovative, and useful tools might be provided through the use of the Web.

2. New Media

Cartographers have always embraced new technology in their quest for more speed, lower compilation and production costs, and an efficient communication system. Developments in printing, and employed by cartography in its various guises have been the norm for over 500 years. When map cartographer/publishers applied printing to map production they used this "New Media" publishing tool to facilitate quicker, more accurate, and cheaper versions of their works. Now digital New Media is being employed by cartography to complement its paper counterparts.

New Media includes a range of new delivery and display platforms; among them are the World Wide Web, interactive digital television, WAP technologies, interactive hyperlinked mapping services, and enhanced mapping packages that are "linked" to large databases – national or global (Cartwright *et al.* 2001). In (very) general terms, it can be argued that New Media has developed over three periods: initially with applications delivered on hard disk drives; then, from the mid 1980s, optical storage applications; and from the mid 1990s, the use of the World Wide Web. It should be noted that the Internet was used to deliver digital mapping applications pre-Web, but the use of the Internet allowed for New Media applications to be published after the popularization of the Web in the latter parts of the 1990s.

Recently, there has been a digital convergence, and relatively inexpensive and accessible tools exist to develop and provide a plethora of (geo)information exploration devices. Tools and techniques are readily available, but methods need to be developed to deliver effective information consumption tools. Different and innovative mapping systems have developed and products have been produced to show two-dimensional, two and a half-dimensional, three-dimensional and three-dimensional plus time (four-dimensional), plus *n*-dimensional data elements.

Pre-Web a number of methods were developed and implemented in making innovative use of the technology of the time. Pre-Web technologies employed by cartography include microformats, film, video, teletext/videotext, videodisc, CD-ROM, and the Internet to transfer data and drawing files. Using computers for the depiction of geography was of much interest to research laboratories in the 1970s. MIT's Media Lab produced *The Aspen Moviemap* in 1978, which was devised and undertaken, used videodiscs controlled by computers to allow users to explore geography using a combination of still images, videos and text (Negroponte 1995; Naimark 1997). Multimedia was born! In 1987 Apple introduced its "HyperCard" and whilst the most used form of hypermedia initially was hypertext, a number of hypermedia mapping applications were developed. Optical storage media like CD-ROM and, later, DVD have been employed for storing massive amounts of information, using an inexpensive and accessible format, which does not rely on Internet access. Today, the World Wide Web has been embraced by cartographic producers and consumers in the same way as an exciting means for facilitating global cartographic publishing, including interactive multimedia applications.

3. The Web and New Media

The revolution in information provision that the Internet, and more particularly the Web, has spawned, has changed forever how information products are viewed. They are now wanted, not demanded, almost immediately. In newspaper terms this would be described as wanting information "before the ink has dried," but for digital information this is probably best described as wanting information "before the data collection sensor has cooled!" Advances in data collection, telecommunications that ensure that collected data is quickly and faithfully transmitted, processing procedures and equipment, map "construction" and "rendering" software and geographical information delivery systems now provide the ability to deliver on demand (geo)information products in almost real time.

We now take for granted this global communication system that links us to the world with simple mouse clicks. The Web is still young, but many users assume, sometimes erroneously, that it can provide all of the information they need, at hand, on demand and immediately. The world of mapping, as noted previously, has always courted new technology to improve partnership in production and delivery. The Web is a most convenient way of "getting to market" cartographic products that can be static or dynamic, archival or current, complete or user built. Now, what is available via the Web are image collections, downloadable data, information sites brimming with maps, online "do-it-yourself" map generation sites, map collections, atlases, and hybrid products. They are being used to inform, educate, and collaborate.

4. Design of Contemporary New Media Products

In 1991 Taylor proposed a model that more appropriately described the true profile of contemporary mapping. He developed the model to indicate the elements that now contribute to what he described as the "New Map." Taylor (1994) further developed his model to expand the "communication" aspects, from just visual display techniques to include both the visual and the nonvisual. The revised model included "interaction" and "dynamics" that lead to improved cognition and analysis, and improved communication, as well as a new range of cartographic products. Later, Taylor (1997, 2003) expanded this model to embrace the use of the Web as a method of communicating New Media cartographic artifacts, and the concept of "cybercartography" was born. The fact that many professionals in the geospatial industry have championed the use of New Media, and that the profession actively supported developing mapping products using some of the "cutting-edge" techniques of New Media, heralded the arrival of interesting and exciting products. New Media information presentation devices, software and techniques on offer today provide the alternative mapping solutions, proposed by Taylor.

The New Map is a device, which aids in producing what Jacobson (1995: 37) has termed Virtual Worlds – produced by the interaction between human cognition (essentially mental maps) and the visual and auditory images that can be produced by computers. Digital cartographic data, unlike alphanumeric data, requires special graphics hardware and software for graphics query, display, and presentation. With technology tied to human endeavor and innovation, the New Map could actually be many maps and provide access to information in ways dictated by the user. The New Map could provide not just a picture of geographical reality, but also a search engine which, besides giving access to geographical data and a means of data selection and display, also allows users to access further data and information plus a background on how things, data systems, data suppliers and facilitators, mapping systems, and so on actually work. The geographically linked "things" are a conglomerate of items, systems, processes, and conventions. So, a complete geographical "picture" may be described by Fig.14.1.

Figure 14.1 depicts the fact that the "real" geographical picture is one that consists of many attributes. If access is denied to support information, that is information that supports how to get data, data standards, data systems, data providers, and data depiction conventions, then the mapping system user is denied a true picture of geographic reality. Also an efficient system would allow users to gain access to the "picture" via a general, surface access mode, or through a rigorous process that Norman (1993) described as "deep interrogation." At the "viewing end" of the electronic mapping process users would be offered depiction methods, which either

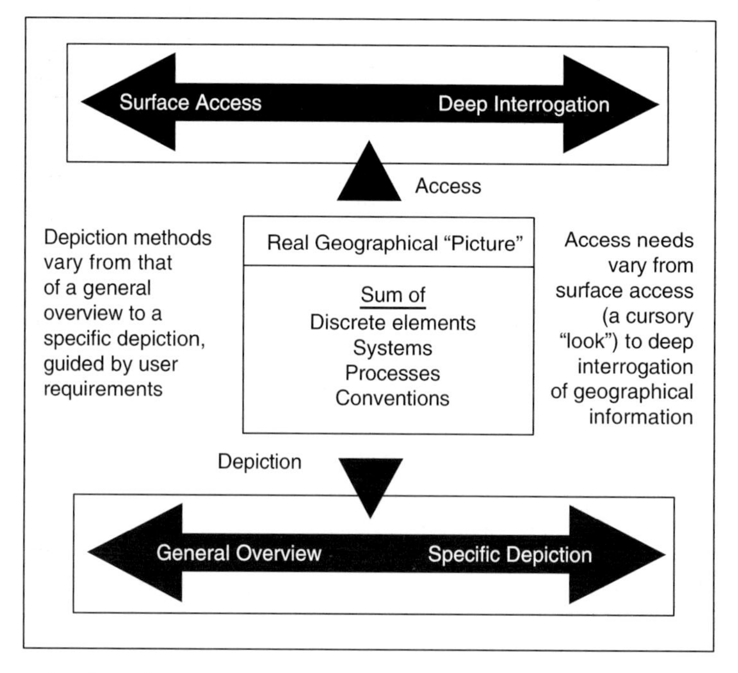

FIG. 14.1. The "real" geographical picture is one that consists of many attributes.

painted a general information overview, or else gave a very specific and precise graphic profile of essential user-defined geographical characteristics. This concept could be employed to facilitate the realization of a system whereby user-defined access to geographical information could be had. Lindgaard *et al.* in Chapter 9 of this volume discuss this in the context of the needs of different types of learners.

5. How Best to Deliver Both Geographical Facts and Cartographic Artifacts so as to Ensure the Provision of an "Exploitable" Information Resource?

Getting visualizations of reality right, and making them work is a difficult task. Sometimes we undertake the design and production of visualizations, and the appropriate access/interaction components somewhat remotely from the user. Getting this "right" takes much time and effort and if users are to appreciate geography, and the elements of geography displayed in visualizations, we need to incorporate their understanding of geography into the products we design and develop. This is not a new phenomenon. For example, Haft (1999) outlined in her discussion about Don Gutteridge's *My Story: Maps* that focused on the Lamberton County verbal map. Haft argued that "a map, and the information from that map only become memorable when a person discovers how to read his own experiences from it" (Haft 1999: 41). How do we allow users of geographical visualizations to achieve this?

Perhaps a solution is geographical storytelling – using mapping packages to tell an "electronic story" (one that is linked to a spatial database to tell a better story based on the movements of peoples and events) more efficiently. For example, *Survivors of the Shoah* (Lang 1995), produced by the Visual History Foundation, California, is an interactive oral history archive of World War II Holocaust survivors. The archive is linked to GIS software that provides view-and-query tools to help users pinpoint videos and time segments relating to topics or persons of interest. The global database comprises alphanumeric, spatial, audio, and video data.

One approach might be to adapt the genre of electronic books that became popular when multimedia was "new" and apply this concept to the production of electronic atlases. About electronic storytelling, and looking forward to publishing using such a medium, Clark (in McIntosh 1990: 42) said:

> These electronic books of the next millennium will not only transport students, but everyone, into realms enabling private study and education exploration unheard

> of today.... The magic of the electronic book will allow more scope and personality, using sound, animated graphics, moving and still pictures and allow the user greater interaction.

Electronic books range from passive entertainment to interactive experiences. Passive entertainment books best described as "electronic page turners," whilst interactive books offer experiences beyond what is first seen on the screen page, whereby digital animation and hypertext are employed for interactive storytelling. Approaches have varied, from the "interactive movie," where theatergoers voted on what action should take place next by pushing buttons attached to their seats, to literary attempts to create hypertext fiction, where the reader chooses on-screen story links and storyline direction (Caruso 1996). Excellent examples of digital "storytelling" products are *The Resident's Bad Day on the Midway* (Wolff 1995) and the CD-ROM book, *Puppet Motel* by Laurie Anderson (Hip Surgery Music 2004), published by the Voyager Company.

Users of interactive books are provided with a different type of access metaphor, that of exploration and discovery. Very little user-guidance is given about how to use the product and the user must explore different means of gaining access to other screens and required information. This was the kind of "seek and find" strategy that was used to develop the highly successful game, *Myst*, where users needed to explore the Virtual Environment to "discover" clues and interaction methods.

As with all forms of interactive media, the quality of the experience rests with the user's willingness and ability to fully engage with the work (Knight 1996). Such products are not all that dissimilar to what an atlas tries to do on the topic of geography!

Electronic books also added dimension of giving "pleasure" to readers. This is one of the elements of using New Media that is very hard to quantify. Also, the "three key pleasures" of electronic storytelling are uniquely intensified in electronic media (Platt 1995), viz:

- Immersion – the sense of being transported to another reality;
- Rapture – entranced attachment to the objects in that reality; and
- Agency – the player's delight in having an effect on the electronic world.

The ideas behind electronic books could be applied to the design of contemporary electronic atlases. If consumers of books are exposed to interactive products they will be most likely disappointed with atlas products that, even though produced and delivered electronically, still merely offer "electronic page turning." Interactive storytelling via atlases may well require a different approach to painting pictures of geography.

6. Geographical Storytelling

"Armchair travelers" have always been able to "build" their own geographical stories using paper maps and other geographical artifacts. Maps like Turgot and Bretez's *Plan de Paris* (1739) and the Manhattan Map Company's *Isometric Map of Midtown Manhattan* allow readers to spoil themselves with a freedom of choice, unavailable on more simple automated maps. This freedom of choice in panorama, vista, and prospect is only available in maps that allow micro readings to happen – a critical and effective principle of information design.

Italo Calvino's *Invisible Cities* (1979: 10–11) gives some insight into what an articulated map reader/interpreter may wish to read into the graphics:

> Relationships between the measurements of its space and the events of its past: the height of a lamp-post and the distance from the ground of a hanged usurper's swaying feet; the line strung from the lamp-post to the railing opposite and the festoons that decorate the course of the queen's nuptial procession; the height of that railing and the leap of the adulterer who climbed over it at dawn; the tilt of a guttering cat's progress along it as he slips into the same window; the firing range of a gunboat which has suddenly appeared beyond the cape and the bomb that destroys the guttering; the rips in the fishnet and the three old men seated on the dock mending nets and telling each other for the hundredth time the story of the gunboat of the usurper, who some say was the queen's illegitimate son, abandoned in his swaddling clothes there on the dock.

Obviously, all readers of microimages will neither have the imagination nor the broad storytelling skills of Calvino, but the opportunity does offer itself when using geographical storytelling.

"Geographical storytelling," if implemented properly, can result in much "Rich Media" being provided to users in formats they are most comfortable using. The image in Fig.14.2 illustrates how this concept has been applied to the development of product.

The author used this concept to build the *GeoExploratorium* (Cartwright 1998), a hybrid CD-ROM/Web resource that enabled users to understand geography by exploring geographical space using metaphors that were user-driven, including a storytelling metaphor (Cartwright 1999). Later, the *GeoExploratorium* metaphors were applied in developing an educational installation for four different user groups – the "Nintendo Generation" (initially described by Ormeling 1993), young users of geoinfomation; the "Computer Generation," older users who are comfortable using computers; the "Video Generation," who grew up with broadcast television and videotapes; and the "Print/Audio Generation," the senior audience who use newspapers, magazines, radio, telephones, and person-to-person conversations. Their reality (shown at the right of Fig 14.3) dictated the choice of metaphor that was applied, from the *GeoExploratorium*

FIG. 14.2. Myst. The example shows a typical interface screen. The user interacts by "clicking" spots on the image. Some lead to more information or other screen pages.

"suite" of metaphors. Geographical storytelling is the idea behind building this second product. In the early conceptual phases it was envisaged that the prime requirement of the package was to tell a geographical story, so the *GeoExploratorium* "storyteller" metaphor was the overarching concept around which the product was built, even though other metaphors were employed. The schema of the product development is shown in Fig 14.3.

The enactment of stories is the centerpoint of how the system works. Stories provided can range from simple statements of facts, to narratives, constructed "own stories," talking a user through a geographic place, investigating a "literate landscape," traveled by a *Literate Traveller* or providing expert opinions about certain geographical spaces. The *Literate Traveller* is a theoretical geographical information user who assembles a range of (geo)information resources so as to be fully aware of a geographical area of study, decision-making or travel (Cartwright and Hunter 1999). For a description of an application built on this concept see Cartwright *et al.* (2003). Geographical stories, based on real and accurate geographical locations and facts, can provide a powerful tool to inform the "new" map consumer.

Stories can just provide statements of facts, where no embellishment is required and the user only wants to know "the facts." These facts can be stand-alone, or supported by "on-line" experts who are able to give expert opinions on the geographical space being explored. It may be a narrative,

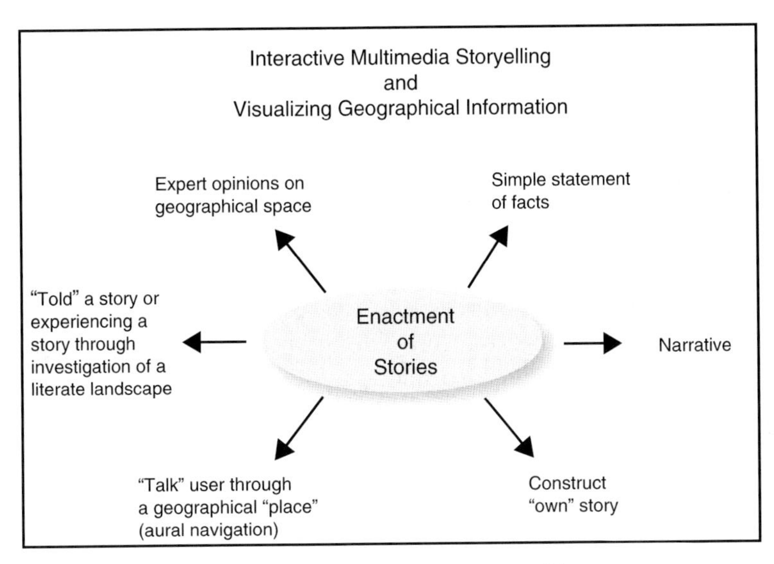

FIG. 14.3. Geographical storytelling.

where a documentary-type video, supported by a comprehensive, and interactive, narrative can "walk" a user through "unknown territory." Users may construct their own story, or be "talked" through an area, where they construct a story using programme support materials and aural navigation aids. Finally, they may decide that they wish to experience a landscape by investigating a "literate landscape" by being told a story.

7. Using Tactile, Discrete and Distributed Resources to Build Installations that Provide Geographical Facts Plus Cartographic Artifacts

It is argued that users of New Media products still require "other" artifacts to complement their use of interactive screen-delivered information and information use tools. The use of tactile resources to supplement the plethora of electronic offerings is seen as a means by which, for the general user, more appropriate information might be provided. Also, it is argued that users do like to have access to many different tools and resources with which to access, explore and discover geographical information. This human way of working with information has been described as "bricolage". Claude Lévi-Strauss in his 1962 (trans. 1966) book *La Pensée sauvage*, (The Savage Mind) described bricolage as a completely different way of thinking. Users have to make-do with what is at hand and use improvisation in problem-solving. However, Papert (1993) said that the user needs time to develop this method of information gathering, and, as Papert (1993: 144) said: "For the true bricolage the 'tools' will have been selected over a long time by a process determined by more than pragmatic utility."

So, using a combination of many media resources, assembled in a digital repository, and then downloaded upon demand, for printing (as hard copy tools), installed in "local" computers and provided "online" via the Web could be a means by which the *bricolage* tools of Strauss are provided. The combination of analog, discrete, and distributed multimedia tools could provision a powerful multimedia resource. Analog multimedia includes all those resources that can be used "close at hand" when exploring geographical information. Along with these "hard copy" elements, analog multimedia can also include audiotape accounts of journeys, oral histories, and comments about other hard copy components. Discrete multimedia provides products from hard disks and optical storage. Distributed multimedia uses communication resources, and particularly the World Wide Web, to link computers locally or internationally (with appropriate "browsers" and "plug-ins" used to access hyperlinked multimedia resources).

Also, an Avatar could be provided as a (virtual) map curatorial guide to provide "support" for geographical story "reading" and access to additional tools that could be employed in a "bricolage" process. For example, the most recent version of the movie *The Time Machine* speculated how a humanlike avatar might assist users looking for specific information. This movie included a scene where, in the mid-21st century, time traveler Alexander Hartdegen (played by Guy Pearce) meets Vox (played by Orlando Jones), a holographic reference source of all human knowledge. The avatar is a repository of all information and this information is made available through an interactive see-through flat screen. All looks real and the curator "human" until the time traveler Hartegen looks behind the screen – and there is nothing there. This concept could be employed (albeit with a less humanlike avatar) to provide multimedia geographical visualization package user support.

The increased access to sophisticated computers by the general public has led to an awareness that resources, like discrete multimedia products and their distributed counterparts on the Internet, the World Wide Web, have revolutionized the way in which information is both accessed and used. Cartographers have embraced the use of interactive multimedia, delivered via discrete or distributed means, as a method of providing products that are easily useable with "everyday" skills, using modest computer platforms and accessible communication resources, like the Web. "Conservative" cartographic tools, including some tools produced and delivered using computers and the World Wide Web, actually disallow users from manipulating the tools provided. Users are unable to modify "the view"; they are only offered what can be called "Frozenscapes" (Cartwright and Heath 2002). The proper use of New Media must allow users to modify the view and the method of tool usage to suit their own "best use" practice.

8. Problems: Inexpert Users of New Media Cartography on the Web – Acceptance, Reliance and Confusion

There now exists a new genre of users. They may never have used maps before, and they consider geographical information in the same way as any other commodity that they can obtain via the Web. These users are attuned to getting information via interactive online resources and they are willing to explore different methods of access other than conventional mapping products. As interactive multimedia cartography products via the Web can be digitally modified between data repository and actual consumption, the delivered product and the user interface can be modified to accord better with individual use preferences and information needs.

Once cartographers knew their users, they knew what they wanted and how they intended to use the cartographic artifact produced. With conventional paper maps, it could be argued that these products were not considered to be "mainstream" information documents, but specialist artifacts to be used by expert users, or users who were willing to "learn the rules" of map use. Now, with almost instant access to geographical data and graphic products via the Web, it is argued that the general public now considers Web-delivered maps to be just part of what New Media delivers. Geographical information delivered through the use of New Media is seen as a part of popular media, rather than scientific documents.

Welsh (1996) argued that the Information Superhighway/Infobahn metaphor had been "overtaken" by commercialism, with its inherent problems and planned biases. He had this to say about the "real" use of the Internet and the World Wide Web (WWW):

> If the Road Movie is the quintessential form of American popular culture, then the notion of surfing or cruising in Cyberspace is, at one level, only a logical extension of this. And the metaphor is actively promoted by advertisers in lifestyle magazines like Wired, or the British.Net, where modems, WWW browser programs and subscriptions to on-line services are marketed like cars or motorcycles (1996: 37).

Welsh saw further problems when access to documents on the Internet provided an unbalanced resource. Something that he saw neither the techno-Utopianists nor the neo-Luddite antitechnology (romantic) intellectuals being able to provide on a useful operational basis. He proposed that what was needed was:

> Something which is thought-provoking, entertaining, stimulating, aesthetically satisfying, life-affirming and revelatory. But then this is merely a description of what constitutes a good book? (1996: 41)

As the Web can be considered as an accepted part of popular media, users could consider its use to be carried out in similar ways as the use of television, video, movies, books, newspapers, journals, radio, and CD-ROM. The delivery of cartographic artifacts via the Web to naive or inexpert users involves different strategies to traditional map delivery and use. Use strategies need to be developed, tested, and implemented. Lindgaard *et al.* discuss the needs of this type of user in Chapter 9.

For cartography the "good book" could be a Web-delivered New Media atlas product that told geographical stories. Designers need to develop skills to design products that allow users to understand the logic behind what is meant by user interfaces (and therefore manipulate things better), to provide feedback that indicates if a procedure has been successful (even though the output may not be immediately apparent), and to design machine "controls" so that they are positioned in a way that correlates with their effects. The designers of spatial products will not be able to merely apply design skills from paper maps, or electronic maps and atlases to multimedia applications. New design methods and approaches need to be developed when applying multimedia for presenting spatial information.

9. Conclusions

Artifacts like maps, if used and presented properly in, say a multimedia system, which is made available through "hard" devices like CD-ROM, or "soft" devices like the Internet and the World Wide Web, would be described by Donald Norman as, metarepresentations: a representation of a representation (Norman 1993). Maps and other devices that articulate four-dimensional spatial information can be seen as cognitive artifacts, or tools – cognitive tools. How these tools interact with the mind and what results they deliver depend upon how they are used (Cartwright 1995).

When one reads prose, like that in Italo Calvino's *Invisible Cities* (1979) the "mind's eye" is encouraged to paint a virtual world. Multimedia and GIS must also allow this to occur, but in this case with the painted virtual world, one which is based on fact and placed somewhere in four-dimensional space. Whichever way interactive mapping eventually establishes itself, the lessons learnt in developing interactive multimedia mapping packages on the plethora of software packages, computers and computer peripherals, will be investments in future mapping and GIS innovations (Cartwright 1995).

Maps, if they are considered to be works of art as well as scientific documents, do have the ability to effectively depict information about geographic phenomena, by combining the communicative effectiveness of well-designed graphics with precise data elements. They provide the opportunity to faithfully show the real attributes of geospatial phenomena, a map attribute depicted by Italo Calvino (1979: 108) in his book *Invisible Cities*, where he describes a conversation between Marco Polo and the Emperor:

> I think you recognize cities better on the atlas that (sic) when you visit them in person, "the emperor says to Marco snapping the volume shut. And Polo answers," Travelling, you realize that differences are lost: each city takes to resembling all cities, places exchange their form, order, distances, a shapeless dust cloud invades the continents. Your atlas preserves the differences intact: that assortment of qualities which are like the letters in a name.

This "new" method of access to and representation of geospatial information, atlases on the Web – Cybercartography, is different to earlier used methods, and therefore, whilst New Media applications can be considered to be at a fairly immature stage of development (compared to paper maps), there is a need to identify the positive elements of the media used and to isolate the negative ones, so as to develop strategies to deliver timely, innovative products.

References

Boswell, J. (2004) *The Life of Samuel Johnson LL.D.*, Adelaide, eBooks@Adelaide http://etext.library.adelaide.edu.au/b/boswell/james/b74l/index.html

Calvino, I. (1979) *Invisible Cities*, Picador, London.

Cartwright, W. E. (1995) "Multimedia and mapping: Using multimedia design and authoring techniques to assemble interactive map and atlas products", *Proceedings of the 17th International Cartographic Conference, International Cartographic Association*, Barcelona, Spain, September, pp. 1116–1127.

Cartwright, W. E. (1998) "The development and evaluation of a web-based 'geoexploratorium' for the exploration and discovery of geographical information", *Proceedings of Mapping Sciences '98, Fremantle: Mapping Sciences Institute*, Australia, pp. 219–228.

Cartwright, W. E. (1999) "Extending the map metaphor using web delivered multimedia", *International Journal of Geographical Information Science*, Vol. 13, No. 4, pp. 335–353.

Cartwright, W. E. and G. Heath (2002) "Geography as seen from the window: Findings about viewpoint-specific images of geography defined by tools of visualization", *Cartography*, Vol. 31, No. 2, pp. 103–117.

Cartwright, W. E. and G. J. Hunter (1999) "Enhancing the map metaphor with multimedia cartography", in Cartwright, W. E., M. P. Peterson and

G. Gartner (eds.), *Multimedia Cartography*, Springer-Verlag, Heidelberg, pp. 257–270.

Cartwright, W. E., J. Crampton, G. Gartner, S. Miller, K. Mitchell, E. Siekierska and J. Wood (2001) "User interface issues for spatial", *Information Visualization, CaGIS*, Vol. 28, No. 1, pp. 45–60.

Cartwright, W. E., B. Williams and C. Pettit (2003) "Realizing the 'Literate Traveller' ", *Spatial Sciences Institute Conference Proceedings 2003, Spatial Sciences Institute*, Canberra.

Caruso, D. (1996) "In the end a game is for entertainment", *The Age*, May 28, pp. C11–C12.

Crane, N. (2003) "Changing our view of the world", *Geographical, The Royal Geographical Society,* April 2003, pp. 33–37.

Haft, A. J. (1999) "The poet and the map: (Di)versifying the teaching of geography", *Cartographic Perspectives*, Vol. 33, pp. 33–48.

Hip Surgery Music (2004) *Puppet Motel*, accessed February 2004 from http://www.hipsurgerymusic.com/AndersonLaurie/puppet-motel

Jacobson, R. (1995) "Virtual worlds: Spatial interfaces for spatial technology", *ESRIARC News*, Fall, pp. 38–39.

Knight, J. (1996) "Honey, don't touch, it's not meant to be touched!", *The Journal of Research into New Media Technologies*, Vol. 2, No. 2, pp. 19–22.

Lang, L. (1995) "GIS supports holocaust survivors video archive", *GIS World*, May, pp. 48–50.

Lévi-Strauss, C. (1962) *La Pensée sauvage*, Plon, Paris.

Lévi-Strauss, C. (1966) *The Savage Mind*, Chicago Press.

McIntosh, T. (1990) "Audio-visual: The word is simplify", *The Australian*, 19 August, 1990, p. 42.

Naimark, A. (1997) "3D moviemap and a 3D panorama", *SPIE Proceedings,* Vol. 3012, Society of Photo-Optical Instrumentation Engineers, San Jose.

Negroponte, N. (1995) *Being Digital*, Rydalmere, Hodder and Stoughton.

Norman, D. A. (1993) *Things that Make us Smart: Defending Human Attributes in the Age of the Machine*, Reading, Addison-Wesley Massachusetts.

Ormeling, F. (1993) "Ariadne's thread – structure in multimedia atlases", *Proceedings of the 16th International Cartographic Association Conference,* 3–6 May, International Cartographic Association, Köln, Germany, pp. 1093–1100.

Papert, S. (1993) *The Children's Machine: Rethinking School in the Age of the Computer*, BasicBooks, New York.

Platt, C. (1995) "Interactive entertainment. Who writes it? Who reads it? Who needs it?", *Wired*, September, pp. 144–149, 195–197.

Taylor, D. R. F. (1991) "A conceptual basis for cartography: New directions for the information era", *The Cartographic Journal*, Vol. 28, No. 2, pp. 213–216.

Taylor, D. R. F. (1994) "Cartography for knowledge, action and development: Retrospective and prospective", *The Cartographic Journal*, Vol. 31, No. 1, pp. 52–55.

Taylor, D. R. F. (1997) *Proceedings of the 18th ICA/PCI International Cartographic Conference,* ICC 97, Stockholm, Sweden, 23–27 June 1997, Vol. 1, Ottoson, l. (ed.), Gavle, Swedish Cartographic Society, pp. 1–10.

Taylor, D. R. F. (2003) "The concept of cybercartography", in Peterson, M. (ed.), *Maps and the Internet*, Elsevier Science Ltd., Oxford, pp. 405–420.
Welsh, J. (1996) "On and off maps: Lost in cyberspace and wasting time on the world wide web, convergence", *The Journal of Research into New Media Technologies*, Spring, Vol. 2, No. 1, pp. 36–41.
Wolff, H. (1995) "Myst", *Australian PC World*, February, pp. 153–154.

CHAPTER 15

Pervasive Public Map Displays

MICHAEL P. PETERSON

Department of Geography/Geology, University of Nebraska at Omaha, Omaha, Nebraska, USA

Abstract

The combination of current computer interfaces and the Web has led to highly interactive and widely distributed map displays. While the computer has increased interaction with maps and the Web has increased their distribution, other types of computer interfaces with maps are possible. Large-screen, computer-driven displays are becoming increasingly common for a variety of applications. To make these displays interactive, new computer interfaces are required and strategies are needed for creating public web map displays that show such things as weather maps, and maps of automobile and flight traffic using large-screen technology. Hand-held Personal Digital Assistants (PDAs) also represent a possible ubiquitous form of information display. Other possible user interfaces for public map displays may be envisioned based on research in the emerging fields of ubiquitous computing and ambient intelligence.

1. Introduction

Adding interactive component to maps has been the central goal of cartography and Geographic Information Systems (GIS) since the development of affordable interactive computers in the early 1980s. This shift was intensified with menu and Windows-based computer systems introduced in the mid- to late 1980s. The melding of mind, map and computer in an "interactive bliss" is still taken as something of a panacea – a way to a higher truth through "visualization". As can be seen from the chapters in this book, cartography is in the process of adapting the interactive and multimedia potential of the computer.

While maps are still being integrated in the classic Windows, Icons, Menus and Pointing device (WIMP) interface, research in computer science is attempting to find alternate forms of computer interaction. A criticism of the WIMP interface is that it requires a considerable amount of concentration by making users sit still and focus their attention on a screen. Other forms of information display are much less demanding. Printed text, for example, surrounds us in many ways but does not require the level of concentration of a WIMP interface. We do not need to "log on" to road signs to use them. Similarly, we do not need to disengage from a discussion with a colleague to jot notes on paper. It is argued here that context-sensitive, computer-assisted interaction with maps could provide us with a better user-experience than an intelligently orchestrated and highly-interactive, computer-based cartographic system.

In an article entitled *That Interactive Thing You Do*, Peterson (1998) points out that our interaction with computers is very shallow. In a conversation, each participant is building up a database of what the other has said. The database is fairly sophisticated at times. For example, you might be able to remember stories that people have told you several years ago although they have forgotten. With the computer, it is somewhat like talking to a person for the first time over and over again. There is no database of the interaction. Because the computer doesn't remember, it cannot raise the sophistication of the interaction. Further, it doesn't challenge us. Imagine if the computer responded with messages like: "I've made this map for you before!"; "Don't you remember where that feature is located!"; "Can we move to a more intelligent level of interaction, please?" Viewed in the perspective of a conversation, the type of interaction that we have with maps on the computer is very simplistic.

The area of research in alternative computer user interfaces is referred to in a variety of ways, including pervasive computing, ubiquitous computing, and ambient intelligence. A common element is the vision that all human-made and some natural products will be embedded with some type of hardware and software. Going beyond the realm of personal computers and the WIMP interface, it is proposed that almost any device, including clothing, tools, appliances, cars, homes, a wall, and even the human body, can be imbedded with hardware for connection to other devices. This area of research combines current network technologies with wireless computing, voice recognition, artificial intelligence and the Internet, "to create an environment where the connectivity of devices is embedded in such a way as to be unobtrusive and ubiquitous" (Webopedia 2004).

In many ways, computational devices are already ubiquitous. Embedded computation controls braking and acceleration in cars, operates medical devices, and controls virtually all machinery. Cell phones, embedded digital cameras, pagers, and digital music players represent widely-used, portable computational devices. Car navigation systems are becoming an increasingly common addition to automobiles.

Large-format displays in public areas, such as airports represent an important type of ubiquitous visual computational device. Using a variety of different kinds of technology, the size of such displays has increased to the size of a large wall and resolutions are in the multiple 1000s of pixels. Plasma displays are now available at up to 84 in. (213 cm) in diagonal measurement. Acrylic screens, designed for rear or front projection with a video projector, are available at sizes up to 135 in. (343 cm). In many cases, multiple monitors or projectors will be arranged next to each other to create a larger display. Organic electronics provide the potential of producing large computer displays with an ink-jet printer and a new fog screen technology allows image projection on tiny droplets of water that are suspended in air. It is anticipated that this latter type of display will be installed at airports or shopping malls and people will be able to interact with the projected "fog" image in various ways. They will even be able to walk through the display. These technological developments are leading to a new medium for the public display of maps. This will help overcome some of the display barriers identified as the "Achilles heel" of cybercartography by Monmonier in Chapter 2.

Large-format displays will also promote collaborative uses of maps that would facilitate their interpretation and analysis. Large electronic map displays in the classroom may serve as a more dynamic version for a standard wall map (Peterson 1995). One of the main problems with current interfaces is that a single, isolated user may not be able to interpret the mapped patterns that are presented to them. A collaborative map-use environment would encourage group interpretation of the mapped patterns.

Alternate forms of computer interfaces have also been developed. Using a cell phone or an electronic key, users are able to interact with computer displays in different ways. Voice-input is becoming more common with computers and this may soon represent an alternative type of user interface with maps. For example, we may be able to walk up to a large electronic map and tell it what we want to see. A video camera can also be used as a user interface device.

Maps would fit well into this new type of computer interface. As a graphic display, maps provide a great deal of information at a short glance. The salient features of a map are recognized in a matter of seconds.

Depending on the complexity, spatial patterns on thematic maps can be comprehended in an even shorter amount of time. Maps that are viewed quickly may be remembered as well as maps that are examined for a longer period of time. This means that interfaces can be developed for maps that require only casual attention. Computer maps would be used in a much less intrusive way than is currently the case.

We are entering a new phase in the marriage of maps and computers. In the beginning, computers were used to help the cartographer make maps on paper. Then, highly interactive computer mapping and GIS systems were promoted. Now, this is seen to be a very intrusive form of map-use that requires an inordinate amount of attention and concentration by the map-user. Less intrusive forms of computer-assisted map presentation may provide a more suitable interface for maps in certain situations. The purpose here is to examine alternative map displays and new interfaces for cartography. First, we look at the concept of pervasive computing and a working example.

2. Pervasive Computing

2.1. The Concept

Pervasive computing is a crossdisciplinary area of research that extends the application of computing to diverse usage models (Mattern and Naghshineh 2002). It is viewed as the next generation computing environment with information and communication technology everywhere, for everyone, at all times. Information and communication technology will be an integrated part of our environment: from toys, milk cartons and desktops to cars, factories and whole city areas – with integrated processors, sensors, and actuators connected via high-speed networks and combined with new visualization devices ranging from projections directly into the eye to large panorama displays. An example of a pervasive computing environment is a weather map display system that has been operating since 1999.

2.2. A Working Example: Public Weather Map Display System

In 1998, an experiment was begun to display updated weather maps in a public setting (Peterson 2001). The purpose was to make weather maps available for viewing in a public place in such a way as to encourage analysis and interpretation (see Fig. 15.1). A large number of frequently

FIG. 15.1. The automated weather map display system at the University of Nebraska - Omaha. Each computer cycles through the display of 4 to 5 weather maps on a continuous basis. Maps are updated hourly on each display.

updated weather maps are available through the Web and these maps are updated on a regular basis, sometimes at intervals of less than 30 min.

Such a public display of maps can serve an important educational function in addition to providing critical information about travel conditions. It can also demonstrate the value of maps for displaying information. The system described here can be used to show any type of image that is frequently updated. A good example would be the display of webcam images. Webcams are available from many parts of the world and depict recognizable landmarks, unique landscapes, or current weather conditions.

The automated display of maps was implemented with a series of older computers. Five were made available for use, with one already 12 years old and limited to 8 MB of memory. The challenge was to create a continuously-operating as well as sequenced display of updated weather maps from the web using this older technology. The solution was to use a simple scripting language and a smaller program that downloaded maps from the web at regular intervals. In the hour between downloads, the series of maps were then displayed in full-screen mode on a continual basis.

Converting these computers to operate in an automated way is not an easy task. The WIMP operating system is based on user interface elements like menus and dialogs that make the operation of the computer dependent on regular user input. Bypassing these user interface elements and creating a computer that works without human input runs counter to the design of the operating system. In order to make a computer download and display images automatically, methods are needed to be developed to supply user input in an automated way. The most difficult aspect of the system, however, was to make the computers run continuously without interruption. Various software components and system enhancements were used in the working system.

(1) ***Server/Client Computer Arrangement*****:** A central master computer/server was used to download all maps at regular intervals. A few minutes after the maps are downloaded by the Master computer, a script running on each of the display or "Client" computers would access and download these files (see Fig.15.2). This approach assures that any problem in downloading images from the Web is limited to one computer. Further, fault tolerance is built into the system by allowing the Client computers to download images directly through the Web if the Master computer goes down. (Fig.15.2)

(2) ***Display Software*****:** This program displays the downloaded images stored in a folder in a continuous loop and must operate flawlessly. Each map is shown for approximately 5–10 s. In addition, the program must

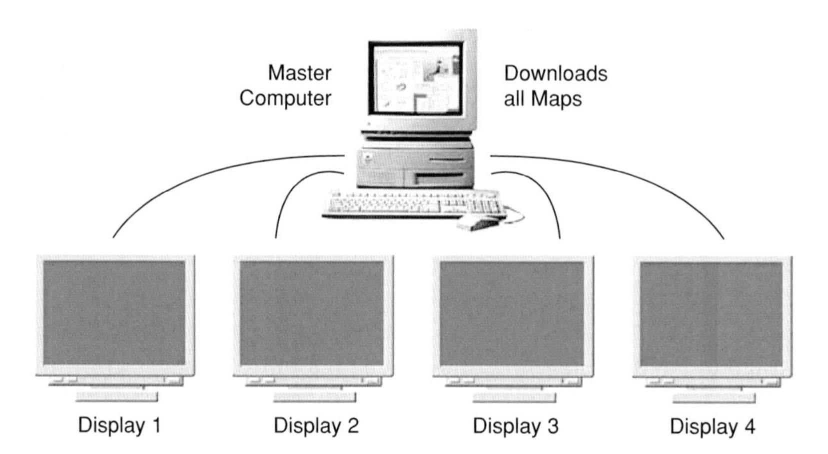

FIG. 15.2. The Master/Client relationship in downloading and displaying weather maps. The "Master" computer downloads and stores all maps. A few minutes after the scheduled download, each client computer downloads the maps from the Master computer.

be able to operate in full-screen mode so that the menus are hidden. A freeware application called JPEG view was used.

(3) ***Timing Software***: Working in the background, this software launches applications at specified times. It is common to use this type of software for many different applications, such as backing-up files in the middle of the night. Cron is the major utility that performs this function. The program scans the file CRONTAB every minute and checks to see if the specified time-date information matches the current time and date. If they are equal, the specified command is executed.

(4) ***Time Synchronization Software***: When downloading images at specific times, it is important that the computer be set to the correct time, especially if multiple computers and separate programs need to synchronize with each other. The Network Time Protocol (NTP) was established to provide the correct time to computers on the Internet. The NTP client that is installed on all the computers in the weather map display system synchronizes the computer's clock to within 50 ms of the atomic clock at the US Naval Observatory.

(5) ***Dialog Handlers***: Dialogs are used to inform the computer-user of an error or some other condition that needs to be brought to the user's attention. The problem with dialogs in this application is that they wait for user input (hitting the OK or Cancel button) and thereby stop all further action by the computer. A method is needed to automatically respond to the dialog. On the Macintosh, a program called Okey Dokey Pro selects the default choice in a dialog box (i.e. the equivalent of pressing the return key) after a preset time has elapsed.

(6) ***Network Administration Software***: A program for network administration is needed if multiple computers are being used to display maps and images. The program can be installed on the same computer that downloads the images. It periodically consults the designated computers to determine if they are still connected to the network and which program is currently active. This type of software is the best way to determine if the display is working properly from a central location.

2.3. The Influence of Pervasive Computing

The weather map display system described here created a regular group of viewers that examined the display of maps on a daily basis. Some would make a special trip to examine the maps while others merely glanced at the display while walking past. Still others would stop to examine the maps before leaving for home. One noticeable aspect of the system is that it influenced discussions about the weather. It represents a "common" display that everyone in the building could see. Most discussions about the weather, a common topic of conversation, used information that could be directly attributed to the display.

The system described here is similar to the flight arrival and departure information provided in airports. Some airports in the US have even implemented a display that depicts the current position of airplanes that have an arrival or departure at that airport (see Fig. 15.3). Certain states in the US have also implemented similar automated weather display computers at rest areas to assist travelers. Referred to in the trade as Digital Visual Messaging (DVM), such systems are designed to be an engaging and dynamic medium that provides up-to-minute information. To make such systems more interactive, we need to examine alternate ways of presenting information.

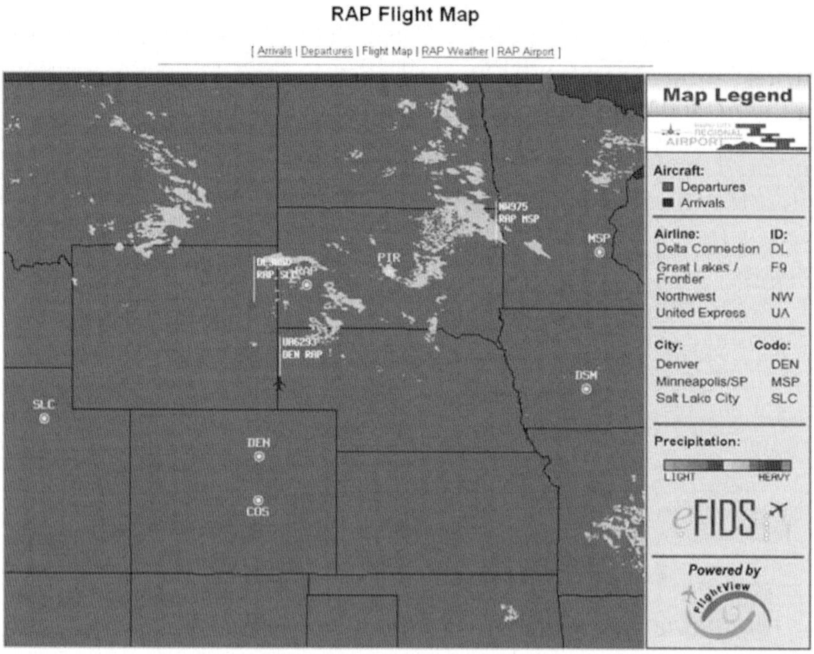

FIG. 15.3. Automated flight information display for Rapid City (RAP), South Dakota, in the US. Airplanes that have departed the airport and those that are en-route are identified on the display. Areas of precipitation are identified through NEXRAD ground-based radar. To access current flight maps, go to http://www.rcgov.com/ Airport/pages/airline_schedules.htm Click on Flight Information. Then click on Flight Map.

3. Alternative Computer Interfaces

3.1. *Ubiquitous Computing*

Originating at the Xerox PARC research centre at Palo Alto, California, ubiquitous computing is a philosophy that aims to enrich our computing environment by emphasizing context sensitivity, casual interaction and the spatial arrangement of computers. It is seen as the third wave in computing. The first wave was based on single mainframe computers that were shared by many people. Presently, we are in the personal computing era with a single user with a mouse and keyboard in front of a single display. The next stage, it is argued, is ubiquitous computing as technology recedes into the background (Weiser 1991, 1994). With the proliferation of mobile devices and the ability to embed large touch screens in public places, a new type of human–computer interaction involving multiple machines and multiple people becomes possible.

The current mode of interaction with a computer involves a single user with a mouse and keyboard in front of a single display. This requires users to sit still and focus their attention, withdrawing from the world around them. The WIMP interface removes the user from the broader environment (Want *et al.* 2004). In contrast, an example of ubiquitous computing is the work of Streitz (2003) in Germany who examines new tools and methods for the embedding of computation in everyday objects. Products, such as Roomware, a DynaWall, an InteraTable, a CommChair and a ConnecTable, are under development by his team. All of these devices attempt to place the computer in the background of the user's experience.

Research in ubiquitous computing has taken two directions. The first involves group interaction with large touch screens in workrooms. This research is divided into those that use the display for collaborative research in a laboratory setting and those that develop more public applications. An example of the latter is the work of Ferscha and Vogl (2002) who examine pervasive Web access through what they call "public communication walls". The method of interaction is a cell phone. The second direction is the examination of user interfaces on mobile devices and ideas, such as geocasting and location aware computers. The purpose is to present context-sensitive information on mobile devices. Chapter 16 by Gartner in this volume discusses such approaches.

3.2. *Communication Walls*

Communication walls, or DVM, are already used in airports to guide passengers between planes. Other types of "digital signage" and wall applications are used in retail, quick-serve restaurants, banking, and command and control environments. Several recent developments in technology have made these types of displays more feasible.

3.2.1. Available Technology

Communication walls are made possible by large screen technology. Four major display types are in use today: the traditional Cathode Ray Tube (CRT), Liquid Crystal Display (LCD), plasma display, and video projectors. Other promising display technology includes the organic display and the fog screen. Resolutions for all of these forms of electronic display are provided in one of the following formats, with cost increasing with greater resolution: (1) SVGA, with "800 × 600" pixels; (2) XGA, with "1,024 × 768" pixels; (3) SXGA, with "1,280 × 1,024" pixels; (4) WXGA, with 1366 by 768 pixels; and (5) UXGA, with "1,600 × 1,200".

The size of each type of display is limited. Standard television or CRT displays range up to 40 in. (101.6 cm) in diagonal measurement while LCD screens are available only up to 30 in. (76.2 cm) at WXGA resolution. Most large screen displays now use plasma technology. These screens are available at up to 84 in. (213 cm) in diagonal measurement at WXGA resolution of 1024 × 768 although 50 in. (127 cm) consumer-oriented plasma screens are much more reasonably priced. The plasma display consists of a lightweight surface covered with millions of tiny glass bubbles. Each bubble, arranged as pixels, has three subpixels – one red, one green, and one blue – that contains a gas-like substance, the plasma, and has a phosphor coating. A digitally controlled electric current flows through the flat screen that causes the plasma inside designated bubbles to give off ultraviolet rays. This light in turn causes the phosphor coatings to glow at the appropriate color. With a wide viewing angle, resolutions up to 1366 by 768 pixels, and a 600:1 contrast ratio, thickness of only 3.5 in. (8.9 cm), and a sharp, bright image, plasma displays are already used for applications, such as digital visual messaging, retail advertising, and boardroom presentations (see Fig. 15.4).

The most promising large screen technology is the LCD projector. These projectors use a very bright halogen lamp to illuminate a small LCD panel. A lens in front of the LCD magnifies the image and projects it onto the screen. In rear or reflective projection, the projector is located behind the screen. Image brightness is a fixed relationship between the projector brightness and the screen surface area. The lumen is a measure of light output. A small increase in the diagonal measure of a screen translates to a big increase in surface area. A projector generates twice the image brightness on a 50 in. (127 cm) screen than on a 70 in. (177.8 cm) screen. LCD

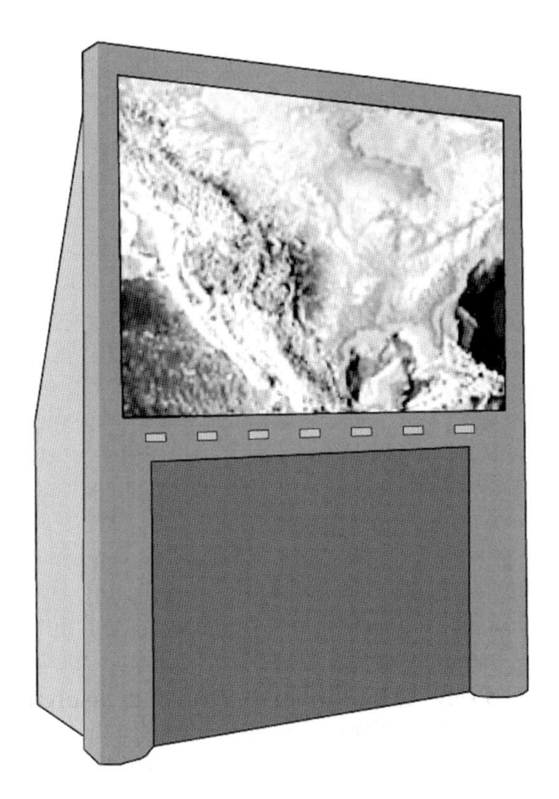

FIG. 15.4. Large screen plasma display mounted in a stand.

projectors require a screen and this is available at various sizes, including acrylic screens for rear projection.

3.2.2. New Display Technology

Two potential technologies are described here that may be used for large map displays in the future: organic screens and fog screens.

Organic electronics is a new technology where organic materials similar to common plastic are used as conductors and/or semiconductors to create optical-electronic devices. This display technology is based on the discovery of conductive polymers in 1977 and the electroluminescent properties of the so-called "conjugated" organic materials (Condottiere 2004). It is now possible to make bright light-emissive diodes in all colors with organic material. These are manufactured by depositing thin films of organic conductors, insulators, and /or semiconductors. Processing temperature is low (<200°C) so that organic devices can be manufactured on cheap, thin and flexible plastic substrates, obtaining flexible electronic devices. Large, flexible displays that use organic transistors for addressing the matrix are presently the focus of several research groups and could challenge silicon-based LCD screens and plasma displays. Organic light emitting diodes can be integrated in large areas to form ultrathin, bright, colored emitting panels that are flexible, thin, can be rolled up and stored or painted on a wall. Organic films can be deposited inexpensively via high-resolution printing. This means that organic displays could be produced with an ink-jet printer (Haskal *et al.* 2002) presenting the possibility that displays could be made easily by anyone with such a printer. Huitema *et al.* (2001) used an organic transistor display with a polymer-dispersed liquid crystal to produce an early example of electronic paper.

The FogScreen is another new type of display screen. A new invention from a Finish company, the FogScreen creates a thin, smooth fog surface and can be used for image projection just like a conventional screen (see Fig. 15.5). The fog, made by ordinary water with no chemicals, evaporates in seconds without any trace. The viewer can walk into the screen and through the projected image. The basic components of the screen are a laminar, nonturbulent airflow, and a thin fog screen injected into a laminar flow. The fog is made within the device using water and ultrasonic waves. The device generates a mild downward air and fog flow. The fog feels dry and cool but does not condense on objects. As a laminar airflow, the fog remains thin, crisp, and protected from turbulence. The FogScreen works very much like an ordinary screen in terms of projection properties. Images are projected onto the fog with a video projector located at a minimum distance of two meters. The FogScreen can be set to produce a translucent or fully opaque screen and can be used for both back- and front-projection. Ambient light is not a problem if a bright projector is used but a projector with at least 3000 lumens is recommended. The next generation of FogScreen will be interactive, much like touch screens. It is envisioned that interactivity will make the FogScreen useful for advertising, in malls for presenting shops and products, in art productions and as interactive exhibits in science and other museums (FogScreen 2004).

3.2.3. Collaborative Multibrowsing

The use of multiple LCD projectors has been investigated by Li *et al.* (2000). He and his team arranged a series of projectors to create a large

FIG. 15.5. Example image projected onto a FogScreen. The device uses droplets of water to create a screen. A high-lumen video projector is used to project the image.

backlit wall display (see Fig. 15.6). Commercial implementations are also available using CRT, LCD, and plasma screens. The advantages of a multiple LCD display are increased image size, increased image brightness, increased resolution, reduced projection distances, and increased depth of focus. All multiple display communication wall solutions face the problem that even minor differences in brightness and/or color characteristics between adjacent displays is often very apparent.

Although ubiquitous computing hardware is widely available, one factor in making ubiquitous computing useful is a way to use multiple heterogeneous displays to view and browse information. Collaborative multibrow sing is proposed to extend the information browsing metaphor of the Web across multiple displays. This involves coordinating control among a collection of Web browsers running on separate displays. The displays may be "public" as in a communication wall or "private" through the use of a series of hand-held Personal Digital Assistants (PDAs). Multibrowsing extends browser functionality by allowing users to collaborate and move existing pages or link information between multiple displays. It also enables the creation of new content targeted specifically for multidisplay environments (Johanson *et al.* 2001, Johanson 2002). Kray *et al.* (2003) also examine a mechanism for distributing coherent presentations to multiple

FIG. 15.6. Multiple projectors arrayed to create a backlit, wall-sized display (photo courtesy of Kai Li).

displays. Another multidisplay arrangement has been proposed that creates a surround effect. Here, the interior part of a room becomes a panoramic environment based on a series of projectors arranged in a circular fashion that displays images on backlit screens (see Fig. 15.7). An alternative arrangement could have the projectors inside a circle projecting to the outside. Or, the panoramic view could be arranged as a walk-through room, like a hallway.

A multidisplay collaborative cartographic system would be useful in a group setting to view a series of maps for analysis. Discerning patterns in maps are highly individualistic and a multidisplay system would bring a group of people together to collectively decipher the displayed patterns. It would also be useful for a variety of educational uses.

3.2.4. Stereo Applications

Stereo communication walls in a laboratory setting use two video projectors, a fast dual-output graphics card, and a computer to display two images simultaneously. The Geowall project, for example, makes use of these projection systems to visualize structure and dynamics of the Earth in stereo to aid the understanding of spatial relationships (Steinwand *et al.* 2002).

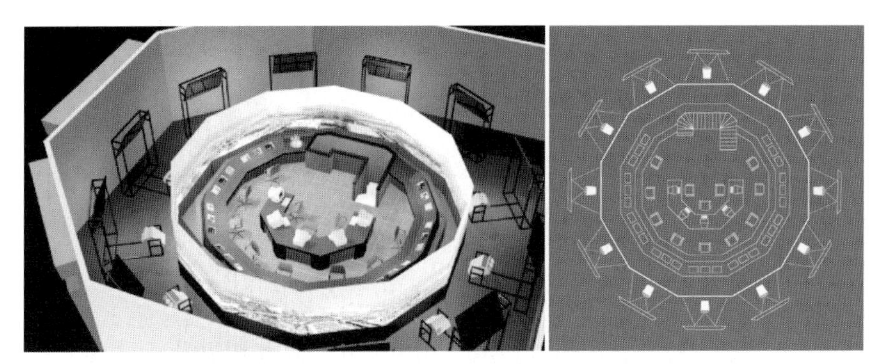

FIG. 15.7. A surround screen arrangement with 12 LCD projectors. The projectors are pointed inward with mirrors and display an image through backlit screens. The room is entered from below (Illustration courtesy Stewart Screens).

The Geowall can be assembled from off-the-shelf components that include, minimally, a computer with a special graphics card, two data projectors, polarizing filters with matching glasses, and a screen that preserves polarization. The graphics card has two monitor outputs to provide separate signals to left and right projectors. The Linear Polization Glasses (LPGs) are like those used in IMAX theaters and work with the display to present the user two separate, slightly offset, images. The lenses of the projectors need to be together as close as possible so that the projected images can be precisely aligned. The simplest way is to physically place the projectors on top of one another using a special stand. Figure 15.8 depicts the arrangement of computer and projectors.

Three-dimensional viewing without the use of special glasses is possible with holographic display. A holographic display system combines the images from two LCD projectors on specially designed, rear-projection opaque screens to create a three-dimensional viewing zone that is visible from certain directions. These screens achieve a striking three-dimensional effect by using special transparent screen material. Holographic systems are compatible with standard three-dimensional graphic and design software, and enable the storage and display of three-dimensional images (Physical Optical Corporation 2004). They accept three-dimensional recorded images, video games or three-dimensional graphics.

3.3. Ubiquitous Mobile Computing

Although efforts have been made to make communication walls more portable, they are essentially tied to a specific location. A true mobile

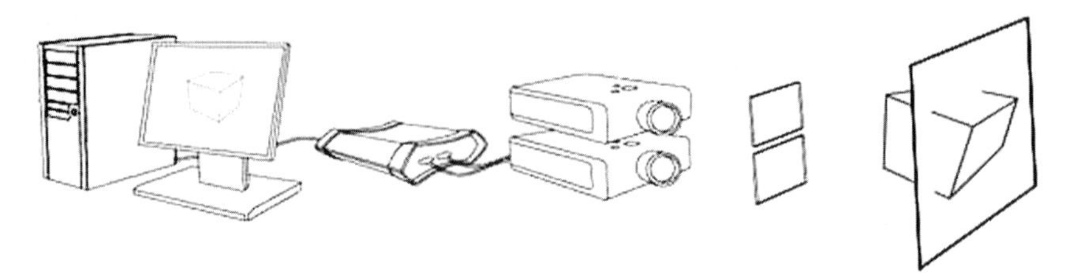

FIG. 15.8. Two projectors used to create a stereo image.

computing model would allow for ubiquitous computing with mobile devices. There has been a great deal of hardware development in this area and dividing line between cell phones and PDAs is blurring. The PDA that combines computing and wireless networking capabilities is the current model used for a mobile device. A number of ideas have been proposed for ubiquitous computing on these devices including location-aware computing and geocasting. Some of these are discussed by Gartner in Chapter 16.

3.3.1. Location-Aware Computing

Many of the electronic devices that we use are affected by location. But, these devices, such as cell phones, laptops, PDAs, have essentially no idea of where they are or what's nearby. Most likely, if you travel a long distance by airplane, the high-tech devices you have would not know that you have switched time zones. Location-aware computing has the potential to integrate locational intelligence into computing devices.

Research in location-aware computing involves location sensing technologies, location representations, location-aware applications, and factors affecting the deployment of location-aware systems in everyday environments. The Global Positioning System (GPS) has added an important new source of location information. As GPS units have gotten smaller, new applications have been developed. For example, a wristwatch-type GPS device is available that can be used to track children. A $200 device from Wherify.com works with an associated online mapping system that tracks a child's location (Wherify 2004). A company called Digital Angel has been developing implants for humans that would monitor a wearer's location, pulse, blood-oxygen level and other vital bodily functions. Internal implants are routinely used for pet identification and Digital Angel has stressed the development of implants for this application, particularly for livestock identification (Digital Angel 2004).

Privacy aspects have become another area of research in location-aware computing (Monmonier 2003). As Monmonier points out, these new "applications have heightened awareness of the map's role as an instrument of both democratic government and social control" (Monmonier 2003: 110). Other social issues in location-aware computing include usability, user acceptance, and implications for Human–Computer Interaction (HCI) and computer-supported collaborative work.

3.3.2. Geocasting

Geocasting is a concept used for sending messages to mobile devices in certain geographic areas where it would be of interest to clients (Dürr and Rothernol 2003: 18). A geocast client is a software component running on a mobile or stationary device like a PDA or a PC. Geocasting forms the basis for a number of location-based services, such as announcement services, advertisement services or friend-finders. A semantic geocast refers to a target area that is specified by its meaning. A sender can broadcast messages to a city center or a specific building, without precisely knowing the physical coordinates.

Location-Based Services (LBS) have been of great interest in Europe where cell phone use far exceeds that in North America. The main goal has been to provide the cell phone user with their location and associated information. Although this service has been implemented on a limited basis, privacy concerns have made the user interface very complicated. A user is queried multiple times to make sure they really want their location computed. The cost involved has also detracted from the implementation of LBS.

4. Ambient Intelligence

Ambient Intelligence (AmI) refers to a development in consumer electronics to create smart electronic environments that enhance the lives of people (Aarts *et al.* 2003). The term is used to denote a new paradigm for user centered computing and interaction. Ambient intelligence represents a vision of the future in which we will be surrounded by electronic environments, sensitive and responsive to people. Ambient intelligence technologies are expected to combine concepts of ubiquitous computing and intelligent systems to put humans at the center of technological developments.

Ambient Intelligence builds on ubiquitous computing and communication. In contrast to ubiquitous computing, which refers to the integration of microprocessors into everyday objects, AmI enables these objects to communicate with each other and to the user by means of wireless networking. An intelligent user interface enables the inhabitants of the ambient world to control and interact with the environment in a natural way through voice and gestures, and personalize based on preferences and context (Ahola 2001).

In an AmI world, people will be surrounded by electronic systems that consist of networked intelligent devices, which provide information, communication services, and entertainment based on location. Technology will become invisible and embedded in our natural surroundings and characterized by effortless interactions that are attuned to all our senses, context-sensitive and adaptive to users. High-quality information and content must be available to any user, anywhere, at any time, and on any device. Furthermore, the devices will adapt and even anticipate our needs. Ambient intelligent environments will merge in a natural way into the environment around us and allow much more natural and human interaction styles. Consequently, the electronics will be integrated into our clothing, furniture, cars, houses, offices and public places (Aarts *et al.* 2003). According to Lindwer *et al.* (2003), the vision of AmI includes:

(1) very unobtrusive hardware;
(2) a seamless mobile/fixed communications infrastructure;
(3) dynamic and massively distributed device networks;
(4) natural human interfaces; and
(5) dependability and security.

Contextual awareness refers to technologies that use sensorial information to adjust the functional behavior of digital devices. To build contextually aware devices, it is required to sense, identify, and interpret relevant contact. The relevance of context depends on the application. From the point of view of AmI, contextual awareness can be viewed as the technology that provides the ambience with environmental senses, which may result in smart and eventually intelligent behavior (Aarts *et al.* 2003). To provide users with information, communication, and entertainment at any desired place and time in an intuitive, efficient, and effective way requires a significant amount of system intelligence (Verhaegh *et al.* 2003).

5. Pervasive Map Interfaces

It is important for us to consider alternate user interfaces for map displays. As technology advances, new methods of map display will augment or displace current methods of map-use by computer. A pervasive map interface will take many forms. On one hand, there will be different methods of controlling the display; on the other hand there will be output types that add to the map display. All of these will be made possible by advances in network technology. It is already possible to walk into a coffee shop and connect to the Internet. The ease with which such connections between devices are established will become much more transparent in future as cities will develop an Internet infrastructure. Three possible examples of a pervasive map interface are presented here.

An example of a different type of input device that can be used as part of the user interface is the video camera. Used for such applications as video conferencing, these cameras have become a standard option with most computers. Some are even built-in to the top of the monitor and are focused on the user. They capture full-color images at 30 frames per second and are used mostly for online chat applications where each user is presented with an updated image of the other user, much like a video phone.

Using a video camera on an Apple Macintosh™ computer, the collection of games called Toysight, are an example of how these cameras may be used. The games are played by standing in front of the camera and using movement and gestures to control virtual sliders, buttons, and menus on the screen. The program performs object and motion detection on the video image to track the movement of the user. The game conveys the feeling of bringing you into the game because you have to move around to make things work (Fothergill and Fothergill 2004). A similar system could be used to control the display of maps on a large-format screen.

On the other end of the user interface spectrum are alternate forms of output. Pictures and video sequences have already been integrated with maps and form the basis of a multimedia cartography (Cartwright *et al.* 1999). While many forms of media for maps have been proposed, most concentrate on the visual sense. An example of an alternate output would be the introduction of smell. Humans can distinguish more than 10,000 different smells (odorants) that are detected by specialized olfactory receptor neurons lining the nose (Alberts *et al.*1994: 722). Scientists believe that smell has the power to unlock memories. A smell machine is very much like the scratch and sniff products of decades ago. These stickers lasted a long time because of their microencapsulation technology. The aroma-generating chemical is encapsulated in incredibly small gelatin or plastic spheres – of the order of a few microns in diameter. When the sticker is scratched, some of the spheres are ruptured and release the smell. Only a few of the tiny fragrance bubbles are ruptured with each scratch. Such encapsulated spheres could be printed onto paper using an ink-jet printer. A similar system has been proposed to place taste on paper. In addition to smell and taste, more advanced methods need to be investigated to integrate sound and touch in maps.

The pervasiveness of the Internet itself is apparent in the city of Cleveland in the US. In a groundbreaking effort, the city is implementing free public wireless access to the Internet for the entire city of nearly 500,000 people by 2006. The OneCleveland project brings together Cleveland's major universities, hospitals, schools, libraries, civic institutions, museums, arenas, research entities, parks, and the public transportation authority to provide blanket broadband Internet coverage. Free high-speed access to the Internet is provided in a concerted effort by more than 1500 institutions and organizations. It will provide Cleveland with more access to broadband Internet than any city in the world.

The project is one of the largest integrated technology efforts in the world and has the potential to transform a city. It is already being used in schools, hospitals, and for public safety in the downtown area. Schools have access from any classroom. Hospitals can instantly exchange medical data of patients. Police officers are able to check national databases for stolen cars and pictures of suspects. The OneCleveland project is a pioneering effort and many other cities are implementing similar systems (OneCleveland 2004). Ubiquitous access to the Internet represents another opportunity for a pervasive map interface and use.

6. Conclusions

The technology of mapping has advanced quickly in the last 40 years. With the help of the Internet and increased international competition, the rate of technical innovation will certainly increase. Along with this technical innovation, alternate methods of map display, interaction, and delivery will also come. Large-format displays described here will represent a major alternative to the current WIMP interface. Maps will benefit greatly from these developments because they will lead to related developments in different cartographic representations and new methods of interaction. As can be seen with quick adoption of maps and the Internet, people seem to readily adapt to these different interfaces with maps.

The current interface for interactive maps would seem to be a serious obstacle to their use. While the WIMP-based, interactive map has been a major advance for cartography, there is still a great reliance on maps that are essentially static. For example, the much-used Perry-Castañeda map collection at the University of Texas only presents static maps. Furthermore, search engines promote the use of static maps by offering an "image search" option that returns links to GIF and JPEG images. Entering any place into this type of search returns hundreds, if not thousands of static maps. Anyone looking for an online map of Africa or Asia, for example, would unlikely find a functional online interactive map and would likely resort to an image search engine to find a static map. For most applications, highly-interactive maps are not needed and few see the need of learning the user interface before being able to view a map. There still seems to be a great reliance on static maps because they are not only easier to find but they may also be easier to use.

If the interface for most interactive maps is too obtrusive, then alternate interfaces need to be developed. The methods described here provide a framework for creating different map displays that would create environments for map-use. The concept of pervasive cartography would integrate maps into more public areas. Ambient intelligence implies that a map display can have an embedded intelligence that can be communicated between devices and conveyed to a user.

Many of the proposals for alternative map interfaces stress collaboration because the interpretation of the mapped patterns can be very difficult, requiring a considerable background in geography, history, and other areas of the social and physical sciences. Most mapped patterns go unnoticed because people lack the background to interpret them. A collaborative map-use environment can bring maps to life by depicting interesting patterns that would lead to analysis and interpretation by a group.

References

Ahola, J. (2001) "Ambient intelligence" *ECRIM News*, No. 47, accessed October 2001 from http://www.ercim.org/ publication/Ercim_News/enw47/intro.html

Alberts, B. *et al.* (1994) *Molecular Biology of the Cell*, Garland Publishing, New York.

Aarts, E. *et al.* (ed.) (2003) *Ambient Intelligence*, Springer-Verlag, Berlin Heidelberg.

Condottiere, E. (2004) "Making electronics unobtrusive with organic materials", MEST Symp in 2004 *Unobtrusive Electronics* (Delft).

Cartwright, W., M. Peterson and G. Gartner (1999) *Multimedia Cartography*, Springer-Verlag, Berlin Heidelberg.

Digital Angel Corporation (2004) *RFID Overview*, accessed 25 March, 2004 from http://www.wherifywireless.com/prod_watches.htm

Dürr, F. and K. Rothernol (2003) "On a location model for fine grained geocast", *Ubicomp 2003*: *Ubiquitous Computing 5th International Conference,* Springer-Verlag, Berlin Heidelberg.

Ferscha, A. and S. Vogl (2002) "Pervasive web access via public communication walls" in Mattern, F. and M. Naghshineh (eds.), *Pervasive Computing*, Springer, Berlin Heidelberg.

FogScreen Inc. (2004) *Projecting Images in the Air*, accessed 25 March 2004 from http://www.fogscreen. com/product.html

Fothergill, A. and A. Fothergill (2004) *Toysight Web Page*, accessed 25 March 2004 from http://www.toysight.com

Haskal, E. I. *et al.* (2002) "Ink jet printing of passive-matrix polymer light-emitting displays", *Journal of the Society of Information Display (SID)*, "Best of SID 2002."

Huitema, H. E. A. *et al.* (2001) "High quality active-matrix displays driven by organic transistors", *Proceedings of SCHARM'01 Science.*

Johanson, B. (2002) *Application Coordination Infrastructure for Ubiquitous Computing Rooms*, accessed 7 February 2005 from http://graphics.stanford. edu/~bjohanso/dissertation/

Johanson, B., S. Ponnekanti, C. Sengupta and A. Fox (2001) "Multibrowsing: Moving web content across multiple displays", *Proceedings of Ubicomp 2001*, September 30–October 2, 2001.

Kray, C., A. Krüger and C. Endres (2003) "Some issues on presentations in intelligent environments", in Aarts, E. *et al.* (eds.) (2003), *Ambient Intelligence*, Springer-Verlag, Berlin Heidelberg.

Li, K. *et al.* (2000) "Early experiences and challenges in building and using a scalable display wall system", *IEEE Computer Graphics and Applications*, Vol. 20, No. 4, pp 671–680, 2000 accessed 7 February 2005 http://www.cs.prince ton. edu/omnimedia/papers/cga00.pdf

Lindwer, M. *et al.* (2003) "Ambient intelligence visions and achievements: linking abstract ideas to real-world concepts", *Proceedings of the ACM Special Interest Group on Design Automation (SIGDA).*

Mattern, F. and M. Naghshineh (eds.) (2002) *Pervasive Computing*, Springer-Verlag, Berlin Heidelberg.

Monmonier, M. S. (2003) "The internet, cartographic surveillance, and locational privacy", in M. P. Peterson (ed.), *Maps and the Internet*. Elsevier Science, Amsterdam, pp. 97–113.

OneCleveland (2004) *The OneCleveland Project*, accessed 25 March 2004 from http://www.onecleveland.org

Peterson, M. P. (1995) "Evaluating the electronic wall map", *Proceedings of the 17th International Cartographic Conference*, pp. 654–662, accessed February 7 2005 from http://maps.unomaha.edu/ mp/Articles/WallMap.html

Peterson, M. P. (1998) "That interactive thing you do", *Cartographic Perspectives*, No. 29, pp. 3–4, accessed 7 February 2005 from http://maps.unomaha.edu/mp/Articles/ThatInteractive.html

Peterson, M. P. (2001) "The automated display of maps and images from the internet", *Journal of South China Normal University*, July 2001, pp. 56–64.

Physical Optical Corporation (2004) *Electro-Optics and Technology Summary*, accessed 25 February 2004 from http://www.poc.com/

Steinwand, D., B. Davis and N. Weeks (2002) *GeoWall: Investigations into Low-cost Stereo Display Systems, USGS Open File Report*, accessed 7 February 2005 from http://geowall.geo.lsa.umich.edu/papers/Geowall.pdf

Streitz, N. (2003) *Interaction Design for the Disappearing Computer*, Springer-Verlag, Berlin Heidelberg pp. 351–355.
Verhaegh, W., E. Aarts and J. Korst (eds.) (2003) *Algorithms for Ambient Intelligence,* Kluwer Academic Publishers, Dordrecht.
Want, R. *et al.* (2004) *The ParcLab Ubiquitous Computing Experiment*, accessed 25 February, 2004 from http://sandbox.xerox.com /parctab/csl9501/paper.html
Webopedia (2004) *Definition of Pervasive Computing*, accessed 25 March 2004 from http://www.pcwebopedia.com /TERM/p/pervasive_computing.html
Weiser, M. (1994) *The World is Not a Desktop in Interactions*, accessed 7–8 January 2005 http://www.ubiq.com/ hypertext/weiser/ACMInteractions2.html
Weiser, M. (1991) "The computer for the twenty-first century", *Scientific American*, accessed 6 January 2005 from http://www.ubiq.com/ hypertext/weiser/ SciAmDraft3.html
Wherify, Inc. (2004) *GPS Locator for Children*, accessed 24 March 2004 from http://www.wherifywireless.com/prod_watches.htm

CHAPTER 16

TeleCartography: A New Means of GeoCommunication

GEORG GARTNER
Research Group Cartography
Department of Geoinformation and Cartography
Vienna University of Technology, Vienna, Austria

Abstract

This chapter aims to analyze elements and the effects of cartographic communication processes in the context of mobile Internet applications. TeleCartography and map-based location based services (LBS) are analyzed by describing their basic elements (positioning, information modeling and presentation, the user and questions of adaptation), the state of the art of current systems and the main areas of research, especially the impact on GeoCommunication. Selected results of experiences in terms of fundamental research (potential of cartographic presentations forms), in positioning (active landmarks), modeling and visualization (guiding and navigation for pedestrians) are given and are brought into the context of GeoCommunication. Telecartography is a specific type of cybercartography.

1. Introduction

Telecommunication infrastructures (mobile networks), positioning methods, mobile in- and output devices and multimedia cartographic information systems are prerequisites for developing applications, which incorporate the user's position as a variable of an information system. Imparting spatial information within such a system in cartographic presentation forms is normally involved. Thus, the resulting system can be called a map-based location based service (LBS). This chapter briefly discusses the elements of a map-based LBS or, more generally, TeleCartography; it outlines main research topics and describes some experiences in the context of conceptual design and developing map-based LBS. A theoretical approach is developed in order to position TeleCartography as a new means of cartographic communication. Telecartography is a specific type of cybercartographic communication as outlined in several chapters in this volume.

2. Telecartography

The main advantages of Internet Cartography are described as gaining better accessibility for the user, enabling higher actuality (van Elzakker 2000), or easier distribution of maps. But, the efficiency of the usage of Internet-based applications, as any other digital mapping application, is determined and restricted by the main attributes of the machines, which are used for accessing and interacting with mapping systems. One of the main issues is that computers are usually not highly mobile. For a lot of cartographic applications this missing mobility and the fact, that the user has to find access to a machine in order to get his/her information or map, is not a major disadvantage. But for enabling cartographic information systems, which could serve products independent from the position/location of a user, in terms of "giving you the information right there where you need the information," the availability of mobile input/output machines and the availability of an infrastructure for wireless submission of information to any location; these are necessary preconditions (Gartner 2002).

The infrastructures and technologies of telecommunication systems are developing rapidly. They have reached a stage where they are judged as a mass market industry. The currently developing new protocols and network environments like WAP, GPRS, or UMTS, can be seen as first test areas for developing TeleCartography applications, because they enable high transfer rates, graphic or even multimedia data transfer, and mobile client/server interactivity. Together with the development of so-called Wireless Information Devices (WID), mobile handheld devices with enriched functionalities (e.g. PDA, PocketPC, Smartphones, Communicators) (Gartner 2002), the fundamentals and preconditions for developing and testing TeleCartography applications are in principle available.

TeleCartography can be defined as the distribution of cartographic presentation forms via wireless air data transfer interfaces and mobile devices (Gartner 2002). Therefore, the development of TeleCartography has to be seen in a close context with the development of telecommunication infrastructures, air interfaces, and related data transfer protocols (e.g. Universal Mobile Telecommunications System, UMTS), and finally the enhancement of mobile devices ("wireless information devices" WID) with functionalities for cartographic information presentation and retrieval (Brunner-Friedrich *et al.* 2001; Gartner 2002).

Cartographic information transmission can benefit from this development because it opens the possibility of serving actual and interactive cartographic products, independent of time and the location of the user. Therefore the disadvantage of the immobility of digital I/O devices by using desktop computers can be overcome, while still taking advantage of the full functionalities and possibilities of digital cartography.

3. Elements of Telecartography and Cartographic Location Based Services (LBS)

A system can be called an LBS when the position of the mobile device and therefore the position of the user is an integral part of an information system. Several types of applications are possible in this context and include simple and text-based applications, which use the cell ID for a rough positioning (e.g. which petrol stations are there around me?), for map-based multimedia applications including routing functionalities. Regardless of the level of complexity of the system's architecture, every map-based LBS requires certain basic elements to handle the main tasks of positioning, data modeling, and information presentation.

3.1. Positioning

The determination of the position of a mobile I/O device is a necessary requirement for every system identified as a LBS. Positioning has to be appropriate to the service and of a nature and level to perform the tasks required. For several applications the necessary level of accuracy needed can be served by the cell ID of a telecommunication network and the position derived from it, which gives a positional accuracy of 50–100 m in urban areas (Retscher 2002). For navigation purposes – in particular in the context of pedestrian navigation – the accuracy demand increases to values of 25 m or less (Gartner and Uhlirz 2001; Retscher 2002). For indoor navigation, the requirements for determining the position are even greater (Gartner *et al.* 2003).

Various methods of positioning are available for different levels of accuracy. These include:

- Positioning by radio network;
- Satellite-based positioning;
- Alternative methods; and
- A combination of these approaches.

Nowadays for outdoor navigation, satellite-positioning technologies (GPS) are most commonly employed. As a result, the achievable positioning accuracies of the navigation system depend mainly on the GPS used, which provides accuracies of a few meters to 10 meters in stand-alone mode or sub-meter to a few meters level in differential mode (DGPS). If an insufficient number of satellites are available for a short period of time due to obstructions, then in a conventional approach, observations from additional sensors are employed to compensate for loss of signal. This is particularly necessary for areas where the satellite signals are blocked, such as indoor or underground environments, or situations in urban areas.

In general, another method of deriving positional information is by using the parameters of a radio network, using the coordinate information of a cell or a base station. This, however, is restricted by cell dimensions. Measuring methods using elapsed time of signals in combination with cell identification, time synchronization, or differences of elapsed time can be used for improved positioning (Retscher 2002).

Alternative methods or improvements of already existing methods are shown by Zlatanova and Verbree (2003). They propose "user tracking" in combination with augmented reality (AR) to improve positioning in bad conditions (indoor/underground). Kopczynski (2003) describes an approach which uses simplified topological relations to determine positions by sketch maps (sketch based input).

3.2. Modeling and Presentation of Information

The limitations of the mobile device used restrict the possibilities of transmitting, refreshing, and displaying location based spatial information. The conditions of the cartographic communication process also have to be fulfilled wherever possible in the context of map-based LBS. The cartographic model has to be clearly perceivable while it is permanently scale-dependent and has to present the task-dependent appropriate geometric and semantic information.

This fact in combination with restrictions in size and format of current mobile devices leads to different types of solutions for presenting information within map-based LBS. These include:

- Cartographic presentation forms adapted to the specific requirements of screen display;
- Cartographic presentation forms without specific adaptations;
- New and adapted cartographic presentation forms; and
- Multimedia add-ons, replacements, and alternative presentation forms.

Rules and guidelines have been developed over the last few years to adapt cartographic presentations to the specific requirements of screen displays (Neudeck 2001; Brunner and Neudeck 2002). A lively discussion about new and special guidelines for map graphics in the context of the very restrictive conditions of TeleCartography and mobile Internet has generated various suggestions and proposals (Gartner and Uhlirz 2001; Wintges 2002; Reichenbacher 2003). In these discussions the main focus will be on questions of graphical modeling, visualization or generally on questions of usability and application navigation (Meng 2002; Wintges 2002). Initial experiences and results have been presented such as the prototyping of the UMTS application LoL@ (Gartner and Uhlirz 2001).

Common rules or standards for cartographic presentations on screen displays are yet to be defined and are influenced by constantly changing technological determining factors. Display size and resolution of state-of-the-art devices are constantly improving; color depth is no longer a restricting factor. Parameters of external conditions during the use of the application (weather, daylight) are hard to model. The needs of an interactive system have to be incorporated into the conception of the user interface, which includes soft keys and functionalities for various multimedia elements. The concept of adaptation has been suggested as a general approach for including the various parameters within a model of map-based LBS (Goulimis *et al.* 2003; Reichenbacher and Töllner 2003). This approach describes links or mutual dependencies between various parameters and these links impact on both data modeling and the cartographic visualization used. New cartographic presentation forms have also been introduced, taking into account that a restricted small screen display has to be used for transmitting cartographic presentations such as a "focus-map" by Klippel and Richter (2003). The challenges of screen presentation for cybercartography identified by Monmonier in Chapter 2 are magnified for telecartography.

For the presentation of spatial information within LBS and on small displays, additional multimedia elements and alternative presentation forms in general are increasingly seen as potential improvements. For example the case of cybercartography in general as illustrated by several chapters in this volume. Methods of Augmented Reality (AR) link cartographic presentation forms, such as three-dimensional graphics to a user's view of reality. Applications in navigation systems are a good example. Cartographic AR applications are based on the idea, that a more intuitive user interface can be created as suggested by Reitmayr and Schmalstieg (2003). Kolbe (2003) proposes a combined concept of augmented videos, which transmit positioning and information transfer by means of video. User interface design is a key issue for cybercartography as illustrated by several chapters in this volume.

3.3. Users and Adaptation

Experiences of developing LBS have led to various suggestions taking into account a more user-adequate approach. The importance of user needs analysis for cybercartography, in general, has been made by Lindgaard *et al.* in Chapter 9, which is critical for effective telecartography. Modeling parameters in the context of the "user" and the "usage situation" are seen as fundaments of more user-adequate approaches, which can be summarized as "concepts of adaptation." Zipf (2003) includes within his understanding of adaptation, components of "adapting to a system," "to a user" and "to a situation." "User needs analysis" is a useful concept in this respect (Lindgaard *et al.* Chapter 9).

The adaptation of cartographic visualizations in this context can include the automatic selection of adequate scales, algorithms for adequate symbolization or even the change to text-only output of information in case of the inadequate graphic potential of output devices. This illustrates an important element of cybercartography in general. Adaptation to the user for telecartography is currently limited to user profiles selected in advance from a list, or entered manually by the user to influence the graphical presentation (size of lettering, use of colors), or to provide predefined map elements. Techniques of user needs analysis developed for cybercartography may prove to be fruitful research areas for telecartography. Adaptation of the visualization to the situation includes the current time of day (day/night), or takes into account the actual velocity of the user. This type of adaptation is used in some versions of navigation software like Tom-Tom, which automatically adapts the map scale to the user's current velocity in vehicle navigation systems.

Various forms of adaptation are summarized by Reichenbacher and Töllner (2003) as "context-adapted Geovisualization." The authors utilize methodologies and algorithms to develop effective cartographic presentations, taking into account the parameters of various output devices and different user needs. This approach is challenging not only the technical developments, but also the questions of how to identify, define, and model the main influencing parameters (e.g. "user," "user situation"). First attempts at empirical studies concerning this subject are made by Zipf and Jöst (2003) and Radoczky (2003), although implementation of experiences from these approaches are rare. The user needs analysis techniques outlined in this volume (Lindgaard *et al.* Chapter 9) have much to offer in this respect and have been used in the implementation of both the Cybercartographic Atlas of Antarctica (Pulsifer and Taylor, Chapter 20) and Canada's Trade with the World (Eddy and Taylor, Chapter 22).

4. Research Questions in the Context of Cartographic LBS and Telecartography

The key elements of map-based LBS, as pointed out in Section 3 of this chapter, are focused on the research questions: Positioning, Modeling, Visualization, and Adaptation.

4.1. Integrative Positioning

Current research on positioning focuses on improving existing methods and in particular finding an integrating concept, including different types of observations and methods, such as satellite-based methods, radio network, and dead reckoning for determining an appropriate position. The combination of the advantages of integrating different methods can lead to a more effective positioning system especially in urban environments. Accuracy of positioning is needed in the low meter range, for pedestrian or indoor navigation. For example, costly preparatory work with network providers is necessary which requires major efforts to synchronize time signals (Retscher 2002) or additional tools for mobile device are essential (GPS-Jackets/Cards). Even assuming that satellite-supported solutions will become much less expensive, some restrictions inherent to the system will remain, especially the blocking of signals of various satellites in specific environments like cities. According to Verbree *et al.* (2003), the combination of GPS- and Galileo-signals will lead to a significant improvement in this context, but will still have the restrictions of all satellite-based solutions (blocking of signals). The range of attainable accuracies will not be significantly improved by combining various satellite-based positioning systems, but there will be an impact on applicability. Alternative approaches like integrating video and augmented reality for improved and supported positioning, as proposed by Zlatanova and Verbree (2003) or Kopczynski (2003), are based on the use of three-dimensional models or special semantic networks, which have to be derived from basic surveying data. The applicability of these approaches in operational conditions differs from "ideal attempted" test areas and has not yet been effectively demonstrated.

4.2. Route Information Systems

Route information systems are of central interest to applications of map-based LBS (Gartner and Uhlirz 2001; Hampe and Elias 2003; Lehto and Sarjakoski 2003; Sarjakoski *et al.* 2003). In the context of modeling routes for pedestrian navigation, a major focus is the question of user-centered approaches (Urquhart *et al.* 2003). A user-centered approach must take into account existing user knowledge about wayfinding, communicating routes, and navigation instructions. This also includes the question of differences in the actions and needs of map users and those map-based LBS users. Does new behavior in map use require new forms of cartographic information transmission development? And is the user of a mobile interactive cartographic information system able to gather all the information the system offers? This leads to the general question of the additional value or benefit of telecartography and cartographic LBS.

In order to answer these questions, knowledge about human behavior in wayfinding is necessary, as well as definitions of the potential of different presentation forms for navigation purposes. In this context potential means the capability of presentation forms to transfer spatial information in a particular situation with geometric and semantic validity and suited to the user's context. First basic studies have been made by Reichl (2003), who proposes a classification of the potential of different presentation forms in the context of route information. In this volume Lindgaard *et al.* discuss the issues of use and usability in the general context of cybercartography.

4.3. Information Presentation and Visualization

Symbolization, visualization, and information presentation in general, in the context of LBS, cannot be discussed without taking into account the restrictions of current mobile devices. These include the size of the display, data transfer rates, terminal access time, and storage capacity, among others. Although a rapid technological development is taking place, the essential restriction of small format screen displays still persists. To cope with this limitation, approaches such as new cartographic presentation forms like "focus-maps" (Klippel and Richter 2003) are developed, or map-related presentations (two-, five-, and three-dimensional presentations), and multimedia cartographic presentation forms (use of photos, videos, or augmented reality) are applied. Prototype applications have proven that there is a potential to integrate new and adapted presentation forms in cartographic LBS (Gartner *et al.* 2003; Radoczky 2003), although solutions for handling huge amounts of data, their automated acquisition, and utilization are lacking. Finally, a discussion of possibilities of "on-the-fly" derivations of high quality presentation forms remains to be answered (Gartner *et al.* 2003). Small screen displays are an extreme case of the presentation challenges facing cybercartography.

Strategies for the optimization of existing cartographic presentation forms in the context of LBS (adaptation of maps for small displays, development of new, and use of alternative presentation forms) may be overtaken by technical innovations. Electronic paper (e-paper) is at the prototyping stage; displays with characteristics of paper (thin, elastic, light) have been developed (Der Standard 31 January 2004), and foldable and virtual keyboards are ready for use (Der Standard 14 January 2004).

5. Experiences and Results from Map-Based LBS Research

5.1. Applicability of Presentation Forms

To communicate route information most efficiently utilizing the concept of adaptation (see Section 2) requires knowledge and analysis of presentation forms, and their applicability and potential for route information systems. In principle various presentation forms could be used to present routes, including all cartographic presentation forms (maps and map-related presentation forms) besides the others that are suitable for spatial-relevant route information, like text (written and spoken), graphics, video, animation, or combinations of these (Reichl 2003). These are integral to cybercartography (Taylor 1997, 2003).

Reichl (2003) analyzes the connectivity between user context and user groups in this respect and proposes a "matrix" that outlines the applicability of different presentation forms. Reichl's observations are based on an empirical study, where a major finding was: communicating route information by maps leads to significantly "more complete" topographical knowledge and understanding than non-map-based procedures. Transmitting route information without maps can enable the user to build up

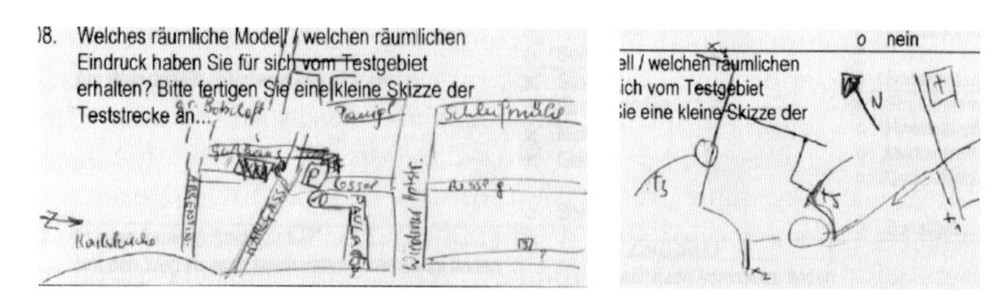

FIG. 16.1. *Sketch Maps* drawn after navigation by textual instructions (left side) and by map-based instructions (right side). Creation of a topological mental map model versus a topological mental map model (Reichl 2003).

topological correct mental representations, but will lack a chorographic meaning of topography (see Fig. 16.1). The study utilized a procedure where participants received navigation instructions in a number of different presentation forms and then had to answer a questionnaire and to draw a sketch map from memory. This study of text-based versus map-based information led Reichl to conclude that map-based information communication leads to a better reproduction of the geometric information and better distance estimation. This conclusion is similar to that of a study performed by Thakkar *et al.* (2001). Further research is necessary to analyze similarities between various presentation forms for human route communication in the adaptation to different styles of learning (Kienberger 2004); and the derivation of route information from one single data set to different presentation forms (Baulch *et al.* and Lindgaard *et al.* in this volume also discuss learning styles).

5.2. Cartographic Support for Wayfinding

Based on the description of the potential of various presentation forms for transmitting route information (Section 4.2), the range of applicability can be tested. In this context the combination of different presentation forms for certain situations (e.g. in mixed indoor/outdoor environments) is particularly interesting. Radoczky (2003) has analyzed the shortcomings of existing LBS applications in this context and proposed – by using the theory of multiencoding (Lovelace *et al.* 1999) – concepts for supporting the communication of route information via map-based LBS in terms of integrating multimedia add-ons and enhancing the user support by various means, including the use of maps, graphics, photos, text, video, panoramas, and animation in various kinds of combinations. The audio channel is used for additional support by means of speech, music, and audio signals. In order to improve the use of the "map" presentation form, various supportive concepts have been proposed, including animations (e.g. for supporting the understanding of a change of map scale), map-relevant presentation forms such as floor plans, bird's eye views, video clips, or VR scenes. Radoczky (2003) focuses especially on the relevance of support while changing presentation forms and proposes some solutions in this regard, based on general theories of multimedia cartography. The relevance of the proposed solutions has been tested by implementing the concept on a mobile input/output device and by a completed feasibility study (Radoczky 2003). The results indicate that the trends are similar to comparable studies (Zipf and Jöst 2003). Radoczky (2003) indicates that some concepts and functionalities have been found to be helpful, independent of sex, age, or educational background for all tested situations (day time, weather conditions).

General Considerations: Overview maps for a general presentation of the whole route and information about distance and estimated time of walking have been found indispensable. An acceptance of 64% of all participants, being "always sure to be on the right way" by using the overview, has been stated. Overview maps in the context of navigation are discussed by Roberts *et al.* in Chapter 10.

Automatic Scrolling: Automatic adaptation of the presented map section to the position of the user has been found indispensable.

Egocentric Mapview: Eighty six percent of all participants preferred a "track up" oriented map, that means the map is always adapted to the user's direction of move, only 9% preferred "north-up" orientation.

Multi-encoded Navigation Instructions: In particular the integration of photos in case of decision points for better decoding, the map's information turned out to be helpful (73%). Panorama photos (73% of male, 36% of female participants), combinations of maps and spoken or written text for route information communication (82% argued that this combination is the most helpful of all combinations) have found high acceptance.

Supported Change of Scale: To enhance the understanding of the effects of a change of scale (in terms of changing from one scale dependent cartographic presentation to another), support can be useful. The change can be done abruptly, step by step, or animated. Indeed, 55% of all participants did not need a supported change of scale.

Presentation Forms for Indoor Environments: For indoor navigation (test case: University of Technology, Vienna), different presentation forms have been used: floor plans (accepted by 82% of the participants as "supporting navigation by finding the way immediately"), bird's eye views (64% acceptance, by 36% acceptance of male persons and 91% acceptance of female persons), or animated three-dimensional graphics (55% acceptance).

The following conceptions have been stated as only partly acceptable/helpful:

Speech Interaction: There has been no confirmation to the theory that mobile users might prefer speech commands instead of graphic–haptic interfaces. Sixty eight percent of the participants preferred a written menu for the selection of destinations instead of speech commands.

Photo Realistic Presentations: The use of photographs or 360°panoramas has been stated as not useful in the context of small displays and pedestrian navigation, although the specific use of photos and panoramas for landmark identification and as support for decision points has found positive acceptance (photos: 73% "yes, helpful for decision making," panoramas: 55%).

Change of Presentation Forms in Mixed Indoor/Outdoor Environments: In the case of changing presentation forms by transit from outdoor to indoor navigation, changes supported by dynamical zooming between maps and floor plans have got the highest acceptance to all other combinations (intermediate steps, changes from two-dimensional graphics to three-dimensional graphics, and vice versa).

Radoczky (2003) has shown with the studies that the chosen presentation form and the effort of supporting cartographic methods are decisive for the acceptance of pedestrian navigation systems. Adequate methods can accelerate the process of wayfinding and can avoid uncertainties of the user.

6. Summary

In this chapter major aspects of the concept of TeleCartography have been discussed. Map-based LBS include aspects of integrative positioning, context-adapted data modeling, and multimedia route communication. As a result the prerequisites for positioning, data modeling, and information communication have been analyzed. Findings and results from research projects, accomplished at the University of Technology Vienna, have been presented.

References

Brunner, K. and S. Neudeck (2002) "Graphische und kartographische Aspekte der Bildanzeige" in Kelnhofer, F., M. Lechthaler and K. Brunner (eds.), *Tele-Kartographie and LBS*, Geowissenschaftliche Mitteilungen, Vol. 58, pp. 77–84.

Brunner-Friedrich, B., R. Kopetzky, M. Lechthaler and A. Pammer (2001) Visualisierungskonzepte für die Entwicklung kartenbasierter Routing-Applikationen im UMTS-Bereich, *Angewandte Geographische Informationsverarbeitung* XIII, Beiträge zum AGIT-Symposium, Salzburg 2001, pp. 72–77.

Der Standard, 14 January 2004.

Der Standard, 31 January 2004.

Gartner, G. (2002) "Telecartography: developing map-based location based services", in *Kartografisch Tijdschrift*, ISSN 0167-5788, XXIX, Vol. 2, pp. 34–41.

Gartner, G. and S. Uhlirz (2001) "Cartographic concepts for realizing a location based UMTS service: Vienna city guide 'LoL@' ", *Mapping the 21st Century - Proceedings of the 20th ICC*, Beijing, pp. 3229–3238.

Gartner, G., A. Frank and G. Retscher (2003) "Pedestrian navigation system for mixed indoor/outdoor environment", in Gartner. G. (ed.), *LBS and TeleCartography*, Geowissenschaftliche Mitteilungen, Vol. 66, pp. 161–167.

Goulimis, E., M. Spanaki and L. Tsoulos (2003) "Context-based cartographic display on mobile devices", in Gartner, G. (ed.), *LBS and TeleCartography*, Geowissenschaftliche Mitteilungen, Vol. 66, pp. 25–33.

Hampe, M. and B. Elias (2003) "Integrating topographic information and landmarks for mobile navigation", in Gartner, G. (ed.), *LBS and TeleCartography*, Geowissenschaftliche Mitteilungen, Vol. 66, pp. 147–157.

Kienberger, A. (2004) *Lerntypenangepasste kartographische Visualisierungsmöglichkeiten für thematische Karten*. Diploma Thesis, TU, Vienna.

Klippel, A. and K. F. Richter (2003) "Chorematic focus maps", in *LBS and TeleCartography*, Gartner, G. (ed.), Geowissenschaftliche Mitteilungen, Vol. 66, pp. 39–44.

Kolbe, T. (2003) "Augmented videos and panoramas for pedestrian navigation", in Gartner, G. (ed.), *LBS and TeleCartography*, Geowissenschaftliche Mitteilungen, Vol. 66, pp. 45–52.

Kopczynski, M. (2003) "Localisation with sketch based maps", in Gartner, G. (ed.), *LBS and TeleCartography*, Geowissenschaftliche Mitteilungen, Vol. 66, pp. 117–123.

Lehto, L. and T. Sarjakoski (2003) "An open service architecture for mobile cartographic applications", in Gartner, G. (ed.), *LBS and TeleCartography*, Geowissenschaftliche Mitteilungen, Vol. 66, pp. 141–147.

Lovelace, K. L., M. Hegarty and D. R. Montello (1999) "Elements of good route directions in familiar and unfamiliar environments", in Freksa, C. and D. M. Mark. (eds.), *Spatial Information Theory: Cognitive and Computational Foundation of GI Science*, *Lecture Notes in Computer Science*, Springer, Berlin.

Meng, L. (2002) "Zur selbsterklärenden multimedialen Präsentation für mobile Benutzer", in Kelnhofer, F., M. Lechthaler, and K. Brunner. (eds.), *TeleKartographie and LBS*, Geowissenschaftliche Mitteilungen, Vol. 58, pp. 99–107.

Neudeck, S. (2001) *Gestaltung topographischer Karten für die Bildschirmvisualisierung*, Vol. 74, Schriftenreihe des Studienganges Geodäsie und Geoinformation der Universität der Bundeswehr, München, Neubiberg.

Radoczky, V. (2003) Kartographische Unterstützungsmöglichkeiten zur Routenbeschreibung in Fußgänger Navigationssystemen im In- und Outdoorbereich, Diploma Thesis, TU, Vienna.

Reichenbacher, T. (2003) "Adaptive methods for mobile cartography", *Proceedings of the 21st ICC*, Durban.

Reichenbacher, T. and D. Töllner (2003) "Design of an adaptive mobile geovisualization service", in Gartner, G. (ed.), *LBS and TeleCartography*, Geowissenschaftliche Mitteilungen, Vol. 66, pp. 17–23.

Reichl, B. (2003) "Kategorisierung des Potentials von multimedialen kartographischen Präsentationsformen für die Verwendung auf kleinen Displays", Diploma Thesis, TU Vienna.

Reitmayr, G. and D. Schmalstieg (2003) "Collaborative augmented reality for outdoor navigation and information browsing", in Gartner, G. (ed.), *LBS and TeleCartography*, Geowissenschaftliche Mitteilungen, Vol. 66, pp. 53–59.

Retscher, G. (2002) "Diskussion der Leistungsmerkmale von Systemen zur Positionsbestimmung mit Mobiltelefonen als Basis für LBS", in Kelnhofer F., M. Lechthaler and K. Brunner. (eds.), *TeleKartographie and LBS*, Geowissenschaftliche Mitteilungen, Vol. 58, pp. 42–58.

Sarjakoski, T., A. Nivala and M. Hämäläinen (2003) "Improving the usability of mobile maps by means of adaption", in Gartner, G. (ed.), *LBS and TeleCartography*, Geowissenschaftliche Mitteilungen, Vol. 66, pp. 79–85.

Taylor, D. R. F. (1997) "Maps and mapping in the information era", in Ottoson, L. (ed). *ICC-Proceedings,* Vol.1, Swedish Cartographic Society, Stockholm, Sweden, pp. 1–10.

Taylor, D. R. F. (2003) "The concept of cybercartography", in Peterson, M. P. (ed.), *Maps and the Internet*, Elsevier, Cambridge, pp. 405–420.

Thakkar, P., I. Ceaparu and Yilmaz, C. (2001) "Visualizing directions and schedules on handheld devices", *A Pilot Study of Maps vs. Text and Colour vs. Monochrome*, University of Maryland, Department of Computer Science.

Urquhart, K., S. Miller and W. Cartwright (2003) "A user-centred approach to designing useful geospatial representations for LBS", in Gartner, G. (ed.), *LBS and TeleCartography*, Geowissenschaftl, Mitteilungen, Vol. 66, pp. 69–79.

Van Elzakker, C. (2000) "Use of Maps on the Web", in Kraak, M. and A. Brown. (eds.), *WebCartography*, Taylor and Francis, pp. 21–36.

Verbree, E., C. Tiberius and G. Vosselman (2003) "Combined GPS-Galileo positioning for location based services in urban environment", in Gartner, G. (ed.), *LBS and TeleCartography*, Geowissenschaftl, Mitteilungen, Vol. 66, pp. 99–107.

Wintges, T. (2002) "Geo-Data Visualization on Personal Digital Assistants (PDA)", in Gartner, G. (ed.), *Maps and the Internet* 2002, Geowissenschaftliche Mitteilungen, Vol. 60, 2003, pp. 178–183.

Zipf, A. (2003) Forschungsfragen zur benutzer- und kontextangepassten Kartengenerierung für mobile Systeme, *Kartographische Nachrichten*, Vol. 1, Kirschbaum Verlag, Bonn. pp. 6–11.

Zipf, A. and M. Jöst (2003) "User expectations and preferences regarding location bases services - results of a survey", in Gartner, G. (ed.), *LBS and TeleCartography*, Geowissenschaftliche Mitteilungen, Vol. 66, pp. 63–68.

Zlatanova, S. and E. Verbree (2003) "User tracking as an alternative positioning technique for LBS", in Gartner, G. (ed.), *LBS and TeleCartography*, Geowissenschaftliche Mitteilungen, Vol. 66, pp. 109–117.

CHAPTER 17

Sound Maps: Music and Sound in Cybercartography

PAUL THÉBERGE

Canada Research Chair in Technological Mediations of Culture, Institute for Comparative Studies in Literature, Art and Culture (ICSLAC) and Music, Carleton University, Ottawa, Ontario, Canada

Abstract

As cartographic objects and processes are increasingly understood as multimodal in character, sound needs to be conceived as an integral element of any cartographic project. Arguing against simple notions of the sonification of data at the level of form, this essay suggests that a more thoroughgoing strategy of "sound design" needs to be pursued. Properly understood, sound design requires that voices, sounds and music, as well as silence, spatial location, signal processing, and interactivity be considered as part of the overall design process and related to other cartographic elements at the level of structure and theme. Furthermore, sounds need to be considered as not simply formal elements but as cultural objects, replete with meanings and associations. Drawing on the literature of sound in cinema and new media, the chapter builds on previous theories of image–sound relations and explores possibilities for new ways of integrating and manipulating sound within an interactive, multimedia context.

1. Introduction

> The words faded away, and without warning, his mind was filled with pictures. A brilliant silvery world flared up before him, as clear as his own echo vision at night. His eyes popped open in surprise.
> "What's happening"?
> "Echo pictures", the elder told him patiently. "We see with echoes, and with practice it's also possible to sing echo pictures..." (Oppel 1997: 34)

In Kenneth Oppel's children's story, *Silverwing*, the main protagonist is a bat, named Shade who embarks on a long journey guided only by the memory of a sound map sung to him by his mother. Unlike other bats who simply use echo location to navigate through the night sky and to capture insects, Shade learns to remember images in sound, to project realistic sound images to other bats, confounding them in the process, and to decipher the echoes of sounds past (the history of his colony is contained in a cave where the echoes of voices from the past continue to resonate for centuries).

Unlike bats (and perhaps a handful of human cultures, such as the aboriginals of Australia, whose history, mythology, and geography are traced out and passed from one generation to another in "songlines"; Chatwin 1987), most of us rely more heavily on our eyes than our ears for navigation and for charting spatial relationships. Indeed, in Western culture, where cartography has long enjoyed the status of a specialized field of study and practice, we do not tend to associate maps with sound at all. Maps are visual representations *par excellence*: they epitomize principles of formal abstraction and processes of visualization and graphic rendering that have developed in our culture over hundreds of years and, throughout this development, their spatial logic has remained essentially silent in character. Even when sound has been used in the mapping process – as in the case of technologies such as sonar – its use does not encourage us to develop new modes of listening; rather, sound information is automatically transformed into visual information, allowing us to "see" things (the ocean floor or schools of fish). In such instances, we are not expected to experience sound so much as witness its effects, visually, through a process of technological mediation.

Shade Silverwing's sound maps, on the other hand, are experienced in an almost visceral fashion, as a series of fleeting sound–images; his maps are thus, temporal, sequential, and emotional in character. And in following this sequence of sound–images, Shade's adventure unfolds; the sound maps giving structure to the narrative as a whole. As the field of cartography turns increasingly to computer and Internet-based multimedia as a format for the presentation of maps and other data as is proposed by Taylor (2003) in the paradigm of cybercartography, sound (among other sensory modes of communication) needs to be considered as an integral part of the mapping process. In so doing, spatiality and temporality, data and narrative structure, image and music, and many other relationships will present themselves as both problem and opportunity.

In this chapter, I would like to examine some of the possibilities and requirements of using sound in multimedia contexts. While there is a small but growing literature on sound applications in cartography and in computer-based multimedia (often referred to as "sonification"), much of it tends to ignore the uses of sound in past and contemporary media – media, such as film, television, and gaming – preferring, instead, to arrive at solutions from a kind of analytic formalism – a formalism based on an objectified, scientific model of sound and aural perception – on the one hand, or relying on a strategy of exploiting the technical possibilities inherent in the computer for generating and manipulating data, on the other. A more cultural-based approach to sound, however, would require that one not only consider historical uses of sound in a variety of media (including perhaps even opera and dance, as suggested by Cook 1998), but also different forms of *listening* practice associated with contemporary media as well. With respect to the latter, it is important to recognize that, as users we come to new media with a host of past experiences – experiences that shape both our perceptions and our expectations.

2. Sound in Media/Sound Design

There are other compelling reasons for investigating the uses of sound in other media, not the least of which is the growing convergence of new and old media forms, especially between computer games and conventional cinema. While there are obvious economic imperatives that lie behind such a convergence, the effects of this particular form of remediation on the content and style of both media are significant. In the case of many action and fantasy games derived from movie titles, such as *The Matrix* and *The Lord of the Rings*, there has been a trend towards the development of realistic animated figures, sometimes rendered from live actors featured in the film itself and, in terms of sound, the use of original soundtrack music and sound effects. Even popular television dramas, such as *Law & Order*, have been turned into game formats and derive a certain cache (and a sense of verisimilitude with the original) from the fact that they make use of the voices of the television cast. Of course, the relationship between these media is not a one-way street: Hollywood producers have long had a fondness for special effects and virtual reality (as a sci-fi theme), and adaptations of video games for the big screen, such as *Lara Croft: Tomb Raider* and *Resident Evil*, with their hyperactive pace and their visual and audio effects, have become increasingly popular. At the end of 2004, there were at least half-a-dozen Hollywood projects based on videogames in the planning stages, or already in production.

The sonic complexity of these media, both in terms of their production and reception, has created a context within which cybercartography must now develop its own approach to sound design. I use the term "sound design," here not in its limited sense, as designating the process of creating individual sound effects for media, but rather, in the larger sense, in which sound is part of an integrated process in which the expressive possibilities of sound are planned as part of the overall thematic or narrative structure of the work. The term, "sound design" has been in use in Hollywood since the 1970s, although it is seldom used in a consistent manner, and the associated designation of "sound designer" – given to the person responsible for the overall approach to the soundtrack, much like the cinematographer is responsible for the overall "look" of a film – has been applied most notably to individuals, such as Walter Murch who, from his earliest work on films such as *Apocalypse Now*, set a new standard for the use of sound in cinema.

As a point of departure, the notion of sound design requires that one consider sound not simply as something that is "added to" or "synchronized with" visual images. As sound designer Randy Thom has argued in relation to cinema, "Sound has to be *designed-into* a movie. That is, the movie has to be designed for sound, not the other way around". (1999: 9; emphasis in the original). With regards to cybercartography, such a demand seems to require that the very concept of a "map" be rethought: How can a map be designed *for* sound? Is it possible to conceive of a map where the visual components do not take precedence over sonic components? What new relationships between sound, image, and other elements need to be forged that are perhaps unique, or at least different from those that exist in cinema, gaming, and other media?

In attempting to answer some of these questions one must first confront the multimodal character of sound itself: in a typical media production, multiple voices, environmental sounds and sound effects, music, and silence, exist simultaneously, draw on different cultural codes and modes of reception, and produce complex relationships both amongst themselves and in conjunction with images and narrative. Furthermore, sound is deployed spatially in most media (ranging from the mono reproduction of sound, where only the illusion of depth can be created to surround sound systems that envelope the listener in multichannel audio) subjected to various forms of signal processing (e.g. reverb, equalization, and the like) and, in interactive media such as videogames, sounds can be made to respond to user input. In these ways, the modern soundtrack produces a kind of plenitude that can be as subtle as it is, at times, overwhelming.

3. Silence

The idea of "silence" needs some qualification in this context and should perhaps be addressed first, if only briefly, before turning to the more familiar categories of speech, sound effects, and music in multimedia. Indeed, it may seem odd to think of "silence" in multimedia at all: so much of our experience with computers and the Internet is dominated by silence that one might think that the whole point of *multi*-media is to create at least some kind of sound, all of the time. Furthermore, in film and television, "silence", at least total silence, can hardly be said to exist: total silence risks being perceived as a technical breakdown and thus, "room tone", or some other kind of ambient, background sound is almost always present on the soundtrack.

In cinema, however, as many scholars have argued, silence only becomes possible, as a formal element, as a result of sound. Silence, in this sense, is "the product of a contrast" and it is neither neutral nor empty, but rather, charged with the sounds that precede it (Chion 1994: 57). As such, relative degrees of silence have been used for specific effects in media, such as cinema, for many years: for example, the general level of environmental noises can be attenuated, allowing for a focus on a particular sound (such as the ticking of a clock; Chion 1994: 57), or ambient sounds can be suppressed entirely, thus giving the impression of entering into the mind of a character (Chion 1994: 89). With regards to the latter, we can think of selected *parts* of the soundtrack that are silenced (producing, in this case, what might be referred to as a "diegetic silence", Théberge, forthcoming), while other parts, such as music, continue to sound, or are given special prominence.

At a larger level of analysis, Claudia Gorbman has suggested that in certain instances the absence of a particular sound or category of sounds, can be understood as a kind of "structural silence": this occurs when a sound that has been part of a particular sequence in a film narrative is no longer present when a sequence of a similar type returns at a later point in the film (1987: 19). By first creating and then taking advantage of audience expectations, a filmmaker can thus call attention to the new sequence, usually with a particular narrative intention: for example, the final death-row song sequence in von Trier's *Dancer in the Dark* (2000), is sung a cappella, in contrast to earlier production numbers within the film that are presented with full orchestrations; in this way, the final scenes are given a kind of stark realism that would be otherwise impossible within the conventional codes of the musical while, at the same time, retaining the expressive character of the song mode (Théberge, forthcoming).

These different kinds of "silence" create a set of possibilities (and challenges) for cybercartography: if one accepts, at least in principle, the possibility that all types of sound – speech, sound effects, and music – can exist simultaneously within any one part of a multimedia production, then the absence of one or more of these elements should be treated not as an arbitrary event, but as the result of a conscious decision-making process – a decision taken within an overall design context. Thus, when and how one uses the various modalities of sound, alone and in combination with one another, should not be solely decided at the local level (at the level at which various sounds might be considered in relation to a particular image), but at the structural level, where thematic elements and the general content of the work are determined. Different kinds of "silence" (for example, an attenuation of music tracks) can be employed to create different kinds of emphasis (an emphasis on a particular sound effect, a visual sequence, a text, or a data set); and, unlike film or television, which often adhere to codes of realism, cybercartographical projects may be able to accommodate instances of total silence – instances where visual elements are allowed to stand on their own and thus, gain a special kind of silence. Again, given the long history of map-making in Western culture, the very idea of a "silent map" may appear odd, but in a multimedia context, a silent map should always be considered as more than simply a map without sound – the lack of sound should have a purpose.

4. The Voice

In turning to the issue of voices in multimedia, it is important to recognize that voices are often accorded a particular kind of deference in most media: in television news, sport casting, and other programs, the voice is often used authoritatively to convey information and, also, to create a quasi-discursive connection with the home audience (Morse 1985); and in fiction genres, dialogue serves a number of functions, not the least of which is to reveal character: the voice is understood to express the unique essence of an individual, their emotions and psychological motivations (Kozloff 2000: 43–47). Recently, in the interactive action game, *Grand Theft Auto: San Andreas* (Released in 2004), character development as manifest through the voice has been given a new dimension: depending on how the user decides to develop the main character during the course of the game, the character will say different things in different situations by drawing on a repertoire of hundreds of individual, recorded vocal statements (in a sense, the interactive dimension suggests that the "voice" the player discovers in the game is their own). Thus, in information, fiction, and interactive genres vocal performance is often taken as a measure of "truthfulness" or authenticity and any discrepancy between the speech of a character and their actions is regarded with some suspicion. Sarah Kozloff has outlined a number of instances in fiction film where narrators appear to adopt an ironic stance in relation to the events depicted or, as in the case of Kubrick's *Barry Lyndon* (1975), narrators whose commentaries appear intentionally misleading, and thus, "unreliable" (Kozloff 1988: 102–126).

The value of positing an unreliable narrator, or the degree to which irony is possible in cartographic presentations is unclear. However, if one can consider cartography as a kind of documentary media, then the issue of truthfulness – or the claim to truthfulness – is an important one. Film historian Bill Nichols (1985) has argued that the use of the voice in documentary film and television has taken on a number distinctly different forms during the course of the past century (Nichols uses the term "voice" both literally, in relation to the voices we actually hear in the documentary, and figuratively, to represent the authorial voice of the director), and, among the most important distinctions to be made between documentary styles, is the manner in which they lay claim to the "truth". Different uses of the voice in documentary are thus related to issues of ideology.

Nichols' typology includes the "voice-of-God" style of voice-over narration, often associated with the Griersonian tradition of documentary film and early newsreels, with their typically male, authoritative, and didactic tone; the *cinéma vérité* style, that emerged during the 1950s and early 1960s and made claim to a direct, unmediated access to reality; the "string-of-interviews" approach to documentary, which has become the stock in trade of much recent documentary filmmaking and television journalism, where individuals speak the "truth" of their personal experience of historical events but seldom can lay claim to a knowledge of larger social forces; and finally, the "self-reflexive" form, where the documentary filmmaker purposefully inserts their own presence and point of view into the film (filmmaker, Michael Moore, is perhaps the most well-known representative of this latter style). Ultimately, for Nichols, each of these forms makes different claims to different kinds of "truth" and always does so based on a particular range of ideological assumptions.

For cybercartography, the important point to be made here is that the choice of a particular voice (a male, female, or child's voice), the way it is produced within the program material (as voice-over, in conjunction with photos, or embedded within video segments), or how it is "staged" (given special treatment through spatial panning, reverberation, or other special effects; Lacasse 2000), the position from which it speaks (as expert, witness, or guide), and the manner in which it addresses the audience, are considerations that have significant aesthetic, social, and political implications. Having said this, however, it is also important to recognize that the use of the human voice presents a range of possibilities that can contribute greatly to the cartographic project. Maps, statistical data, and other forms of information are too often presented as objective, disembodied, and socially neutral when, in fact, all knowledge is situated knowledge, produced by individuals for other individuals, within institutional settings, and with particular goals or agendas in mind. The social and cultural context of cybercartography is critical, as is argued by Martinez and Reyes (Chapter 5) and by Taylor (Chapter 1). The human voice can be used to personalize and embody knowledge, the identification of social actors can clarify the origin (and bias) of information, and access to multiple voices can illuminate not only different points of view, but also the social, cultural, and political stakes that are implicit whenever notions of geography, territory, identity, economy, and the nation state are invoked.

A technical and formal issue that must also be addressed with the use of voices is their duration and their relationship to images, printed texts, and other data. Recorded voices are, by their very nature, temporal in character and, as such, they can sometimes slow down the pace at which the user may progress through a multimedia program. Allowing users to choose, if and when to listen to a recording (or video segment) is the most obvious way to deal with the problem, but the implicit message in such strategies is that sound is somehow optional or secondary in importance (the corresponding option, to dim the screen of the computer because it distracts one from listening closely to the audio is seldom offered). Careful editing of voice recordings into short, content-related, or thematic file segments that can be accessed when the user clicks on a related, part of an image, map, or relevant set of data might be a more useful strategy for dealing with the problem of duration – one that also allows the user to more effectively "manage" the order and flow of the voice segments.

The relationship of the voice to visual and other elements is a more difficult one for which there is no easy solution. In general, it is important to make use of verbal information that complements, rather than simply repeats information that is presented in some other form: too often in multimedia presentations, from PowerPoint lectures to educational CD-ROMs, spoken and written (or graphical) elements are virtually identical, rendering both speaker and text redundant. While information overload is certainly an issue to be concerned within multimedia, most cartographic products tend, if anything, to be overly dense with regards to visual content; this is not surprising given the history and dominant practices of cartography, but a multimedia product may need, in some instances, to thin out the visual dimension, that is, to design both images and textual elements for sound. Making use of dynamic maps – maps where graphical and textual elements change over time – or animations, also present themselves as possible solutions to the problem of combining sounds, images, and texts: in effect, "temporalizing space", as Lunenfeld has suggested (2000: 160–161).

Finally, one needs to consider whether the content of recorded voices always needs, as is often assumed, to be also rendered in textual form. Verbal expressions, accent, pace, cadence, and emphasis are characteristics of spoken language that should be valued and are, in many instances, difficult to separate from the "content" of speech (or are at least difficult to translate into written form). Where it is deemed important for users to be able to access printed material related to recorded voices, short summaries or complementary texts that elaborate or amplify, in some way, the spoken segments might be an alternative to verbatim transcriptions that tend to diminish the unique characteristics of vocal performance. From the opposite side of the coin, the voice synthesis capabilities of computers are sometimes proposed as a means of rendering printed texts into speech. However, while such strategies may be useful (especially for the visually impaired), the expressive quality of synthesized voices often leaves much to be desired and they become tiresome very quickly. The relationship of spoken and written texts must ultimately be considered as part of the overall design of the multimedia program, taking into account the amount and nature of the material in question, the intended audience, and other factors.

5. Sounds and Soundscapes

The inclusion of ambient, or "environmental" sound textures, sound effects, "ear-cons", data sonifications, and the like offer an entirely different set of possibilities to cybercartography. Furthermore, these possibilities can be exploited and layered in ways that are quite different from the ways in which the human voice can be treated: sound effects can exist in the background or foreground, or both simultaneously; they can be layered, creating dense and diffused textures, or isolated as singular, focused events that are initiated by user input; they can add atmosphere to visual images, offer specific forms of information, or guide the user; and finally, they can be deployed spatially, enveloping the listener, or they can be made to move fluidly through a designated area, supplying spatial cues or tracking user movement.

A relatively small and sporadic literature on the uses of sound in computing and multimedia has been in existence since at least the mid 1970s, and attempts to bring together some of the findings of individual researchers date back a decade or more: for example, the journal, *Human-Computer Interaction*, devoted a special issue to nonspeech audio in 1989; a useful attempt to summarize and theorize some of the main issues concerning sound and cartography can be found in an essay by Krygier, published in 1994; and, in their summary of the field, Kramer *et al.* 1998; make an explicit call for the formation of a coherent discipline of sonification. Ironically, one of the most extended and wide-ranging efforts to explore the uses of sound in computing remains unpublished (see Buxton *et al.* 1994). Given the scattered development of this field of study (and to some degree, sound studies in other media suffer from a similar lack of cumulative research), it is perhaps not surprising that much of the recent literature on sonification continues to address many of the same issues that preoccupied its predecessors: these include discussions of the psychological and cognitive aspects of sound, the mapping of data onto sound variables (i.e., representing changes in data as changes in pitch, loudness, timbre, or other parameters of sound), using sound to provide feedback to enhance user interaction, processing sound to provide spatial cues, and developing sonic resources for the visually impaired, among others.

While such work is important in exploring the various possibilities offered by sound and is not without theoretical insight, much of it has remained essentially functional in character: that is, the role of sound is often regarded as simply one more way of adding redundancy in the presentation of data, or as a means of providing user feedback. Furthermore, much of the case study material deals primarily with the lower levels of structural relations between sound, image, and data, for example, in the one-to-one relationship between a set of data and a particular sound attribute. Surprisingly, within such studies the choice of sounds is sometimes quite arbitrary – in some cases, choosing sounds simply because they are "pleasant" in some way. Furthermore, the higher levels of program structure – the relations between different data sets, different "pages" of information, the relationship between thematic elements of the overall program and different categories of sound – are seldom addressed in any meaningful way.

Within the overall approach to sound design suggested here, larger structural relationships are important because they can supply both a rationale for the many choices of sound material that need to be made at the local level and, also because they can provide a strategy for using sound to draw attention to relationships between elements of data and other program material that might be otherwise only implicit in nature. Even at the lowest levels of decision-making, however, the quality of sounds chosen is important: sound designers in film and other media have long recognized that audiences remember sounds that are distinctive in some way much more readily than they remember differences in more "abstract" sound qualities, such as pitch and loudness.

At this lowest level of sound design, it is thus essential to recognize that all sound effects – both individual sounds and sound textures – are complex events; they have characteristic shapes and tonalities, a distinct grain or texture; they always come from some kind of source (although that source may be only more or less recognizable); they may have particular cultural associations for the listener and, finally, they bear the marks of the spaces and conditions in which they are recorded and processed (Lobrutto 1994), and are influenced by the spaces and conditions in which they are reproduced. Thinking in such terms is not unusual in the world of cinema, where the creation of sound effects is an important part of the postproduction process, but to a large degree, goes against common practice in the world of computers: anyone who uses a computer is familiar with the range of alert and status sounds – various chimes, bells, beeps, electronic sound effects, and "orchestra hits" – that accompany operating systems and many applications, sounds that are often lifeless, occasionally humorous, but mostly banal in character. Early video and computer game sounds were similarly constrained by limitations of memory and chip design and each succeeding generation of games has had its characteristic sound. Listening, again, to the curious, synthesized sound of early game action, such as *Pac-Man* devouring dots, is part of the charm game fans experience when playing older games today (Herz 1997). Even an average contemporary sound card for the PC has much greater sound capability – producing multiple channels of synthesized or digitally sampled audio – than anything manufactured a decade, or more ago (and recent gaming consoles typically possess even greater capability and sound quality).

Partly in response to these technical developments and, more importantly, as game developers have increasingly sought to emulate the quality and feel of cinematic experience, users have come to expect much more from computer audio. For sound designers, this requires a greater attention to the ways in which sounds are chosen, edited, and processed for multimedia. Commercial sound effects libraries – containing literally thousands of recordings of human, animal, machine, and environmental sounds – have existed for decades and are a useful resource for prerecorded sounds. However, while such libraries are in widespread use in media production, they are almost never relied upon exclusively as sound sources and, equally important, the sounds are seldom used without first being carefully layered and processed so that they fit into the overall sonic context of the production.

The quality of sound effects is important, in part, because they contribute so much to the way in which movies, television programs, and computer games create the illusion of "real" worlds. This has been evident in cinema ever since the introduction of synchronous sound in the late 1920s and even more so since the arrival of Dolby sound in the 1970s; indeed, the whole notion of "realism" in cinema is closely tied to the advent of synchronous sound and further developments in sound recording, editing, and mixing practices. In gaming, the first generation of highly successful, action-oriented and "immersive" games – games such as *Doom* and *Myst*, both released in 1993 – relied similarly on sound as a way of increasing the realism of their three-dimensional visual environments. The *Myst* series in particular, which now exists as a set of four separate games and various other spin-offs released over the course of the past decade, is striking in its use of grainy, squeaky sound effects that lend an almost tactile sense to the many ancient, rusted machines that populate its otherwise deserted worlds, encouraging users to interact with them.

More importantly, *Myst* also made use of sound effects (and music) as integral parts of its overall game plan: paying attention to sound and, in some instances, remembering specific sounds or sequences of sound become important strategies for solving *Myst's* puzzle-like structure. Compared to *Doom* and its many imitators, with their emphasis on hyperactive movement *through* space and engagement with a variety of demons and other foes, *Myst's* attention to sonic and visual detail forces a slower pace upon its users and encourages a different type of "interactivity". While action and first-person shooter games have certainly become the dominant genres in contemporary game culture, such differences in the design of games may be important for cybercartography insofar as they provide different models for how sound, image, and movement relate to one another at a structural level and, also, how they can sometimes influence the character of user interaction (for a discussion of these and other differences in the structure and gameplay of *Doom* and *Myst*, see Manovich 2001: 244–253; Smith 2002: 494–496).

The production of immersive sound textures offers other possibilities for the engagement of users in cybercartography. In recent decades, the relationship between geography and sound has been most thoroughly explored by researchers concerned with the "soundscape", a term coined by R. Murray Schafer in his book, The *Tuning of the World* (1977). For Schafer, the most important sounds in any landscape are those related to geography, climate, animal, and human life: the sounds of the ocean, forests, wind, insects, and the din of city traffic, are referred to as "keynote" sounds – sounds that are not necessarily noticed, consciously, but form the pervasive background against which other sounds are heard (Schafer 1977: 9–10).

Furthermore, keynote sounds, such as the ocean, are often archetypal in character and thus, have deep resonances within culture. Within Schafer's lexicon, "sound signals" are foreground sounds that we consciously listen to, such as warning bells, sirens, and the like; and, in much the same way as landmarks distinguish particular locales, "soundmarks" are sounds that are unique to particular places and are especially regarded or recognized by local communities (following Schafer, an extensive vocabulary of sound terms is developed in Truax 1978).

This figure–ground relationship in environmental sound offers a variety of possibilities for creating analogous relationships with images and maps. For example, environmental sound textures could be used in conjunction with large-scale maps to provide a parallel stream of information about a given geographical area, with different keynote sound textures cross-fading from one to another as the user moves a pointing device over different regions of a map. Similarly, in more close-up maps or even urban street maps, individual soundmarks or other effects might be sounded, against a more diffuse background texture, in specific spatial locations and in response to user input. Topographical maps could be enhanced through the dynamic use of sound effects – changing from agricultural sounds, to forest sounds, to the sound of windswept mountain tops – in relation to topographical information.

In examples such as these, it should be stressed that the sound tracks need to be constructed according to their own sound logic and not simply reduced to the structure of data or other map variables: that is, the link between sound and map should be one of a parallel, analogous movement that can be more or less general in character rather than the result of a mechanical, one-to-one correspondence between sounds and different types of information depicted in the map. If sound is to function as a partially independent mode or channel of information, then the logic of sound must be arrived at through a careful analysis of both natural and artificial sound environments and the possibilities offered by any given map or program.

As discussed above in relation to the design of individual sound effects, the production of diffuse, ambient, or environmental sound textures is a painstaking, highly detailed process: although the casual listener is seldom aware of it, dozens of sound tracks are typically mixed together to create the background sound texture in a film or game production. Even something as apparently simple as the sound of a gurgling brook might be made up of multiple layers of water sounds: from individual droplets to the sound of a rushing current. Conversely, in their account of a studio workshop on the creation of virtual spaces, More *et al.* (2003) advocates the use of actual recorded environments over those constructed out of individual sounds or those supplied, premixed, in commercial sound libraries, in part, because of their greater acoustic and spatial coherence. In some instances, it may be only after the various keynote textures and other sounds have been produced that one can determine the number and types of linkage that can be made with the visual material. Even where such linkages are implicit, however, the detail and quality of the sound environments, and the degree to which individual sound events can be made to respond to user input, may have an influence on the level and type of user engagement that can be achieved with the program material as a whole.

In some cases, working by way of analogy rather than strict imitation may be necessary because visual and audio modes of reproduction are very different from one another. In viewing a film, television, or multimedia program, one is always looking at a "frame", a "cut-out" of greater or lesser dimension. Sound, by comparison, extends beyond the frame, filling the space in which it is reproduced, thus creating a disjuncture between visual and aural space – between what is "on" and "off-screen"; the way in which sounds and images are then recombined is a process that is both psychological and ideological in character (Doane 1980). At a simple, functional level, however, it follows that there will be a certain lack of precision in the spatial location of sound in relation to images, especially where the issue of scale between the size of the monitor screen and the potential space of mono, stereo, or surround audio reproduction is concerned (even within the audio track itself, there are discrepancies between different parts of the audio spectrum with regards to the degree to which sounds can be located accurately by the listener: low frequency sounds cannot be located as precisely as high frequency sounds). And while the spatial dimension of audio reproduction is greater, in some ways than the visual dimension, it too is limited: while objects on the screen can be located in a matrix made up of left–right and up–down axes, there is very little sense of a corresponding up–down axis in audio – audio, as presently reproduced in conventional speaker arrays, exists primarily on a single plane between a left–right axis and a front–back axis.

Despite such discrepancies, the potential for audio to offer a rich, immersive experience for the user is an important design consideration for cybercartography. In cinema, "realism" dictates that the surround channels are generally used very sparingly so as not to distract the audience from the screen itself. Even in television programs that are broadcast in simple, two-channel stereo, however, there is a noticeable broadening of the sound "image" when moving from indoor scenes to outdoor ones. In gaming, with its emphasis on action, special effects, and interactivity, there tends to be a greater use of surround channels – for "flybys" and other effects – regardless of the potential disjuncture between audio and visual coherence (in this regard, gaming has a looser, more tenuous connection to cinematic codes of realism). How the relationship between visual and aural space will be organized in cybercartography, and what its implications are for the ways in which users make sense of these spaces, will be largely governed by the nature of the program material, the number of channels available for audio reproduction, and the governing aesthetic of the production design overall.

6. Music

Aesthetic considerations will perhaps come to the fore even more explicitly in considering the relationship between music and multimedia. Music has a long history of involvement with other art forms. Indeed, in the traditional languages of some African cultures there is no precise word or concept for "music", as such: music is inseparable from instruments on which it is played, from dance, and from the ceremonial contexts in which it plays a part. Even in Western culture, where music has achieved the status of an autonomous art form, it nevertheless enjoys a centuries-old association with religious ceremonies, with theater, opera, and popular forms of entertainment; more recently, it has become an integral part of our experience of cinema and television drama, as well as documentary genres and even television news programming. In many ways, music is central to our concepts of multimedia: music was a key element in Wagner's notion of the *Gesamtkunstwerk* – a notion whose influence on the history of film music has been both recognized and critiqued (Paulin 2000), and even regarded as something of a point of departure for recent theorizing about virtual reality (Packer and Jordan 2001).

With music, however, one is also faced with the most complex type of sound to be used in cybercartography: music has its own structural logic, often thought of as a "language" in itself; it operates according to own modes of expression and reception; and, despite its apparent autonomy, it is always laden with cultural associations and meanings of various kinds. The structural logic of music is something that is too often ignored in the uses of musical sounds in the field of sonification: sonification strategies usually begin with the structure of data and attempt to "map" it on musical sounds, reducing them, in the process, to the structure of the data itself. While this may be a useful strategy in some cases, using timbre or other musical elements much as one might use gradations of color in illustrating quantitative changes in data, it functions most successfully when the changes in data are relatively simple in character: for example, where simple increases and decreases in quantities might be represented as analogous changes in pitch or timbre, or where the amount of change or the number of variables is relatively small; more complex changes in data can certainly be translated into changes in musical variables, but the ability of users to remember those changes with any degree of precision may be limited. For listeners, memory in music tends to be more closely related to distinctive melodic, rhythmic, harmonic, or other musical motifs – motifs that exist at a slightly higher level of musical structure than that of simple pitch or duration. Creating relationships between visual elements or data and real musical structures (rather than sonic attributes) may offer a different kind of strategy for sonification in cybercartography.

The use of musical motifs as meaningful entities within multimedia contexts has been most thoroughly explored in music drama and in film. Again, Wagner is of central importance here, in particular, his technique of leitmotifs: the *leitmotif* is a short, recognizable musical phrase that can be linked to words, characters, or particular events within narrative (the leitmotif is most often melodically distinctive but it can also be primarily rhythmic in character or have specific associations with musical instrument timbres); more importantly, the leitmotif is not static, it can change and develop as characters develop, revealing subtle variations in their psychological motivations, and two or more leitmotifs can be combined, underlining the implicit relationship between different characters or events in the narrative. Thus, a distinct form of *musical* logic runs parallel to, and informs the narrative in both its moment-to-moment movement and in its larger structures.

While the leitmotif technique is only one way of creating relationships between music and narrative, and the structural "unity" that it seeks is only one possible aesthetic goal, among many others, for music, its use has been remarkably persistent in the history of film music. Indeed, many film scores – from the classic scores of Max Steiner (*King Kong* 1933) and Erich Korngold (*The Adventures of Robin Hood* 1938) to the recent blockbuster, multifilm scores of John Williams (*Star Wars* 1977 and its subsequent releases), and Howard Shore (*The Lord of the Rings* series, beginning in 2001) – have made use of leitmotifs (with some variations in the application of the technique), as the primary means for integrating music and large-scale narrative form.

The point here is not to advocate the use of the leitmotif, as such, as a particular technique for cybercartography (for all its possibilities, it seems ridiculous to posit computer-based multimedia as some kind of *Gesamtkunstwerk* of the future – a gesture that would be, at best, overly romantic at this point in history). What the leitmotif technique suggests, however, is that it is possible for music to maintain its own formal and structural characteristics while entering into significant associations with, and indeed, contributing to and informing other types of artistic or organizational logics. While the complete unity and integrity of music and drama sought after by Wagner is quite likely impossible – one form or the other will always tend to predominate, especially in film (Paulin 2000) – the many possible ways in which musical structures might be intertwined within the structures of multimedia have only begun to be explored. And perhaps the lesson for the field of sonification in this regard is that, whatever relationships are to be drawn between data and other levels at which multimedia information is organized, a more thoroughgoing understanding of music and its formal properties (as opposed to its physical attributes) needs to be considered, at the outset, as an essential part of the sonification project.

The significance of music lies not only in its forms, but also in its relationship to culture: music, in this sense, is closely tied to notions of individual and collective identity and to our sense of geography and place. In this way, music, as an expression of cultural geography, has much to contribute to cartography in a multimedia context. However, the relationship between music and culture needs to be handled with some care if it is not to slip into a kind of musical stereotyping or cliché on the one hand, or a form of easy exoticism, on the other. Certainly, the residue of musical clichés runs deep in Hollywood film music, where the simple use of the pentatonic scale and a few percussion instruments can often signal the movement to an exotic, Oriental locale. Indeed, the stereotyping of ethnic music has been a feature of Hollywood scoring from the earliest days of cinema, when Erno Rapée included excerpts of the music of various "Nationalities" in his book on piano and organ music designed for use in early, silent films (1924), to the present, where digital samples of musical instruments from around the world are readily available for use in the evocation of foreign lands and cultures (Théberge 2003). This stereotyping is far from static; however, Claudia Gorbman (2000) has outlined a history of film music in relation to Hollywood's portrayal of North American Indians and what is remarkable in her account is the ways in which the uses of music have changed over time, reflecting on the one hand, changes in the status of Indians within American culture and on the other, changes in musical style (e.g. the arrival of so-called World Music). What Gorbman's study illuminates is the subtle ways, in which music and the ideological representation of "Otherness" are intertwined, and how such an entanglement can change and take on new forms.

In the present global context, two phenomena (among others) – the proliferation and circulation of popular music via media and technology, and the large-scale displacement of ethnic populations – have loosened the ties of musical style with specific cultures and places, creating a vast range of hybrid musics and a variety of multicultural complexities that make it increasingly difficult to "map" music onto specific places in any easy, singular, or straightforward fashion. To what Arjun Appadurai (1996) has called, in describing these phenomena, the "mediascape" and the "ethnoscape", we might add another term, the "genrescape", to designate that complex mix of musics that exists in any locale, and are made up of a variety of local traditions and global influences, indigenous musical instruments and electronic modifications, playing techniques, and media practices.

If cybercartography is to remain true to the cultural geography of the present and not simply drawn on the stereotypes and clichés of the past, then a subtle, historically and culturally informed approach to music must be adopted at the outset when designing sound for multimedia projects. Surprisingly, one of the more interesting attempts to break out of the confines of stereotypical uses of music in recent years has come out of the world of gaming: in *Grand Theft Auto*: *San Andreas* (2004), Rockstar Games uses a wide range of popular music styles in an attempt to portray the "genrescape" of West Coast radio programming, circa 1990. Unlike similar action games that have relied on the conventions of Hollywood *film noir* or on a simple reference to rap music to situate the player within the thematic of urban crime, *GTA* employs musical styles as diverse as country music, rap, pop, and alternative rock, accompanied by an elaborately produced collection of satiric radio commercials and DJ patter reminiscent of the early 1990s. In this way, as the player moves through the various cities (there are three in all) and rural areas depicted in the game, they are situated in a subtle and complex manner within a "genrescape" that is at once general and specific, temporal and spatial in character.

GTA, through the metaphor of radio, also introduces a kind of mediated user interaction with music. However, the question of how music might be made interactive in a more direct fashion is not addressed. In keeping with the ideas and concerns outlined earlier, it is important to consider the unique formal characteristics of music if a viable strategy for user interaction is to be achieved, that is, if interactive possibilities are to be *designed for* music rather than simply designing music for interactivity. There are, no doubt, many ways in which interactivity might be proposed with regard to musical textures, however, pursuing concepts or practices that already have some degree of commonality in both music and in technical media may offer the most fruitful possibilities. The ideas of "layering" and "looping" are perhaps two such concepts.

"Layering", as a common technique in graphic design, is sometimes found in cartography in the form of overlays. It can be considered as analogous to "parts" in music (or "tracks" in a multitrack sound recording): soprano, alto, tenor, and bass lines in a choir composition; the various instrumental sections of an orchestral score; the individual bits of noise that make up the dense clouds of sound in electronic textures, the drums, bass, guitar, and vocal tracks of a pop music recording; the individual recordings that are combined to make a dance remix or forced together in a "mashup", can all be thought of as forming the "layers" of a musical composition. These layers or parts, always possess a certain level of independence and autonomy, but are also intimately integrated – rhythmically, harmonically, and thematically – with the other parts of the composition in which they are embedded.

Breaking music down into its constituent parts, either through multitrack recording or Musical Instrument Digital Interface (MIDI), and making these parts available to the user, may be a fruitful strategy for introducing interactivity into multimedia programs. The various musical parts could be linked to a series of specific images (or fragments thereof), overlays, or other types of information to create multimedia, puzzle-like structures. Technologies that allow for this type of multitracking, in music at least, have been in existence for many years and have even been adapted for the Internet, for example, during the 1990s, former pop musician, Thomas Dolby, created a browser plug-in called, "Beatnik" that allowed users to experiment with multitrack mixing. Dolby has recently developed similar technologies for cell phones. One important design consideration in such strategies is to consider the type of graphic or information that might be most usefully linked to this particular level of musical structure – a level that exists much higher in the compositional scheme than that of the musical motif discussed earlier.

"Looping" is also a technique that holds many possibilities for musical and multimedia interactivity. Lev Manovich has discussed looping in relation to both the early days of cinema and decades later, early computer games, as partly a response to the technical limitations of these media, as well as a unique formal property – a property that may allow for new kinds of narrative structure or, possibly, a bridge between narrative and interactivity (Manovich 2001: 314–322). Looping in music, certainly shares some of the qualities suggested by Manovich. During the late 1940s and early 1950s, composers of *musique concrète* exploited the possibilities of looping on phonograph records and magnetic tape, as a central part of their compositional strategy, and during the 1980s, looping became an essential technique in the digital sampling practices associated with dance and rap music.

In computer games, however, the looping of music has been too often the result of technical limitations rather than the exploitation of new possibilities. Early versions of computer games such as *Sim City*, for example, for all their sophistication in allowing users to create complex urban environments, made use of only a handful of sound signals (sirens and the like) against a background of looped music so as to maximize computer power and memory for purposes of graphic rendering. Games have come under much criticism for this repetitious use of music, but at least one critic has suggested that the use of multiple loops of different lengths might allow for a more complex and seamless approach to the limitations imposed by technology (Bessel 2002).

The problem in computer games, however, is not so much looping per se, as with the level of musical structure where the looping is imposed: by looping entire musical arrangements, especially where the arrangements are of short duration (often only a few bars of music), the loops become obvious and tiresome. In this regard, the multiple-loop solution proposed by Bessel is intended more to disguise the loop points than to use them for creative purposes. In conjunction with the idea of layers or parts, as suggested above, however, one can come to a very different understanding of the role of loops in music. In popular music, some parts of the musical texture – typically drums, bass lines, and rhythm guitar parts – are characterized by greater levels of repetition than other, more featured parts, such as vocal lines (although, here too, repetition is also built into the structure of verse and chorus). In dance music, numerous percussion tracks of various durations might be looped and layered together to form an overall rhythmic pattern, itself looped at an even higher level of the musical structure. This technique of creating music of layers, and loops within loops, has become exceedingly popular and are the driving forces behind the release of computer programs, such as Ableton's "Live" that permit users to combine loops in real-time performance contexts.

Understood in this way, loops (and layers of loops) could be embedded within an overall musical structure that is dynamic and under the control of the user. In combination with sets of data or information that is equally dynamic, or linked to narrative or animated sequences, music could be integrated with multimedia in new ways and, perhaps as suggested by Manovich, become one of the links to a new kind of interactivity.

7. Conclusions

It is unlikely that humans, even with the powerful technologies at our disposal, will ever become as adept at navigating with sound as the bats in Kenneth Oppels' children's stories. However, a more sensitive approach to the modalities of speech, sound, and music may indeed allow sound to play a more significant role in multimedia productions and, especially in cybercartography. The technical requirements of using sound in a more integrated fashion within multimedia should not be underestimated. Making multiple tracks of sound available to users in a dynamic context – cross-fading between sound files, moving sound in a stereo or multichannel spatial field, making multiple sound loops available to the user at the click of a mouse–will inevitably make demands on computers in terms of bandwidth, memory, and processing power – demands that will have to be shared with the creation of images and other requirements. The issue of designing multimedia for sound is thus imperative, and must be addressed at the outset if sound is to be considered as an equal player in multimedia productions. However, if lessons can be learned from the cultural history of cinema, gaming and other media, and combined with the unique data processing and interactive capabilities of computers, then the potential benefits for cybercartography are indeed great.

References

Appadurai, A. (1996) "Disjuncture and difference in the global cultural economy", *Modernity at Large: Cultural Dimensions of Globalization*, University of Minnesota Press, Minneapolis, pp. 27–47.

Bessell, D. (2002) "What's that funny noise?: An examination of the role of music", *Cool Boarders 2, Alien Trilogy and Medievil 2*, in King, Krzywinska, G. and T. (eds.), *Screenplay: Cinema/Videogames/Interfaces*, Wallflower Press, London, pp. 136–144.

Buxton, W., W. Gaver and S. Bly (1994) Auditory Interfaces: The Use of Non-speech Audio at the Interface, Unpublished manuscript, accessed August 11, 2004 from http://www.billbuxton.com/Audio.TOC.html

Chatwin, B. (1987) *The Songlines*, Viking, New York.

Chion, M. (1994) *Audio-Vision: Sound on Screen*, Claudia Gorbman (trans.), Columbia University Press, NY.

Cook, N. (1998) *Analysing Musical Multimedia*, Oxford University Press, Oxford.

Doane, M. A. (1980) *The Voice in the Cinema: The Articulation of Body and Space*, Yale French Studies, No. 60, pp. 33–50.

Gorbman, C. (1987) *Unheard Melodies: Narrative Film Music*, Indiana University Press, Bloomington.

Gorbman, C. (2000) "Scoring the Indian: music in the Liberal Western", in Born, H. G. and D. (eds.), *Western Music and its Others: Difference, Representation, and Appropriation in Music*, University of California Press, Berkeley, pp. 234–279.

Herz, J. C. (1997) "Ditties of the apocalypse", *Joystick Nation*, Little, Brown and Co., Boston, pp. 101–111.

Human-Computer Interaction (1989) *Special Issue on Nonspeech Audio*, Vol. 4, No. 1.

Kozloff, S. (1988) "Invisible Storytellers: Voice-over Narration", *American Fiction Film*, University of California Press, Berkeley.

Kozloff, S. (2000) *Overhearing Film Dialogue*, University of California Press, Berkeley.

Kramer, G. *et al.* (1998) *Sonification Report: Status of the Field and Research Agenda*, a report prepared for the National Science Foundation, International

Community for Auditory Display (ICAD), accessed 6 July 2004 from http://www.icad.org/websiteV2.0/References/nsf.html

Krygier, J. B. (1994) "Sound and geographic visualization", in MacEachern, A. and D. R. F. Taylor (eds.), *Visualization in Modern Cartography*, Pergamon Press, Oxford, pp. 149–166.

Lacasse, S. (2000) *'Listen to My Voice': The Evocative Power of Vocal Staging in Recorded Rock Music and Other Forms of Vocal Expression*, PhD thesis, University of Liverpool.

Lobrutto, V. (1994) *Sound-on-Film: Interviews with Creators of Film Sound, Westport*, Praeger, CT.

Lunenfeld, P. (2000) *Snap to Grid: A User's Guide to Digital Arts, Media, and Cultures,* MIT Press.

Manovich, L. (2001) *The Language of New Media*, MIT Press, Cambridge, MA.

More, G., L. Harvey, J. M. and M. Burry (2003) *Implementing Nonlinear Sound Strategies within Spatial Design: Learning sound and Spatial Design within a Collaborative Virtual Environment,* Vol. 17, No. 8, August, Melbourne, DAC, pp. 128–134.

Morse, M. (1985) Talk Talk Talk—The Space of Discourse in Televison News, Sportscasts, Talk Shows and Advertising. Screen Vol. 26, No 2, pp. 2–15.

Nichols, B. (1985) "The voice of documentary", *Movies and Methods*: *Vol. 2, An Anthology*, University of California Press, Berkeley, pp. 258–273.

Oppel, K. (1997) *Silverwing*, HarperCollins, Toronto.

Packer, R. and K. Jordan (eds.) (2001) *Multimedia: From Wagner to Virtual Reality*, W. W. Norton & Co., New York.

Paulin, S. D. (2000) "Richard Wagner and the fantasy of cinematic unity: the idea of the Gesamtkunstwerk", in Buhler, J. *et al.* (eds.), *The History and Theory of Film Music, Music and Cinema*, Wesleyan University Press, Hanover, NH, pp. 58–84.

Rapée, E. (1924) *Motion Picture Moods*, G. Schirmer, New York.

Schafer, R. M. (1977) *The Tuning of the World*, Alfred A. Knopf, NY.

Smith, G. M. (2002) "Navigating myst-y landscapes: killer applications and hybrid criticism", in Henry, J., T. McPherson and J. Shattuc (eds.), *Hop on Pop: The Politics and Pleasures of Popular Culture*, Duke University Press, Durham, NC, pp. 487–502.

Taylor, D. R. F. (2003) The Concept of Cybercartography in Peterson, M. P., ed., Maps and the Internet, Elsevier, Amsterdam, pp. 403–418.

Théberge, P. (2003) "Sampling the world: cultural commodification and world music", in Lysloff, R. and L. Gay (eds.), *Music and Technoculture*, Wesleyan University Press, Middletown, CT, pp. 93–108.

Théberge, P. (forthcoming) "Almost silent: the interplay of sound and silence in contemporary cinema and television", in Beck, J. and A. Grajeda (eds.), *Lowering the Boom: New Essays on the History, Theory, and Practice of Film Sound.*

Thom, R. (1999) *Designing a Movie for Sound, iris*, Vol. 27, pp. 9–20. Also available online in a somewhat different version at http://www.filmsound.org/articles/designing_for_sound.htm, accessed 04 April 2000

Truax, B. (ed.) (1978) *Handbook for Acoustic Ecology*, ARC Publications, Vancouver.

CHAPTER 18

Interactive Mapping for People Who are Blind or Visually Impaired

REGINA ARAUJO DE ALMEIDA (VASCONCELLOS)
Department of Geography, University of São Paulo
São Paulo SP, Brazil

BRUCE TSUJI
Human Oriented Technology Laboratory (HOTLab)
Department of Psychology, Carleton University
Ottawa, Ontario, Canada

Abstract

Static and interactive tactile maps are discussed in the context of providing survey and mobility information to people who are visually impaired or blind. The heterogeneous nature of visual impairment is examined, as is the nature of tactile perception. Technologies associated with tactile maps are reviewed and the application of interactive tactile maps for populations, in addition to those who are visually impaired, is also considered. Cybercartography has considerable potential in this respect.

1. Introduction

Common sense might suggest that vision is a prerequisite for the acquisition and development of space perception. Certainly, the introspections of people with unimpaired vision point to visual perception as fundamental to an understanding of the environment and it is difficult to imagine how people with severe visual impairment or blindness can rely on alternative senses to gather information about the world and acquire geographical knowledge. To make matters worse, maps as graphic representations of space are usually unavailable to users who are visually impaired unless they are translated into an alternate language or medium. Tactile maps are rendered in a format to be read by touch. They can also be combined with other images and sounds as multimedia representations of geographical space.

This chapter will explore the use of tactile and multimedia maps for users who are blind, visually impaired, as well as those with no visual impairment. (Although color blindness has an effect on the use of maps, this chapter will not address that particular impairment.) An overview of the etiology and nature of visual impairment will be followed by a summary of the current state of tactile and multimedia maps. The chapter will then continue with a description of psychology's contribution to map-use by blind and sighted people. Finally, the chapter will explore some of the technologies associated with assisting the map-reader who is visually impaired as well as the map creator who may or may not have a visual impairment.

2. Visual Impairment and Blindness

The World Health Organization estimates that 180 million people worldwide are visually disabled and 40 to 45 million of these individuals cannot walk about unaided (World Health Organization 2000). While "walk about unaided" may be a meaningful operational measure of the severity of deficit, it makes objective comparisons difficult. In order to understand the number of people who are afflicted with visual impairments one must rely on the statistics maintained by various national organizations. For example, Heller (1991) cites the data of the US and Canadian advocacy groups indicating that approximately 10% of their clients report themselves to be totally blind. Similarly, the American Foundation for the Blind (2004) estimates that 10% of visually impaired individuals are totally blind amongst a population of 1.3 million Americans who reported themselves legally blind in 1994–95. The definition of "legal blindness" is central visual acuity of less than 20:200 with correction or a visual field of 20° or less. These two criteria have different impacts on visual perception. For someone whose corrected vision is 20:200 what many people are capable of seeing at 200 meters they are only capable of seeing at 20 meters. The visual field impairment means that such an individual must "build up" a visual environment from many smaller parts or pieces. In comparison, a typical nonimpaired visual field spans approximately 180°. For Canadians aged 40 years and over in 2001, Campbell *et al.* (2004) estimate that approximately 21% of the population of people who are visually impaired have visual acuity of less than 20:200. Similarly, the US Census Bureau data (McNeil 2001) indicates that approximately 18% of those aged 21 to 64 years who reported a visual impairment were "unable to read newspaper print".

Typical causes of congenital total blindness include retinopathy of prematurity (ROP, caused by excessive oxygen administered to premature infants) onchocerciasis, and trachoma. However, as indicated above, the incidence of adventitious or late blindness is far greater than congenital blindness. Adventitious blindness is caused by a variety of diseases including: macular degeneration, cataracts, glaucoma, diabetic retinopathy, retinal detachment, retinitis pigmentosa, AIDS/HIV-related cytomegalovirus and Karposi's sarcoma as well as burns, accidents, and other traumas (Canadian Opthalmological Society 2003).

There is a significant correlation between age and visual impairment (McNeil 2001; Cossette and Duclos 2002). One consequence of this fact is that many visually impaired users of maps will have had some prior experience with sight and possibly even with visual maps. Certainly the definition of "legal blindness" means that many such individuals retain some vestiges of sight. The user who is visually impaired may have primarily a central field loss wherein the center of the visual field is hazy or unclear, which is sometimes a consequence of macular degeneration. Alternately, a multiple field loss may involve patches of darkness or haziness. This type of loss is often associated with diabetic retinopathy. Glaucoma is often associated with a severe loss of peripheral vision leaving the individual with vision in only a very narrow 10 or 20° cone. As well, there are many other types of impairments that may result in loss of contrast, difficulties associated with changes in lighting conditions, blurring, or distortion. Indeed, lighting conditions and glare are often cited as the most common problems encountered by people with visual impairment in attempting to navigate the environment (Smith 1990; Geruschat and Smith 1997). There is a common misconception that visual impairment is binary – present or not – but it is not as homogenous as many think. The map-reader who is visually impaired may have no vision whatsoever or they may have various degrees of visual input under different lighting conditions. They may have had no visual input at any time in their lives or their impairment may have developed over years or decades of otherwise normal visual input.

In summary, visual impairment is considerably more complex than common understanding would attest. Furthermore the development of maps that are usable and useful to users with visual impairments must be consistent with the capabilities of this population. The next section will describe the current state of tactile maps.

3. Tactile and Multimedia Maps

Tactile maps have raised symbols on a flat surface (see Figures 18.1 and 18.2 in the accompanying CD-ROM) and differ from models (see Figures 18.3, 18.4, 18.5 and 18.6 on the CD-ROM), which are three-dimensional, and are usually perceived through the hairless (or glabrous) skin on the underside of the hand. Tactile perception works with active kinesthetic movement to enable people to sense raised images, surfaces and objects with their fingers. Research is also being conducted using other body parts, such as the eyes or head. For example, Tsuji *et al.* (2004) investigated navigation through a virtual space. Hagen compared head movement controlled azimuth combined with a joystick to determine forward/backward movement versus a joystick alone to control both (Tsuji *et al.* 2004). They found that the proprioceptive feedback associated with head movement directed navigation led to significantly better memory for landmarks in their virtual space compared to the joystick alone.

Maps of course have been made since ancient times, from materials as diverse as stone, wood, ceramics or sand. With the invention of Braille in the mid-19th century, (see the review by Foulke 1991) it became possible to describe map features using a font that blind persons could read. Embossed images and Braille text have been used for 150 years in maps. During the last decades, tactile maps have been produced using a wide variety of techniques and materials, such as plaster, embroidery, aluminum, embossed paper, molded plastic and silk-screen printed fabric (see CD-ROM Figures 18.7 and 18.8).

While there is still much room for progress in terms of tactile symbol vocabularies (Rowell and Ungar 2003), some of the greatest opportunities appear to exist with the advent of multimedia technologies. Visual and tactile images, sounds, and even smells or taste can be combined to transform mapping techniques and conventional maps into an interactive multimedia cybercartography that is much more than simply digital cartography and GIS (Taylor 2003, and the major elements of Cybercartography are discussed in the chapters of this book). What was once a solely tactile cartography aimed at map-users who were blind and visually impaired can now take advantage of all five senses and many innovations in telecommunications and computer technology. An example of the latter is the notion of "telementoring" in which haptic devices can be linked across a network to permit an individual's hand in one location to be guided by another elsewhere (Ballantyne 2002). Clearly Cybercartography has considerable potential for communication with, and by, individuals who are blind or visually impaired.

The Internet has had a significant impact on maps and mapping. Quite recently, Peterson (2003) had estimated that over 200 million maps were distributed via Internet per day. This number continues to increase each year. From one site alone (www.mapquest.com) daily map downloads increased from 700,000 in 1997 to 20,000,000 in 2001 (Peterson 2003). However, the Internet per se has had a mixed impact on tactile maps. On the one hand, Internet map distribution has made it possible for a great many different and very unique tactile maps to be created while on the other when such visually oriented maps are turned directly into tactile form without consideration of the unique properties of tactile perception, the results are often less than spectacular. In a pilot study Tsuji *et al.* (2004) found that their tactile maps of a university campus were very poor for assisting the mobility of blindfolded, sighted participants. Tsuji *et al.* concluded that they needed to attach their tactile maps to a firm substrate; include landmarks that might be considered irrelevant on purely visual maps (such as sources of smell and sounds); and provide a series of maps that communicated overall survey knowledge plus mobility maps for specific paths.

For a brief period of time a tactile feedback mouse made it possible for an individual to get haptic feedback when the mouse encountered lines and different textures (see http://www.utoronto.ca/atrc/reference/tech/altmouse.html). Unfortunately, at the time of this writing the Logitech iFeel mouse has been discontinued and there do not appear to be any equivalent devices. Taylor (Chapter 23 in this volume) argues that market forces have limited the availability of odor displays for smell. Market demand also has an impact on the availability and affordability of haptic devices.

Nonetheless, there are four aspects of Internet maps that must be noted here. First, the interactive attribute of Internet maps implemented using scalable vector graphics (SVG) means that the user, who is visually impaired, can zoom or magnify a given map to accommodate their impairment. This would obviously be of benefit to the map-reader who has impaired acuity. Secondly, the SVG format also permits output of maps to devices such as touch pads (Bulatov and Gardner, 2004; Rotard and Ertl 2004). Thus equipped, a visually impaired map-reader might be able to get overall survey information but the limited portability of such equipment would lessen its value for assisting mobility. Third, the directions output by sites such as www.mapquest.com could be transformed into verbal output either by the Internet site itself or via screen reading applications like JAWS for Windows (see http://www.freedomscientific.com/fs_news/

nr_JAWS401.asp). While this capability is very promising, currently most Internet-based directions are optimized for the automobile driver as opposed to the pedestrian who is visually impaired. Handheld mobile devices are however developing rapidly and this is discussed in relation to Cybercartography by Gartner in Chapter 16 of this volume.

Finally, an emerging capability is for the map to be output into speech or nonspeech audio directly. In its simplest form this is already being widely deployed in GPS-based car navigation systems. While many such systems are now available (see http://www.gps-practice-and-fun.com/car-gps.html) the speech output is typically restricted to very simple phrases such as "turn right" or nonspeech tones to indicate that a turn should be accomplished immediately. At least one system is optimized for the pedestrian who is visually impaired (see http://www.visuaide.com/gpssol.html) but the reliance on GPS technology means that inside buildings or shopping malls or even in built environments where the height of buildings may hinder GPS satellite line of sight, the system may not be able to provide effective navigation assistance. Elsewhere, experiments with nonspeech audio are being conducted to determine its efficacy for the communication of graphs. For example Flowers and his colleagues (Flowers *et al.* 1997; Flowers and Grafel 2002) have utilized nonspeech audio to effectively communicate the degree of correlation in scatterplots and also to communicate long-term weather patterns. If the two-dimensional information contained in statistical graphs can be communicated using sound, might it be possible to imagine an extension to two-dimensional maps? Ferres (2004) has created speech output of simple line graphs and we can hope that a similar speech-based sonification of maps would be a logical extension of that research. Théberge in Chapter 17 and Trbovich, Lindgaard and Dillon in Chapter 11 discuss sound for Cybercartography from a variety of approaches.

In addition to the potential of providing better information to visually impaired users, this aspect of cybercartography can also address the latent needs of much larger populations. Potential beneficiaries include the sighted but intellectually impaired, and children, as well. The basic concepts associated with multimedia cartography are simply too valuable to relegate to the category of Universal Design exclusively. As Theofanos and Redish (2003) have argued, by addressing the needs of the visually impaired audience, all audiences benefit. The extensive research on sound and touch in traditional tactile cartography has implications for the design of cybercartographic products for all users and is an extension of the arguments made by Monmonier in Chapter 2.

Sounds, touch and vision can be easily put together to depict the physical and social environment. Current technologies can already add smell and research will find ways to include taste in the future. Visualization techniques are improving and expanding to virtual reality and three-dimensional images to depict geographical space, also using remote sensing techniques to collect, process, save, represent and update information as illustrated in several chapters in this book.

Independently of impairment, our objective should be a multisensory and multicultural cartography similar to what has been proposed by the lead author in previous work (Almeida/Vasconcellos 1993, 1995, 1996, 2001). As opposed to "special needs" cartography, this new paradigm addresses a cartography which can be used by anyone, including children of all ages and indigenous people with their unique cultures. This new paradigm would include tactile maps and all new technologies related to this area, such as immersive virtual environments and TDG (see below). Conventional cartography will still live in the future and there is space and need for all maps. Tversky (1993) talks about "cognitive collages" as opposed to "cognitive maps" as a reflection of the multidimensional and multisensory nature of our internal representations of the external environment. Cybercartography can get closer to encompassing the many realities inhabiting our minds, memories and dreams, representing the real and the hidden worlds with all its dimensions including the physical, cultural and social. Related arguments have been made to this effect by several authors in this book and are central to the concept of cybercartography (Taylor 2003).

Considering these new approaches in cartography, all map-users (including people who are visually impaired) can learn to obtain information through previously underutilized channels. Those exposed to computer games today may be better prepared to deal with multisensory resources because they are already using touch, sound, and tactile/vibratory feedback to gather geographical information and move around virtual spaces. The implications of gaming for Cybercartography are outlined by Cartwright in Chapter 14, Greenspan in Chapter 13 and Taylor in Chapter 23. People use their five senses, if they have them, in different ways and situations, many times without being conscious of it. What is required are the techniques and strategies necessary to use the other senses in conjunction with, or in addition to sight. While technology in the 21st century will provide us with multimedia and multisensory maps, it will remain our responsibility to deploy those in a manner which helps rather than hinders us in our consumption of information (Stanney *et al.* 2004).

4. Tactile and Spatial Perception

Effective tactile perception is an obvious prerequisite for effective tactile map-usage. Among others, Burton and Sinclair (1996) have noted that touch is unique among the senses. Touch has a dual nature that arises in the touched object (the distal stimulus) as well as the sense perceptions of the body part that does the touching. The best example of this duality is experienced by running a hand over part of one's body and alternately attending to the skin of the hand doing the touching or the touched skin. The skin is both highly sensitive and highly variable in its ability to sense. Greenspan and Bolanowski (1996) report that the skin exhibits an electrophysiological response to a stimulus as small as a 5.5 μM dot etched onto the surface of glass. At the same time, the human threshold for distinguishing between one and two points applied to the skin (called the two-point limen) varies 20 times between the shoulders and the fingertips with the greater sensitivity exhibited by the more distal fingertips (Stevens and Green 1996). When people exert control over their touching (by actively moving the hand for example) they appear to exhibit a greater degree of sensitivity than when the touching is done passively or without active movement (Heller and Myers 1983). Like the visual and auditory senses there appears to be a decline in tactile sensitivity with age. Stevens and Patterson (1995) have reported a 1% per year decline in tactile acuity between 20 and 80 years but according to Greenspan and Bolanowski (1996) there appear to be no gender differences in tactile acuity, unlike the case with spatial abilities (Tapley and Bryden 1977).

Although Greenspan and Bolanowski (1996) report that spatial stimuli are aggregated or summed across the two hands (this is called spatial summation) and Foulke (1991) reports that faster Braille readers utilize two index fingers instead of just one, tactile perception is fundamentally serial in nature. People cannot take in and analyze tactile stimuli in the massively parallel way they analyze visual scenes. Instead, tactile stimuli are perceived more slowly and more sequentially than visual stimuli (Heller 1991). In part, this explains why the Braille reading rate for senior high school students is 80–90 wpm compared to visual reading rates of 250–300 wpm (Foulke 1991).

Applying what is known of tactile perception to tactile maps, it is clear that it will take more time for people to apprehend a tactile map than one that is visual. And, while the skin is wondrously sensitive, people may not be able to distinguish different textures to the same degree that they can distinguish different colors or objects in the visual world.

In Chapter 10, Roberts, Parush, and Lindgaard have summarized the literature concerned with the spatial cognition of sighted users. However, we are interested here in the spatial cognitive behavior of users who are visually impaired. Trbovich, Lindgaard and Dillon (Chapter 11) evaluate this research in the context of multimodal interface design. Ungar (2000) distinguishes between tasks associated with Near Space versus Far Space. The former are smaller scale and can be performed without changing body location whereas the latter are associated with navigation in the world around us. Ungar reports that congenitally and early-blind participants perform Near Space tasks at a level equivalent to participants who are late-blind or sighted. Similarly, Heller and Kennedy (1990) in their drawing tasks found that their blind participants exhibited an understanding of spatial perspective. This is also consistent with Kennedy *et al.* (1991) who found that their congenitally blind participants were able to interpret line drawings from different perspectives. As well, mental rotation effects (slower response times to identify one object as the same when it is rotated further out of alignment with the other) have been reported by Hollins (1986) and others. These results imply that spatial abilities (at least those sufficient for Near Space tasks) develop even in the total absence of visual input. Ungar (2000) goes on to argue that the Far Space performance of users who are visually impaired is inconsistent and (compared to sighted users) may be due to the selection of poor strategies. Ungar states that "lack of visual experience does not prevent the acquisition of spatial representation" (Ungar 2000: 233). This is an important statement because it demands an entirely different research agenda from one in which the focus is on understanding the nature of the visual impairment and associated spatial deficit (e.g. Warren 1977). Instead, a more fruitful approach may be understanding the less-effective strategies adopted by users who are visually impaired and facilitating the learning of more effective ones.

Similarly, Jones (1975) has argued that movement deprivation of people who are visually impaired (by well-meaning parents intending to protect their children, for example) has led to less practice and therefore a poorer understanding of space. Facilitating the understanding of space using tactile maps seems to be one way of overcoming the lack of prior learning. In fact, participants who are congenitally blind perform the same as, or even better than participants who are sighted or late blind in judgments of texture and tactile form perception (Heller and Kennedy 1990; Heller 1991). Again, this implies that utilizing tactile maps could be an effective strategy for helping people who are visually impaired, acquire a "cognitive collage" (Tversky 1993) of their world. The development of such a collage is obviously further enhanced by the addition of sound and other elements of cybercartography as discussed earlier in this chapter.

5. Tactile Cartography: A Different Kind of Map

People learn about geography and gather spatial information through direct exploration, verbal explanations, sounds, smells, movements and gestures, models, visual and tactile images, and maps. In fact, maps and images are far more important to the visually impaired than to the sighted person. For blind people cartography can be the bridge to reality, while to the sighted, maps and graphic representations in general are an abstraction of reality. We consider tactile maps to be of two kinds:

Orientation and Mobility Maps: Orientation and mobility maps depict indoor and outdoor environments and both are almost always large-scale representations of buildings, routes, streets, transportation networks, neighborhoods, and recreation areas. People with visual impairments use such maps to navigate in space and to be aware of the physical environment, its obstacles and dangerous features (see CD-ROM Figure 18.8). An orientation map gives an overview of the area, while mobility maps provide the kind of detailed information that is needed for an individual to move safely through space. Mobility maps show relevant information for the traveler, elements to be used such as sidewalks, or elevators. Verbal or sequential maps can replace tactile mobility maps to explain routes to blind people. Golledge and his collaborators (Loomis *et al.* 1993; Golledge *et al.* 1996) have developed auditory maps using GPS technology. There has been much research combining sound and tactile graphics and this will be presented later in this chapter. Gartner in Chapter 16 on telecartography discusses similar applications.

Geographical Maps: There are two main types of geographical tactile maps that represent larger areas in much smaller scale than 1. mobility maps, and 2. thematic maps, representing qualitative and quantitative geographical information in tactile form. They portray a large variety of themes such as economics and trade, population, education, environment, relief, geology, flora, and fauna. Usually there is a great diversity of scales from local, regional, national to global. They show location, pattern, and distribution of data, using a tactile graphic language in both choropleth and topographical fashion. The tactile atlas of Latin America (Taylor 2001) and the map reproduced in CD-ROM Figures 18.2 and 18.6 is an example of such an approach.

General reference maps include topographical maps or relief models similar to those produced for the sighted user. However, in this case all symbols are produced in raised format with Braille text. Likewise political borders and other features are rendered in tactile format (Fortin 2003).

6. Tactile Map Design, Production and Use

Design and production of tactile maps is a task often performed by non-cartographic specialists (Tatham 2003). International standards can facilitate design but there are still works in progress with little consistency in size, format, labels, or even the existence of scales (Rowell and Ungar 2003). The greatest element of consistency reported by Rowell and Ungar (2003) seems to be a convergence on thermoform and microcapsule paper.

Orientation and mobility maps are more likely to have a system of conventions established at the international level. In any case, it is important to prepare users and mapmakers to deal with tactile cartography. There is a need to produce and disseminate tactile graphics, especially maps, not only for the blind but also as teaching materials for all schools (Vasconcellos 2001).

The International Cartographic Association has given much support to this field of cartography and in 1984 created a Commission on Maps and Graphics for Blind and partially sighted People aimed at the study and dissemination of tactile cartography. It has members from over 20 countries and has organized four major international conferences in the USA, UK, Japan, and Brazil, and several regional meetings on all continents, with important achievements and a large number of publications. Even prior to the creation of this commission however, cartographers and geographers had studied and developed and tested tactile maps.

Many authors have made significant contributions to tactile cartography. Research on tactile map design, production and use has been carried out by a large number of specialists (Wiedel 1972, 1983; Kidwell and Greer 1973; James 1982; Schiff and Foulke 1982; Barth, 1987; Tatham and Dodds 1988; Ishido 1989; Edman 1992; Vasconcellos 1996; 2001; Perkins and Gardiner, 2003). Guidelines for map design, production and use can be found in these studies, including issues such as scale, standardization, degree of generalization, and choice of symbols.

As stated by Vasconcellos (1996, 2001), the design of a tactile map involves a cartographic language with tactile symbols such as raised lines, points and textures. The nature of geographical data, either quantitative or qualitative, has to be considered while selecting the correct variable to represent the particular data element. Graphic variables (e.g. size, value, texture, color, orientation, and shape) can be expressed in point, linear and area modes. Color is not a part of the tactual graphic language, but elevation (height) can be used alone or combined with other variables in multiple ways, to represent features in addition to height of the terrain. Color can however be useful for some categories of visual impairment to increase the contrast amongst visual elements.

Clarity and simplicity are important qualities in tactile mapping, as is the use of a map legend or key. A collection of maps or a multilayered tactile map can help avoid complexity and excessive clutter, making tactile perception easier (Vasconcellos 2001; Tsuji *et al.* 2004). Tactile maps when designed for low vision users should combine visual and tactile symbols, preferably in color as suggested above. In this case, hues should be used carefully, as subtle differences may not be adequately discriminated by the visually impaired user. Many studies have been done on legibility and discriminability of symbols from the perspective of the visually impaired as map users. In general, choice of symbols, scale, and degree of generalization greatly depends on the production method used, including master construction and reproduction techniques.

There are many techniques and materials used in the construction and reproduction of tactile maps. The most common methods are:

- *Handmade collage maps* using a large variety of inexpensive materials, usually prepared by nonspecialists. At the CNIB, these maps are created by tactilists (see Figure 18.8);
- *Molded plastic maps* done in vacuum-forming machines (e.g. thermoform and Brailon plastic), and made from a master map prepared in a variety of materials such as embossed aluminum, collage, paper or plaster carved manually or driven by computers (see Figure 18.7);
- *Embossed paper maps* made by conventional Braille printing machines, done by conventional or digital modes;
- *Silk-screen maps* using puff ink, usually printed on paper;
- *Microcapsule paper*, compatible with conventional printers and raised by heat, largely used with digital maps;

The advantages and limitations of these production methods depend on a variety of factors such as cost, available technology, number of copies needed, or the user's ability to read maps (Vasconcellos 2001). Considering these methods, choices are not easy to make, for instance, collage has very good properties of discriminability because of the richness of different textures, but it has low precision and costly reproduction. On the other hand, silk-screen and microcapsule paper maps have low discriminability, but high reproduction capabilities. Besides these more common techniques, there are others such as German film, virktyping or thermocraft.

Mapping for blind and visually impaired users gives rise to theoretical discussions about cartographic communication. Research has shown that tactile cartography has different rules and, in general, more training is necessary for both producers and users. The tactual graphic language has to be introduced to the user prior to tactile mapreading through exercises which help to explain physical space and its representation. There is also the need to teach basic geographical concepts such as scale, point of view, relative positioning, orientation, location, and distance (Vasconcellos 1996, 2001; Fortin 2003).

It is important to consider the user's degree of sensory impairment, previous experience, knowledge about the environment, and cartographic skills. The tactile sense is coarser than vision and it demands that the user assemble the parts actively to form a whole map which differs from vision that is synthetic with high resolution. Despite the limitations of the medium, people who are visually impaired can develop an excellent ability to decode map keys because they are able to use both hands simultaneously, one on the legend and the other on the map. The use of redundancy is advisable in this case, as the repetition of the same information using more than one medium can improve the decoding and understanding of the information depicted on the map.

In most cases, the visually impaired user might need personal assistance or good instructions to understand the message conveyed on a tactile map. Successful cartographic communication depends on innovative approaches and efficient training since many of the problems this group encounters in mapreading are different from those encountered by the sighted user. The combination of different media and sensory channels, used in an interactive way, is important determinant in the cartographic communication process.

7. Tactile Audio Approach to Virtual Spaces: Digital Technology Discovering New Paths

Touch and sound are used as substitutes for sight by people who are visually impaired and sighted people in the dark. If synthesized speech, digital speech, sounds, and pure tones can be synchronized with the touching of a tactile image, in this instance a map or plan, the disadvantages of Braille text might be overcome. Sounds can also be used alone as substitutes for the visual and the tactual languages or combined in different ways. Trbovich, Lindgaard and Dillon consider these modalities elsewhere in this volume and Théberge discussed sound in Chapter 17.

Maps are usually said to represent some segment of the real world at one specific point in time. As a result, Golledge *et al.* (1996) argue that static maps completely miss out on one of the most characteristic dimensions of the real world – its dynamic nature. After bringing up the issue of real maps versus virtual maps and their nature, Golledge *et al.* turn to a consideration of auditory maps and examine how these can be related with tactual maps and visual maps (1996). Golledge *et al.* discuss the advantages and disadvantages of virtual auditory map display compared to visual maps arguing that audition can be incorporated into computer cartography in the form of initiation and completion alerts, duration filters, and auditory icons. The contribution of sounds in these contexts depends on whether the sound is used as a symbol, a metaphor, or is nomic in character. In Golledge's auditory maps different types of sounds are incorporated, as he developed a Personal Guidance System (PGS) combining GPS technology and auditory maps directed to orientation and mobility of blind users. Golledge argues that auditory maps can complement tactile maps and may eventually replace them, having even greater potential as an en route navigation device to overcome the movement restrictions of blind travelers (see also www.visualprosthesis.com for Meijer's complementary view).

Tactile maps with Braille text have been produced and used all over the world. Because the resolution of the tactual sense is much lower than sight, map size has to be large and interpretation of shape is much harder. To solve a few of these problems, a tactile-audio system called Nomad was developed in the late 1980s. This system combined touch and sound so that a tactile map, when touched, spoke the written text under the finger. Much more information could now be placed on a tactile map and Braille was no longer required on the map. During the past 15 years there have been refinements of the Nomad system, most recently with a system called Tactile Graphics Designer or TGD, also by the Australian inventors of the original Nomad system. This new system allows the blind and visually impaired to draw their own maps and graphics using a combination of software and hardware, printer and embosser (Parkes 1998). This is an interesting example of the map-user becoming a map-maker as discussed by Taylor in the concluding chapter of this volume. Research in many countries has also been focusing on dynamic refreshable surfaces using a matrix of solenoid-driven pins, piezo-electric cells and possible use of dynamic shape memory to create tactile surfaces (Vasconcellos 2001).

In 1987 a prototype system called ATMAPSIT (Audio-Tactile Mapping and Spatial Information Tool) was developed by Parkes (1988). The initial work owed much to the encouragement of Dr. Tony Heyes of Royal Guide Dogs Association in Melbourne, Australia where there was clearly an interest in the linking of spatial information to personal travel by blind people. This first tool was described as a grid of touch sensitive wires within the board with a microprocessor where maps, plans and diagrams are registered. A compass with directions is considered important, and is to be added. The tool would be a stand-alone device, not requiring an interface with any other equipment that needs input and output of voice or other audible signals. ATMAPSIT is to be designed in a way to enable a visually handicapped user to have a dialogue with a tactile map. It goes without saying that the design intent was quite ambitious, and not all features were incorporated in the first implementations.

Within 2 years an Australian company (Quantum Technology) had been licensed to develop an A3 size touch sensitive board with an integrated synthesizer and linked to a personal computer. Software development continued (Parkes and Dear 1989) and pure tone sound painting was added to the tactile graphic's capability to respond with speech or digital sounds, linked to point, linear or two-dimensional map features. Sound painting was to act as a sort of choropleth shading in which quantitative and qualitative differences over an area could be linked to changing sound frequencies. For instance, high elevation was mapped onto high frequency sounds. They wrote, "with this facility of sound painting it is possible to paint a sound into any shape". Nine frequencies were used, more could have been added but there was insufficient understanding of the ability of users to discriminate the sounds beyond eight or nine (Parkes 1994). This is an interesting extension of the concept of "earcons" as described by Trbovich *et al.* in this volume.

By the early nineties, the Nomad (see Figure 18.11 on the CD-ROM) system, as it was known, was used in 14 different languages with a range of functions that included direction, distances on any path, a guidance system, the ability to include text files of lower level information, describing the details of a map or plan feature, the recording of digital speech so that any language could be used.

Later, two programs were developed that enabled the making of tactile graphics, either for printing to a capsule paper such as the American system, flexi-paper or to embosser output on to "Braille" paper. Thermoformed substrates could also use the QikTac and TraceMe output. The latter enabled a color image, map or plan to be converted to a dot system suitable for embossing. Having been embossed, the tactile image could have sounds and other audio linked features added. AudioCAD was one of these programs (Parkes 1998). Its essence was to emulate the sighted user QikTac program and enable a blind person to draw a map (or any other graphic) using a range of computer aided functions, such as circles, lines of any length and angle, rectangles and so forth. The CAD component of AudioCAD was described, not as Computer Aided Design in fact but as "Can't Anyone Draw?" (Parkes and Brull 1997). The program is now under redevelopment to use keyboard and voice recognition commands and it is hoped that the blind person's ability to create their own maps will encourage their acceptance of a wider range of spatial information in map or plan form.

AudioPIX and AudioTrip were two other programs in the systems that were developed after Nomad in the mid-nineties and continue today within TGD Workshop. AudioPIX is the platform for all tactile audio graphics, now called "graphics" in the TGD system and provides the functionality whereby information appropriate to mapmaking can be applied to tactile graphics. A TagPad, a touch sensitive pad of US Letter/A4 dimensions holds the tactile image as an overlay, in essentially the same way that this was done 15 years earlier, but there have been improvements in touch and sound access, as technological developments in computer hardware (disk storage and speed, memory access and so forth) have improved.

AudioTRIP (presently under redesign) uses the coordinate based information from AudioPIX and the TagPad (Tactile Audio Graphics Pad). The system allows any map or plan (so long as it has a scale) to be placed on the TagPad. An intended route through the map space is simply traced with a finger or soft tipped stylus. The route features that are a prespecified distance from the "path" are then spoken to the user – as though on a tape recording. The distance of the trip, the time it will take to complete based on prespecified travel speed – are all provided to the intending traveler. Parkes however has also described the trip as a virtual travel system enabling blind persons to take a trip through any "famous" space and learn about its scale and peripheral features. The latter is a "fun" feature and it has been a tenet of the systems developed since 1987 that they should be easy to use and inexpensive, Taylor (2003) argues that "edutainment" is an important element of Cybercartography. The virtual travel system is an interesting example of this.

Currently, TGD Workshop is a graphics tool kit which includes a barcode reader and microphone for speech recognition. There are other systems about to appear in the market and this is very encouraging, as ever more effective hardware and software enables graphics to be brought into the use of blind people.

Overall, the future for people who are blind to have access to and to create map-based information systems is looking brighter. The importance of being able to understand one's spatial environment goes well beyond its general interest value. It can become a key factor in the employment and successful accomplishment of the duties of a person who is blind or visually impaired, in the workplace.

8. Conclusions

Tactile cartography holds great promise for people who are visually impaired and blind. It can help people overcome relevant information barriers and to "see" graphics with their hands and ears. Tactile maps can also help the blind user to navigate through space, inside buildings, and in the environment generally. It is very important that we understand the best and most effective way to design, reproduce, and distribute tactile maps. All kinds of maps and graphics should become available in tactile form, facilitating the orientation and mobility of visually impaired users to a world that is otherwise dominated by images.

Technology has transformed tactile map production and use, bringing new challenges to the field. We have the ability to combine images, sounds and tactile graphics. However, only by intelligently combining these can we avoid a "buzzing booming world of confusion". The importance of avoiding cognitive overload as discussed by Trbovich, Lindgaard and Dillon in this volume is important to cybercartography as a whole and to tactile mapping in particular. Touch has dominated tactile mapping until recently when sound has become increasingly important. In the future tactile maps may only be "tactile" in the sense that touch is a major part of the user interface.

More than ever before, cartographers have to weigh the existing advantages and limitations of different methods prior to producing tactile maps and deciding about components of map design and graphic language to be employed. Scale, degree of generalization, and choice of symbols greatly depend first, on the production methods and techniques and second on the media through which tactile maps are to be presented to the user.

Future research should move from tactile cartography specifically for visually impaired users to a broader field of "niche" cartography that would deliver the appropriate combination of visual, tactile, sounds, smells, and tastes to the appropriate user for the appropriate purpose. This Cybercartography would be multisensory, multimodal, and multicultural in terms of design, production and use.

References

Almeida (Vasconcellos), R. (2001) "Cartography and indigenous populations: a case study with Brazilian Indians from the Amazon region", *Proceedings of the 20th International Cartographic Conference*, Beijing, China.

American Foundation for the Blind (2004) "Statistics and sources for professionals", accessed 5 December 2004 from http://www.afb.org/Section.asp?SectionID=15&DocumentID=1367&Mode=Print

Ballantyne, G. H. (2002) "Robotic surgery, telerobotic surgery, telepresence, and telementoring", *Surgical Endoscopy*, Vol. 16, No. 10, pp. 1389–1402.

Barth, J. L. (1987) *Tactile Graphics Guidebook,* American printing house for the blind, Louisville, KY.

Blum, E. K. and J. Kuchmister (1996) "Using the sound-coding method as an aid for blind people", Slovenia accessed 9 February 2005 from http://www.surrey.ac.uk/~pss1su/intact/TMC/blum1.html

Bulatov, V. and J. A. Gardner (2004) "September 4–7, making graphics accessible," in *Proceedings of SVG Open: 3rd Annual Conference on Scalable Vector*

Graphics, Tokyo, Japan, accessed 5 December 2004 from http://ptolemy.unomaha.edu/~peterson/svg2004/proceedings_en.html

Burton, H. and R. Sinclair (1996) "Somatosensory cortex and tactile perceptions" in L. Kruger (ed.), *Pain and Touch*, Academic Press, New York, NY, pp. 105–177.

Campbell, J., R. Hodge, G. William and R. R. Buhrmann (2004) "CNIB symposium, the cost of blindness: what it means to Canadians", *Estimates of Current and Projected Vision Loss in Canada*, Paper presented at CNIB, Toronto, Ontario, Canada, 31 January–1 February.

Canadian Opthalmological Society (2003) *Diabetes and the Eye*, accessed 5 December, 2004 from http://www.eyesite.ca/english/public-information/eye-conditions/pdfs/Diabetes.pdf

Cossette, L. and E. Duclos (2002) *A Profile of Disability in Canada, 2001*, Statistics Canada, Housing, Family and Social Statistics Division, Catalogue no. pp. 89–577-XIE. Ottawa, ON, Canada, Minister of Industry.

Edman, P. K. (1992) *Tactile Graphics*. American Foundation for the Blind, New York, NY.

Ferres, L. (2004) "Generating natural language descriptions of line graphs: the inspectgraph system", CLIN 2004: *Computational Linguistics in the Netherlands*, Leiden, Netherlands.

Flowers, J. H., D. C. Buhman and K. D. Turnage (1997) "Cross-modal equivalence of visual and auditory scatterplots for exploring bivariate data samples", *Human Factors*, Vol. 39, pp. 341–351.

Flowers, J. H., and D. C. Grafel (2002) "Perception of sonified daily weather records", *Proceedings of the Human Factors and Ergonomic Society 46th Annual Meeting*, pp. 1579–1583.

Fortin, A. (2003) "MA thesis, department of geography", *Mapping for a Visually Impaired Audience: A Case Study on the Legibility and Cognition of Tactile Maps for Education*, Carleton University, Ottawa.

Foulke, E. (1991) "Braille" in Heller, M. A. and W. Schiff (eds.), *The Psychology of Touch*, Lawrence Erlbaum Associates, Hillsdale, NJ, pp. 219–233.

Geruschat, D. and A. J. Smith (1997) "Low Vision and Mobility" in Blasch, B. B., W. R. Wiener and R. L. Welsh (eds.), *Foundations of Orientation and Mobility*, 2nd edn., American Foundation for the Blind, New York, NY, pp. 60–103.

Golledge, R. D., J. M. Loomis and R. L. Klatzky (1996) "Auditory maps as alternatives to tactual maps", *Proceedings, GeoDigital '96 Symposium, Department of Geography, FFLCH*, University of São Paulo, 25–28 November, pp. 129–136.

Greenspan, J. D. and S. J. Bolanowski (1996) "The psychophysics of tactile perception and its peripheral physiological basis" in L. Kruger, (ed.), *Pain and Touch*, Academic Press, New York, NY, pp. 25–103.

Heller, M. A. (1991) "Haptic perception in blind people" in Heller M. A. and W. Schiff (eds.), *The Psychology of Touch*, Erlbaum Associates, Lawrence, Hillsdale, NJ, pp. 239–261.

Heller, M. A. and J. M. Kennedy (1990) "Perspective taking, pictures, and the blind", *Perception and Psychophysics*, Vol. 48, No. 5, pp. 459–466.

Heller, M. A. and D. S. Myers (1983) "Active and passive tactual recognition of form", *Journal of General Psychology*, Vol. 108, No. 2, pp. 225–229.

Hollins, M. (1986) "Haptic mental rotation: more consistent in blind subjects?", *Journal of Visual Impairment and Blindness*, Vol. 80, No. 9, pp. 950–952.

Ishido, Y. (ed.) (1989) *Proceedings of the Third International Symposium on Maps and Graphics for the Visually handicapped People*, Yokohama Convention Bureau, Yokohama, Japan.

James G. A. (1982) "Mobility Maps" in Schiff, W. and E. Foulke (eds.), *Tactual Perception: A Source Book*, Cambridge University Press, New York, NY.

Jones, B. (1975) "Spatial perception in the blind", *British Journal of Psychology*, Vol. 66, pp. 461–472.

Kennedy, J. M., P. Gabias and A. Nicholls (1991) "Tactile pictures" in M. A. Heller and W. Schiff (eds.), *The Psychology of Touch.*, Erlbaum Associates, Lawrence, Hillsdale, NJ, pp. 263–299.

Kidwell, A. M. and P. S. Greer (1973) "American Foundation for the Blind", *Sight, Perception and the Nonvisual Experience: Designing and Manufacturing Mobility Maps*, New York, NY.

Loomis, J. M., R. L. Klatzky, R. G. Golledge, J. G. Cicinelli, J. W Pellegrino and P. A. Fry (1993) "Nonvisual navigation by blind and sighted: assessment of path integration ability" *Journal of Experimental Psychology,* General, Vol. 122, No. 1, pp. 73–91.

McNeil, J. (2001) *Americans with disabilities: Household economic studies 1997, US Census Bureau*, accessed 5 December 2004 from http://www.census.gov/prod/2001pubs/p70–73.pdf

Parkes, D. N. (1988) "Conference of the Australian and New Zealand association of educators of the visually handicapped", *Audio-Tactile Mapping for the Visually Handicapped,* Ormond College, University of Melbourne, Melbourne Australia, January, pp. 18–22

Parkes, D. N. and R. J. Dear (1989) "A new approach to graphics processing for the blind, can graphics be useful, interesting, amusing, educative to those who are blind and to those who are partially sighted? We think so", in *Australian and New Zealand Association of Educators of the Visually Handicapped,* Newsletter Vol. 24, No.2, pp. 7–8.

Parkes, D. N. (1994) "Paper presented to international tactual mapping commission symposium", *Multi-Media Audio-Tactile Maps and Plans: a Sound Space for Blind Users with the 'Touchblaster' Nomad System*, University of São Paulo, Brazil, São Paulo, Brazil, February, pp. 20–26.

Parkes, D. N. (1995) "Access to complex environments for blind people: multimedia maps, plans and virtual travel", *Proceedings of the 17th International Conference, International Cartographic Association*, Vol. 2, Barcelona, Spain, pp. 2449–2460.

Parkes, D. and M. Brull (1997) "It may not be easy but it is possible: a new form of literacy", *The World Blind*, Vol. 14, August, pp. 48–50.

Parkes, D. N. (1998) "Tactile audio tools for graphicacy and mobility: a circle is either a circle or it is not a circle", The *British Journal of Visual Impairment*, Vol. 16.3, 99–104

Perkins, C. and A. Gardiner (2003) "Real world map reading strategies", *The Cartographic Journal*, Vol. 40, No. 3, pp. 265–268.

Peterson, M. P. (2003) "Maps and the Internet: an introduction" in M. P. Peterson (ed.), *Maps and the Internet*, Elsevier Science, pp. 1–16.

Rotard, M. and T. Ertl (2004) "Tactile access to scalable vector graphics for people with visual impairment", *SVG open 2004: 3rd Annual Conference on Scalable Vector Graphics*, Tokyo, Japan, accessed 5 December 2004 from September, pp. 7–10 http://www.svgopen.org/2004/papers/TactileAccessToSVG.

Rowell, J. and S. Ungar (2003) "The world of touch: results of an international survey of tactile maps and symbols", *The Cartographic Journal*, Vol. 40, No. 3, pp. 259–263.

Schiff, W. and E. Foulke (eds.) (1982) *Tactual Perception: A Source Book*, Cambridge University Press, New York, NY.

Smith, A. J. (1990) "Mobility problems related to vision loss: perceptions of mobility practitioners and persons with low vision", *Dissertation Abstracts International*, Vol. 51 No. 5, University Microfilms No. 9026646.

Stanney, K., S. Samman, L. Reeves, K. Hale, W. Buff, C. Bowers, D. Nicholson and S. Lackey (2004) "A paradigm shift in interactive computing: deriving multimodal design principles from behavioral and neurological foundations", *International Journal of Human-Computer Interaction*, Vol. 17, No, 2, pp. 229–257.

Statistics Canada (2004) *Population Pyramids*, 1901–2001 accessed 5 December 2004 from http://www.statcan.ca/english/kits/animat/pyone.htm

Stevens, J. C. and B. G. Green (1996) "History of research on touch" in L. Kruger (ed.), *Pain and Touch*, Academic Press, New York, NY, pp. 1–23.

Stevens, J. C. and M. Q. Patterson (1995) "Dimensions of spatial acuity in the touch sense: changes over the life span", *Somatosensory and Motor Research*, Vol. 12, No. 1, pp. 29–47.

Tapley, S. M. and M. P. Bryden (1977) "An investigation of sex differences in spatial ability: mental rotation of three-dimensional objects", *Canadian Journal of Psychology*, Vol. 31, pp. 122–130.

Tatham, A. F. (2003) "Tactile mapping: yesterday, today and tomorrow", *The Cartographic Journal*, Vol. 40, No. 3, pp. 255–258.

Tatham, A. F. and A. G. Dodds (eds.) (1988) *Proceedings of the Second International Symposium on Maps and Graphics for Visually Handicapped People,* University of Nottingham, King's College, London, UK.

Taylor, D. R. F. (2001) *Tactile Atlas of Latin America*, Carleton University, Ottawa, Ontario, Canada.

Taylor, D. R. F. (2003) "The concept of cybercartography" in Peterson, M. (ed.), *Maps and the Internet*, Elsevier, Amsterdam, NL.

Theofanos, M. F. and J. Redish (2003) " November-December, bridging the gap: between accessibility and usability", *Interactions*, pp. 36–51.

Tsuji, B., L. Hagen, C. Herdman, J. LeFevre, and G. Lindgaard (2004) "Paper presented at the Canadian Association of Geographers of Ontario", *Two Steps into a Cybercartographic Landscape*, Waterloo, Ontario, Canada, October, pp. 29–30.

Tversky, B. (1993) "Cognitive maps, cognitive collages, and spatial mental models" in Frank, A. U. and I Campari (eds.), *Spatial Information Theory: A Theoretical Basis for GIS*, Springer-Verlag, Berlin, Germany pp. 14–24.

Ungar, S. (2000) "Cognitive mapping without visual experience" in Kitchin, R. and S. Freundschuh (eds.), *Cognitive Mapping, Past, Present, and Future.*, Routledge, London, UK, pp. 221–248.

Vasconcellos, R. (1993) "Representing the geographical space for visually handicapped students: a case study on map use", *Proceedings of the 16th International Cartographic Conference,* Berlin, Germany, Vol. 2, pp. 993–1004.

Vasconcellos, R. (1995) "Tactile mapping for the visually impaired children", Proceedings 17th International Conference, ICA, Barcelona, Spain, Vol. 2, pp. 1755–1764.

Vasconcellos, R. (1996) "Tactile mapping design and the visually impaired user" in Wood, C. and P. Keller (eds.), *Cartographic Design: Theoretical and Practical Perspectives*, John Wiley & Sons, London, UK.

Vasconcellos, R. (2001) "Tactile maps in geography" in Hanson, S. and F. E. Weinert (eds.), *International Encyclopedia of Social and Behavioral Sciences*, Elsevier, Boston, MA.

Warren, D. H. (1977) "American foundation for the blind", *Blindness and Early Childhood Development*, New York, NY.

Wiedel, J. W. and P. Groves (1972) *Tactual Mapping: Design, Reproduction, Reading and Interpretation*, University of Maryland, College Park, MD.

Wiedel, J. W. (ed.) (1983) *Proceedings of the First International Symposium on Maps and Graphs for the Visually Handicapped,* Association of American Geographers,Washington, DC.

World Health Organization (2000) *Blindness: Vision 2020 - The Global Initiative for the Elimination of Avoidable Blindness, Fact sheet #213,* February 2000, accessed 5 December 2004 from http://www.who.int/mediacentre/factsheets/fs213/en/

CHAPTER 19

Exploring Conceptual Landscapes: The Design and Implementation of the Georgia Basin Digital Library

ROB HARRAP

Queen's University GIS Laboratory, Queen's University, Kingston, Ontario, Canada

SONIA TALWAR

Department of Geography, University of British Columbia and Geological Survey of Canada, Natural Resources Canada, Vancouver, British Columbia, Canada

MURRAY JOURNEAY

Geological Survey of Canada, Natural Resources Canada, Vancouver, British Columbia, Canada

BOYAN BRODARIC

Geological Survey of Canada, Natural Resources Canada, Ottawa, Ontario, Canada

RYAN GRANT

Institute for Resources, Environment and Sustainability, University of British Columbia, Vancouver, British Columbia, Canada

JOOST VAN ULDEN

Geological Survey of Canada, Natural Resources Canada, Vancouver, British Columbia, Canada

SHANNON DENNY

Geological Survey of Canada, Natural Resources Canada, Vancouver, British Columbia, Canada

Abstract

The design of Web-based digital libraries for thematic education, concept exploration, and community decision support represents a fundamental challenge in cybercartography. The incorporation of ideas from knowledge representation, education theory and practice, geomatics, cartography, and hypermedia into a consistent and usable suite of tools represents a challenge to both the conceptual architect and the system engineer. The conceptual architecture is hampered by the radically different interface metaphors each of these component approaches embody. The technical architecture is hampered by contrasting and often incompatible system architecture requirements for operation in a Web browser, and by the emphasis on ephemeral trends and approaches in the Web design community. A workable compromise requires balancing the two sets of requirements in a flexible way that honors the audience's core needs.

The Georgia Basin Digital Library (GBDL), which supports sustainable development education, community development, and a rich interaction between social and scientific modes of understanding, attempts to accomplish this balance by using interface techniques that directly expose concept semantics and interconcept relationships to users. In GBExplorer, the test implementation for GBDL, cartographic and hypermedia tools are driven by a concept–network interface element, although in principle any of the three interface elements can be used for direct engagement. Associated online community support tools provide the ability for users to directly add to the digital library knowledge base at both the information asset and community atlas site-of-interest level. Ongoing research examines two-dimensional versus three-dimensional modes of interaction, supporting technology from artificial intelligence and geomatics, and especially the social impact and role of the digital library.

1. Conception of the Georgia Basin Digital Library (GBDL)

The Georgia Basin Digital Library represents both a stand-alone project – a digital library resource and information-sharing environment for communities in the Vancouver, B.C. area – and also a component of a larger system of tools and set of objectives that significantly predate it. The design, implementation, and subsequent modification of the Georgia Basin Digital Library (GBDL) need to be understood in terms of the context in which these activities took place. The approach to the design of the initial digital library, in fact, was based on methods explicitly developed to envelop the context of use in the act of design. These methods and their application to GBDL are detailed below, followed by a discussion of the actual implementation details, and in turn followed by a discussion of ongoing modifications to the system. While the implementation details of the GBDL are somewhat peculiar to the specific application and target audience, the design framework followed and described herein is useable in a wide range of contexts and is sufficiently general to be applicable to cybercartography problems.

As part of ongoing research and community-involved decision-making efforts addressing issues of long term sustainability for the Vancouver area and surrounding district – collectively, the Georgia Basin physiographic and ecological region – a group of researchers centered at the University of British Columbia's Sustainable Development Research Initiative constructed an integrated back-cast modeling tool (Robinson *et al.* 1996) populated with spatial-temporal data covering the last few decades of the regions growth. The model, called Quest (see Fig. 19.1), allows a user, often led by a professional facilitator, to examine the consequences of policy

FIG. 19.1. Output screen for Lower Fraser Basin Quest, a stand-alone implementation of the Quest engine. Note the integration of maps, graphical displays, and text.

decisions about the governance of the Georgia Basin region and lifestyle choices made by citizens. The Quest model uses the initial choices made by the user or group of users to show possible futures of the region based on rigorous scientific models and the available data; these are typically staged as 10, 20, 30, and 40 year in-the-future views and are reported back in the form of charts, graphs, maps, and simple text descriptions. The Quest system, generalized to urban/rural zones, has been commercialized by Envision, Inc. and applied to a number of other urban centers as well.

The emphasis on modeling, on the importance of social context, and on the involvement of user groups described herein are conceptually similar to the approaches taken by CentroGeo in Mexico reported earlier in this volume in the chapters by Reyes and Martinez, although the implementation is different (Fig. 19.1).

Given the number of possible initial choices in policy and lifestyle available to the users of Quest, and given the number of technical results that the system might then report, a critical problem with Quest emerged early in its use. Without a skilled facilitator, of whom there are very few available, the system is sufficiently complex and intimidating that the desired result – some synergistic understanding of how complex decision making affects the future of a region – is unlikely. Users of Quest, for the most part, are overwhelmed by the number of policy issues and number of outcomes possible. Choosing to have very few policy decisions, in the form of general pre-sets, alternatively, tends to remove the entire point of the system, which is support for the fluid exploration of a range of ideas and outcomes. The overwhelming problem is one of facilitation. There are few skilled facilitators, and this dramatically limits the reach of the tool. Furthermore, the Quest tool should in principle be useable on the World Wide Web, and hence, capable of reaching a very wide audience, but human facilitation on the Web is technically difficult and logistically impossible for the Quest team to achieve.

The GBDL was conceived as an accessory tool to be used at a standalone kiosk or on a Web site, in conjunction with the Quest modeling tool. The original intent was for the Digital Library to act as an extended learning resource adjunct to Quest, such that users confronted with new terms or new concepts in Quest could choose to shift to GBDL, wherein they would encounter supporting material explaining the concept, giving examples, and encouraging a return to Quest for further concept exploration. The resulting configuration of interfaces, data sources, and flows is shown in Fig. 19.2.

The instrumental purpose of GBDL, in other words, was originally to support users, who have become blocked in their use of Quest. The human context of GBDL was the range of possible users, who might become

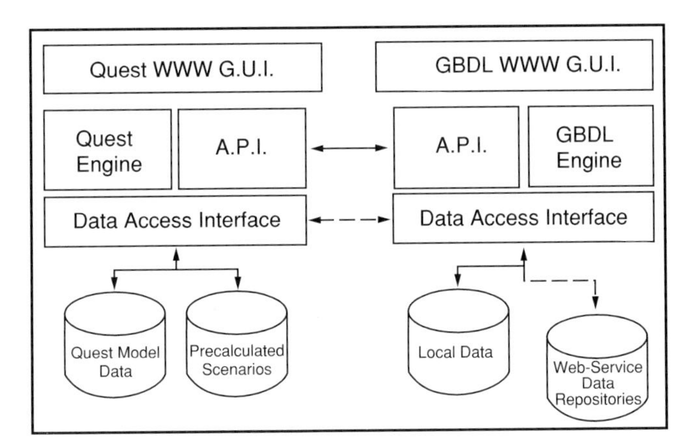

FIG. 19.2. Interfaces, Engines, and Application Programming Interfaces (API) in the envisaged GBDL-Quest toolset.

blocked in this way. The tool interoperability context of GBDL was how the different tools – GBDL and Quest – might exchange information about user interaction events and share underlying data and knowledge resources in order to better address the user's needs.

When the design team initially began scoping the requirements for GBDL, these three issues were at the forefront. The interoperability issue seemed, at the time, to be easily resolved by collaborating with the team members from Quest. The instrumental needs issue, and the scope of all possible users, on the other hand, seemed daunting. As a result, the design team turned to two design techniques well-known in software development circles but less well-known in geomatics, cartographic, and Web design circles – persona-based design and scenario-based design.

2. Design Process for Georgia Basin Digital Library

The design of GBDL is based around a series of metaphors that capture the essence of the design process in different ways. By combining a series of metaphors that have different strengths and weaknesses, the overall design process is guided and strengthened. The key to this layering of metaphors is that each of them captures a different perspective on the purpose of the system, on the corresponding interface elements and affordances, and on the requisite underlying architecture (Fig. 19.3). While software developers may be expert at such architectures, they are typically less adept at conceptualizing the interface layer and may be completely unaware of the user's intents in accessing the system. Domain experts, knowledgeable about the task that the system users are engaged in at a technical level, are often unaware of the technical issues in implementing a tool, and more importantly may be naïve in their conceptualization of what the user is actually doing (Fig. 19.3).

The designers of the GBDL included domain experts, knowledgeable in both digital library techniques and the scope of the intended library vis-à-vis the Quest model, and software experts, charged with making a functional and useable tool. Early design sessions for the GBDL revealed that the team had a simplistic understanding of the user's needs, skills, and desires at best. The metaphor of scenario-based design (Carroll 2000) and the related metaphor of persona-based design (Cooper and Reimann 2003) were used to fill this design gap.

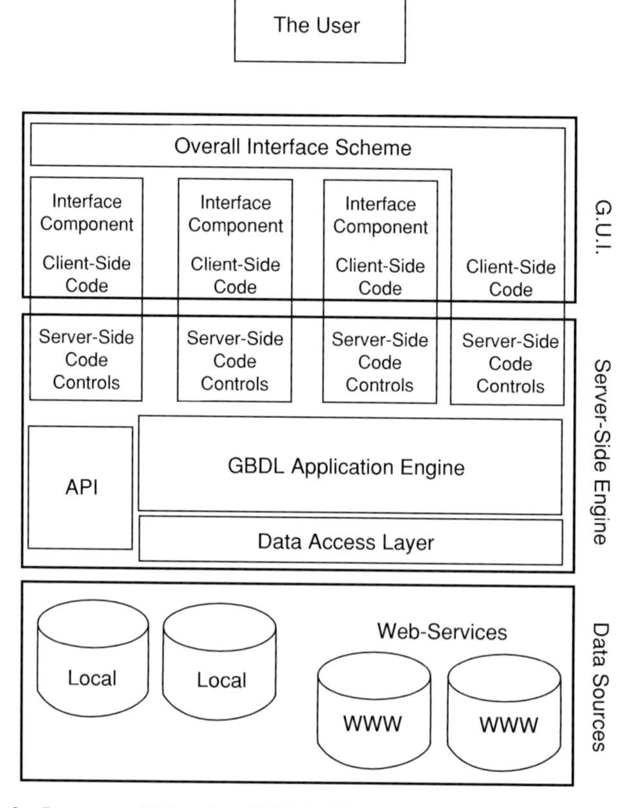

FIG. 19.3. Layers within the GBDL. Note that the user sees the overall interface scheme and subordinate interface components, while the application programmer is fundamentally concerned with client- and server-side codes, application engines, and data abstraction.

2.1. Scenario-Based Design

The basic tenet of scenario-based design is that design should begin with a human-centric telling of a situation in which a user interacts with a tool to achieve a result. Scenario-based design has been applied to a wide range of software development projects but is relatively unknown in the Digital Library and especially, geomatics communities.

Scenarios, constructed by usability experts as a result of interviews with typical users, are story based rather than being formal structures; they differ from technical use cases in that they attempt to capture the mood and intent of a process rather than the details of how the tool will achieve the desired result. For example, a scenario for a regional environmental digital library might be:

> Sam has been wondering if any environmental fieldwork has been done in his neighborhood of North Vancouver. He has a few minutes to spare before going to class, so he accesses the Digital Library Website. The page loads a little slowly – Sam is using a dial-up connection – but after a couple of minutes, he has a map page of his neighborhood. He searches for field data access tools in the interface, and discovers that he can show field work by individual, by group, or as an entire collection. He tries to load the map of the entire collection, but it is just taking too long on his slow connection. He sees that he can save the process and resume it later, so he does this, intending to restart the process from a faster machine at school.

The scenario captures what the user wants to do, some of the realistic constraints that the user faces, and some ways that a system might alleviate constraints. There are no real technical details here, but a software designer seeing this knows at least some of the core functionality that the library tool will need. More importantly, the system interface designer knows that tool functions will need to be prominent, which ones might be secondary, and why.

By building a suite of such storytelling scenarios, as an ongoing process in collaboration with a wide range of users, a system designer can perform the equivalent of a complete release–test–revise cycle before writing any code. Furthermore, the scenario stories are accessible. Users, seeing a badly designed interface will generally become frustrated and offer relatively little useful feedback simply because they are in fact confused! A suite of scenarios sets the proper stage for development early in the design process.

2.2. Persona-Based Design

While scenarios situate the design process around a particular task faced by a user or user group, it is often useful to understand exactly what predispositions users bring to the task of using a system. In other words, we can design for a user who is focusing on the task and has no historical predispositions about tools, or we can try to capture a more detailed view of a user's needs, skills, and behavioral styles. The process of developing a suite of user profiles, whether from interviews of real users or by design team collaboration, was pioneered by Cooper (Cooper and Reimann 2003); the user profiles are called personas.

An example of a persona for the GBDL is Jennie, a high school student assigned to use GBDL and Quest to investigate the impact of large box stores on the urban environment.

> Jennie is a high school student, in Grade 11. She is currently taking a regional economic geography module in her High School Geography course. She isn't academically oriented: her favorite subjects in school are Drama and English, because she is thinking of becoming an actress. She keeps tabs on the local film and television industry and has managed to work as an extra a couple of times, once even in the foreground of a shot. She isn't as obsessed with fashion as most of her friends, mostly because she simply can't afford to compete – her parents are high school dropouts and money is tight around the house. She has decided that, homework or no homework, she's at least going to get a college diploma to fall back on if the acting thing doesn't pan out. She's seen enough waiters and waitresses in downtown Vancouver reading scripts to know that the supply is generous and the demand is light, to use her Economics teacher's way of speaking. One thing Jennie does care about is commuting time: the bus is a drag and the Skytrain doesn't come near her house. She's heard that some communities are working to put shopping and parks throughout a region, which would help her a lot. Jennie comes to the GBDL because she has to work on a project on this topic – urban planning – for her economic geography course and . . .

In the case of a digital library, such as GBDL that must both serve as a stand-alone tool and also interface with another tool such as Quest, the range of personas that capture the realistic range of users will be quite large and have dramatically different skills, knowledge, desires, and interaction styles. The process of capturing these divergent users at first appears to be an exercise in futility: is it conceivable to build a tool that will in fact address the needs of a highly divergent audience? Though this may, in fact, be an insurmountable design problem, the persona process is vital in that it demonstrates this difficulty before any actual development takes place, allowing multiple tools to be designed as necessary, or changes to the project scope to be made in a proactive rather than reactive manner. For each of these tools, then, the corresponding suite of personas serves to keep the designers designing for a specific audience.

2.3. *Integrating Personas and Scenarios*

With the personas taking the part of synthetic or aggregated "actors" and the scenarios describing representative situations that arise around use of a system, a situated design process can be started. In the case of the GBDL, a white paper (Harrap *et al.* 2001) was used early in the design process and framed the dialogue that followed. One key innovation in the GBDL, the use of a design metaphor called Pattern Language Theory (Alexander *et al.* 1977) was framed in the white paper and situated in design scenarios and in terms of real personas, emerged in the initial implementation of GBDL as a working system. Ironically, a *Pattern Language for Web Site Design* has since been published (Van Duyne *et al.* 2003): the GBDL design can thus, be regarded as using pattern languages in both design and in implementation. Pattern languages have since emerged as a major force in software design and the field of "Web Cartography" is ripe for a fully fleshed out pattern language to be developed.

2.4. Pattern Languages and Design

The core idea of pattern languages is that some design choices are both pivotal and recurrent in many situations, and so can be treated as the elements of a grammar of design. Patterns are both situated in a culture and to some extent transcendent of culture: the original *Pattern Language* (Alexander *et al.* 1977) was a cross-cultural study of the folk-utility of architecture, with patterns ranked by their universality in world cultures. Patterns are published in a manner that is structured and so amenable to formal use in a knowledge representation system, and yet with a combination of imagery and narrative style that allows a highly personal expressiveness to emerge from a pattern collection.

The overall structure of an individual pattern is:

(1) Title, with a ranking of universality (not, somewhat, universal);
(2) Illustration, usually a photo, but sometimes a map or figure;
(3) Things that would lead you to this pattern;
(4) Rhetorical statement of the problem;
(5) Discussion, often illustrated with maps, figures, charts, and photos;
(6) Proposed solution statement; and
(7) Things that could follow from this pattern.

The pattern titles are numbered, with the original language for urban planning and architecture having about 250. Items 3 and 7 in a pattern use this numbering system as references to interrelate the sections. Each section is 3–5 pages long. The result is that the original book, published in 1977, is one of the first examples of a large and effective hypertext, long before the World Wide Web.

Since 1977, a number of works both on the theory of pattern languages (Alexander *et al.* 1977; Alexander 1979) and on applications of this approach to computer science (Coplien and Schmidt 1995; Buschmann *et al.* 1996; Larman 2002), user interface design (Borchers 2001), and Web site design (Van Duyne *et al.* 2003) have been developed.

In the case of the GBDL, an early design objective that arose from the persona and scenario driven process was to use a pattern language (see Fig. 19.4) for sustainable development and natural hazards as a framing mechanism for learning about these issues in the site. Since the individual patterns are both formal and rich in narrative, they represent a knowledge representation framework suitable for educational purposes but supportive of assisted knowledge discovery tools developed in a rich Web environment. Web-based research revealed that a pattern language for sustainable development already existed (Ecotrust 2000) and a collaborative design effort was established with partners from both the GBDL team and the Ecotrust team (Fig. 19.4).

In the original *Pattern Language* the layout of the book is scoped from the very large scale – national to international – to increasingly local scale as the book "progresses". The final patterns deal with individual elements of pieces of furniture. This scale-dependent ranking is both intuitive and natural in book form and for dealing with urban planning. As the GBDL design team considered the relevant metaphor for using sustainability patterns on a Web site, it became clear that some similar framing mechanism would be needed or the personas we envisaged as representative of the user community would clearly not find the site intuitive. This is, in fact, the power of scenarios and personas, that during design sessions people would argue as if these are real people with real concerns and even personality quirks. The debate centered around whether the patterns should be part of the underlying architecture of the system or also part of the visible knowledge representation – in the *Pattern Language* book, the user knows the patterns are there and the point of the system as a language is to encourage

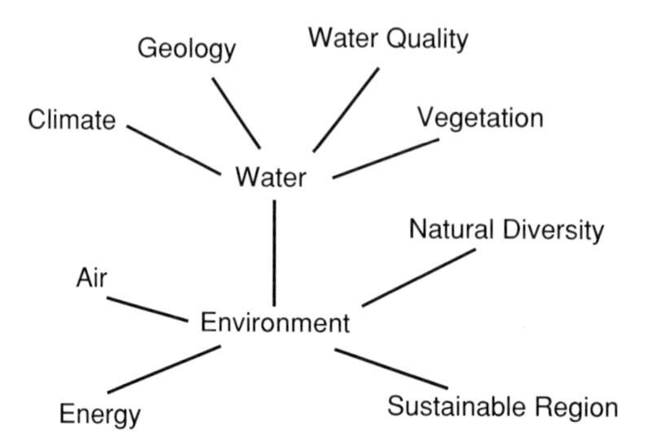

FIG. 19.4. Fragment of a Sustainability Pattern Language Network.

users to capture the system into their personal approach to solving problems. In a Digital Library, for sustainability intended to support the use of a modeling tool, like Quest, making the language visible, seemed appropriate for similar reasons, and yet a Web site lacks the physical affordances of a book to capture the essence of the overall language and make the user understand the relevance of the entire holistic system.

Information architecture (Wurman 2000; Shedroff 2001) aims to resolve these types of issues by combining visual representations of material with rigorous empirical studies of how these representations work in practice. One method that has evolved since the mid 1990s for Web site information architecture, is the use of network diagrams to represent connected concepts, an outgrowth of the use of such diagrams in artificial intelligence, cartographic concept mapping, and graphic design. In the case of a pattern language for sustainability being made visible, linking the relationships between patterns that the original text performed by numbering schemes by hyperlinks instead, as a prototype developed by Ecotrust (2000) did, and furthermore showing the relationships as a dynamic diagram that adapted to the user in a fluid way seemed like an appropriate metaphor for imparting the structure of the pattern collection as a whole. Visual representation of knowledge of this type have been used by Web developers to show museum collections (Smithsonian 2000), related words in a thesaurus (Plum Design 2000), and Web site relationships (Touchgraph 2004).

This approach to blending an understanding of the content of a pattern with the architecture of the pattern language as a whole, by mixing visual representations of content with text and other media, was an early design decision that fits the overall scenario and arguably would be appealing to the personas developed. A full usability study using a trial interface was not carried out; instead a trial implementation of a pattern language browser with GIS linkages was carried out (Grant 2001).

One significant concern with the approach taken was that the design might be overly complex for novice users, expecting them to think in detail about their ongoing interactions. This violates one of the main design points of Web sites – users do not in general want to think (Krug 2000)! Testing this concern became a priority of the research group.

2.5. Linkages to Related Projects

Another significant issue with the process of design for the GBDL was the hope that some of the interface and Web engine components of the system would be reusable in a related project. This project, a design and implementation of a kiosk for the Science World museum in Vancouver, British

Columbia, was paired to a theatre-based multiuser version of the Quest modeling tool. As with the Web-based GBDL, the purpose of the kiosk was to support user knowledge about regional sustainability issues both before and after Quest theatre sessions. In addition, the kiosk was intended to be used by children; the scenarios and personas for this design were thus, significantly different. In the end, an implementation using a helicopter body, animation-based interfaces, and photographic content was chosen: the attempt to integrate the projects at a technical level was limited to reuse of underlying spatial data. In retrospect, the personas and scenarios for this secondary project are sufficiently different from those of GBDL that any attempt to strongly link development should have been discouraged. Scenarios and personas are thus, useful not only to suggest what to do, but also to suggest what not to do.

2.6. Design Process Summary

The process described above notably does not involve any traditional software design or even requirements formation, until much of the conceptual design is finished. This reflects the emphasis on user-centered story-based design. In summary, the steps taken in this process are:

(1) Idea generation;
(2) Development of persona and scenario sets;
(3) Open discussion of scenarios, with role-playing of personas; participation of the user community is ideal;
(4) Background research, including inspection of the strengths, weaknesses, opportunities, and threat of similar project approaches (SWOT analysis);
(5) Revision of the core scenarios to reflect discussion and background research. Preparation of a design document centered around realistic use cases;
(6) Prototyping of the first deliverable on paper or with simple implementations;
(7) Open discussion of the core scenarios, design document, and prototypes;
(8) Implementation of the full system as a working prototype;
(9) Over time, development of a refinement model, involving recasting of personas, rewriting of scenarios; and
(10) Revision of the full system.

This approach, and notably the circular refinement method in items 8–10, has strong similarities to the Agile or Extreme (Cockburn 2001) programming style that arose from within the Open Source software movement. However, unlike most Agile software development approaches, the persona and scenario-based approach described here involves substantial development before code is written. The Agile process emphasizes on the development of prototypes, and corresponds to the following steps 6–10 above without the essential, context-setting exercises in steps 1–5.

3. Implementation

The design of the Gorgi Basin Digital Library followed the steps outlined above. The first five steps resulted in the creation of a design white paper (Harrap *et al.* 2001) that was accompanied by prototyping of portions of the system (Grant 2001) before a full implementation was attempted. Feedback from open discussions of the design document and prototypes (step 7) resulted in revised plan for action, and in 2000–2001 a full implementation of the Georgia Basin Digital Library was carried out by staff at Natural Resources Canada.

Notably, background research and brainstorming outside of the persona- and scenario-process described resulted in additional components being included in the first full implementation. The results of this parallel development are discussed in detail later. The components that originated outside of the process are identified below.

3.1. Overall Environment

The original implementation of the Georgia Basin Digital Library, called GBExplorer, was constructed as a Web-based tool with six core modules. The underlying architecture is discussed in some detail below, but in essence GBExplorer presents content as pages akin to those in a tabloid newspaper using dynamic HTML code and visual interface elements generated by an underlying Web engine using technologies, such as Web Services, Java applets, a database engine, and a Web-mapping toolkit. The pages are for the most part created in response to user interactions – they are dynamic rather than static – and this means that the user interface that fulfills the needs of a persona interacting along the arc of a scenario emerges from the interactions in a fluid way rather than being entirely prescripted. This has the advantage of affording flexibility to the system, but at the cost of requiring significant Web engineering efforts to build the first working system. GBExplorer was constructed out of reusable parts, on the other hand, so efforts to build follow-up systems were less onerous.

3.2. Modules

The six modules of GBExplorer are:

(1) A newspaper like "Front Page".
(2) Ideas and Perspectives, a semantic network and pattern language based concept exploration interface.
(3) Local Stories, a Web-mapping interface that allows community members to map assets relevant to them.
(4) News and Information, a Web news filter service that uses keywords to build custom newspaper-like summaries relevant to the GBDL community.
(5) Library Collections, a metadata-based spatial data discovery tool.
(6) Future Scenarios, a link to the online version of Quest (not active).

Of these, the Front Page, Ideas and Perspectives, and Future Scenarios were in the original design scenarios, while Local Stories, News and Information, and Library Collections were added during development in response to ongoing discussions with the community and to ongoing research about effective Web-based information architecture.

3.3. Front Page

The entry point for using GBDL (See Fig. 19.5) is a tabloid-newspaper styled home page typical of many current Web sites. A multicolumn format (Fig. 19.5) and the use of color and font hints divide the conceptual page, and the familiar metaphor lets novice users make immediate sense of the arrangement. A top navigation bar links to other components of GBDL: this style of navigation structure is common on the Web (Nielson 1999) and most users will find it self-obvious (Fig. 19.5).

Georgia Basin Digital Library makes extensive use of Web Service feeds, drawing content from a range of other sites. In addition, the page that the user sees is dynamically generated by a Web engine rather than being precisely authored. This allows content to be fluidly altered, and in addition allows the use of timed content that is scheduled to appear in the future, allowing the page to appear to be more commonly updated than it is. In principle, this design also allows the site to be modified to fit a user profile in the case of a returning user.

3.4. Ideas and Perspectives

As discussed above, one of the fundamental roles for the GBDL is to provide a novice user with the ability to learn about issues before they

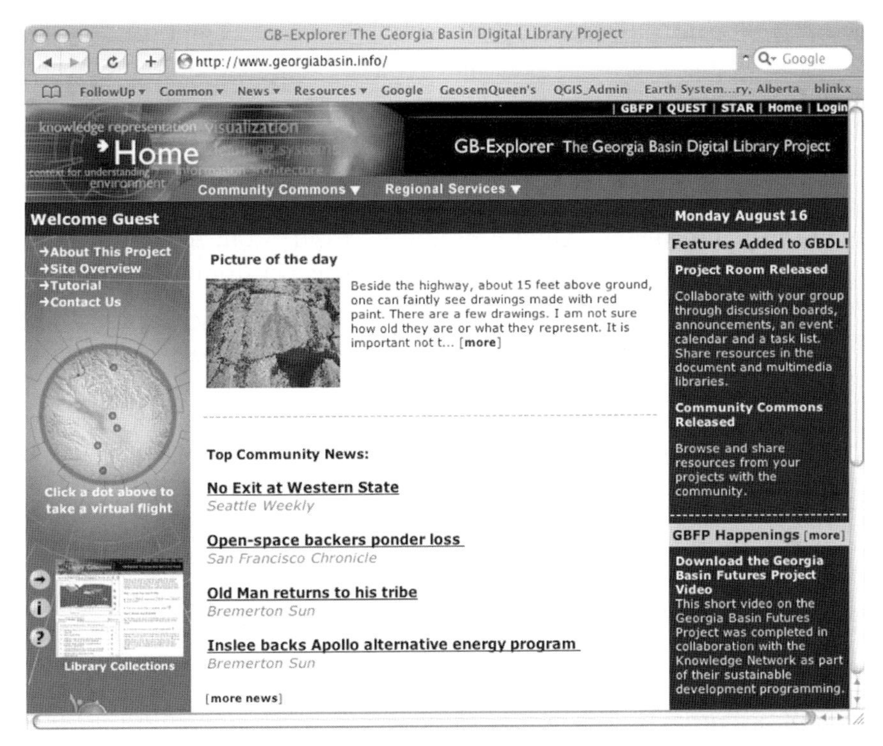

FIG. 19.5. Front page of the GBDL site. Note the similarity to tabloid newspapers.

know what they are called. Through the use of navigation metaphor – the diagrammatic semantic network construct – and a rich and detailed content structure – the pattern language – the Ideas and Perspectives module of GBDL accomplishes this. Ideas and Perspectives (Fig. 19.6) contains three thematic elements: the semantic network, a dynamic diagram that adapts to user interaction (based on the ThinkMap Web programming toolkit, see Thinkmap 2004), a hypermedia display area for patterns, and a map display (based on the Open Source MapServer toolkit, see MapServer 2004).

Interaction in Ideas and Perspectives is driven by the Thinkmap semantic network. Choosing a new node linked to the current one shifts the diagram focus, revealing new nodes and new opportunities for browsing. In most cases, the networks become increasingly detailed as one "clicks in". In principle, multiple types of links, perhaps color coded, could be implemented, perhaps showing alternative interpretations or cross-linking the networks. The browser supports zooming back to the starting node and backing up.

Clicking on a new node causes the pattern language fragment authored to correspond to that node to load in the hypermedia browser. Images, video, sound files, text, and figures are loaded, and if necessary a scroll bar is generated. Many patterns are long enough to require significant scrolling, but the scroll bar concept is familiar to most users as it is ubiquitous in desktop software.

The pattern language fragment has an embedded instruction string, built using a custom-purpose programming language that allows a script to assume control of the other elements of the interface, which loads a custom map in the map browser, symbolizes it as necessary, and zooms on a target area as desired. Using this, an author can choose a custom map to illustrate the point made in a pattern, highlighting our geospatial focus in GBDL's design and construction.

Ideas and Perspectives (see Fig. 19.6) is capable of not only showing a single network–pattern–map set, but of hosting several. In the case of the GBDL, we chose to support both a local network and also the *Ecotrust Sustainability Pattern Language* described above, to illustrate that multiple perspectives can easily coexist in the system (Fig. 19.6).

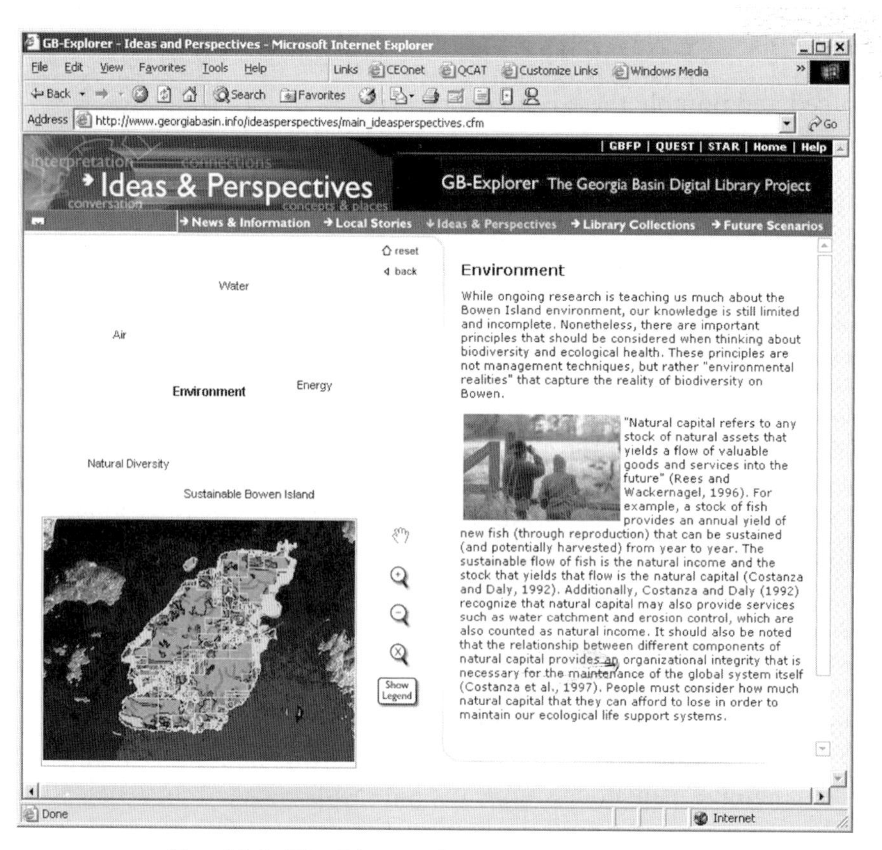

FIG. 19.6. The Ideas and Perspectives interface.

Interaction in *Ideas and Perspectives* is entirely driven by the user. The user can click to more detailed topics or stay at a simple, overview level as desired. In principle, this provides a user-controlled level of complexity that allows GBDL to scale from novices to expert users fluidly. Alternatively, multiple parallel collections can be used to reach significantly different audiences; in principle, some content could be shared, acting as boundary artifacts between multiple ways of seeing the world. This issue is discussed in detail in the conclusions.

3.5. Local Stories

The Local Stories module of GBDL was added during the implementation stage, and was not scoped in the original persona and scenario process. It consists of a map- and form-based interface for community mapping. Groups within a community collectively assume the responsibility to manage a collection of geographically coded notes; these are, when vetted by the community, available for all to see. The overall format is broadly similar to the pattern language style used in *Ideas and Information*, but the style is not enforced and notably there is no overarching network of content elements. The principal access method for notes is via the map interface: selecting a map element brings up the associated notes.

Given the range of possible note topics we decided to implement an icon-based coding system, wherein an author of a new item in Local Stories would pick a representative icon that would thereafter appear on the map. The icons used are from a collection of sustainability mapping symbols called the Green Map system (Brawer 1995), which is familiar to at least some of the community groups that the module was intended to serve. The system includes 125 icons covering a wide range of site types, including museums, cultural sites, markets, spiritual sites, and so on. A hypermedia help system associated with the map allows users to browse the library of symbols. The symbols provide a simple search interface, since they can be turned on and off, effectively highlighting certain classes of features.

Geospatial referencing in Local Stories (see Fig. 19.7) is relative: when creating a new piece of content the author clicks on the map display to site the story. Since the user can zoom to the required level of detail, this allows reasonable resolution providing the map layer used as a backdrop has sufficient resolution. The stories themselves are entered into a simple form and added to an underlying database. They are not indexed or cross-referenced in any way, and in fact, cannot refer to each other except indirectly (Fig. 19.7).

FIG. 19.7. The Local Stories viewing interface. Groups are selected in the bottom left, and the Green Mapping System icons provide a simple mechanism for limiting the display contents.

Local Stories provides a critical element to the GBDL: user community-driven content. Within the bounds of the community vetting process, users can add content that takes the overall Library in directions we did not even consider. The result, and the implications for the use and subsequent redesign of the system, are discussed below.

3.6. News and Information

Like Local Stories, Ideas and Information was added during development. It consists of a simple information access page that is built via hidden queries to Web Service content providers on the World Wide Web.

The queries in Ideas and Information are grouped by scope and by topic. The scope is chosen via a tabbed interface (see Fig. 19.8) and includes international, national, and local levels of interest. The topic is chosen from lists built by the system architects to reflect the sustainable development focus of GBDL. Finally, the user can choose to look for News providers or alternatively, for Web sites (Fig. 19.8).

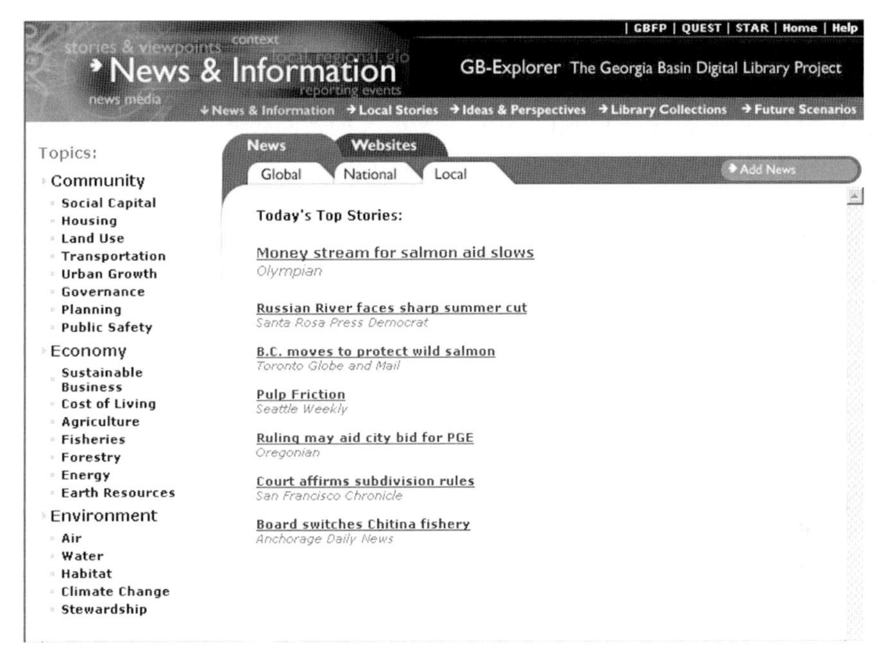

FIG. 19.8. The News and Information Interface. The content is dynamically generated using prebuilt queries to Web Service information providers.

News and Information builds out the content of the newspaper analogy for GBDL. It provides a vital link between the geospatial, focused content of GBDL, and the wider world of communicated information. By using online news providers rather than locally constructed content, it does this without creating the need for substantial and ongoing content creation by the site authors.

3.7. Library Collections

The GBDL components, discussed till this point, are addressed to relatively unsophisticated users interested in sustainable development, natural hazards, and urban growth in the Georgia Basin region. In fact, the users are considered to be unaware of technical terms, formal methods, and mapping techniques to the point that Ideas and Information specifically guides them to learn these. Library Collections, unlike the other modules, is not for this audience, but for more sophisticated users and system content authors who wish to find spatial data on the World Wide Web, accessible via Web Services-enabled Web engines. Like Local Stories, Library Collections was added during development; unlike Local Stories, it was as much for the creators of the system as for the envisioned users, who were captured in the personas generated in the early design stage.

Library Collections allows a user to browse metadata that identifies spatial data resources. It allows spatially-based, source-based, and metadata keyword-based queries to be performed, and returns results as maps and lists as needed. When a query return list is displayed, items can be selected and the entire metadata entry for the item is displayed in turn (see Fig. 19.9).

Library Collections is capable of finding resources from across the Web because these resources adhere to a uniform standard for metadata and information retrieval. In Canada resources of this type are championed by the Canadian Geospatial Data Infrastructure (CGDI) program, while in the US they are championed by the Federal Geographic Data Committee (FGDC). These organizations have created standards for metadata, provide access to framework mapping data for North America, and are working to standardize Web-based information access via Web Services.

3.8. Architecture

Like most dynamic web sites, GBDL consists of a mix of static content, code, and database content. The key components of this approach are

FIG. 19.9. Library Collections, a more advanced interface for information discovery from within GBDL.

briefly summarized here; this is not intended to be a technical review of the implementation of GBDL or similar systems.

The GBDL makes use of code that, interacting with a Web server, creates the overall structure of the pages the user sees. This includes the menus, screen layout, and most aspects of the interaction the user can undertake. The page the user actually sees is thus, created on the fly using the rules in this code: the code retrieves content from a database and generates a page that is sent to the client's browser. A key and often overlooked issue here is that not only is the content of the page generated, the form of the page is as well. In principle, a page design could be a result of a dynamic interaction between the user, their use history, and GBDL. Although this is a very interesting area for research, the GBDL team is just beginning to scratch the surface of such dynamic interface approaches.

Much of the code in GBDL acts to call for content from databases, Microsoft SQL-Server in the case of GBDL as it now exists. Since database calls are somewhat generic, using the standard SQL language as a basis, any database with the capacity to interact with a Web server would do as well, and the programming team has experimented with other databases such as My-SQL and POSTGRES.

The Web server handles the actual transaction with the user, taking the code- and database-generated page and handing it off to the user's Web client. The GBDL is standardized on Microsoft Internet Explorer 5, though it works for most modern browsers.

The GBDL makes limited use of Java, a programming language that allows tools to build, which execute on the users computer rather than on the home server: Java applets are called client-side because they operate autonomously once downloaded, whereas transactions that must be handled by the main server are called server-side. The Thinkmap applet in GBDL is an example of a Java applet that, nevertheless, is capable of requesting server actions; the computer graphics techniques that display the network of connections, on the other hand, are handling client-side.

The GBDL makes use of Web Services (OpenGIS 2004), a relatively new approach for allowing code and Web servers in particular to make formal requests for information from other servers: these are much like the Web servers that generate the pages a user sees, but in the case of Web Services the transaction is in the background and the interface is between software systems. Web Services standards provide for resource discovery, content acquisition, and support the requisite metadata to make these operate. A webserver approach is also being used in the Cybercartographic Atlas of Antarctica and the Cybercartographic Atlas of Canada's Trade with the World described in this volume (Pulsifer *et al.* in Chapter 20; Eddy and Taylor in Chapter 22).

Underlying GBDL is a data model that provides conceptual organization to the database content. Much of the database content consists of the suite of concepts used internally by GBDL and externally by users, for example, the semantic network in *Ideas and Perspectives*. The data model also links concepts to spatial objects to enable thematic indexing of maps and their components. The database thus, acts as a concept repository and management system (i.e. an ontology system), whereas the interface acts as a client to the concept services provided by the database. In effect, GBDL is ontology-driven. This provides a consistent storage framework and moreover allows for reasoning about relationships between geographic datasets at both the data and concept level. It supports interoperability between divergent user data standards and especially, classification systems. It is also capable of storing both a geographic data set and the classification taxonomies that the user employed in constructing the dataset, and so comparisons between differently classified datasets are possible. Discussion of the application of the data model to GBDL can be found in Brodaric *et al.* 2003.

3.9. Summary of Implementation

As discussed above, the original implementation of GBDL included components that were in the scenario- and persona-scoped design, and elements that were added during the early implementation stage upon consultation with early users. The planned tools included the entry page, the Ideas and Perspectives discovery page, and the linkages to the Quest system. Added tools included the News and Information tool, the Library Collection tool, and most importantly, the Local Stories tool. The modular design of the underlying code made adding new components relatively simple. This raises the question, discussed below, of whether stepping outside of the scenario- and persona-process is a good thing.

3.10. Results

The original implementation of GBDL, as shown in the screenshots above, went online in 2001. The system was shown at conferences, in talks with community groups, and received attention in the online media. Users began to visit the site and make use of the tools. Most importantly, they began to provide feedback to the designers. A PhD thesis by one of the authors (Talwar) is in progress to examine the role of GBDL in community-based decision making. Some simple conclusions from the early user feedback follow.

First, users were drawn to the Local Stories component of the Georgia Basin Digital Library. User groups began constructing local, thematic content in Local Stories as soon as the tool was made available. The content ranged from nature observations to personal recollections, and there were requests for more functionality in this component almost immediately.

Second, community users tended to avoid using the Ideas and Perspectives module, but this module received the bulk of attention in academic and software circles. The relationship between pattern languages, semantic network idea navigation, and narrative tools in particular was of great interest in the Geographic Information Science (GIS) community.

Third, there was an immediate perceived need for a group-based tool, wherein a personalized version of GBDL could be used by a community with little outside intervention or even access. This was alternatively conceptualized as being redeployments of the entire system on private servers, or the creation of semiprivate or private components within the main server environment. This is discussed in detail below.

Finally, the planned relationship to the Quest project, which required significant codevelopment of interfaces between Quest and GBDL, did not fully materialize as the effort to make Quest available online was postponed. GBDL, as delivered supports the Quest approach to understand community development through time, but Quest is presently available only through sessions facilitated by Envision Sustainability Tools Inc.

4. Repair

In a continually evolving tool environment, the redesign and redeployment of a toolset to address feedback from the community represents repair of the tool (Brandt 1994). In the case of a scenario- and persona-driven process, this involves several forms of repair:

(1) Repair of the concept of the user (i.e. of the personas, and how representative they are).
(2) Repair of the content of the scenarios (i.e. of what we conceive of users as wanting to accomplish).
(3) Repair of the interface of the tool toward original flaws or the changed conception of the user and the scenarios.
(4) Repair of the underlying architecture of the tool to support new interface architectures, to provide sufficient response time, and to match changing technologies.

As noted above, we realized early in the repair process that there are really three divergent communities using GBDL: local communities, who focus on Local Stories; groups, who were interested in a project- and group-based approach; and the academic designers of the tool, who were interested in testing out theories of design! The repair process for GBDL needed to recognize at least the first two of these groups explicitly.

In response to feedback from the community, the team worked to improve the usability of the Local Stories module. This included the introduction of a cover-page that allowed entry to different geographic locales with different community interests being represented.

In response to the need for project-based tools, a separate project on geological data management and field data integration was carried out using a broadly similar architecture to GBDL. The Geosemantica project (Geosemantica 2004) involved the construction of project-rooms, wherein a team could build shared documents, maps, and other resources and to which access was controlled by a password system. Geosemantica also incorporates significant field data and data management tools, and is multilingual. It is the focus of a multicountry effort to share geological data between Andean nations, funded by the Canadian International Development Agency (CIDA).

Finally, with the repair of GBDL itself, the underlying architecture was modified to make use of open-source tools wherever possible. This simultaneously gave the development group access to source code for tools and allowed for the redeployment cost of the entire system to be significantly reduced. An open-source approach is also being used by the Cybercartography and the New Economy Project reported earlier in this volume (Pulsifer and Taylor in Chapter 7).

5. Second Stage Repair

The repair of GBDL and the realization of the diversity of the user communities led to the development of a broader follow-up project, internal to Natural Resources Canada. The *Pathways Project* (http://sdki.nrcan.gc.ca/p_path_e.cfm) focuses on the development of decision support science and technology, to help bridge the gap between community policy and decision-makers, such as planners and emergency responders, and natural scientists modeling the environment. Scientifically, *Pathways* is involved in the development of assessments related to water quality and quantity, and geohazard risk (i.e. earthquakes and landslides) to natural and human infrastructure. Technologically, *Pathways* is concerned with (1) web-based interoperability of data and software tools; (2) customizable web interfaces to geospatial and nongeospatial data and related software tools; and (3) development of "what-if" scenario modeling (integrated assessment) tools to aid community decision making.

Pathways explicitly recognizes the need for tools to support technical users who, though GBDL and Geosemantica did endeavor to support them, fell well outside the persona- and scenario-scope of the original GBDL development. The *Pathways* development process also recognizes that building tools for the construction of digital libraries, rather than building specific instances of digital libraries, is a necessary goal to move the overall effort of fluidly supporting diverse users in a digital library forward. The *Pathways Project* runs from 2003 to 2006.

6. Lessons Learned

The design and implementation of the Georgia Basin Digital Library, as described above, led to a series of realizations about both the process of carrying out such projects and also about the specifics of digital libraries in particular.

First, the use of scenarios and personas focused the design process on pragmatic deliverables for a real audience. The repair process, where user feedback reframed some of the questions, led to a real evolution of the tool and of our appreciation of who our audience is. The realization that there are in fact several distinct audiences with discontinuous needs was accentuated by the persona process, which leads designers to constantly frame their work in terms of candidate people rather than check list items.

Second, the design of effective spatial digital library interfaces is significantly impacted by the general lack of map handling skills in the user community. While advanced users gave feedback that suggested the development of an online GIS analytical tool, beginner users saw a map interface as a backdrop that allowed them to situate themselves in a community, and nothing more. In particular, the attempt to build an idea exploration tool was largely lost on the intended users, but appreciated by advanced users familiar with abstract representation methods such as symbolic diagrams.

Third, at least some of the difficulties with the first iteration of GBDL resulted from applying the framework inconsistently: some components were designed within the persona- and scenario-framework and some were not. In particular, the implementation of the Local Stories module as a separate entity from Ideas and Perspectives led to the latter being regarded as a static tool with little community content: staying focused on user stories would have led us to see these as really one system, sooner.

Finally, a more technical framework for viewing user- and task-based perspectives is provided by ontologies, formal structures – more formal than pattern languages – that organize concepts, tasks, and capabilities and allow computation to be performed on or between these (Berners-Lee *et al.* 2001; Decker 2004). Although the GBDL project developed initial approaches to the use of ontologies to manage the content accessed in a web-based digital library environment, these are now more of a focus in the ongoing *Pathways Project*. Integrating more formal approaches earlier in the project may have been beneficial in terms of supporting the development of more powerful interface elements.

7. Conclusions

The Georgia Basin Digital Library demonstrates how a user-centered digital library can be designed via software development techniques that address the user – persona-based design – and the task at hand – scenario-based design. These are well-established techniques in the software-design community but have not been documented within the spatial digital library community until now. These approaches address the common dilemma that software engineers, who must build the system, have little appreciation of what the system is for. As seen in the case study presented above, large projects tend to take on a life of their own, and a composite of formal, informal, and reactive design methods were in the end incorporated into the process for the GBDL. Although this might seem less-than-ideal from an academic perspective, from a pragmatic perspective this is inevitable as a study of the software-design literature reveals (Hunt and Thomas 2000). Cybercartography as a discipline requires software constructs; the case of the GBDL illustrates approaches and limitations of formal, semiformal, and informal design for these, and complements the approaches and experiences discussed in other chapters of this volume.

The lessons learned through the development process, including both the successes and failures, strengthen the view that the persona- and scenario-techniques provide valuable focus for the design of a toolset for a real community. While the degree to which the design process was enforced was somewhat mixed, the resulting tool does meet real needs to a great extent and has resulted in a groundswell of calls for the application of the GBDL engine to other communities. This particular approach to cybercartography, in other words, has been accepted by a real community. This is also seen in the Cybercartographic Atlas of Lake Chapala described by Reyes and Martinez earlier in Chapter 6 in this volume.

The ongoing *Pathways* development project continues to develop spatial digital library tools in support of community based spatial decision support, with an emphasis on underlying technology that will enable the use of more sophisticated analytical approaches in future interfaces, and to abstract the particulars of data models away from analytical processes, leading to greater sharing of data, information, and knowledge resources.

References

Alexander, C., S. Ishikawa and M. Silverstein (1977) *A Pattern Language: Towns, Building, Construction*, Oxford University Press.

Alexander, C. (1979) *The Timeless Way of Building*, Oxford University Press.

Berners-Lee, T., J. Hendler and O. Lassila (2001) "The semantic web, a new form of web content that is meaningful to computers will unleash a revolution of new possibilities", *Scientific American*, 284, May pp. 34–43.

Borchers, J. (2001) *A Pattern Approach to Interaction Design*, John Wiley & Sons, New York.

Brandt, S. (1994) *How Buildings Learn: What Happens After They Are Built*, Penguin Books.

Brawer, W. (1995) *The Green Map System,* accessed July 2004 from http://www.greenmap.org

Brodaric, B., M. Journeay, R. Harrap, S. Talwar, J. Van Ulden, R. Grant and S. Denny (2003) *The Architecture of the Georgia Basin Digital Library: Using Geoscientific Knowledge in Sustainable Development*. Geological Survey of Slovenia, Geologija, Vol. 46, No. 2, pp. 343–348.

Buschmann, F., R. Meunier, H. Rohnert, P. Sommerlad and M. Stal (1996) *A System of Patterns: Pattern-Oriented Software Architecture*, John Wiley & Sons, New York.

Carroll, J. M. (2000) *Making Use: Scenario-Based Design of Human-Computer Interactions*, MIT Press, Cambridge, Massachusetts.

Cockburn, A. (2001) *Agile Software Development*, Addison-Wesley, Boston, Massachusetts.

Coplien, J. A. and D. C. Schmidt (eds.) (1995) *Pattern Languages of Program Design,* Addison-Wesley, Boston, Massachusetts.

Cooper, A. and R. M. Reimann (2003) *About Face 2.0: The Essentials of Interaction Design*, John Wiley and Sons, New York.

Decker, S. (2004) *Markup Languages and Ontologies*, accessed July 2004 from http://www.semanticweb.org/knowmarkup.html

Ecotrust (2000) *Conservation Economy Net: A Pattern Language for Sustainability,* accessed July 2004 from http://www.conservationeconomy.net/

Geosemantica (2004) *Geosemantica: Su Ventana a la Geologia de Cordillera*, accessed 7 February, 2004 from http://node.geosemantica.net.

Grant, R. (2001) *Representation of Geological Knowledge: Development of On-Line Tools for Learning with Geographic Information Systems*, BSC. Thesis, Department of Geological Sciences and Geological Engineering, Queen's University.

Harrap, R. M., M. Journeay, S. Talwar and B. Brodaric (2001) *Discovering The Community of Our Future: Decision Support and Digital Library Tools Supporting Informed Community Decision-Making in the Georgia Basin Region, White Paper,* accessed at http://www.gis.queensu.ca

Hunt, A. and D. Thomas (2000) *The Pragmatic Programmer: From Journeyman to Master*, Addison-Wesley.

Krug, S. (2000) *Don't Make Me Think: A Common Senses Approach to Web Usability*, Que Books, Indianapolis, Indiana.

Larman, C. (2002) *Applying UML and Patterns: An Introduction to Object-Oriented Analysis and Design and the Unified Process*, 2nd ed., Prentice Hall, Upper Saddle River, New Jersey.

MapServer (2004) *MapServer Homepage*, accessed June, 2004 from http://mapser ver.gis.umn.edu

Plum Design (2000) *Plum Design, Visual Thesaurus*, accessed July 2004 from http:// www.visualthesaurus.com/index.jsp

Nielson, J. (1999) *Designing Web Usability*, New Riders Press, Berkeley, California.

OpenGIS (2004) *The Open GIS Consortium*, Website, accessed August 2004 from www.opengeospatial.org

Robinson, J. B. and the Sustainable Development Research Institute (1996) *Life in 2030: Exploring a Sustainable Future for Canada*, UBC Press, Vancouver, B.C.

Shedroff, N. (2001) *Experience Design*, New Riders Press, Berkeley, California.

Smithsonian (2000) *Revealing Things Online Exhibit*, accessed June 2004 from http://www.si.edu/revealingthings/

Thinkmap (2004) *Thinkmap Visualization Software Homepage*, accessed June 2004 from http://www.thinkmap.com

Touchgraph (2004) *TouchGraph LLC Website*, accessed July 2004 from http:// www.touchgraph.com

Van Duyne, D. K., J. A. Landay and J. Hong (2003) *The Design of Sites: Patterns, Principles, and Processes for Crafting a Customer-Centred Web Experience*, Addison-Wesley, Boston, Massachusetts.

Wurman, R. S. (2000) *Information Anxiety 2*, Que Books, Indianapolis, Indiana.

CHAPTER 20

The Development of the Cybercartographic Atlas of Antarctica

PETER L. PULSIFER
Geomatics and Cartographic Research Centre (GCRC), Department of Geography and Environmental Studies, Carleton University, Ottawa, Ontario, Canada

AVI PARUSH
Professor of Psychology, Human-Oriented Technology Lab (HOTLab), Department of Psychology, Carleton University, Ottawa, Ontario, Canada

GITTE LINDGAARD
Professor, Natural Sciences and Engineering Research Council (NSERC) Industry Chair User-Centred Design, and Director Human Oriented Technology Lab (HOTLab), Department of Psychology, Carleton University, Ottawa, Canada

D. R. FRASER TAYLOR
Geomatics and Cartographic Research Centre (GCRC) and Distinguished Research Professor in International Affairs, Geography and Environmental Studies, Carleton University, Ottawa, Ontario, Canada

Abstract

This chapter discusses the ongoing development of a Web-based Atlas of the Antarctic region entitled the Cybercartographic Atlas of Antarctica (The Atlas). An overview of the design, development, and ongoing implementation of The Atlas is presented. Central to The Atlas development approach is the extensive analysis of the needs of a general public target user group. Through user needs analysis (UNA), specifications are established and prototypes used to test concepts before expensive development tasks are carried out. Constant feedback between process stages is built into this iterative User-Centred Design (UCD) approach. The UCD results provide specifications for many aspects of cartographic design, including interface elements, usage context, and information architecture. The chapter presents preliminary interface designs built on content supported by a prototype mediator-based system architecture. The concluding sections discuss emerging research challenges and directions in terms of cartographic representation, atlas design, and systems development.

1. Introduction

1.1. Digital Atlas Development

Significant changes have taken place in cartographic theory and production over the past two decades as a result of developments in computer and communications technology (Taylor 2003, Taylor Chapter 1 in this volume). Since the mid 1990s, the Internet has emerged as a key element in transforming the discipline and process of cartography providing a faster method of map distribution in comparison to paper or CD-ROM formats, different forms of mapping and new areas of research (Peterson 1997; Cartwright *et al.* 2000).

A number of new possibilities emerge, including online interactive visualization (Cammack 1999), maps built from distributed data sources, maps updated with near–real-time data precisely georeferenced using satellite positioning systems (Monmonier 2000), and potentially collaborative map use utilizing telepresence.

Until recently, the ability to produce maps for broad public consumption was limited to those with access to the requisite tools (i.e. photogrammetric stereoplotters, large format printers) and data (topographic survey data, census information). With the advent of relatively inexpensive computers and software and easy access to the Internet, many citizens of the world's wealthier nations can now produce and readily distribute maps to a large audience with relative ease. This emerging reality presents new possibilities for different groups in society through activities, such as public participation Geographic Information System (GIS) (Jankowski and Nyerges 2001) and counter mapping (Peluso 1995; Caquard Chapter 12 in this volume). However, these new possibilities do not eliminate the well-established challenges of effectively communicating and exploring geographic information traditionally addressed by cartographers. Indeed, they create new challenges.

As the creation of Internet maps increases (Peterson 2003) and online analysis of geographic information becomes a reality, it is reasonable to expect that maps will increasingly become integrated with other forms of information (Taylor 2003). A broad question emerges: where, in the unbounded domain of "information" are maps situated in terms of conveying data or explaining concepts (Cartwright 1999)? This question is particularly relevant with respect to the creation of online, multimedia atlases. Lobben and Patton (2003) suggest that while maps are still the primary focus in a digital atlas, various supporting media also provide a significant amount of information.

The effective integration of maps and supporting media is a central challenge of the atlas development process, however, there are others including:

- The balance between developing a "closed" atlas, where content and every possible interactivity is programmed by the author, or a more "open" extended database atlas that provides exploratory interactivity (Borchert 1999);
- Integration of, and navigation through, various orders (high to low) of information content; and
- Designing an interface metaphor that enables the user to better access atlas content and functionality (Howard 1999).

Several of the chapters of this book discuss these topics and, the Cybercartography and the New Economy Project (CANE), as described by Taylor in Chapter 1 of this volume, is designed to conduct fundamental research on these and other key issues in cybercartography.

This chapter discusses the ongoing development of a Web-based Atlas of the Antarctic region entitled the Cybercartographic Atlas of Antarctica, henceforth termed The Atlas. The Atlas is an applied research component of the CANE project. The chapter provides an overview of the design, development, and ongoing implementation of The Atlas and concludes with a discussion of emerging research challenges and directions.

1.2. The Cybercartographic Atlas of Antarctica Project

1.2.1. The Antarctic Region

Antarctica is the coldest continent on Earth with the record low temperature of −89.2 °C recorded at Russia's Vostok station. The continent has no permanent inhabitants and is primarily covered by ice. The Antarctic ice sheets cover an area of 14 million sq km that translates to a volume of

30 million cu km. The closest continent to Antarctica, South America, is more than 1000 km away. Considering Antarctica's remote location relative to most of the world and its limited capacity to support human life, one may ask why is Antarctica of relevance? There are a number of reasons. Antarctica is increasingly being seen as a unique laboratory for studying global processes, like climate change. The ecological sensitivity of the poles makes them useful as early warning indicators to detect global environmental change trends and effects. The most obvious relationships are related to the hydrosphere (global ocean currents) and atmosphere (climate change). Less obvious are the lessons we can learn using the Antarctic as a case study of human exploration, resource exploitation, territoriality, resource management, the role of science in society, and developing concepts of a global commons (Berkman 2002).

1.2.2. Project Overview

The Cybercartographic Atlas of Antarctica Project aims to develop an online atlas portraying, exploring, and communicating the complexities of the Antarctic continent for education, policy, and research purposes. Data from a number of international sources are being incorporated into The Atlas. In collaboration with experts from different fields of science, these data are used to develop theme-specific modules. The development of Atlas modules for pedagogical purposes and for use by the general public is an integral part of The Atlas. Learning modules incorporated into The Atlas will provide a thematically based synthesis of information (Baulch *et al.* Chapter 21 in this volume). To develop this component, a partnership has been established with an Ottawa area group of educators, Students on Ice (http://www.studentsonice.com). Researchers are working with staff at Students on Ice to evaluate user needs.

1.2.3. Project History

The concept for a cybercartographic atlas of Antarctica was first suggested by Dr Daniel Vergani (Argentina) following work on a cybercartographic atlas of Latin America (Alviar *et al.* 2001). In collaboration with Dr Vergani, development of The Atlas has been led by The Geomatics and Cartographic Research Centre (GCRC) at Carleton University, Ottawa, Canada. The Canadian Committee for Antarctic Research (CCAR) discussed and approved the project in 1999. The work was then presented to the Scientific Committee on Antarctic Research (SCAR) Working Group on Geodesy and Geographic Information (WG-GGI) meeting in Tokyo, in July 2000. The project was formally adopted by SCAR WG-GGI (presently the Geographic Information Group of Experts) at its meeting in Siena,

Italy in July 2001. A development workshop in Puerto Madryn, Argentina in December 2001 brought together a number of key stakeholders to discuss the conceptualization and initial design phases of the project. A second workshop in Ottawa (May 2002) further developed The Atlas concept and design by identifying specific atlas elements.

1.2.4. Partnerships

It is not the intention of The Atlas project to collect substantive new data, but rather to bring together selected existing datasets in a new multimedia form. Source data are being accessed through partnerships with members of the SCAR community. A variety of data sources are being used. Framework data layers include remote sensing data, such as those collected as part of the Radarsat Antarctic Mapping Project (Jezek 1999). Primary Topographic data is provided by the Antarctic Digital Database project. This vector database, compiled under the direction of researchers at the British Antarctic Survey, is constructed from source maps with scales primarily between 1:100,000 and 1:1,000,000. A number of other databases are being made available from sources such as: The United States Geological Survey's Atlas of Antarctic Research, the King George Island GIS (KGIS), Australian Antarctic Division, Wuhan University (PRC), and many others.

2. Atlas Design

2.1. User-Centred Design

Central to The Atlas project is a move away from development that is focused on the capabilities of the technology – to one that involves an analysis of the user for whom these technologies will be applied, an approach that is discussed in several chapters in this volume. Figure 20.1 illustrates a "waterfall" approach to product development. In this approach, product specifications are established by developers. Testing is usually carried out late in the process after a great deal of investment has been made.

In contrast, Fig. 20.2 illustrates a user-centered design process, where users are identified and involved from the start of a project. Through user needs analysis (UNA), user interface requirements are established and prototypes used to test concepts before expensive development tasks are carried out. Constant feedback between process stages is built into the process. The result is a style guide or design specification that is based on

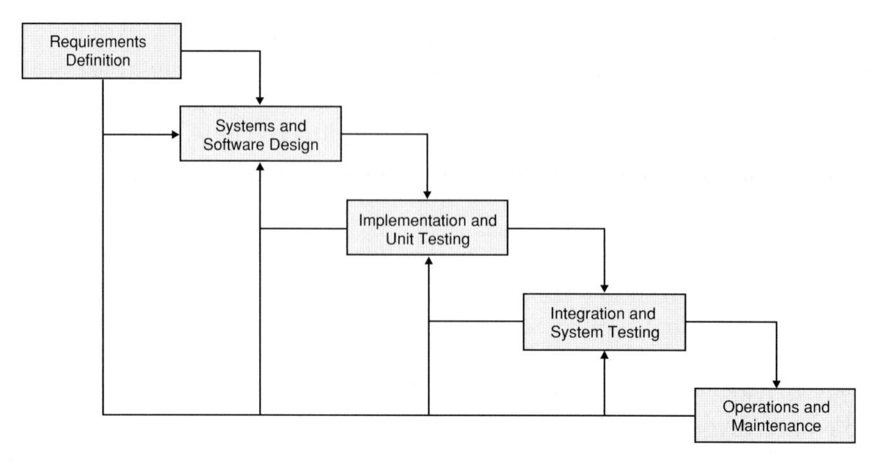

FIG. 20.1. A "waterfall" design approach. Users are typically involved towards the end of the process (adapted from Royce 1970).

user requirements. These specifications are then used during the development stage.

While a detailed discussion of user-centred design is beyond the scope of this chapter, it is important to note that this development strategy is being used as an integral part of the research and development related to The Atlas. Although this approach is centred on the user, all stakeholders,

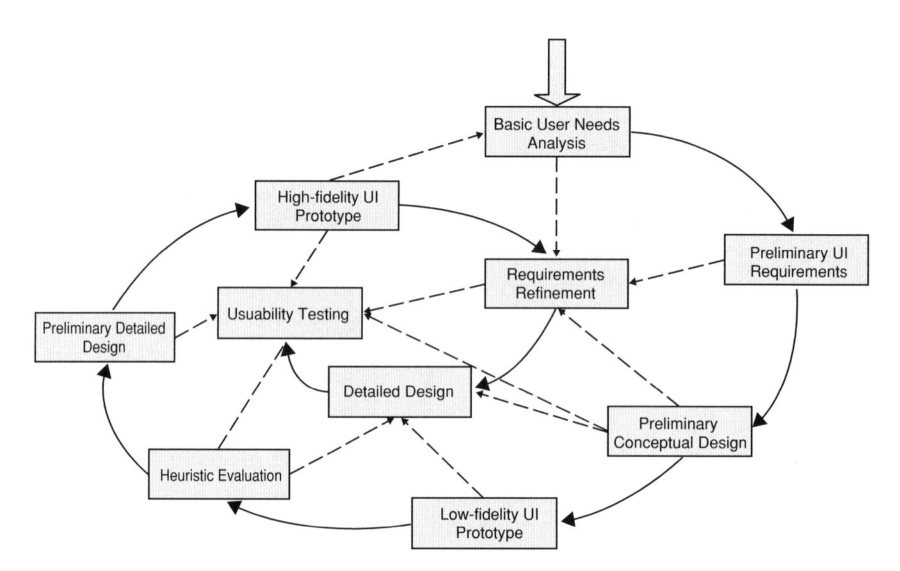

FIG. 20.2. A user-centred design approach based on user needs from the outset of the project (Image created by A. Parush).

including developers and content providers are involved in the needs analysis process. This inclusive method facilitates better cybercartographic production that involves multidisciplinary teams, partnerships, and multiple stakeholders. A similar approach has been taken to the creation of the cybercartographic atlases produced by CentroGeo (as described by Reyes and Martinez in Chapters 5 and 6).

2.2. Design Process

Atlas research and development is being carried out by a team drawn from several disciplines including cartography, psychology, cognitive science, and cultural mediation. Figure 20.3 illustrates research clusters that have emerged from these disciplines to deal with various theoretical and applied research aspects related to the development of The Atlas. These are described more fully by Lauriault and Taylor in Chapter 8. Note that stakeholders, such as the Scientific Committee on Antarctic Research and Students on Ice are included in the development domain as part of the

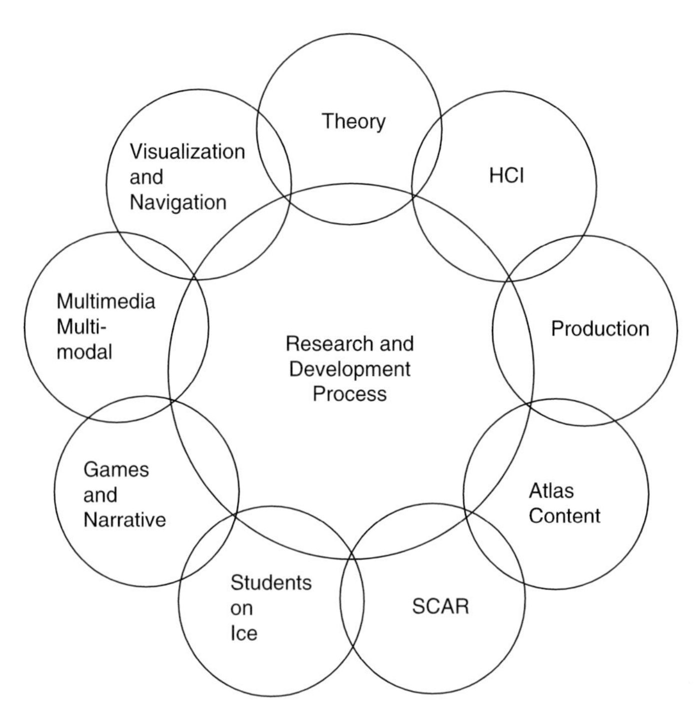

FIG. 20.3. A graphic representation of interaction between groups within The Atlas development team.

integrative, interactive, and inclusive nature of cybercartography (as described by Taylor in Chapter 1).

2.3. User Needs Assessment (UNA)

The initial user requirements specifications were based on a UNA process that included interviews with potential users, teachers, content developers, technical developers, and other stakeholders (Philp *et al.* 2004). The initial development of The Atlas predated the CANE project and thus, some research and development activities were carried out before the formal UNA process commenced. The overall Atlas development process commenced with several stakeholder meetings. The first workshop, held in December 2001, involved members of the Antarctic Science Community. The result was an agreement on the high-level objectives of the project, potential participants, and establishment of a preliminary list of themes that could be addressed in The Atlas. Two follow-up workshops with members of the Antarctic Science Community were held in May 2002 and October 2003. During these workshops, content themes were further developed along with potential data sources. Broad technological approaches were proposed and evaluated. The results of these workshops in the form of papers and reports were incorporated into the formal UNA process when it began early in the CANE project.

In addition to involving the science community, representatives from Students on Ice were involved in a stakeholders meeting held in May 2003. During this meeting, a variety of high-level issues were discussed and visions shared, including a description and purpose of the proposed atlas, specifics of the target audience, and major functions.

The stakeholder workshops resulted in two high-level statements of goals as established by the Students on Ice group and members of the Antarctic Scientific Community, respectively.

Students on Ice:

- A cybercartographic atlas portraying, exploring, and communicating the complexities of the Antarctic continent for education, research, and policy purposes, highlighting the global importance of Antarctica as the continent of science and peace.

Scientific Committee on Antarctic Research:

- Create an innovative new product and methodology to complement discovering, utilizing, presenting, and distributing existing informa-

tion and data about Antarctica to a wide variety of users, including scientists, decision makers, and the general public;
- Facilitate increased cooperation and information exchange between Antarctic stakeholders under the terms of the Antarctic treaty; and
- Through international cooperation, to develop and link National Atlases of Antarctica.

These statements are considered throughout the UNA, interface design, and content development process.

The user component of the UNA process began with student and educator interviews and the evaluation of existing curricula, and learning modules. A Hierarchical Task Analysis (HTA) identified specific tasks that users would perform with the atlases resulting in user requirements specification. These specifications established, for example:

- Content topics commensurate with the province of Ontario's curriculum;
- Criteria for structuring content;
- User activities;
- Need for tutorials instructing users how to use site functions (i.e. interactive visualizations, map viewers);
- Interface design elements;
- Navigation strategies; and
- Use of multimedia.

The user requirements specifications do not provide the cartographers and content developers with a "how-to" on developing the Atlas, but rather inform those developing the Atlas about usage context, technological constraints, curriculum requirements, and best practices based on research (Khan *et al.* 2004b).

2.4. Interface Design and Testing

User requirements and the HTA provided guidance for the user interface design (Bronsther *et al.* 2003; Narasimhan and Bronsther 2004). Two user interface teams developed in parallel conceptual designs underwent several iterations resulting in the proposal of three distinct designs. A familiar Web-based interface was developed as a low-fidelity prototype for an initial heuristic evaluation (Fig. 20.8, Khan *et al.* 2004a), whereby "experts" (either from the human computer interaction or cartography researcher groups) are asked to review and evaluate the interface. As tasks are given, these experts perform a "walkthrough" of the interface, and using a list of usability heuristics (rules-of-thumb for good design), they identify issues and usability problems and make comments about the interface. These problems can apply to one of the heuristics identified from the list. For example:

- Problem: You indicate/report that you are feeling lost at a certain point in the interface... you cannot tell what you did to get to the screen you are at, where you are in the interface, or what to do next...
- This problem applies to the heuristic visibility of system status.
- The system should always keep users informed about what is going on, through appropriate feedback within reasonable time. So, you report this problem in the heuristic evaluation.

After the prototype was tested with the heuristic evaluation, formal usability commenced.

Research questions and directions related to interface design are discussed further in Section 4.1.

2.5. Information Structure

A primary objective of any atlas is clarity of communication. From a review of research, Lobben and Patton (2003) suggest that simplicity in design and representation is desirable in achieving clarity of communication. Organizing and structuring information contained within an atlas is a key element of designing for clarity of communication. The UNA results for The Atlas concluded that the topic category hierarchy should be no more than three levels deep, arguing that an excessive number of categories may overwhelm users. At the same time, a detailed treatment of a topic should be divided into subtopics, or "chunks" to facilitate information synthesis. Respecting such design specifications presents a challenge for the cartographer when dealing with complex geographic subjects. As does including other suggestions from the psychological literature outlined in several chapters of this book.

Global environmental issues involve many interrelated domains with complex interrelationships (Berkman 2002). Data related to these domains are mediated to form information categories or concepts (see Pulsifer and Taylor, Chapter 7 of this volume). In addition to organizing information based on conceptual or domain "space," geographic information also has spatial and temporal components that need to be considered.

There is no single "best" method for organizing such complex information that can be viewed from a variety of perspectives.

In developing The Atlas and in keeping with the UNA results, a modification of the familiar paper atlas table of contents metaphor was established for initial prototypes. Topics and subtopics are organized by Volume, Chapter, and Content Module (the concept of a content module is discussed further in Section 2.5.1). Topics are hierarchically nested and ordered from high to low in terms of conceptual relationship and depth of treatment. While the structure is based on a traditional table of contents metaphor, the digital implementation of The Atlas allows us to depart from the predefined "one-to-one" relationships inherent in the paper atlas approach. There is essentially a parent–child relationship between the topic levels. For example, Volume 1 can be associated with a number of Chapters (one to many). Conversely, a chapter can be associated with more than one volume. At a lower level, Content Modules can be associated with many chapters (and in turn, volumes). Box 20.1 presents a textual listing of a table of contents for The Atlas based on the UNA, author, and science community perspectives.

BOX 20.1

Preliminary Table of Contents for the Cybercartographic Atlas of Antarctica

Volume 1: Antarctica in a Global Context

Chapter/Theme: Antarctica and Global Environmental Change

Marine Ecology: *The Effects of Climate Change on Top Predators: Declining Elephant Seal Populations* (X. Liu, Z. Stanganelli, D. Vergani, Argentina, Canada)

Cryology: *Glacier Dynamics: Glacier Extent in the McMurdo Dry Valleys (Ross Sea) region* (B. Woods, N. Short, Canada)

Chapter/Theme: Human Aspects

Geopolitics: *Territorial Claims in Antarctica Territoriality* (P. Pulsifer, S. Caquard)

Exploration: *Voyages of the Explorers* (S. Caquard)

Volume 2: The Continent of Antarctica (A Regional Analysis)

Chapter/Theme: The Antarctic Peninsula

Content Module: Hope Bay

Chapter/Theme: The Ross Sea Region

Content Module: McMurdo Dry Valleys

Chapter/Theme: Sub-Antarctic Islands

Content Module: King George Island

(*continued*)

BOX 20.1
Preliminary Table of Contents for the Cybercartographic Atlas of Antarctica (continued)

Volume 3: Antarctic Science
Chapter/Theme: Geoscience
Content Module: Subglacial Antarctic Lake Exploration
Chapter/Theme: Life Science
Content Module: Antarctic Seals
Chapter/Theme: Physical Science
Content Module: Environmental Research

Volume 4: Human Activity in Antarctica
Chapter/Theme: Protecting the Antarctic Environment
Content Module: Environmental Protection in the Context of an International Treaty:
The Protocol on Environmental Protection to the Antarctic Treaty ('The Madrid Protocol')
The Antarctic Protected Areas System
Chapter/Theme: Exploring Antarctica
Content Module: Exploration of Antarctica in the 20th Century
Chapter/Theme: Representations of Antarctica
Content Module: The Art and Music of Antarctica
Chapter/Theme: Human Impacts on the Antarctic Environment
Content Module: Tourism in Antarctica
Chapter/Theme: Living in Antarctica
Content Module: Human Response to Harsh Environments

For the initial prototype, the table of contents has been converted to a series of interface element, such as graphic menus (Fig. 20.8). Alternatively, an information map could be presented to the user allowing zoom and pan operations to navigate conceptual space. Figure 20.4 displays a proof of concept prototype for such an interface. Figure 20.5 demonstrates the zoom function of the prototype. Given the modular nature of information structure framework, the potential exists to establish more than one table of contents or views of the content.

An illustration of the table of contents (Fig. 20.6) suggests that The Atlas content can be situated in terms of space, time, and depth of discussion and related to different perspectives: Science to Policy/Natural to Human. While this is a model that is deemed useful for development, it is recognized that it is but one possible model.

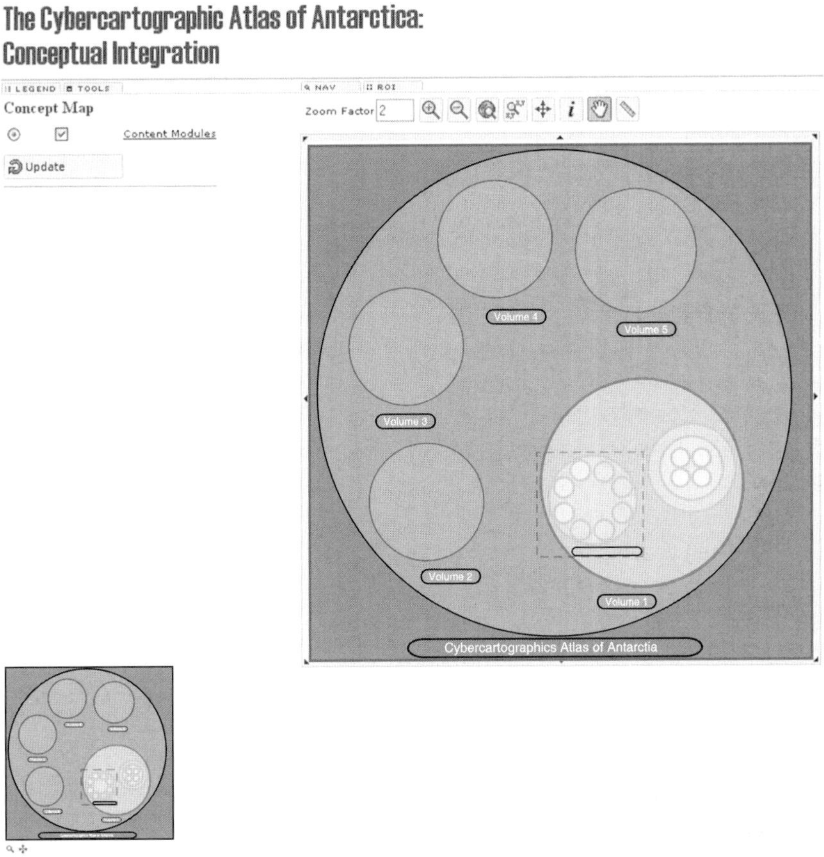

FIG. 20.4. Using Venn diagrams as a concept map allows users to navigate content using a graphical metaphor. This approach highlights the interrelationships between content modules established by The Atlas authors. (content developed by P. L. Pulsifer) (adapted from Berkman 2002).

Working from within the model, decisions are made in terms of the techniques and methods used to represent data (e.g. text, video, interactive visualization), establishing relationships between concepts (content structure), interface navigation design, and cartography (e.g. symbolization). Experience has shown that there are interrelationships between these elements and, for example, a decision related to content structure has implications for interface navigation design.

Although various members of the development team bring different expertise to the process, it is clear that overall development of The Atlas must occur as an integrated whole. While this statement may seem self-evident to an experienced cartographer, it nonetheless has fundamental implications for current interoperability models that are based on the concept of separating content from presentation (i.e. eXtensible Markup Language).

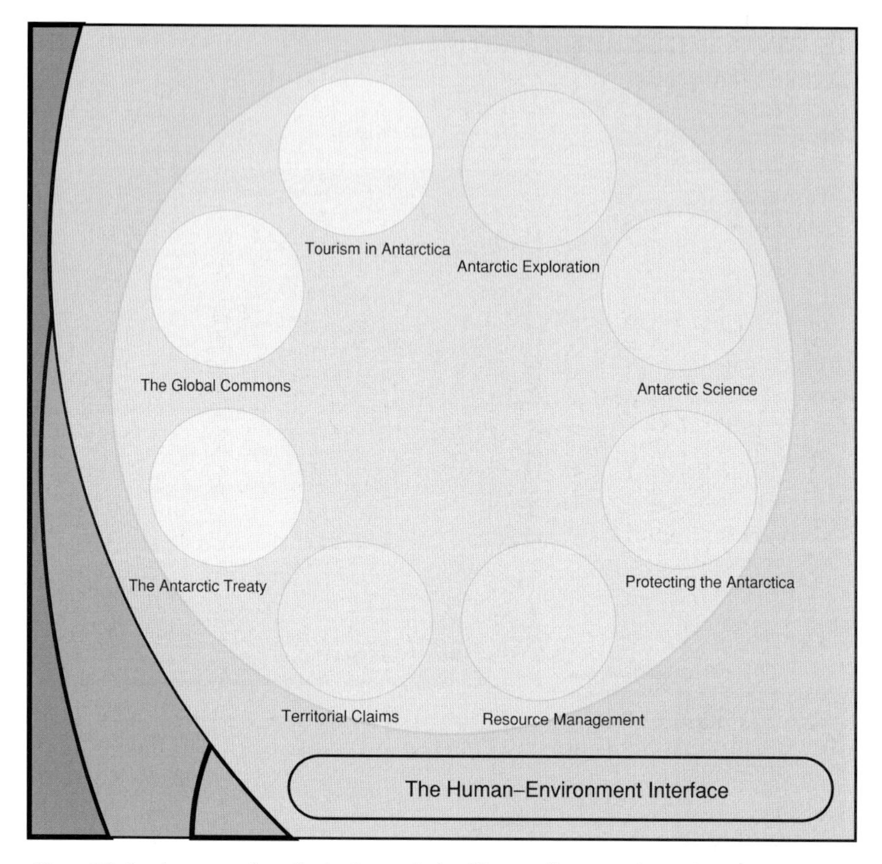

FIG. 20.5. A more detailed view of the Venn diagram-based concept map used to navigate The Atlas (content developed by P. L. Pulsifer).

2.5.1. Content Modules

Content Modules are Web-based representations of Atlas topics or concepts. A Content Module is:

- A component of the atlas containing cartographic, narrative, and multimedia elements for the purpose of examining a particular question, topic, area, or phenomena related to the Antarctic region;
- Associated with one or more Chapters and Volumes (broader concepts);
- Evaluated for quality;
- Owned;
- Developed for a particular audience; and
- Described by a well-defined set of properties.

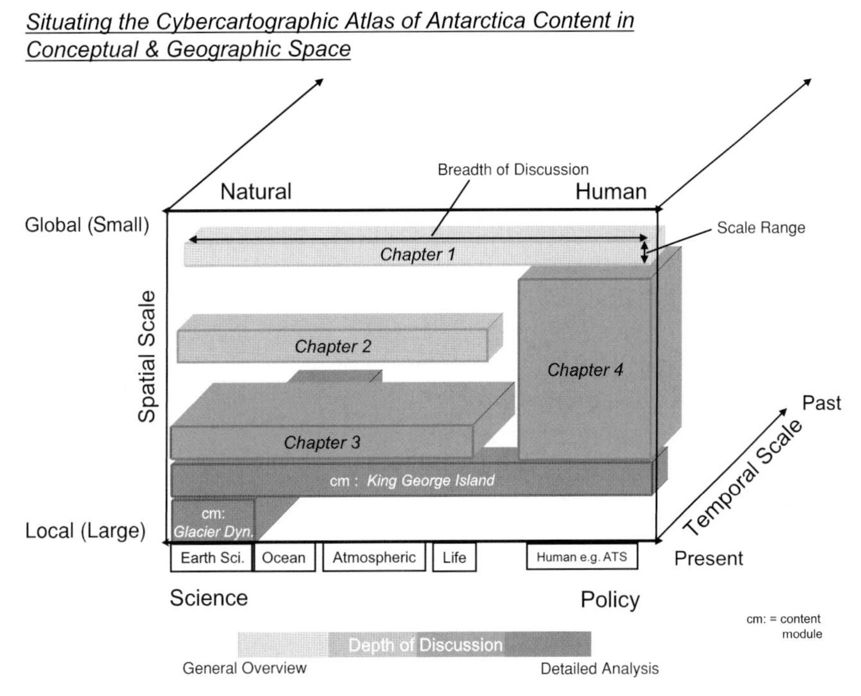

FIG. 20.6. Situating elements of The Atlas information structure in space and time with an indication of depth of discussion and domain relationships.

Depending on position within the conceptual structure, a content module would typically have a primary Web page. However, more complex modules may, for example, involve linking to subwindows to view multimedia elements. A variety of representational elements may be used within a given module. Box 20.2 lists some of the elements that may be used in a given content module.

2.6. *Interoperable Distributed Systems*

Recent specifications proposed by the Open Geospatial Consortium (OGC) and the International Organization for Standardization (ISO) are enabling the development of the "spatial Web" (Moses 2003). This spatial Web is making it possible for users to discover, access, and process digital geospatial data over the Internet. The Atlas is based on the Web services architecture that underpins the spatial Web. Figure. 20.7 illustrates how image, vector, three-dimensional terrain, and scientific model data support applications using the Internet. In keeping with the project's Open Cartographic

BOX 20.2

Representational Artifacts that may be used in a Content Module

Selected Geographic Information Elements:

- Maps
  - Topographic (i.e. shorelines, elevation contours)
  - Thematic (i.e. magnetic anomalies, food distribution)
- Georeferenced imagery
  - Passive sensor imagery (i.e. aerial photography)
  - Active sensor (Radar images – require interpretation)
- 3D/4D Terrain rendering
  - Topographic/Thematic drape
  - Georeferenced imagery drape

Hypermedia Elements:

- Text
  - Static narrative
  - Hyperlinked
  - Georeferenced (place name)
  - Glossary/Keyword
- Digital images
  - Photographs
  - Illustrations
- Sound
  - Earcons
  - Environmental
  - Narration
  - Music
  - Sonification
- Animation
  - Process
  - Time series
  - Augmented
- Video
  - Real-world
  - Simulated
  - Augmented
- In Situ Sensors
  - Webcams

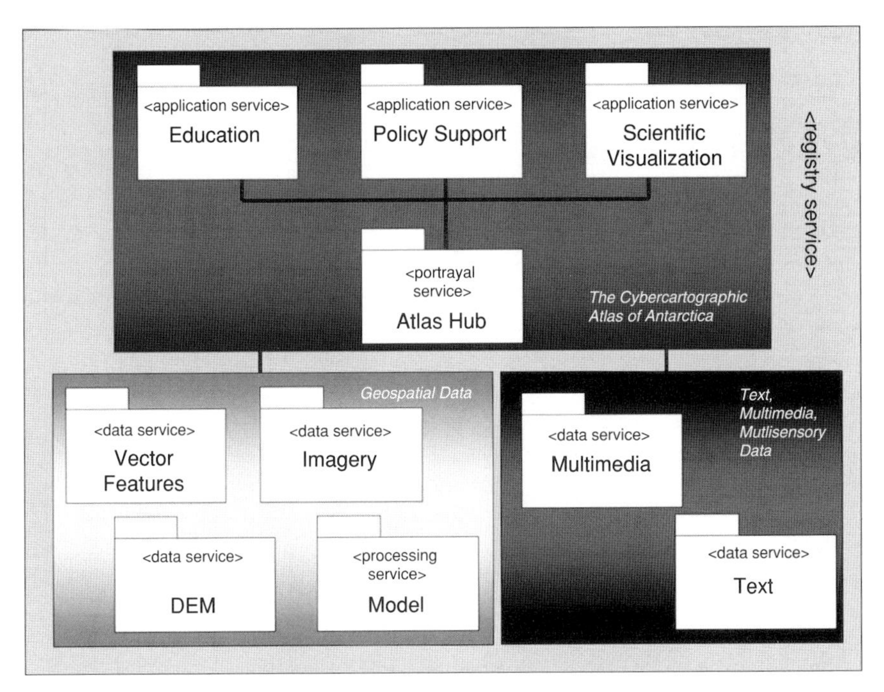

FIG. 20.7. A conceptual diagram illustrating elements of an interoperable, distributed system being developed to support The Atlas and other Antarctic science applications.

Framework, text, and multimedia content will be delivered using the same architecture (See Pulsifer and Taylor, Chapter 7 of this volume). The ability to search for and discover data resources is integral to developing Atlas content modules. To improve access to data, members of the Geomatics and Cartographic Research Centre (GCRC) have recently agreed to assist in the development of the Antarctic Data Directory System developed by the Joint Committee on Antarctic Data Management. This directory system, hosted by NASA's Global Change Master Directory can facilitate data sharing between The Atlas and the global Antarctic Science Community.

3. Implementation

3.1. *Interface Designs*

As part of the iterative User-Centred Design process, a number of low-fidelity prototypes have been developed. Figure 20.8 illustrates the first

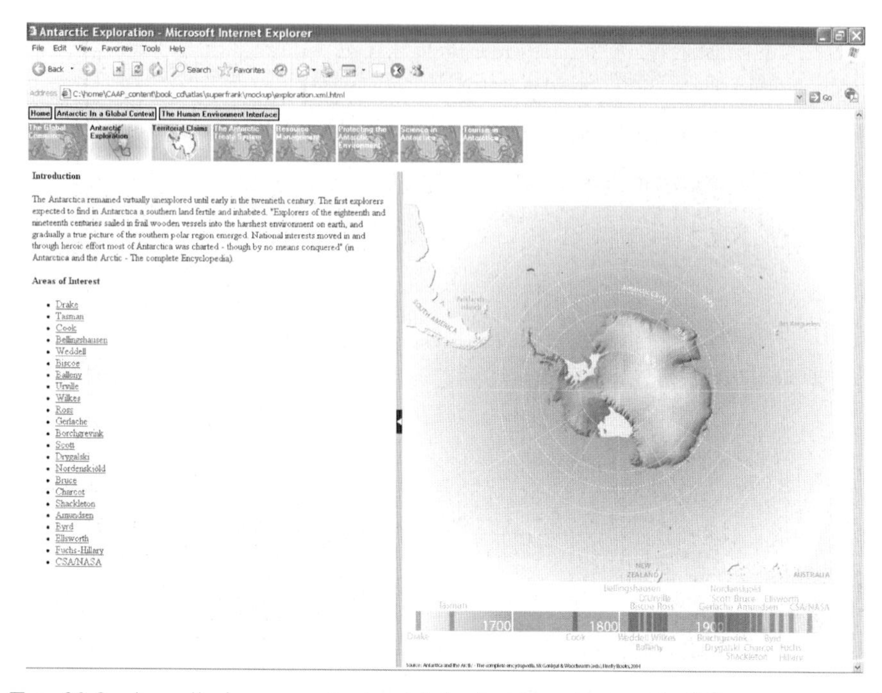

FIG. 20.8. A preliminary content module developed as part of the user-centred design process that aims to teach students about exploration of the Antarctic region (content developed by Caquard, S., P. Pulsifer, J. P. Fiset and A. Hayes). Results of formal testing of this content module will inform the next iteration of the design process (Rasouli *et al.* 2004).

iteration interface intended for use in preliminary formal usability testing with high school aged students.

In addition to atlas interfaces that use a more typical Web interface design, researchers are examining the use of alternative approaches that allow users to access content based on topic or media type (Fig. 20.9).

3.2. Prototype Content

Several project researchers are developing content modules for The Atlas. In Chapter 21 of this volume, Baulch *et al.* describe some of the theoretical and pedagogical considerations for designing a content module for The Atlas, specifically dealing with the topic of climate change. Given the conceptual complexity of the topic of climate change, a decision was taken to first develop several modules of smaller scope. Thus, the consid-

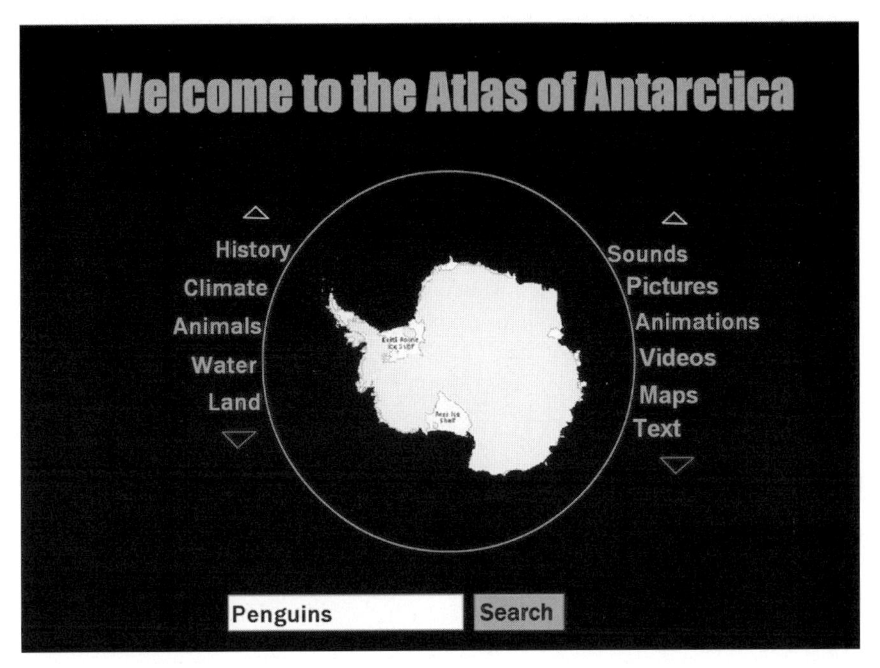

FIG. 20.9. Radial menu user interface linking between different contents and media for combined browsing (developed by Parush, A. and S. Narasimhan).

erable development effort required for the climate change module will benefit from more extensive user input, analysis of user needs, and interface evaluation resulting from initial formal user testing. The preliminary content modules include treatments of topics ranging from physical and biological processes (e.g. marine ecology; Fig. 20.10), and human aspects (e.g. territoriality; Fig. 20.11), and the aforementioned module discussing climate change. In addition, modules are being developed by international partners. Included in these modules are treatments of the History of Human Activity on King George Island and Glacier Dynamics, King George Island (Vogt 2004).

3.3. *System Architecture*

3.3.1. *From Data to Atlas*

As stated, The Atlas is based on a distributed data architecture whereby geospatial information and associated information (multimedia elements etc.) are retrieved from servers provided by project partners and other publicly available data resources. Retrieving a map from a remote server

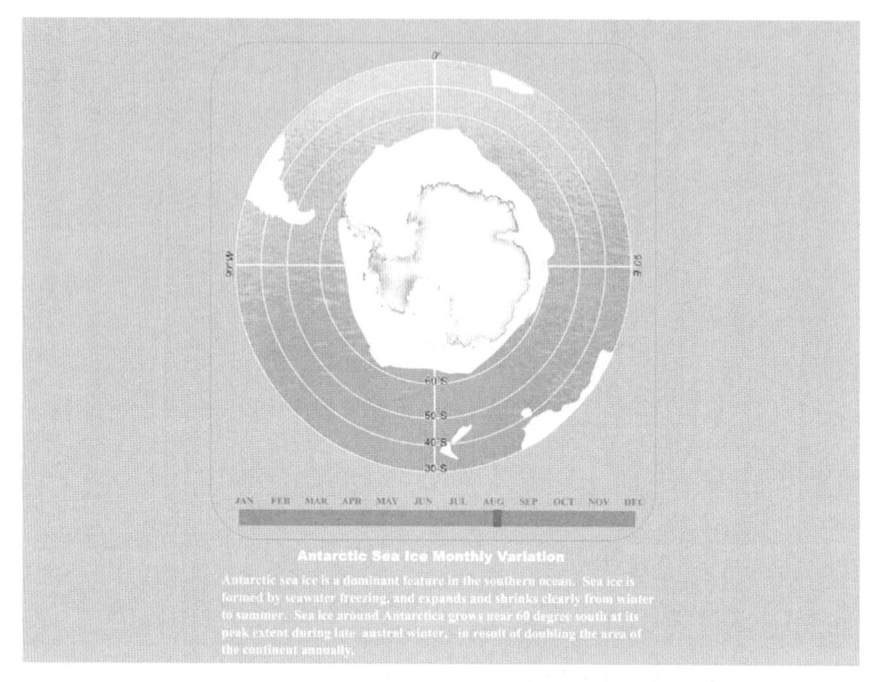

FIG. 20.10. As part of a content module focused on marine ecology, a Web-based animation conveys the annual variation in sea–ice extent in the Antarctic region (content developed by X. Liu in collaboration with scientific advisors Vergani, D. and Z. Stanganelli) (Liu *et al.* 2004).

does not however constitute the creation of a digital atlas. Atlas development requires that these maps, spatial features, and other information elements be placed in a conceptual and usage context (as discussed in Section 2.5). This is critically important to effective learning as Lindgaard *et al.* outline in Chapter 9 of this volume.

To place the distributed information in a particular context requires that information resources be well understood by atlas authors and end users. This goes beyond the need for basic metadata stating projection, Internet connection URLs, and responsible party contact information.

In a situation, where the information resources (databases, media repositories, etc.) used to develop an Atlas are managed by people close to the development process, the nature, and structure of information resources from a syntactical, schematic, and semantic perspective are typically well known. This supports effective and appropriate use and integration of information resources. However, when attempting to populate an atlas using distributed resources, such knowledge is limited to that, which is exposed by the distributed source. For example a standards based "Web

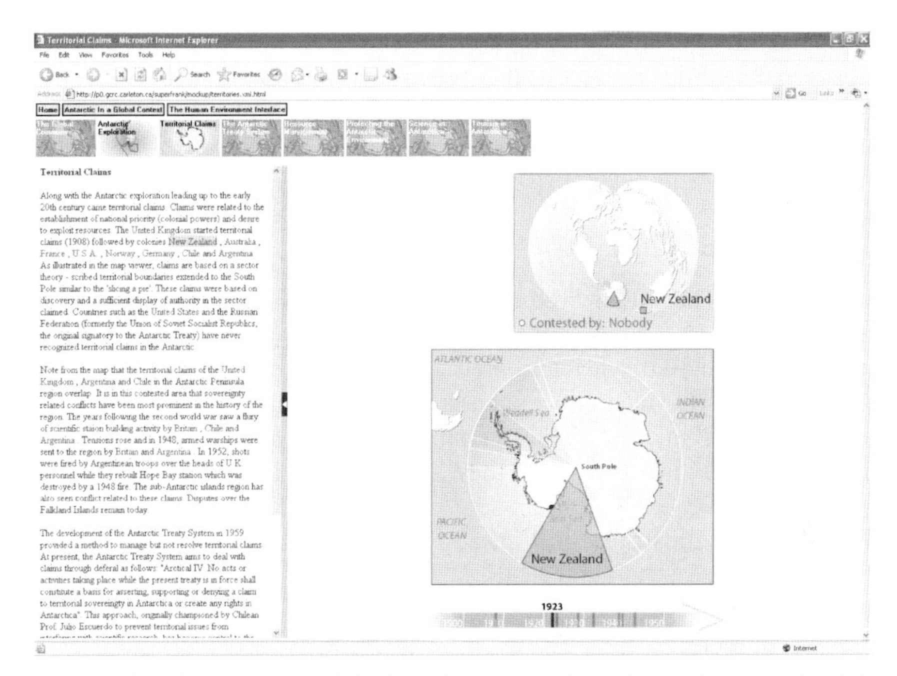

FIG. 20.11. A content module that aims to teach students about territorial claims in the Antarctic region using interactive vector-based graphics (content developed by Caquard, S., P. Pulsifer and J. P. Fiset).

Feature Server" will provide the name of features returned to a client, however, at present a semantic description of that feature is not necessarily returned.

To start addressing the issue of semantic interoperability for The Atlas, researchers are participating in a project that aims to establish a spatial data model for the Antarctic Science Community. The importance of modeling for cybercartography, in general, is outlined by Reyes in Chapter 4 of this volume. The first step in this process is the development of a Feature Catalogue that establishes known physical features that exist in the Antarctic region (SCAR 2004). Feature names are accompanied by a precise definition. While this effort is very useful in providing semantics for well-defined physical features, the effort is just starting to address the difficulties of integrating features related to human–environment interactions. For example, in relation to environmental protection, integrating spatial data that maps the risk to species as a result of a particular human activity can be difficult. If the concept of risk is not well defined within the environmental protection regime, then the entire risk model may need to be examined if resulting maps are to be integrated. This type of application may go beyond the establishment of a feature catalogue.

The challenges of integrating various forms of information are of central concern in The Atlas project in theoretical and practical terms. Some of the theoretical issues have been reviewed. To work towards implementing systems that may facilitate more effective Atlas development, a technical architecture is being developed and evaluated.

3.3.2. Mediator Based Architecture

Conceptually, The Atlas architecture is based on the three-tiered Open Cartographic Framework, described in Chapter 7 of this volume. In this instance, the mediator level models and associates the spatial features, maps, information elements, such as multimedia, narrative, metadata, and the area of interest. The described architecture is being built with reference to and aims to extend the Open Geospatial Consortium Discussion Paper 01-037 (OGC 2001).

Each element illustrated in Fig. 20.12 is briefly discussed here:

(a) *Feature* – In general, a feature is defined as an abstraction of a real-world phenomenon with a geographic feature being one that is associated with a location relative to earth. In this architecture, features are specifically defined using the Geographic Markup Language with the semantics being further defined using a feature catalogue developed using the ISO TC211 series of standards (19110 in particular). In this context, we are typically referring to a collection of features (layer) rather than individual feature instances. Layers are associated with lineage information (how the features were processed), metadata using an ISO 19115 profile (currently being defined as part of the Antarctic Spatial Data Infrastructure program), and cartographic style information. Typically, style information is stored in the OGC/ISO Styled Layer Descriptors Format. These associated elements comprise a map layer complete with metadata description and styling information.

(b) *Map Context* – As described above, feature collections define layers. In turn, maps are constructed by compiling related layer. Maps are defined and stored as a Web Map Context document. The map context defines service connection information as well as style information that may be associated with the map. This style information may take priority over feature styling. While the current specification is designed for Web Map Services (WMS), extension of the specification to include Web Feature Servers (WFS) and other Web Services is being investigated.

(c) *Information Element* – This class includes atlas elements that would typically be considered nonspatial, but would often be associated with

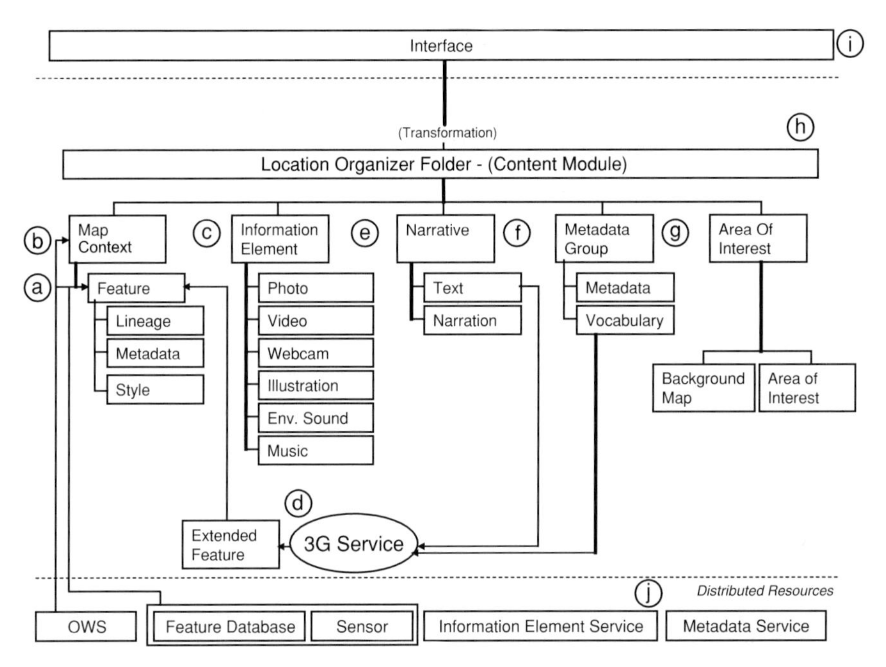

FIG. 20.12. Diagram depicting the prototype system architecture of the Cybercartographic Atlas of Antarctica.

a map context or feature. In traditional terms, we might consider these elements as marginalia. Where georeferencing of an information element is desired, the element must be associated with a feature, map context, or the area of interest.

(d) *3G Service* – In this case, three coupled services (GeoParser, Gazetteer, and Geocoder) are used to dynamically parse and interpret textual information and subsequently geocode the text using a gazetteer. The parsing and interpretation process can use vocabularies associated with the application domain. The aforementioned feature catalogue can be considered a vocabulary that may be used in conjunction with a gazetteer to establish the usage context of a particular term (i.e. Ronne Ice Shelf).

(e) *Narrative* – A text-based explanation of the content portrayed in the content module. This narrative is typically authored by a person with expertise in the particular domain discussed. The narrative is typically associated with a map context, feature, or information element. Narratives can be divided into smaller segments, which in turn can be associated with various elements in the model.

(f) *Metadata Group* – The metadata group provides a high-level description of content module as a whole. This particular form of metadata is analogous to a Catalogue entry. For The Atlas, a metadata profile, defined by in the Antarctic Master Directory has been adopted. Shared vocabularies, feature catalogues, glossaries, etc., are stored here.

(g) *Area of Interest* – This element defines the spatial bounds of the content module using GML constructs. The AOI includes a background map, which would typically be generated by a Web Map Server. Although this background map can be interrogated using WMS GetFeatureInfo request, for performance reasons it is typically used as a rich, image-based backdrop to interactive map content. Interactive content would typically be delivered using a Web Feature Server.

(h) *LOF* – The Location Organizer Folder (LOF) is a container used to organize various elements in the model. Effectively the LOF is a content module as previously defined.

(i) *Interface* – The Interface is the point at which a human or machine-based user interacts with the system.

The preceding points provide an overview of The Atlas architecture at the time of publication. Development is ongoing.

4. Research Directions

Research related to development of The Atlas is being structured to support knowledge development in a number of ways. Individual research projects are examining various specific aspects of cartographic representation, while the collaborative effort of theorizing, designing, and implementing The Atlas is informing the way in which we integrate geographic information. As a result of these processes, new insights are being developed in terms of how theory meets practice in the form of implementation. The interaction between theory and practice is integral to the paradigm of cybercartography, as discussed by Taylor *et al.* in this volume.

4.1 Cartographic Representation

A number of ongoing research initiatives are examining various methods of cartographic representation.

A content module, addressing glaciology examines the effective use of three-dimensional and four-dimensional visualization for exploring and communicating dynamic processes (Woods *et al.* 2004). Specifically, this research examines approaches to online, interactive terrain rendering.

A content module addressing the affects of climate variability on southern elephant seal populations examines the use of animation to communicate scientific research results (Liu *et al.* 2004). More specifically, this research examines how such animation can be supported by a spatiotemporal, object-relational model used to support the research being communicated.

The content module, addressing territoriality and exploration is a component of research that examines how geographic information is integrated in the context of human activity (the human–environment interface). Specifically, this research looks at how we negotiate meaning between high-level concepts (such as acceptable risk, significant degradation, etc.) and disparate sources of geographic information exposed by Web services.

4.2. *Atlas Design*

A central challenge facing the project is the integration of Atlas content. Organizing Atlas elements in an efficient and effective manner not only at the interface level, but at lower levels of the system – system mediators, data infrastructure, etc., are critical to the usability and usefulness of The Atlas. The design challenge is to create an atlas that is clear and simple and can deal with complex issues while meeting the evolving needs of the user.

Cartwright *et al.* (2000) point out that experienced users of geospatial information possessing in-depth domain knowledge may benefit from customized tools for advanced exploratory analysis (DiBiase 1990). A user from the general public can benefit from simpler, more standardized tools. However, more understanding is required in terms of how user needs evolve through use of the product. Advanced exploratory analysis entails some level of domain knowledge and the ability to use more advanced analytical methods and tools. While a user from the general public may not possess such knowledge at the outset of her interaction with an atlas, domain knowledge will conceivably increase through use of the atlas. Similarly, the provision of effective training and support within an atlas can assist in increasing user competence in terms of methods and tools for manipulating geospatial information.

This iterative process may necessitate the move from Bochert's (1999) "closed atlas" (focused on communication) to a more open "extended database" atlas. As user knowledge increases, can the system adapt to allow progressive movement towards exploratory analysis? Research related to The Atlas is examining how such a system might be implemented. Lindgaard *et al.* (Chapter 9) distinguish three kinds of users: knowledge seekers, feature explorers, and apathetic users; and the challenges of designing products and to integrate these for each user type is challenging. They argue that cybercartographic atlases must be cognitive learning tools, so The Atlas is designed with this in mind. For example, let us consider an Atlas narrative discussing Southern Ocean fisheries at a high level. A user from Australia would like more information about this topic for territories associated with her country. From within The Atlas, a visual search engine allows the user to search and organize information related to the topic of interest. The Atlas itself was not authored to include such detailed information, however, a connection to a distributed information resource has been established; in this example, the Antarctic Master Directory.

This system provides the user with a familiar Atlas structure, in addition to providing the possibility of agency as their knowledge develops. The example below provides the user with a map of information space that offers an overview of a particular topic, in this case, Southern Ocean/Sub-Antarctic fisheries (Fig. 20.13). The resulting information map provides the ability to evaluate the amount of information available from this particular information resource. Upon identifying a useful resource, the user can also

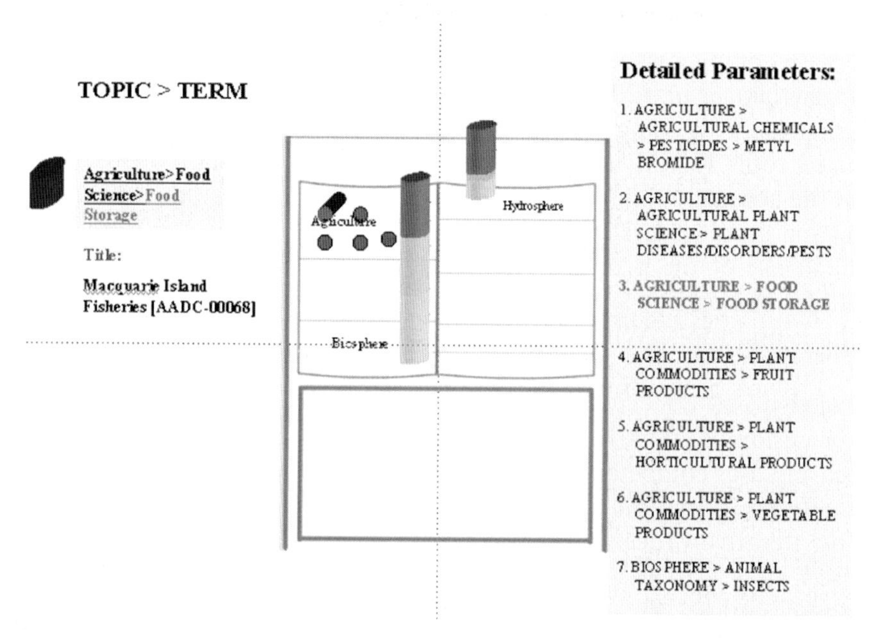

FIG. 20.13. An information map provides the Atlas user with a graphic representation of the type and volume of information returned using particular search parameters. The relative location of symbols indicates the semantic similarity (developed by Zhou, Y., S. Roberts and A. Parush).

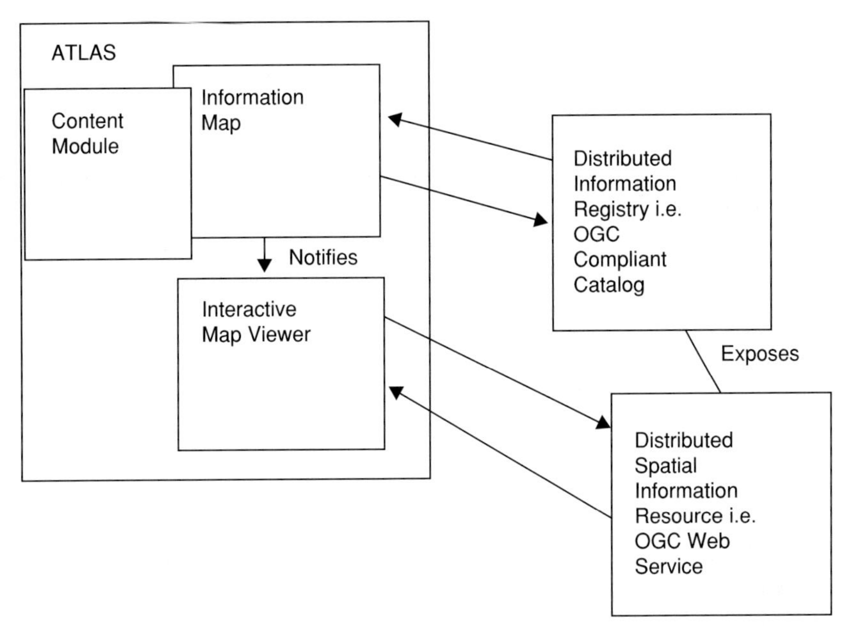

FIG. 20.14. The information map is used within The Atlas to support the integration of distributed information while allowing users to maintain orientation within the existing Atlas interface.

establish if there are spatial information services related to the resource (Fig. 20.14). In the case of interoperable resources (i.e. OGC/ISO compliant), the maps, features, coverages, etc., are then viewed or explored using the Atlas interface. Although the information map plots information from an external resource, there is no need for the user to leave the Atlas interface due to the distributed architecture of the system. The present scenario provides the user, the ability to maintain orientation while extending the content of the Atlas to meet evolving information requirements. This supports a meaningful interactive cognitive learning experience controlled by the learner but within a structure, which does not confuse the users or disorient them in cyberspace.

It is important to note that this type of user activity is distinct from viewing an Atlas content module. In effect, the user is required to mediate the information – that is to abstract, integrate, and to some extent make representation decisions (i.e. which layer will be displayed from the distributed spatial information resource). This brings us back to fundamental usability issues (e.g. how appropriate is this type of user activity? Does the user have suitable domain knowledge to draw meaningful inferences? Does the user have the skills to manipulate data from an external spatial information resource? How does the atlas designer evaluate these user qualities in a Web-based environment?). These questions and others are being considered as part of ongoing research related to The Atlas.

4.3. *Implementation*

Research related to cartographic representation and atlas design has resulted in a number of innovations related to implementation. In summary, some of the issues being addressed are:

- The move towards a widely available form of spatiotemporal cartographic representation using browser supported Scalable Vector Graphics;
- Establishing methods to deal with latency issues related to distributed databases. While the concept of distributed databases is powerful in theory, in practice delays related to data volume and bandwidth can result in delays for the user, thus degrading the user experience. A virtual database approach is being developed whereby data are stored locally to support high performance while maintaining a dynamic link to distributed resources. The inclusion of near–real-time data in The Atlas presents a special challenge in this respect; and
- Providing innovative implementation models for supporting data access and integration to support atlas development (see Section 3.3.2). To promote wide accessibility to a wide range of users and providers, these models are being developed using open standards and specifications, such as those developed by the World Wide Web Consortium (W3C), Open Geospatial Consortium (OGC), and the International Organization for Standardization (ISO).

5. Conclusions

When considering digital atlas production, many new possibilities are presented through the use of the Internet, distributed databases, location based services, and other emerging technologies. Experiences till date in designing and implementing The Cybercartographic Atlas of Antarctica suggest that many theoretical and practical challenges remain for providing a usable and useful open system (high-level of user control) with analytic capabilities that fully take advantage of these new developments. The strong, interdependent links between content, presentation, and the user context in which these systems are used, suggests that ongoing research aimed at supporting effective integration and representation of geographic information in a way that is meaningful to end users, is required.

References

Alviar, M. L., F. Lopez, C. Reyes and D. R. F. Taylor (2001) "Cybercartography and the Environment: The Chapala Atlas", Paper presented at the 20th International Cartographic Conference, Mapping the 21st Century, Beijing, PRC.

Berkman, P. A. (2002) Science into Policy: Global Lessons from Antarctica. Academic Press, London, UK.

Borchert, A. (1999) "Multimedia atlas concepts", in Cartwright, W., M. P. Peterson and G. Gartner (eds.), *Multimedia Cartography*, Springer-Verlag, New York, pp.75–86.

Bronsther, A., S. Roberts, P. Trbovich and G. Lindgaard (2003) *Cybercartography and the New Economy: Preliminary HCI design guidelines for the design of interactive multimedia multimodal application,* Human Oriented Technology (HOTLab), Carleton University, Ottawa, Canada.

Cammack, R. G. (1999) "New Map Design Challenges: Interactive Map Products for the World Wide Web", in Cartwright, W., M. P. Peterson and G. Gartner (eds.), *Multimedia Cartography*, Springer-Verlag, New York, pp. 155–172.

Cartwright, W. (1999) "Extending the map metaphor using web delivered multimedia", *International Journal of Geographic Information Science*, Vol. 13, No. 4, pp. 335–354.

Cartwright, W., J. Crampton, G. Gartner, S. Miller, K. Mitchell and E. Siekierska *et al.* (2000) "Geospatial information visualization user interface issues", *Cartography and Geographic Information System*, Vol. 28, No. 1, pp. 45–60.

DiBiase, D. (1990) "Visualization in the earth sciences", *Earth and Mineral Sciences, Bulletin of the College of Earth and Mineral Sciences*, PSU, Vol. 59, No. 2, pp. 13–18.

Howard, D. (1999) "Metaphors and user interface design", in Craglia M. and H. Onsrud (eds.), *Geographic Information Research: Transatlantic Perspectives*, Taylor and Francis, Bristol, PA, USA, pp. 565–576.

Jankowski, P. and T. Nyerges (2001) *Geographic Information Systems for Group Decision Making: Towards a Participatory, Geographic Information Science*, Taylor and Francis, London.

Jezek, K. C. (1999) "Glaciological properties of the Antarctic ice sheet from RADARSAT-1 synthetic aperture radar imagery", *Annals of Glaciology*, Vol. 29, pp. 286–290.

Khan, S., S. Narasimhan and A. Parush (2004a) *Heuristic Evaluation of Cybercartographic Atlas of Antarctica Prototype*, Human Oriented Technology (HOTLab), Carleton University, Ottawa, Canada.

Khan, S., M. Rasouli, A. Parush and G. Lindgaard (2004b) *Requirements Specification*, Human Oriented Technology Lab, Ottawa, Canada.

Liu, X., P. L. Pulsifer and D. R. F. Taylor (2004) May 27–29, *The Cybercartographic Atlas of Antarctica: Visualization for Antarctic Science,* Paper presented at AntGIS04, Chinese Antarctic Centre of Surveying and Mapping, Wuhan University, Wuhan, PRC.

Lobben, A. K. and D. K. Patton (2003) "Design guidelines for digital atlases", *Cartographic Perspectives*, Vol. 44 (Winter 2003), pp. 51–62.

Monmonier, M. (2000) "Webcams, interactive index maps, and our brave new world's brave new globe", *Cartographic Perspectives*, Vol. 37, pp.51–64.

Moses, R. (2003) July, 2, *OGC's Mandate,* accessed 7 February 2005 from http://www.directionsmag.com/article. php?article_id=384

Narasimhan, S. and A. Bronsther (2004) *Interim Report: Cybercartography and New Economy Project (user interface design/prototyping)*, Human Oriented Technology (HOTLab), Carleton University, Ottawa, Canada.

OGC. (2001) *Location Organizer Folder Specification* (Draft Candidate Implementation Specification, Discussion Paper No. 01–037): Open GIS Consortium, accessed 7 February 2005 from http://www.opengeospatial.org/docs/ 01-037.pdf.

Peluso, N. L. (1995) "Whose woods are these? Counter-mapping forest territories in Kalimantan, Indonesia", *Antipode,* Vol. 27, No. 4, pp. 383–406.

Peterson., M. P. (1997) "Cartography and the internet: introduction and research agenda", *Cartographic Perspectives*, Vol. 26, pp. 3–12.

Peterson, M. P. (2003) "Maps and the internet: an introduction", in Peterson, M. P. (ed.), *Maps and the Internet*, Elsevier Science, pp. 1–16.

Philp, K., M. Rasouli, S. Narasimhan and G. Lindgaard (2004) *User Needs Assessment Report For Cybercartography and New Economy Project "Canada's Trade with the World" Product* (CD-ROMs, Human Oriented Technology (HOTLab), Carleton University, Ottawa, Canada.

Rasouli, M., K. Philp, S. Khan, G. Dunn, G. Lindgaard and A. Parush (2004) October 29–30, *User-Centred Design of Educational Cybercartographic Atlases*, Paper presented at CAGONT (Canadian Association of Geographers—Ontario Division) Conference, University of Waterloo, Waterloo, Canada.

Royce, W. W. (1970) "Managing the development of large software systems", Paper presented at the IEEE WESCON.

SCAR. (2004) *SCAR Feature Catalogue*, accessed 28 November 2004 from http://www.aad.gov.au/default.asp? casid=6259

Taylor, D. R. F. (2003) "The concept of cybercartography", in Peterson. M. P. (ed.), *Maps and the Internet*, Elsevier, Cambridge.

Vogt, S. (2004) *Personal Communication in Bremerhaven, Germany on the afternoon of July 23rd, 2004* at the Alfred Wegener Institute for Polar and Marine Science.

Woods, B., P. L. Pulsifer and D. R. F. Taylor (2004) July 25–31, *Visualization of Earth Science Spatial Data for Antarctic Research as Part of the Cybercartographic Atlas of Antarctica Project,* paper presented at Scientific Committee on Antarctic Research, Open Science Conference, Bremen, Germany.

CHAPTER 21

Cybercartography for Education: The Case of the Cybercartographic Atlas of Antarctica

SAMANTHA BAULCH

The Delphi Group, Ottawa
Ontario, Canada

RONALD MACDONALD

Kawartha Pine Ridge District School Board
Lakefield District Intermediate Secondary School
Lakefield, Ontario, Canada

PETER L. PULSIFER

Geomatics and Cartographic Research Centre (GCRC)
Department of Geography and Environmental Studies
Carleton University, Ottawa, Canada

D. R. FRASER TAYLOR

Geomatics and Cartographic Research Centre (GCRC),
and Distinguished Research Professor in International Affairs,
Geography and Environmental Studies
Carleton University, Ottawa, Ontario, Canada

Abstract

This chapter examines the educational design of the Cybercartographic Atlas of Antarctica for high school students. It examines different learning methods and establishes a need to create a learning environment that appeals to Howard Gardner's Multiple Intelligences Theory, while incorporating elements of edutainment to capture the attention of teenagers. Working in collaboration with Students on Ice, an organization that leads learning expeditions to the Antarctic and the Arctic, this chapter describes a prototype learning module, based on climate change, and a prototype gaming component for the Atlas.

1. Introduction

The use of maps, atlases, globes and more recently remote sensing has always been an important part of high school curricula, particularly in the subjects of geography, history and science. Maps not only help students learn about their world, they also help students to develop spatial and problem solving skills. The *Geography for Life: National Geography Standards*, a well-used document for geography curriculum, stresses the use of these resources in the classroom because "knowing how to identify, access, evaluate and use all these... resources will ensure students a rich experience... and the prospect of having an effective array of problem-solving and decision-making skills for use in both their educational pursuits and their adult years" (American Geographical Society 1994: 64).

With advances in technology in the realm of CD-ROMs and the Internet, maps and atlases are becoming very dynamic learning tools. For example, atlases can now incorporate many forms of multimedia, such as photos, sound and video, creating a much more appealing learning experience for students. Many examples of these digital atlases exist such as the *Canadian Geographic Explorer* and *Microsoft Encarta World Atlas*. More recently, the provincial government of Quebec is funding the production of an *Internet Atlas of Quebec*. Its mandate is to support Quebec's Social Science program for the primary and senior levels (a prototype can be viewed at: http://www.Atlasduquebec.qc.ca/ scolaire/) (Anderson *et al.* 2002).

With the development and increased use of these digital atlases in the classroom, it is essential that they create an effective learning experience. To help achieve this goal, a team of researchers at Carleton University is examining the creation of a Cybercartographic Atlas for Antarctica, as described in this volume by Pulsifer *et al.* in Chapter 20.

The research team prefers the term Cybercartographic Atlas, rather than Internet atlas or digital atlas, because it better encapsulates the goals. The concept sees a cybercartographic product as (Taylor 2003):

- Multisensory;
- Multimedia;
- Interactive;

- Comprising an information package;
- Developed by interdisciplinary teams;
- Forming new partnerships; and
- Applicable to a wide variety of subjects.

The Cybercartographic Atlas of Antarctica (from now on referred to as the Atlas) is meant to be a multifaceted atlas, allowing both Antarctic research scientists and the general public to learn more about this remarkable continent. The research scientists and the general public will have their own interfaces in order to meet each group's specific needs. For example, for the general public a teaching focus has been identified, specifically targeting the high school learner. Therefore, this chapter will discuss the educational component of the Atlas for high school students.

1.1. *Cybercartographic Atlas for Antarctica and Education*

As the coldest and most desolate place on Earth, Antarctica serves as a giant laboratory to look at the Earth's past and predict its future. This concept is not lost on *Students on Ice*, an organization that orchestrates expeditions for students to both Polar Regions. Their mandate is to provide students from around the world with inspiring educational opportunities at the ends of our earth, and in doing so, help them foster a new understanding and respect for our planet" (Students On Ice 2004).

The research team is fortunate to have teamed up with Students on Ice to develop an educational plan for the Atlas, for the mutual benefit of both parties. Using their experience with teaching students about Antarctica, and the research team's experience with cybercartography, the following questions are discussed in this chapter:

- How should learning with a cybercartographic atlas be approached?
- What requirements are needed to make the Atlas useful for both teachers and students?

2. Themes of Technology and Education

Cybercartography, with its use of multimedia, has numerous potential benefits for education. Multimedia in education allows for self-paced, interactive learning. Many researchers (Litke 1998; Harasim 2000) also assert that multimedia allows for a greater emphasis on student centered learning rather than teacher centered learning, which in recent years has been viewed as the preferred method of instruction. The use of cybercartography in education can play an important role in teaching students about the world around them. It also provides students with an opportunity to use and interact with computers – a major need in this information age. While studies show that students perform equally well using conventional learning methods as they do with computer assisted methods (Hiltz 1990; Matthew 1996; Urven *et al.* 2000), a growing body of literature suggests that students prefer computer assisted methods (Hurley *et al.* 1999; Urven *et al.* 2000; Neuhauser 2002; Thirunarayanan and Perez-Prado 2002).

Another theme in the literature is the debate over teaching *with* and teaching *about* technology. One of the problems with using computers in the classroom is that many teachers feel uncomfortable with the hardware and software due to lack of experience and training (Ross 2001). With cutbacks in education budgets, this situation is only exacerbated. Students who are not familiar with computers will also need to feel comfortable with the hardware and software before other learning takes place. Therefore, using a computer in the classroom as a teaching tool often begins with learning *about* the technology before learning *with* the technology.

There are many theories in academic circles about the best methods to approach teaching and learning with multimedia. Early computer-based materials were more influenced by *behaviorist* concepts while current materials are more founded on *cognitive* models of information processing (Deubel 2003). The primary tenant of "behaviorism is that there is a predictable and reliable link between a stimulus and the response it produces" (Deubel 2003: 64). Thus, learning with multimedia in a behaviorist format is very linear. Constructivists, on the other hand, believe that learning occurs when the student constructs new ideas based on their current or past knowledge. Multimedia learning in this format is multifaceted and emphasizes knowledge discovery.

2.1. Gardner's Multiple Intelligences Theory

One of the most widely distributed cognitive learning theories is Howard Gardner's *Multiple Intelligences Theory*. In the early 1980s, Gardner questioned not only the typical uniform style of teaching but also our concept of a singular intelligence. From his research he discovered that students learn in very different ways and outlined seven intelligences to describe these differences. These intelligences were presented in his widely read book *Frames of Mind* (1983). In this book Gardner argued that we all exhibit each of the seven intelligences, however we have strengths and weaknesses in different ones. The seven intelligences are:

Linguistic intelligence: "involves the sensitivity to spoken and written language, the ability to learn languages, and the capacity to use language to accomplish certain goals" (Gardner 1999: 41).

Logical-mathematical intelligence: "involves the capacity to analyze problems logically, carry out mathematical operations, and investigate issues scientifically" (42).

Musical intelligence: "entails skill in performance, composition, and appreciation of musical patterns" (42).

Bodily-kinesthetic intelligence: "entails the potential of using one's whole body or parts of the body (like the hand or mouth) to solve problems or fashion products" (42).

Spatial intelligence: "features the potential to recognize and manipulate patterns of wide space as well as the patterns of more confined spaces" (42).

Interpersonal intelligence: "denotes a person's capacity to understand the intentions, motivations, and desires of other people and, consequently, to work effectively with others" (43).

Intrapersonal intelligence: "involves the capacity to understand oneself, to have an effective working model of oneself – including one's own desires, fears, and capacities – and to use information effectively in regulating one's own life" (43).

Later, Gardner added an eighth intelligence, which is presented irregularly in the education literature. While presented here, it will not be discussed in detail in this chapter. This eighth intelligence is:

Naturalist intelligence: "involves capacities to recognize instances as members of a group (more formally, a species); to distinguish among members of a species; to recognize the existence of other, neighboring species; and to chart out the relations, formally or informally, among the several species" (49).

Gardner claims that educators tend to focus on teaching to the first two intelligences (*Linguistic Intelligence* and *Logical-mathematical intelligence*), neglecting students who exhibit the other forms. Therefore, he sees multiple intelligences (MI) theory as a basis of education reform that will encourage educators to look for ways to reach these other intelligences. Gardner continues to research MI theory and its use in education through *Project Zero* (http://www.pz.Harvard.edu/Default.htm), an educational research group at Harvard University.

2.2. Multiple Intelligences in the Classroom

Multiple intelligences theory was introduced just over 20 years ago, and in that time an extensive volume of literature, both books and academic articles, has been produced on how it can be incorporated into the traditional classroom. Examining how MI theory is used in the regular classroom will be useful in creating learning modules for the Atlas. Campbell *et al.* (2004) in their book, *Teaching and Learning through Multiple Intelligences*, suggest the following activities for appealing to the multiple intelligences as seen in Table 21.1. Howard Gardner has reviewed *Teaching and Learning with Multiple Intelligences* (2004), which is now in its third edition.

2.3. Multiple Intelligences and Multimedia

With the advent of the World Wide Web in 1993, and increase in use of educational CD-ROMs, multimedia has become another tool for educators

TABLE 21.1
Teaching and Learning Through Multiple Intelligences, Suggest the Following Activities for Appealing to the Multiple Intelligences (Cambell et al. 2004)

Linguistic Intelligence -Use of story telling -Writing a newsletter, booklet, letter, poem, play etc. -Giving presentations	**Spatial Intelligence** (known as **Visual Intelligence** in Campbell *et al.*) -Chart, map cluster or graph -Designing a poster, bulletin board,. mural, architectural drawing, etc. -Color code the process of an event, etc.
Logical-Mathematical Intelligence -Use of story problems -Creating timelines -Describing patterns or symmetry -Designing and conducting an experiment -Using code	**Interpersonal Intelligence** -With a partner use "out loud problem solving" -Paticipate in a group -Address a local or global problem -Give and receive feedback -Teach someone else about a topic, event, etc.
Musical Intelligence -Examining rhythmical patterns -Use of background music to enhance learning -Making an instrument and using it to demonstrate something -Relating the lyrics of a song to a topic, event, etc.	**Bodily-Kinesthetic Intelligence** -Role playing -Building or constructing something -Creating a series of movements to explain an event, topic etc. -Field trips
Interpersonal Intelligence -Explaining your philosophy about a certain topic or subject -Writing journal entries about a topic, event, etc. -Creating a personal analogy for a topic or event, etc. -Set and pursue a goal	

to appeal to the intelligences. Connecting multimedia to MI theory usually resides in the realm of educational literature as it is largely ignored by the psychology literature. (see Section 2.4).

The education literature has looked extensively at the use of MI theory in multimedia or e-learning, for example, the following researchers suggest that using multimedia will appeal to all seven intelligences (Veenema and Gardner 1996; Nelson 1998; Freundschuch and Hellvik 1999; Chen and Chang 2000; Passey 2000; Witfelt 2000).

Freundschuch and Hellevik contend that by providing a variety of mediums by which we can convey information, learners who posses different strengths and abilities in each of the different intelligences have multiple mechanisms that will encourage them to learn to their fullest potential" (Freundschuch and Hellevik 1999: 276). Table 21.2, developed by Freundschuch and Hellvik to link the various intelligences to multimedia, illustrates this.

Where this academic literature is lacking is in specific, classroom-tested examples. The literature tends to focus on ideas of how multimedia incorporates MI theory; it rarely describes how they were actually used. There is, however, a rich resource of material describing classroom-tested scenarios by teachers, online teachers, and operators of online learning institutions. This material is typically presented in less academic journals (i.e. Multimedia Schools) or education conference papers that are also freely available online. As the majority of classroom teachers do not have the means to access more expensive databases or journals, this is the mode by which teachers can communicate with each other. While the bulk of this material has not been academically scrutinized in the traditional sense, it can be argued that these data have been practically tested in the classroom. Therefore, these practical classroom experiences may be helpful in illuminating MI theory in computer assisted learning approaches. For example, Martin and Burnette (2000) explain how they use an electronic portfolio, a repository of a student's work in an electronic format, to incorporate MI theory. Shiratuddin and Landoni (2001) explain how they built a children's E-book based on MI theory.

2.4. Criticisms of MI Theory

While MI theory is widely accepted by the education community, it does have its critics. For example Klein (1997) critiques Gardner's idea of "multiple intelligences." Gardner (1993) argues that the existence of groups

TABLE 21.2
Medium and Potential Multiple Intelligences

Medium	Multiple Intelligence
Sounds (music and narrative)	Musical
	Logical-mathematical
	Linguistic
	Musical
Video (including sound, images and movement)	Logical-mathematical
	Linguistic
	Spatial
	Kinesthetic
Text	Linguistic
	Musical
	Kinesthetic
	Spatial
Animation	Musical
	Logical-mathematical
	Linguistic
Interactivity	Kinesthetic
	Spatial
	Spatial
Hypermedia	Kinesthetic
	Logical-mathematical

such as geniuses support MI theory as they illustrate exceptionality in one specific intelligence. He cites examples such as Babe Ruth being high in bodily-kinesthetic intelligence and Barbara McClintock being high in logical-mathematical intelligence. Gardner concedes that people, even geniuses, can excel in more than one domain, for example Albert Einstein is considered to have exhibited both high spatial and logical-mathematical intelligences (1993). However, Klein (1997) argues that this is where MI theory fails because if "Gardner fails to show that most achieve excellence in one specific domain, then his claim that the intelligences are independent is threatened" (381). Klein suggests that this shows instead that "the system as a whole is one single intelligence, and specific abilities, such as spatial reasoning, are mere components of this intelligence" (380). However, one of the points behind Gardner's theory is to examine our idea of what intelligence means. Is our concept of singular intelligence correct? Or is it possible that there are multiple forms of intelligence?

Sternberg (1994) argues that MI theory has no specific tests, experiments or otherwise that can be conducted on it. He questions the wide use of a theory that does not have direct supporting evidence. However, Gardner contends that MI theory "represents a critique of the standard psychometric approach accordingly, having a battery of tests is not consistent with the major tenets of the theory" (1999: 80). He proposes that intelligences must be assessed in an *intelligence fair way* "in ways that examine the intelligences directly rather than through the lenses of linguistic or logical intelligences" (1999: 80). For example, to assess spatial intelligence "one would allow people to explore a terrain for a while and see whether they can find their way around it reliably." (1999: 80).

Another criticism of MI theory is that it is based more on intuitive reasoning than on the results of empirical research studies (Aiken 1997: 196). However, Gardner contends that MI theory is based wholly on empirical evidence (Gardner 1999). He claims that hundreds of studies were reviewed and the "actual intelligences were identified and delineated on the basis of empirical findings from brain science, psychology, anthropology, and other relevant disciplines" (1999: 85).

These criticisms are valid. However, if MI theory injects new life into lesson planning, for the education benefit of all students, then it has done its job. It helps to explain why some students do not learn well under traditional logical or linguistic teaching methods. It seeks to find other ways to help students obtain a deeper understanding of their subject material. It also allows teachers to recognize that students may have different means of exhibiting intelligence, making the classroom far more egalitarian.

Multiple intelligences theory has the potential to be a useful framework in cybercartography because of its encompassing and egalitarian nature. Just as cybercartography attempts to include different disciplines in seeking knowledge, MI theory seeks to include different means of learning in seeking knowledge. Multiple intelligences theory also gives a voice to the educational perspective, rather than just the psychological perspective.

To sum up, MI theory is a practical model for the Atlas for number reasons:

1. Gardner's theory has been discussed in the literature of geographers interested in multimedia in cartography, such as Freundschuh and Hellevik (1999);
2. The principles behind MI theory closely emulate the principles behind cybercartography;
3. Teachers are very familiar with the MI theory, and a cybercartography atlas that utilizes it will be very familiar to them. This will help promote the use of the Atlas;

4. As mentioned previously, some teaching practices already presume that multimedia integrates the multiple intelligences;
5. Many researchers believe that multimedia appeals to the multiple intelligences;
6. It leads to a more egalitarian learning tool; and
7. It provides an educational perspective on learning.

3. Educational Requirements for the Atlas

To ensure that the Atlas is a justifiable and useful product, it must meet the needs of both teachers and students. The research group for the Cybercartographic Atlas of Antarctica has conducted a User Needs Assessment (UNA) with high school students and high school teachers. The information obtained from the UNA will help to establish the educational dimensions of the Atlas. However, based on the current literature on multimedia and online learning, the educational requirements outlined below should also be considered.

3.1. Accuracy and Quality

The most important concern for creating a dynamic learning tool like the Atlas is not the interactive technology, but the content and accuracy of the material used. Weston *et al.* (1999) stress that the content should be "relevant..., reliable..., and up to date" (37). The importance of this statement cannot be stressed enough as a well-designed Web site does not equate to a well-designed learning tool if the content has not been thoroughly examined for quality and accuracy. The research team at Carleton University is fortunate to have the expertise of several Antarctic research scientists and Antarctic educators, like Students on Ice, to examine the content.

3.2. The Ontario Curriculum

One way to ensure the value of the Atlas is to use quality curricula. As the research team has determined that high school students will be one of the primary user groups for the Atlas, it is necessary to find ways of making it both useful and attractive to this group. One possible way of attracting students is to develop learning modules that can be used in the high school curriculum. If the learning modules are connected to the high school curriculum, teachers will be more ready to use the Atlas. As the research team is located in Ontario, and therefore has access to Ontario schools and teachers, it is reasonable to use the Ontario high school curriculum to develop the modules.

Another push for using the Ontario curriculum comes from Students on Ice. Students on Ice is currently in the process of creating curriculum to be accredited by the Ontario Ministry of Education. If accredited, students who participate in one of their expeditions will receive one Ontario high school credit. Students on Ice would, therefore, like to utilize the learning modules as part of their program.

3.3. Appealing to the Audience

Researchers (Weston *et al.* 1999; Vrasidas 2002;) in the literature on online and multimedia learning stress the importance of creating learning environments that appeal to the students using it. A detailed analysis of the users will be presented in the UNA, however, for the moment two important factors should be considered when designing the Atlas: the age of the user and their experience with the Internet.

As the Atlas is designed for the high school learner (grades 9 to 12), it can be assumed that the majority of the users will range in ages 13 to 19. The Atlas should be designed specifically for this age group.

An important critique presented by Schrum and Hung (2000) about using the Internet in education is the challenge faced by many students in learning how to use the various tools available to them. These tools must be mastered before they can begin to construct new knowledge of the subject material. While Canada did become the first country in the world to connect its public schools online (Canada's School Net 2004), it is important to understand how familiar students are with the Internet. According to a Statistics Canada study by Dryburgh (2000), 90% of Canadian teenagers (aged 15–19) have used the Internet. This suggests that Canadian high school students are familiar with this medium and the percentages have increased since 2000.

3.4. Computer Use in Schools

To ensure the use of the Atlas in schools it must be usable in schools. Most researchers (Weston *et al.* 1999; Reeve *et al.* 2000; Skaalid 2001; Vrasidas 2002) stress the importance of examining the Internet speed and computer hardware capabilities of users to ensure a site's success. To achieve a picture of what these capabilities are in Ontario schools, the Ottawa-Carleton District School Board (OCDSB) will be used as an example. The OCDSB is the seventh largest school board by student population in Ontario. It serves approximately 80,000 students across 118 elementary and 27 secondary schools. The computer capabilities of the OCDSB's high schools are as follows:

Internet Speed: In OCDSB high schools receive approximately 10 MB per second speed rate or 1 MB per second over their connection (Goerz 2002).

Flash Player plug-in: The programs installed on all high school computers in the OCDSB range from version 4.x to 6.x (Hawes 2003).

Common computers used by high school in OCDSB include:
Windows Computers

- Northern Micro Spirit PII Celeron: 853 exist in high schools.
- Digital Venturis 575: 555 exist in high schools.

Macintosh Computers

- I-Mac G3: 574 exist in high schools.
- PowerMac 5260: 428 exist in the high schools (Hawes, 2003).

The OCDSB uses a computer lab infrastructure, meaning that teachers who want to integrate computer use in their classes must book a computer lab. This can affect accessibility to the computers as the labs are in high demand from either a regular scheduled computer class or from other teachers also wishing to integrate computer use into their lessons. Accessibility to quality computers may also be affected by a teacher's subject area. For example, the best computers in the school are typically reserved for computer classes; subject areas that traditionally have not used computers in the classroom have to book labs with the less superior computers.

As one of the goals of this project is to increase the potential of cybercartography, the design of the Atlas should only take into account these issues, but it should not be limited by them.

3.5. Students with Special Needs - Accessibility

As students with a range of accessibility issues will use the Atlas, it is essential that it be built so that every student in a class will be able to use it. Almeida and Tsuji consider this issue in *Interactive Mapping for People Who are Blind or Visually Impaired* in Chapter 18 of this volume. Another source to examine is The World Wide Web Consortium's (www.w3.org), *Web Accessibility Initiative* (WAI). The WAI is committed to promoting a high degree of usability for people with disabilities. Through the distribution of its *Web Content Accessibility Guidelines 1.0*, it hopes to promote accessibility to the Internet for all people. Adopting these guidelines will help to ensure that the Atlas will be useful to all students.

4. The Learning Experience of the Cybercartographic Atlas of Antarctica

The goal of the educational component of the Atlas is to create an accurate and quality learning experience, while at the same time encompassing the needs of both teachers and students as well as following the framework of MI. To do this, the Atlas is broken down into learning modules and a gaming component.

However, it will be difficult to continuously apply the intelligences throughout the Atlas. Even Gardner recognizes the challenges of applying all the intelligences to a single topic (Gardner, 1999). Multiple Intelligences theory should, therefore, be used as a guideline rather than a prescriptive set of rules. Using the research presented by Freundschuh and Hellevik (1999) and Campbell *et al.* (2004), what would a Cybercartographic Atlas of Antarctica look like using MI theory? Table 21.3 illustrates the potential that the Atlas has at appealing to the multiple intelligences.

The prototype learning module will be described in detail first, followed by a brief discussion of the gaming component.

4.1. A Climate Change Learning Prototype

To understand what a learning module would look like in the Atlas, one topic will be developed so it can be examined and critiqued. The finalized version of the learning module can then be used as a guideline and template for subsequent learning modules.

Through consulting with *Students on Ice* and examining the Ontario high school curriculum, climate change was chosen as the topic for the prototype learning module. Climate change refers to any change in climate over time, whether due to natural variability or as a result of human activity. It was picked as a prototype topic for several reasons:

1. It has strong connections to both the Ontario high school geography and science curriculum.
2. It is a major component of Students on Ice's Antarctica learning expedition.

TABLE 21.3
Multiple Intelligences and Cybercartographic Atlas of Antarctica Activities

Multiple Intelligences	Sample Activities
Linguistic	• Listening to an interview of an Antarctic research scientist on a particular topic • Reading text on a topic related to Antarctica
Logical-mathematical	• Constructing or viewing a historical timeline of Sir Ernest Shackleton's journey to and from Antarctica (See Fig. 21.1) • Working with climate data • Conducting an experiment on the floatability of icebergs
Musical	• Identifying Antarctic fauna (penguins, seals) by their various sounds • Building a shelter to withstand Antarctica's climate
Bodily-kinesthetic	• Engaging in an activity where the student controls the mouse, keyboard or joystick (i.e. an interactive activity where a student digitally dissects a penguin) • Examining maps or satellite images to develop an understanding of Antarctica's terrain
Spatial	• Viewing animation to explain a problem or situation, such as Antarctica's role in global climate
Interpersonal	• Examining and debating the Antarctica Treaty • Working with another student online on a project
Intrapersonal	• Writing a journal entry as a research scientist far away from home

3. It fits well with a Cybercartographic Atlas of Antarctica as there is a great deal of research currently being conducted on the continent's climate change and its role in global processes. This research has lead to the development of a variety of multimedia resources. In particular, NASA's Scientific Visualization Studio (http://svs.gsfc.nasa.gov/) contains many MPEGs examining climate change, such as decreases in ice. Videos like these can be incorporated into the Atlas given its multimedia component.
4. It is a major topic in the media and something about which students would have knowledge and interest in.
5. It is applicable to a wide variety of subjects, for example, geography, science etc.

4.2. Case Studies for the Learning Module

One of the major topics that *Students on Ice* addresses in their online curriculum to Antarctica is climate change (http://www.studentsonice.com/index.php ?content=antarctica_unit_9_global_climate_change). There

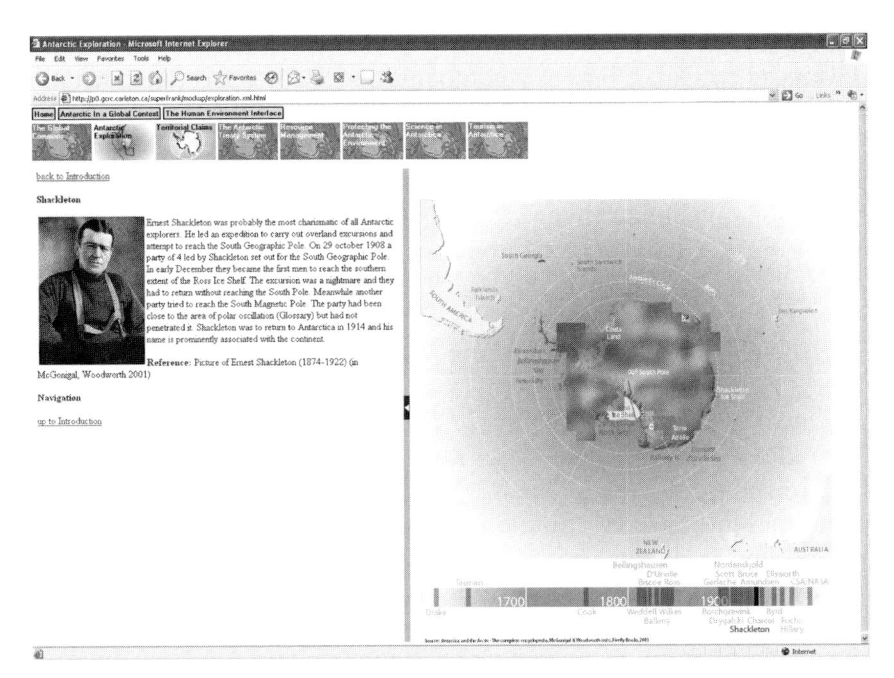

FIG. 21.1. A prototype learning module plotting Ernest Shacklenton's South pole expedition both cartographically and in the form of a timeline.

are many ways in which this information can be presented. As this section of the Atlas is designed to be a public site, with an emphasis on the high school learner, it would be ideal to create a general interactive site that both high school students and the general public could enjoy, with links to specific tasks for students to complete. An excellent method to make this site interesting is to follow an interactive, hands-on format, which attempts to appeal to the multiple intelligences. This is a format that is proposed by Veenema and Gardner (1996) when using technology with MI theory.

The basic design of the climate change learning module, uses examples from three excellent educational resources. Two of these resources are Websites that best exemplify Veenema and Gardner's (1996) idea of using multimedia in education. The other is the online learning curriculum designed by *Students on Ice*. These resources are as follows:

Green Learning:

> Recently launched by the Pembina Institute for Appropriate Development, this site is dedicated to teaching about the environment using the curriculum from the province of Alberta. This site provides information to both students and the general public, but contains additional exercises that students can complete in the classroom.

War of 1812:

Produced by Gala Films as a companion to their 4-part documentary on the War of 1812, this site is intended to teach about the War of 1812.

Students on Ice: Global Climate:

Part of Students on Ice's curriculum for Antarctica, this module contains information on climate change as well as student exercises to be completed while on their Antarctic expedition.

4.3. The Climate Change Learning Module

Both the *Green Learning* and *War of 1812* sites divide the overall topic of a subject into several subtopics – this is often referred to as "chunking" in the online learning literature. This is a practice many researchers believe help Web users synthesize information (McCampbell 2001; Skaalid 2001). The topic of climate change should therefore be divided into several subtopics as follows.

4.3.1. The Wonder of Antarctica

To help students appreciate the importance of Antarctica, it is necessary that they first understand its impact on the planet. Students on Ice addresses the topic of climate change in Antarctica by first describing why the nature of Antarctica gives it such a major role to play in global climate change. Their site points out that:

- It is affected by and impacts on some of the earth's most powerful wind patterns.
- It stores up to 70% of the world's fresh water.
- Incredibly powerful high and low pressure zones interact in the lower latitudes.
- Though not well understood, oceans and atmosphere redistribute surplus heat energy from the tropics to the poles.
- It receives more radiation at any point in its summer than the equator does.

One method to illustrate this information is to use a clickable map that highlights some of these key points, in the same sort of format that the *Green Learning* electricity lesson uses landscape to illustrate the different uses of electricity. This particular lesson was obviously designed for a much younger audience, but the idea can be expanded and developed for high school students. The control of the mouse and keyboard would appeal to students who exhibit bodily-kinesthetic intelligence, while the map format would appeal to students with a spatial intelligence. Text describing such information as the impact of the southern oceans on global climate will appeal to linguistic intelligence. This particular subtopic could impress upon students the importance of protecting Antarctica.

4.3.2. Climate Change and Human Activity

At this point, it would be logical to describe to students what climate change is and why many researchers believe it is being exacerbated by human activities. Students on Ice describes some of the human activities that contribute to global warming; for example the burning of fossil fuels. An animation illustrating these activities and the greenhouse gases they produce would be appropriate. This would also be an excellent area to include some science, such as the chemical substances and reactions of these gases. This could be described with text and accompanying animations. Animations would appeal to spatial intelligence while the logic of chemical reactions would appeal to mathematical-logical intelligence.

4.3.3. Impact of Antarctic Climate Change on the World

The Ontario curriculum, in both science and geography, requires students to understand the effects of climate change on the environment. There are many effects that can be detailed in this section. One topic would be to look at the effect of severe Antarctic ice melt on ocean levels, which would give it a global focus. Nova, the science television series on PBS, has an excellent site (http://www.pbs.org/wgbh/nova/vinson/ice.html) illustrating what would happen to locations around the world if the ice were to melt. Maps showing the effects of ice melt on Canada would help Canadian students better understand impacts of radical climate change.

4.3.4. Climate Change – Natural Process versus Human Activities

Students need to be introduced to the concept that, while widely accepted that, human activities are leading to global climate change, it is a debated topic among some circles. Some scientists contend that changes in temperature are due to natural rather than human activities. This could involve reading articles presenting the two sides of the debate. The War of 1812 site does an excellent job of this type of situation by including textual information on the different perspectives of the Americans and the British at various battles, such as the battle of Queenston Heights. Therefore, an obvious way to present this debate would be to read textual information. This would appeal to linguistic intelligence, but the debate could appeal to interpersonal intelligence. The War of 1812 site also contains paintings of the period to further illustrate their textual information. Simple graphics could be used to help students better understand the complicated debate on global warming.

4.3.5. Research on Climate Change in Antarctica

The Student on Ice site describes the research currently being conducted in Antarctica on climate change. It describes methodologies such as ice core sampling and remote sensing, which help to provide information about past air temperatures and the relationship with changes in carbon dioxide and methane levels in the atmosphere. A description of these processes would be necessary. Green Learning contains an interactive activity in its electricity unit where students build a virtual circuit. Something similar could be done to allow students to virtually operate an ice core sampler or see how a satellite operates. Students could then examine changes in ice through looking at several remote sensing photos (Fig. 21.2), or examine the data

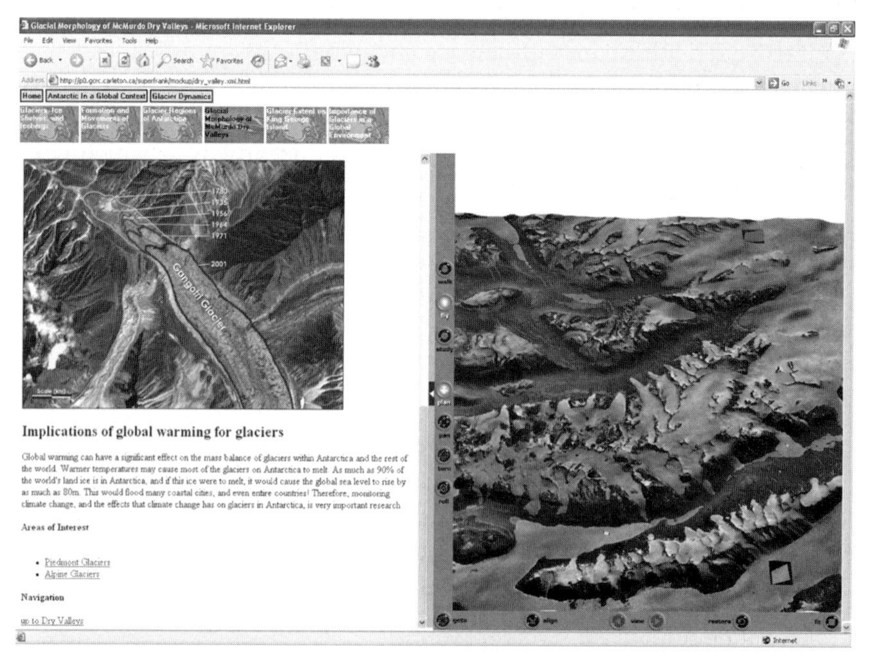

FIG. 21.2. Learning module using satellite imagery and terrain surface rendering to illustrate the effects of environmental change on the glaciers of the world in general and the Dry Valleys region of Antarctica in particular (image compliments of Y. Zhou *et al.* 2004).

obtained from ice core samples. Such an exercise would appeal to mathematical-logical, spatial intelligences and bodily-kinesthetic intelligences.

4.3.6. Action Against Global Warming

Global warming is a bonafide atmospheric phenomena. Therefore, it is important to leave students with an understanding of its most negative consequences and what they can do to prevent them. This last section should review the importance of Antarctica to the world using awe-inspiring video. It should then let students know about organizations (government and non-governmental organizations) that are involved in the issue and how students can become active in promoting their goals. In particular, it should describe the organizations Canada is involved in, such as the Kyoto Protocol, as per the Ontario curriculum. Finally, it should describe some of the simple things students can do in their own lives to reduce the level of greenhouse gases. For example, the Green Learning site has an excellent activity where students can turn on and off household appliances to see how much energy they use. Using a method such as this, students will understand the simple tasks they can do such as turning off a light when not in use to reduce energy use and thereby help the environment.

4.4. Student Activities on Climate Change

As previously mentioned, the learning modules are expected to appeal to both high school students and the general public, with separate activities that high school learners can do in the classroom. This is the format adopted by the Green Learning site. The climate change activities should expand on what the learning modules have already taught, and lead to a deeper understanding of the material. Examples of classroom activities include:

- Plot the mean annual temperature of Antarctic over a period of years. Explain why determining a temperature trend is difficult.
- Conduct experiments on greenhouse gases, such as determining carbon dioxide.
- Debate the nature, extent and mechanisms of global warming
- Write a report on how students can help protect Antarctica.

4.5. A Prototype Gaming Component

To complement the learning experience of the Atlas, the cybercartographic team has added a gaming component. This component was included to provide a fun, but educational way to learn more about Antarctica.

In the prototype game, the student starts by creating and naming a new character. This character is a new arrival to an Antarctic research station and learns about Antarctica through speaking to other characters, participating in lectures and going on quests, such as collecting ice core samples (Fig. 21.3).

Gaming has the potential to incorporate many elements of MI theory. For example:

- Use of a joystick or keyboard controls will appeal to the bodily-kinesthetic intelligence.
- Navigating in a gaming world will appeal to the spatial intelligence.
- Communicating with characters will appeal to the linguistic and interpersonal intelligence.

FIG. 21.3. Prototype gaming component where the main character is communicating with penguins (From Game Protype created by Dormann *et al.* 2005).

4.6. *Climate Change and the Ontario Curriculum*

As mentioned previously, it is important that the learning modules and the gaming component be connected to the Ontario curriculum. The UNA carried out at the HOTLab incorporated the curriculum in their analysis.

5. Discussion

This is an early discussion of preliminary conceptual and functional design of the Atlas. However, before building and testing of the Atlas occurs, it is essential that the UNA is reviewed and its findings taken into account. It is also important to incorporate research from psychology as discussed in various other chapters in this volume. This will ensure that the Cybercartographic Atlas of Antarctica holds to the principle of being multidisciplinary.

Based on this and the information presented in this chapter, a prototype-learning model has been created. This model can then help to answer some very essential pedagogical questions:

- Is MI theory valid for this type of learning resource?
- How do the various multimedia elements appeal to learning and MI theory?
- How should the Atlas be designed so that teachers find it useful in the classroom?
- How should the site appear so that it appeals to high school students?
- How should the site be designed so that it is accessible to all students?

Answering these questions will help to refine the module. The finalized version of the climate change module will then be used as a guideline for subsequent learning modules.

6. Conclusions

In this age of dynamic technology, a Cybercartographic Atlas with a strong educational focus can be a useful learning tool. It has the potential to incorporate many aspects of MI theory and, therefore, to potentially reach more students, something that can be challenging in the traditional classroom. The approach we have described has the ability to teach to all intelligences using integrated forms of multimedia, and will leave students with a deeper understanding of this fascinating continent.

References

Aiken, L. R. (1997) *Psychological Testing and Assessment* (9th edn.), Allyn and Bacon, Boston.

American Geographical Society (1994) *Geography for Life: National Geography Standards 1994,* National, Geographic Research and Exploration, National Council for Geographic Education and National Geographic Society Washington, DC.

Anderson, J., J. Carrière, P. Pitre and J. Le Sann (2002) "A prototype internet school atlas for Quebec, Canada", *Society of Cartographers, Bulletin,* Vol. 36, No. 2, 7–14.

Atlas du Quebec, accessed November (2004) from http://www.atlasduquebec.qc.ca/scolaire/

Campbell, L., B. Campbell and D. Dickinson (2004) *Teaching and Learning Through Multiple Intelligences*, Pearson Education, Inc., Boston.

Canada's SchoolNet, accessed November (2004) from http://www.schoolnet.ca/home/e/whatis.asp

Chen, P. C. and T. H. Chang (2000) *Multimedia as a Teaching Tool for Multiple Intelligences,* Presented at 6th annual International Conference on Engineering Education, 14–16 August 2000 Taiwan, Taiwan, accessed 7 February 2004 from http://www.ineer.org/Events/ICEE2000/Proceedings/papers/WD5-1.pdf

Dryburgh, H. (2000) *Changing Our Ways: Why and How Canadians Use the Internet,* Statistics Canada, cat no. 56F0006XIE, accessed 7 March from http://www.statcan.ca/english/IPS/Data/56F0006XIE.htm

Deubel, P. (2003) "An investigation of behaviorist and cognitive approaches to instructional multimedia design", *Journal of Educational Multimedia and Hypermedia,* Vol. 12, No. 1, pp. 63–90.

Dormann, C., J-P. Fiset, S. Caquard, B. Woods, A. Hadziomerovic, E. Whitworth, A. Hayes and R. Biddle (2005) *Computer Games as Homework: How to Delight and Instruct*, Home Oriented Informatics and Telematics Conference, 13–15 April 2005, University of York, United Kingdom (forthcoming).

Freundschuch, S. M. and W. Hellevik (1999) "Multimedia technology in cartography and geographic education", in Cartwright, W., M. P. Peterson and G. Gartner. (eds.), *Multimedia Cartography*, Springer-Verlag, Berlin, pp. 15–172.

Gardner, H. (1983) *Frames of Mind: the Theory of Multiple Intelligences*, Basic Books, New York.

Gardner, H. (1993) *Multiple Intelligences: the Theory in Practice, Basic Books,* New York, New York.

Gardner, H. (1999) *Intelligence Reframed: Multiple Intelligences for the 21st Century*, Basic Books, New York.

Goerz, J. (2002) (email from Jessica Goerz, Educational Technology Leader, Ottawa-Carleton District School Board, Ottawa, Ontario, 13 December 2002).

Green Learning, accessed November (2004) from http://www.greenlearning.ca

Harasim, L. (2000) "Shift Happens: Online Education as a New Paradigm in Learning", *Internet and Higher Education*, Vol. 3, pp. 41–61.

Hawes, T. (2003) (e-mail from Tim Hawes, Secondary Consultant, Ottawa-Carleton District School Board, Ottawa, ON, 21 February 2003).

Hiltz, S. (1990) "Evaluating the Virtual Classroom", in Harasim, L. (ed.), *On-Line Education: Perspectives on a New Environment*, Praeger, New York, pp. 133–183.

Hurley, J. M., J. D. Proctor and R. E. Ford (1999) "Collaborative Inquiry at a Distance: Using the Internet", *Geography Education, Journal of Geography*, Vol. 98, pp. 128–140

Klein, P. D. (1997) "Multiplying the Problems of Intelligence by Eight: A Critique of Gardner's Theory", *Canadian Journal of Education*, Vol. 22, No. 4, pp. 377–394.

Litke, C. Del (1998) Virtual Schooling at Middle Grades: a Case Study, Thesis (PhD), Graduate Program in Educational Research, University of Calgary.

Matthew, K. (1996) "The Impact of CD-ROM Storybooks on Children's Reading Comprehension and Reading Attitude", *Journal of Educational Multimedia and Hypermedia*, Vol. 5, pp. 379–394.

Martin, G. P. and C. Burnette (2000) *Maximizing Multiple Intelligences Through Multimedia: a Real Application of Gardner's Theories Multimedia Schools,* accessed 7 March 2004 from http://www.infotoday.com/MMSchools/oct00/martin&burnette.htm

McCampbell, B. (2001) E-learner = Self-motivated?, *Principal Leadership,* Vol. 2, No. 2. pp. 63–65.

NASA's Visualization Studio accessed November (2004) from http://svs.gsfc.nasa.gov/

Nelson, G. (1998) "Internet/Web-based Instruction and Multiple Intelligences", *Education Media International*, Vol. 35, No. 2, pp. 90–94.

Neuhauser, C. (2002) "Learning Style and Effectiveness of Online and Face-to-Face Instruction", *The American Journal of Distance Education*, Vol. 16, No. 2, pp. 99–113.

Nova - Mountain of Ice: If the Ice Melts accessed November 2004 from http://www.pbs.org/wgbh/nova/vinson/ice.html

Passey, D. (2000) "Developing Teaching Strategies for Distance (Out-of School) Learning in Primary and Secondary Schools", *Education Media International,* Vol. 37, No. 1, pp. 45–57.

Project Zero, accessed November 2004 from http://www.pz.harvard.edu/

Pulsifer, P. L., D. R. F. Taylor, B. Eddy and T. P. Lauriault (2003) *Cybercartography and the New Information Economy: Canada's Trade with the World and the Cybercartographic Atlas of Antarctica,* Presented at 21st International Cartographic Conference, 10–16 August 2003 Durban, South Africa: pp. 1093–1101 [CD-ROM].

Reeve, D., S. Haedwick, K. Kemp and T. Ploszajska (2000) "Delivering Geography Courses Internationally", *Journal of Geography in Higher Learning*, Vol. 24, No. 2, 228–237.

Ross, H. (2001) *Can computers make a difference? Edutechnot,* accessed 4 February 2005 from http://www.edtechnot.com/notross601.html

Schrum, L. and S. Hung (2002) "From the Field: Characteristics of Successful Tertiary Online Students and Strategies of Experienced Online Educators", *Education and Information Technologies*, Vol. 7, No. 1, pp. 5–16.

Shea-Schultz, H. and J. Fogarty (2002), *Online Learning Today: Strategies that Work*, Berret-Koehler Publishers Inc., San Francisco.

Shiratuddin, N. and M. Landoni (2001) *Multiple Intelligence Based E-Books*, The 2nd Annual Conference of the LSTN Centre for Information and Computer Sciences, accessed 1 October 2004 from http://www.ics.ltsn.ac.uk/pub/conf2001/papers/Shiratuddin.htm

Skaalid, B. (2001) "Web Design for Instruction: Research-Based Guidelines", *Canadian Journal of Education Communication*, Vol. 2, No. 3, pp. 139–155.

Sternberg, R. (1994) "Commentary: Reforming School Reform: Comments on Multiple Intelligence: the Theory in Practice", *Teachers College Record*, Vol. 9, No. 4, pp. 561–569.

Students on Ice, accessed November 2004 from http://www.studentsonice.com

Taylor, D. R. F. (2003) "The Concept of Cybercartography", in Peterson, M. P. (ed.), *Maps and the Internet*, Elsevier, Amsterdam.

Thirunarayanan, M. O., and A. Perez-Prado (2002) "Comparing Web-Based and c: A Quantitative Study", in *Journal of Research on Technology in Education*, Vol. 34, No. 2, pp. 131–137.

Urven, L. E., L. R. Yin, B. D. Eshelman and J. D. Bak (2000) "Presenting Science in a Video-Delivered, Web-Based Format: Comparing Learning Settings to Get the Most Out of Teaching", *Journal of College Science Teaching*, Vol. 30, No. 3, pp. 172–176.

Veenema, S. and Gardner, H. (1996) "Multimedia and Multiple Intelligences", *The American Prospect*, Vol. 7, No. 29, accessed 7 March 2004 from http://www.prospect.org/print/V7/29/veenema-s.html

Vrasidas, C. (2002) "Systematic Approach for Designing Hypermedia Environments for Teaching and Learning International", *Journal of Instructional Media*, Vol. 29, No. 1, pp. 13–25.

War of 1812, accessed November 2004 from http://www.galafilm.com/1812/e/intro/index.html

Weston, C., T. Gandell, L. McAlpine and A. Finkelstein (1999) "Designing Instruction for the Context of Online Learning", *The Internet and Higher Education*, Vol. 2, No. 1, pp. 35–44.

Witfelt, C. (2000) "Educational Multimedia and Teachers Needs for New Competencies: a Study of Compulsory School Teachers Needs for Competence to Use Educational Multimedia", *Education Media International*, Vol. 37, No. 4, pp. 235–241.

Woods, B., R. Biddle, S. Caquard, C. Dormann, J. P. Fiset and A. Hayes (2004) *NeverWinter Nights in Antarctica: Developing a Video Game for Affective*

Learning, CAGONT (Canadian Association of Geographers - Ontario Division) Conference, Waterloo, Ontario, 29–30, October 2004.
World Wide Web Consortium, accessed November (2004) from http://www.w3.org
Zhou, Y., X. Liu, B. Woods, P. Pulsifer and D. R. F. Taylor (2004), *Visualization in Cybercartography*, CAGONT (Canadian Association of Geographers - Ontario Division) Conference, Waterloo, Ontario, 29–30 October 2004.

CHAPTER 22

Applying a Cybercartographic Human Interface (CHI) Model to Create a Cybercartographic Atlas of Canada's Trade with the World

BRIAN G. EDDY

Geomatics and Cartographic Research Centre (GCRC),
Department of Geography and Environmental Studies, Carleton University,
Ottawa, Ontario, Canada

D. R. FRASER TAYLOR

Geomatics and Cartographic Research Centre (GCRC),
and Distinguished Research Professor in International Affairs,
Geography and Environmental Studies,
Carleton University, Ottawa, Ontario, Canada

Abstract

The Cybercartographic Human Interface (CHI) model proposed in Chapter 3 is used as an organizing framework for the development of a Cybercartographic Atlas of Canada's Trade with the World (CTW). The model is used to differentiate the respective domains and their characteristics, and to examine cybercartographic issues involved in their integration and implementation in interface design. "Content-context" dependencies suggest human needs and content analysis must proceed in parallel before considering interface design options. A consideration of these reveals a number of organizational and societal issues that often lie beyond the control or influence of the cybercartographer. Practical, organizational and institutional requirements, as well as classical cartographic issues are discussed in relation to options for implementation.

1. Introduction

It has been argued that to produce a cybercartographic atlas, a model is required (Reyes Chapter 4). The Cybercartographic Human Interface (CHI) model discussed in Chapter 3 is presented here as a framework for the development of a cybercartographic atlas of Canada's Trade with the World (CTW). Our first challenge is to rethink the concept of atlases in terms of how we see their construction and use in the information era. One contrast with conventional paper atlases is that cybercartographic atlases may be considered "living atlases" in a number of ways. Cybercartographic atlases can be continuously updated. The concept of an atlas is no longer static; it is no longer a "fixed product" in the conventional sense. Users can interact with the content in a variety of new ways, and information can be provided that are specific to user needs and learning styles (Baulch *et al.* Chapter 21). Additionally, there is no distinction between "producers" of cybercartographic atlases and their "users." As a result, users can contribute to cybercartographic atlases enriching their content and increasing interaction at each of the primary, secondary and tertiary levels of the CHI model.

Building and maintaining a cyberatlas, creates significant new challenges which have not been met before in conventional atlas construction. In Chapter 1 and several subsequent chapters of this volume, the interactive relationship between theory and practice as a key element in the paradigm of cybercartography has been emphasized. It is in this new context that we examine the potential usefulness of the CHI model as basis for "praxis"; defined here as "theoretically informed practice." The major elements of the CHI model are used to establish goals for atlas development, as well as to highlight the challenges and constraints that may be expected in attempting to apply theory in practice. The development of the CTW starts from the theoretical concepts embodied in the CHI model and moves to practice. Once again, we view the relation between theory and practice as a set of iterative feedback loops.

The application of the CHI model for the CTW atlas is presented here in three sections. First, we begin by outlining a general approach to project design utilizing the principles of the CHI model. Figure 22.1 shows a simplified CHI model that builds upon Taylor's (1997) three-part schematic. This highlights the cybercartographic, human and interface

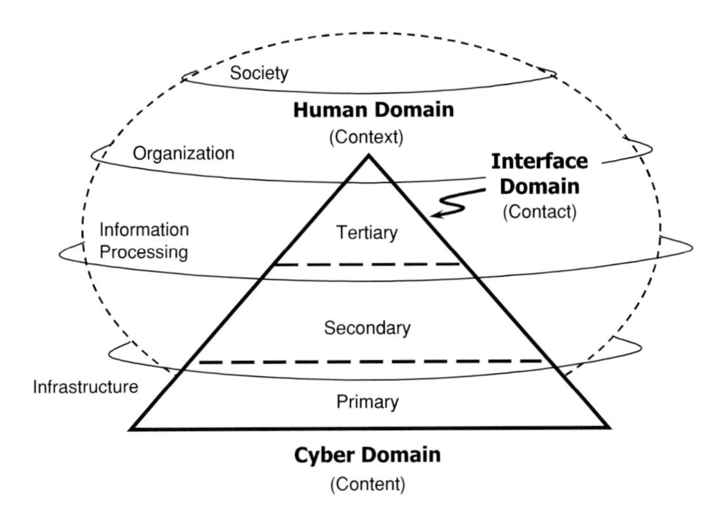

FIG. 22.1. Simplified Cybercartographic Human Interface (CHI) model, adapted for application to the development of a cyber atlas of Canada's Trade with the World.

domains, but with important distinctions for how these interact in each of the primary, secondary, and tertiary levels. Second, we examine some important issues and experiences encountered in both the human and cybercartographic domains. Finally, we outline options for technological infrastructure and interface development, and highlight a number of related issues associated with the development of cybercartographic atlases (Fig. 22.1)

2. Project Design

In project design and implementation, initial attention is given to the three principle domains of the CHI model. Information gathered about each domain (e.g. characteristics, constraints, requirements etc.) is used in project planning and throughout all phases of the development cycle. Figure 22.2 simplifies elements of the CHI model in terms of differentiating the user, interface, and the content, and is used as a general project planning framework. As suggested by human factors psychologists, the approach taken includes and extends a "user-centred design" approach as we utilize user needs analysis (UNA) prior to the development of the atlas (Fig. 22.2).

Although UNA captures critical information that must be fed into the development process at the earliest stages, additional considerations are also required with respect to broader societal and organizational levels

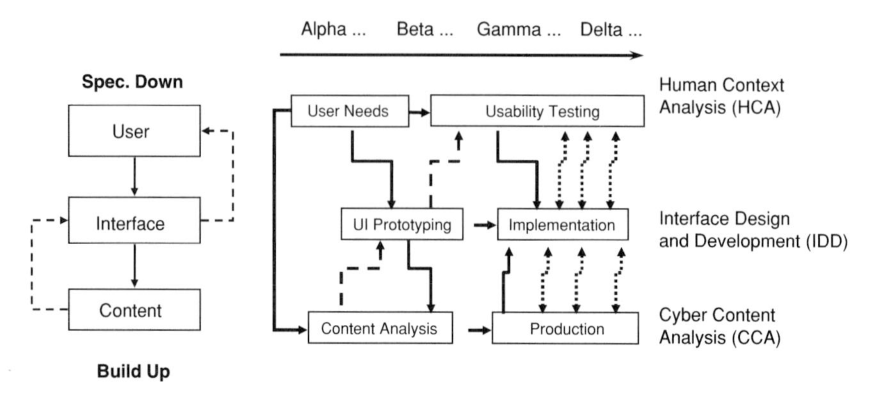

FIG. 22.2. Adaptation of the CHI model in a macro-level project design.

within the human domain. All aspects of the human domain are considered under the umbrella of a "Human Context Analysis" (HCA) that continuously feeds the development of the cyberatlas at different stages of the project. This is an example of the socio-cybernetic feedback process described in Chapters 1 and 5.

In parallel with the HCA, it is equally important to proceed with a "Cyber Content Analysis" (CCA). One tenet of integral theory is that "the lower sets the possibilities of the higher", indicating that it is important to determine early in the project what content is available (or possible) to meet specific user needs. As we illustrate below, this involves considerable effort from a number of areas of expertise involving subject experts, database administrators, Geographic Information System (GIS), and geomatics professionals, as well as the identification of economic, legal, organizational, and political issues and constraints associated with the acquisition and use of specific content sources.

The third component of the CHI model suggests that Interface Design and Development (IDD) proceeds only after sufficient information is collected from the HCA and CCA. The manner in which cybercartographic applications are considered to be dynamic, nonlinear, interactive, and dialogical is reflected in the development process itself, wherein the HCA and CCA represent the demand and supply dimensions respectively. A useful principle illustrated in this approach is that we "specify from the top-down" and develop or "build from the bottom-up." The interaction takes place at the IDD stream of activities. IDD presents significant challenges for people involved in reconciling and mediating conflicting constraints and requirements between the CCA and HCA. A number of issues that have emerged in applying this approach to the CTW are elaborated in subsequent sections.

This project design also incorporates several distinct phases, or stages, during which cybercartographic applications "emerge." Figure 22.2 summarizes these phases as Alpha, Beta, Gamma, and Delta. The terms are used here in a different context than in other formal system designs and development methodology. We argue that a cybercartographic atlas, being an "emergent" product that results from efforts of people across organizational and disciplinary boundaries, cannot be developed in any rigid, step-wise fashion (as suggested by many formal systems development approaches). We do recognize, however, that distinct and timely breakpoints can be identified within the development and production cycle allowing assessment of progress in relation to the overall production process.

One of the central principles of the CHI model is that cybercartographic applications are considered holonic artifacts that mimic the human mind, but aim to overcome physical constraints in information processing and communication. Cybercartographic applications are multiauthored and reflect a broader human need for extended information processing capacity. An ecological view suggests that applications may be viewed as emergents in response to broader human-societal needs for adaptation to ever changing environmental conditions. It follows that the manner in which they are developed also mirror psychosocial development patterns. In short, this approach views the cycles and phases of development to mirror similar cycles and stages in human adaptation to changing environmental conditions.

There are several psychosocial development models that may be used, Wilber (1999) identifies over thirty of these, and the one used here is based on Spiral Dynamics (Beck and Cowan 1996). In this view, applications emerge, in part, as an artifact of a human development process. The central feature of the spiral dynamics model situates human development as a series of nested stages, each of which has a corresponding cycle of development that must be followed through to completion. Here we use their phases of the cycle of development as a framework for identifying distinct phases of development. The phases are summarized as follows:

The Alpha Phase – the initial phase of a development characterized by a broad human concern and societal context, a need for "change" that also issues a "need to know" calls for better information support on some issue. It is partly characterized by an information gap which is understood in relation to limitations in the existing information-base. Alpha phase activities begin with an HCA and CCA, and identify critical requirements across content/context relational scales: from the societal to the organizational, individual/interface functionality, information processing and flow, and infrastructure requirements. Alpha Phase outputs are various specifications and/or requirement documents that attempt to describe and remedy the perceived problem as best as possible.

The Beta Phase – involves a more thorough analysis of similarities and differences in the HCA and CCA, and ends with initial attempts to produce rudimentary IDD prototypes. It is critical at this stage to be able to "differentiate" and cross-correlate across the HCA and CCA elements. The primary aim of initial IDD prototypes is to expose and explore more critical issues in reconciling HCA requirements with CCA constraints. This phase is the beginning of an increasingly iterative phase of interface design, development and experimentation, where the team proceeds in small steps in consultation with both HCA and CCA team members. If successful, a better understanding of the dynamics typically unfolds, and an initial reconciliation is reached on how to proceed with more formal development. This includes critical choices in infrastructure and interface development tools, content processing plans and pathways, and an overall production plan.

The Gamma Phase – although all phases of the project cycle are equally important, the Gamma Phase is perhaps the most critical. Significant effort in time, money, and resources is invested in the initial development of a system. Although the focus is on the physical development of the system, it requires continued participation of people representing the three CHI domains at all levels. Whereas beta-level prototypes may offer various "conceptual" type interfaces and options, Gamma phase development proceeds with the real interface and atlas development process. Here, critical challenges may be encountered. Gamma phase work may discover significant variances between what was anticipated from the beta-design, and what the technology actually delivers. In a cyber development environment, where programming environments are open and flexible, interface developers may be surprised by the significant variation between what third-party providers say a technology can do, and what can actually be implemented.

The Gamma phase is the most critical phase of the development cycle. If a sufficient number of things go wrong, the development process can end up in a "Gamma Trap" (which is the most common point of failure in systems development). It is therefore critically important to have significant iteration and refinement, involving changing directions in infrastructure and interface design if necessary, based on interactions among individuals representing each of the CHI domains. This leads to the creation of an initial "operational prototype" that can be launched for external review and feedback.

The Delta Phase – in this phase the operational prototype developed during the Gamma phase is extended to include more iterative refinement of both content and interface functions, and features based on feedback from beyond the immediate user base of those involved in the development process (test subjects used in design stages). For cyber environments, such as the Internet (on which the majority of atlases will reside), this may involve considering new information from users and perspectives that were not captured during previous phases. If the operational prototype succeeds in meeting a number of essential requirements of a majority of users, then such iterative feedback will result in a "Delta Surge", and a publicly available cyberatlas will emerge.

Cyberatlases, as "living, dynamic atlases", are not fixed, one-time products. They mirror many aspects of Internet sites and are constantly updated with new information. Users from a variety of backgrounds may not only use the atlas as an information source, but may also contribute to it in many ways. Applying the holonic tenets further, it can be argued that if cyberatlases are not used, then they will "self-dissolve." In the Delta phase, a point of stability in the cycle is reached, but does not guarantee longevity. Maintaining a cyberatlas will depend largely on the level of user interactivity. We may also anticipate that new human contexts and societal issues will emerge, which in turn will lead to new Alpha phases, creating a continuous cycle. This model illustrates the iterative feedback processes between theory and practice characteristic of the paradigm of cybercartography, and we will now explore these issues for the development of the CTW.

3. Human Context Analysis

Our HCA proceeds on a number of levels that include the four general levels outlined in Chapter 3 and illustrated in Fig. 22.1: (1) societal; (2) organizational; (3) information processing; and (4) infrastructure. The aim is to gain an understanding of the characteristics and requirements associated with each level as well as to examine the nested (holonic) nature of the linkages among levels. The approach "envelops the process" within a constructively critical perspective from which we can return to for iterative review and refinement as required. The HCA attempts to make clear the relationship between the dictae and the modis, following the approach of Latour (1999), and adds additional detail across the four general levels.

3.1. The Societal Level

At a societal level, we can explore the circumstances that create the need for a Cybercartographic Atlas of Canada's Trade with the World. Why is trade important? Why focus on Canada's relationship with the world? Why is there no information source to fill this need? Is an "atlas" an appropriate medium for filling this information gap, and if so, is a "cybercartographic" atlas necessary? Who should develop it and what should it contain? Who will use the CTW and for what purpose? Clearly, these are just a few of many questions that can be asked at this level; questions that we return to throughout the development cycle.

The motivation for the development of the CTW stems from an information gap at the national level in Canada. In the emerging new economy, Canada continues to be increasingly dependent upon foreign trade, and in particular, trade with the United States. While experts debate these important and complex issues for Canada, the educational system and the general public are not fully aware of their significance, and the impact these will have on society as Canada is increasingly absorbed into a global economy. There is no shortage of data and information available on Canada's Trade with the World, but this is found in a variety of disparate source and forms more suitable to experts rather than the general public. Data and information on trade are not easy to access. In addition, there are few sources that give a synoptic view of trade involving spatial patterns and trends over time. The geography of Canada's trade with the world is not readily available from existing source in an easy to use format.

The CTW is designed to fill this gap, and to explore the international and national context of Canada's trade with the world. Innovative cartographic design, the provision of data, facts, statistics and formal knowledge, and interactive interfaces are all important components of the CTW, and a synoptic approach is taken in order to give a better understanding of the complex issues and their interrelationships with other issues. Latour argues that how information is constructed in society depends upon the players involved, and the broader societal context. This mirrors the arguments made by Martinez and Reyes earlier in this volume and is an important facet of the paradigm of cybercartography. Developing the CTW has involved the formation of partnerships with a number of key organizations within the Federal Government of Canada, university and the educational system (this has been described by Lauriault and Taylor in more detail in Chapter 8). The partnerships created to produce the CTW bring us to an analysis of the organizational level, and its requirements and constraints.

3.2. *The Organizational Level*

Figure 22.3 illustrates how the organizations involved relate to the different levels of the CHI model. Two divisions within Statistics Canada (Geography

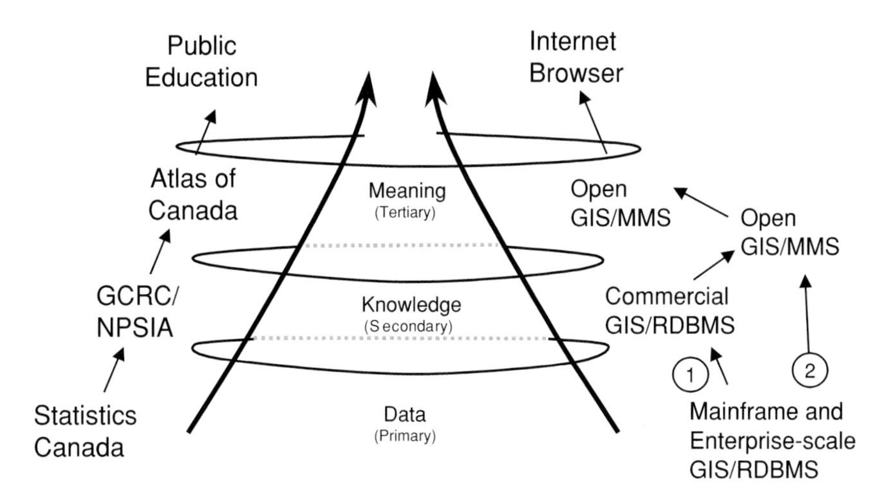

FIG. 22.3. Key organizations involved in the CTW in relation to the CHI model and corresponding geospatial technology infrastructures. GIS = Geographical Information Systems, RDBMS = Relational Database Management Systems. Note: see text for description of technology interface options 1 and 2.

and International Trade) represent the primary level, and the Atlas of Canada serves a public outreach capacity at the tertiary level. The secondary level is facilitated by researchers in the Department of Geography and Environmental Studies and the Norman Patterson School of International Affairs (NPSIA) at Carleton University, with the help and contributions of subject experts in both Statistics Canada and the Atlas of Canada. In Latours' terms, these stakeholders represent the alliances that form among organizations, with the higher level support of the Government of Canada.

In following a user-centred design approach, an initial task involved the organization of a stakeholders meeting to identify a "common vision" for the CTW (Lindgaard 2003). The common vision that emerged was stated as:

> The development of CTW is an integral part of a Social Sciences and Humanities Research Council (SSHRC), which supported multidisciplinary research project intended to extend the Atlas of Canada to enable an exploration of Canada's trade at three nested scales: with the US, with the continent, and with the world. The possibility of considering interprovincial trade exists.

Additionally, two other high level objectives were also identified:

(1) The product is a service to the Canadian public initially geared to teachers, and students (grade 8–12). This sets the minimum, likely to be expanded later, and

(2) Exploring trade "relationships and issues", for example, environment, security, quality of life, resources for the purpose of a variety of educational functions.

Having made these decisions, the next step in the HCA involved conducting a more formal User Needs Assessment (UNA) (Fig. 22.3).

3.3. The Information Processing Level

We situate the UNA (Philp *et al.* 2004) at the information process level of our model, where the focus is on the information processing at the immediate human–computer interface. The UNA in this study involved interviews with high school students and teachers, as well as a thorough analysis of trade related content in textbooks currently used in grades 8–12 of the Ontario high school curriculum. The general results of the UNA reveal some interesting characteristics in relation to the CHI model, one of which is how it clearly differentiates needs in terms of content requirements from functional interface design issues. Content requirements are summarized in Fig. 22.4. Part of what the UNA uncovered was that trade was not taught as a subject on its own, but within a number of subject areas, and at different levels of depth in grades 8–12. Given the complex nature of

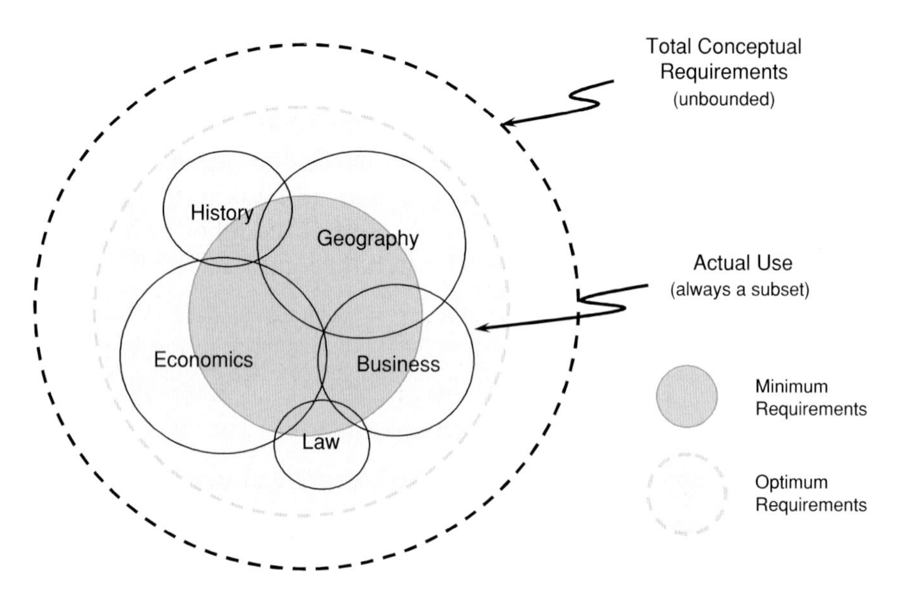

FIG. 22.4. Content requirements shown as content–context relations among multiple, but interrelated subjects.

trade, and its linkages with many issues, this is not surprising. Trade is taught more centrally in geography, economics, and business, but is also an important aspect in understanding history and law. It can be inferred from these results that for the broader general public, the subject of trade is likely to be discussed, analyzed, critiqued, or better understood in a variety of similar contexts.

Many functional interface requirements were identified, several of which provide critical information for atlas design (Khan *et al.* 2004). Students make extensive use of Internet-based resources at both home and school, and work both individually and in groups on the completion of assignments. Group activity and discussion around trade issues, as well as the use of meaningful visual aids, are important learning strategies. Students need to print maps and tables, and interrogate statistics for a broad number of economic sectors and countries. Significant challenges are created by time and resource constraints for both teachers and students. Although the textbooks provide a plethora of information on trade, very little of it was actually used in the real time classroom setting. On average, only six of one hundred and thirty hours was spent on teaching some aspect of trade (less than 5%).

Situating these results in CHI terms reveals a number of elements that were significant for the cybercartographic and interface domains. Although the "possibilities" for the need of a plethora of information support are truly there (as indicated by the broad range of coverage in the textbooks), the actual usage pattern is significantly less, dictated by constraints in time and resources at the classroom level. It is likely that the general public will spend equal or less time in learning about trade than students and teachers. These aspects of "actual versus potential" are important to differentiate in deciding which content is most appropriate in meeting specific users' needs (Fig. 22.4).

Figure 22.4 illustrates a plurality of contexts for which the same primary trade content (data, maps, tables, and knowledge) may be used in a tertiary level environment, and can be viewed as a "top-down" perspective of the CHI model. In information entropy terms, there are conceptually an infinite number of "possibilities" for the use of trade information, but in reality, only a limited number of "probabilities" in which the CTW will be actually used. The challenge for the CTW team was to identify the most likely usage patterns in order to narrow the scope of requirements to a manageable level in terms of both the initial development and implementation of the atlas, as well as its ongoing iterative updating and maintenance. Figure 22.4 also illustrates that a reasonable approach targets a minimum set of requirements that address as wide a potential user base as possible (in this case, partial coverage of five different subject areas). An optimum approach would aim to meet all the requirements of all subject areas, but would still be less than the total conceptual requirements space because these are boundless or "virtually infinite."

4. Cybercartographic Content Analysis

As mentioned above, the CHI model offers a general working principle that cybercartographic applications are specified from the top-down, but are built from the bottom-up. Top-down and bottom-up interactions characterize the complexity of the nonlinear dynamic interaction between the human and cyber domains. Although they are complex, it is helpful to differentiate these interactions at primary, secondary, and tertiary levels. Human needs and use, technology, organizational culture, and the nature of content are all fundamentally different at each level of the model. In this section, we explore how the CHI model was used to differentiate and organize activities associated with these general levels of the process.

4.1. The Societal Level

A preliminary alpha analysis explored a content–context relationship at the societal level by explaining the extent to which trade "maps and data" would help give a better understanding of trade "issues." This question arose from an examination of existing online trade mapping and data interfaces that provide a wealth of basic or primary data (Industry Canada 2004; US Dept. of Commerce 2004). This assessment concluded that users could not derive meaningful synoptic-level information from these sources without already having some substantial knowledge of trade. Of fourteen important contemporary trade issues identified, it was determined that maps and data would be helpful in understanding only six of these (Eddy and McDonnell 2003). Part of the reason for this is that primary (or basic) dollar-value maps are not contextualized around specific issues. If we consider an issue as a type of "storyline," then it might be worth exploring how tertiary level maps may augment storylines to reinforce a better understanding of trade issues.

This alpha phase analysis revealed the magnitude of the task involved in translating large volumes of primary data into meaningful cybercartographic syntheses. How will the CTW atlas help to derive meaningful (contextual) information from primary data? How could cybercartographers and interface designers mediate this process? Should the content of the trade atlas be "contextualized" around specific trade storylines? And to what extent should we allow users to interrogate the primary data and develop their own interpretations and storylines? These questions are explored further in the subsequent sections, but first some other realities of the cybercartographic process, involved in acquiring and making meaningful use of primary trade data at the organizational level, are considered.

4.2. The Organizational Level

As mentioned above, the CHI model positions our two key government partners, Statistics Canada and the Atlas of Canada, at the primary–secondary, and secondary–tertiary level boundaries respectively. The team was already familiar with the capabilities and capacities of the Atlas of Canada site, and its internal process for content publication. Early in the project, it was assumed that the CTW would initially be developed as an "additional content" for the existing Atlas of Canada site (including its current interface, server technology, and internal content delivery process) (Atlas of Canada 2004). The site is built upon the open source Minnesota Map Server (MMS) technology, and includes over one hundred existing themes and issues on the geography of Canada. The majority of content has had substantial synthetic-level processing and contextualization around specific issues, by cartographers and experts from a variety of disciplines.

In contrast, our partners in Statistics Canada do not provide such a high level interface to their data. In fact, their general working climate is significantly more specialized, using a mainframe-based system for recording, extracting, analysis and output of trade information. The Atlas of Canada is public-oriented in its content delivery (facilitated by open source technology). Statistics Canada, on the other hand, is much more private-oriented in terms of its client base, and in its mix of proprietary and commercial-based information technology infrastructure. Statistics Canada also has highly legalized, formal protocols surrounding its information processing. Although Statistics Canada does have a public-level information site, the vast majority of their services are in the provision of highly specialized analysis and reporting to government, industry and subject-specific research clients. They are a public agency, but direct access by the public to trade data in particular is difficult.

Accessing and using content from the Atlas of Canada site are essentially free, whereas accessing and using Statistics Canada data requires purchasing and licensing agreements. Even within a project that had already established a formal working relationship in which Statistics Canada was an official partner on the CTW project, formal agreements still had to be negotiated and settled according to standard government and cost-recovery protocols. Completing these arrangements took more than six months during the latter part of the alpha phase of the project, which added an unexpected delay in gaining access to the primary data.

These issues are highlighted here to indicate their importance for cybercartographic applications that aim at working across multiple organizations and with large volumes of authoritative data, which may be directed toward a virtually unlimited user base. Dealing with different working cultures, operational protocols, and information processing requirements across these organizations presents significant challenges in atlas development. Many of these challenges were less significant in the past when cartographers aimed to produce onetime products. But for a living cybercartographic atlas, where information will be continually updated from different institutional sources, overcoming these barriers is critical.

Such challenges include political, legal, and institutional factors, above and beyond strictly information processing requirements. For organizations, such as Statistics Canada, ensuring responsible use of their data is of legal significance. In fact, the formal agreement made between the CTW project and Statistics Canada clearly differentiates specific protocols that must be followed depending on the level of processing beyond the primary to secondary content boundary (GCRC 2003). We return to these issues in following sections, and simply introduce them at this stage as being of critical importance in arranging access to primary data for using in a cybercartographic context. The creation of new partnerships between government, industry and academia, which is one of the seven elements of cybercartography (Taylor 2003), is relatively easy in theory, but much more difficult in practice (Lauriault and Taylor Chapter 8). The CTW experience has shown that high level partnership agreements can quickly be reached in principle, but that is often "the devil is in the details." Once again, theory and practice involve iterative and often unexpected feedback loops.

4.3. The Information Processing Level

Whereas the information processing level in the HCA is captured in a UNA, a CCA focuses on the internal information processing requirements for meeting the UNA. Our experience has been that fulfilling organizational requirements is closely related to the information processing across the primary, secondary, and tertiary levels. The aim of the CTW is to provide the most up to date, relevant, and authoritative trade information in contexts that are meaningful to the intended user base. This can include the use of primary data, but more often requires syntheses of various combinations of primary (data), secondary (knowledge), and tertiary (meaning) level content.

A major challenge was to make the leap from the very large volume of primary data to tertiary level presentation. A major multidisciplinary research effort was required to reach a synthesis suitable for the subject areas as shown in Fig. 22.4. First, the content requirements from the UNA were extracted as a starting point for formal analysis. By the time the project entered beta phase, a Trade Atlas Content (TAC) team was organized (as part of the overall project matrix) to analyze the requirements against the characteristics of the data available through the International Trade Division (ITD) (Eddy *et al.* 2004). It was decided that a synoptic level atlas would contain the following components:

(1) Introductory chapters that cover both general trade theory (and, related economic theory and terms), geopolitics and the emergence of regional trade agreement (RTA) blocks, and international trade organizations (e.g. WTO, IMF). This component represents the more knowledge-oriented content providing users with what is "currently known" about trade.
(2) General historical trends of Canada's international trade, and more detailed trade data and maps covering specific economic sectors. A special chapter was included on Canada's trade with the US using data recorded at respective provincial and state levels. This component is more data-oriented providing users with statistical content concerning trends, patterns and flows of Canada's trade.
(3) A third section highlights a number of contemporary trade issues at both local and global scales which are currently topics of public debate. This component represents the more meaning oriented component of the atlas, and is much more subjective and contextual. These arise from first understanding the content of the first two sections of the atlas.

The content has been created at a general level for education and information purposes in response to the UNA. Quizzes and learning modules for students have been incorporated in various places throughout the chapters. In relation to trade data (and trade mapping), the requirements specified the need to provide users with an ability to select various trade sectors across various time periods and trading partners. There are a number of trade classification systems in use that had to be analyzed to support this need. Current standard systems include the North American Industrial Classification System (NAICS) and the international level Harmonized System (HS). The analysis and discussion among trade subject experts concluded that neither of these systems provides the capacity to look at longer term trends necessary for a synthesis of Canada's evolving trading patterns with the world. A third classification system in use with Statistics Canada is the Standardized Imports/Exports Grouping (SG) system. The SG classification is a normalized approach that reads data from various classification systems across both time and origin, and was therefore, deemed most appropriate for the CTW.

A key finding is that the number of possible content elements that can be generated by either of these classifications is virtually unlimited. Each classification system contains a hierarchy of sector or commodity groupings. At the most detailed level, there are several hundred groupings, recorded on monthly intervals over many years. Providing maps at yearly intervals for the higher levels in the SG system (at level three there are 71 sectors) would require several hundred thousand possible "maps," not including the many ways of classifying or symbolizing trade in each map. Each map contains a different distribution of data values (in $CAN) from which various derivatives may also be used (e.g. % of GDP), adding to the number of possible representations.

As outlined in Chapter 3, if information entropy is understood to be the number of possibilities available in a content–context relationship, then a huge number of possibilities result from applying the primary data against many possible contexts. Choices, therefore, had to be made to ensure that requirements were met, while respecting realistic constraints imposed by primary data, organizational and resource requirements, and most importantly, in meeting a production timeline. In CHI terms, this involved diminishing the probability space within the total range of possibilities. This mediation process is a fact of life in information production whether this is done by a human or by a more automated process such as agent based computing. (The role of the cartographer as mediator is more fully considered by Pulsifer and Taylor in Chapter 7.)

A key lesson learned in developing the CTW is that because of the constraints identified among all levels, "choices" always need to be made. Such choices are subjective and qualitative and are thus inevitably subject to critique. Critical theory and the societal implications involved are discussed in later sections of this chapter. For now, we turn to our experience in applying the CHI model in the interface domain.

5. Interface Design and Development

Examination of the information processing requirements in relation to the systems or infrastructure level was intentionally omitted from the cybercartographic content analysis. There are infrastructure and interface issues that set additional constraints upon information processing and flow at each level, but are more appropriately considered together in the Interface Domain of the CHI model. As Table 3.2 in Chapter 3 illustrates, there are different technologies, infrastructure, and interface requirements at each level of the CHI, and our experience in discovering these patterns across the participating organizations and information processing levels closely fits with those listed.

5.1. *Interfaces and Technologies*

Interface design and development in a CHI context involves two aspects:

(1) Interfacing both existing and emerging technological infrastructures that promote internal information processing and flow (internal interface); and
(2) Interface development at each level of the model, including that of the CTW "atlas interface" (external interface).

In general, working from the bottom-up, the full range of technologies encountered include legacy mainframe environments used within Statistics Canada, to GIS/RDBMS used in the Geomatics and Cartographic Research Centre (GCRC), to open source mapping and Web-based Internet technologies used by both GCRC and the Atlas of Canada; in their case, Minnesota Map Server (MMS) (Fig. 22.3). Interfaces associated with all of these technologies are important when considering the possible range of user needs. Figure 22.3 illustrates two possible technological pathways for information flow and interfacing technologies from the primary to tertiary levels. The option of using GIS/RDBMS software as "intermediary" technology between mainframe data storage, and retrieval and open source web-based mapping environments exists at the secondary level. Standard commercial GIS/RDBMS software was used during the alpha and beta phases of content analysis, and production to explore the characteristics of the data, as well as to identify analytical and cartographic issues in meeting tertiary level requirements. For the CTW, the possibility for bypassing the GIS/RDBMS option depended in part on the level of data processing and analysis functionality available in open source technologies with respect to the functional needs of the intended users. Both options were considered. There are advantages and disadvantages to each approach.

Using commercial GIS/RDBMS at the secondary level provides a greater analytical and data integration capacity. Although open source technology avoids the need for commercial GIS/RDBMS, its analytical and data manipulation functionality is limited, and in addition requires a more significant learning curve for content developers (only people with sufficient programming experience can work directly with the technology). The approach taken was to implement the trade data in both environments, using commercial software for data integration tasks in preparation for migration to the MMS environment. Custom data integration products were generated using the commercial software and uploaded into the MMS environment (e.g. integrating the primary trade data tables with standard international and Canada-US political boundary maps and auxiliary data sets).

As Pulsifers' (Chapter 7) Open Source Cartographic Framework (OCF) illustrates, the use of open source technologies allow a number of options to be implemented at the tertiary interface level. These principles were used in the development and implementation of the CTW. A significant influence on the choice of technology used was a shift toward the use of open source and SVG and Flash-based interface environments that extend the geospatial capacities of the MMS/OpenGIS technologies. Following this approach, a beta phase prototype was developed with these technologies to explore some of the issues discussed throughout this chapter. These new technologies provide significantly greater possibilities for user interaction in exploring, discovering, visualizing and analyzing data. Before presenting the results of the beta phase prototype, it is necessary to discuss a key issue that surfaced concerning user freedom and responsibility.

5.2. User Freedom and Responsibility

The capacity of tertiary level interfaces to provide various degrees of interactivity across multiple levels of content is central to the concept of cybercartography and many of the issues discussed in previous sections of this chapter. Interactivity extends well beyond navigating through existing, structured content. Higher degrees of interactivity allow users to explore patterns in both the data and knowledge, from which they can derive their own meanings. The capacity for technology to enable this pluralistic approach is one of the most distinguishing features of more sophisticated cybercartographic environments.

The general approach taken was to provide users with an authoritative, structured framework to learn about trade, including contemporary issues, through the presentation of geographical patterns and statistics of Canada's trade with the world. Adopting a critical theory perspective, users are made aware that the initial content provided is just "one way" of presenting a view of Canada's trade with the world. Many trade issues are open to different interpretations, and users are encouraged to "navigate, explore and discover" for themselves alternative trade patterns, statistics, knowledge and issues, and possibly to construct their own storylines or modules that they may add to the content of the CTW.

A number of questions emerged in adopting this more open and dialogical approach. How should the functionality and content of the atlas be structured so that users do not get lost in "too much information", and what level of guidance is appropriate in this respect? Additionally, government partners and subject experts expressed legal and scientific concerns about how users might interpret and use the data. Giving users the freedom to interpret the data in any way they wish (as is the case with much Internet content) is not a straightforward matter. For example, liability issues arise if users cite our partners (i.e. Statistics Canada or the Atlas of Canada) as the "source" supporting their interpretation. This presents the creators of cybercartographic atlases with two choices at either end of a continuum:

(1) Present content in a manner that implies it is the "only" view of reality, which users must accept without modification;
(2) Present content in a completely open access environment where users are free to draw their own conclusions and interpretations without any formal guidance or support.

It can be argued that none of these two extremes are either realistic or desirable, but signify a continuum within which cybercartographers must mediate information processing and flow. All forms of cybercartographic mediation involve compromises, and the CTW adopted the following compromise and rationale. In CHI terms, the proposed content offers users "what is currently known" by trade experts using the supportive data and maps (the authoritative element). As an educational product, the atlas also contains a series of quizzes and learning modules. One compromise is to incorporate a user profile that not only serves as a storage mechanism for students work (either formal assignments, or their own explorations), but also as a basis for formal evaluation of how much the student has learned about trade. As students (or other users) successfully complete a series of quizzes or learning modules, points are added to their profile that allow users to enter into a deeper levels of the atlas, providing more advanced functional tools to explore, analyze and interpret the data more freely. This approach might alleviate the concerns of content providers and partners by at least ensuring that users analyzing the content (and maybe generating new content) have acquired an appropriate level of knowledge and expertise on the subject. In general, the "degrees of freedom" provided to users should have corresponding levels of "responsibility."

The general relation between the degrees of freedom and responsibility can be set at the different levels of the CHI model. As users enter at a tertiary level, and become more advanced subject experts, they may then enter a secondary level where they can carry out their own analyses and interpretations. This general principle was used in the development of the CTW placing it at the open access end of the continuum, but arguing for the need of guidance and structure. The psychological literature on learning described in several chapters in this volume supports the need for some structure within which interactive learning can take place. The complexities and nature of trade do not lend themselves to completely unstructured and unguided approaches. We now turn to exploring interface issues in the provision of analytical cartographic functionality that aim to compromise these concerns.

5.3. A CTW Beta Prototype

As discussed above, the OCF model was used as a guiding framework for integrating a suite of open source technologies that not only provide greater flexibility in visually appealing and interactive interface design, but also allow a variety of analytical functions. A beta prototype was developed to explore and test how this level of functionality would help resolve some of the conflicts between HCA requirements and CCA constraints. Figure 22.5 shows a snapshot of the CTW beta prototype built using a third-party utility GeoClip (www.geoclip.net) (a functional copy of this prototype appears in the CD-ROM accompanying this book). GeoClip runs as an extension in ArcGIS 8.x, and provides content developers the ability to generate compact interactive interfaces for a particular map theme. Figure 22.5 summarizes some of the main features incorporated in GeoClip that are similar to those used in the development of the CTW. The focus of attention here is on the interactive and analytical functionality rather than on interface design.

The GeoClip prototype presents a number of very important features, as illustrated in Fig. 22.5. The key feature is the dynamic map display, where the data is handled in a thin-client configuration (i.e. data are cached to the client application from a server depending on the selection made by the user). Standard features common to many Internet mapping sites include controls for zooming, panning, and searching for features, selecting layers (or themes) and labeling, and dynamic textual and spatial queries of map elements. The more advanced cartographic elements include an "interactive legend" which allows users to select the number and range of class intervals, the color palette, and other classification tools. Not shown in

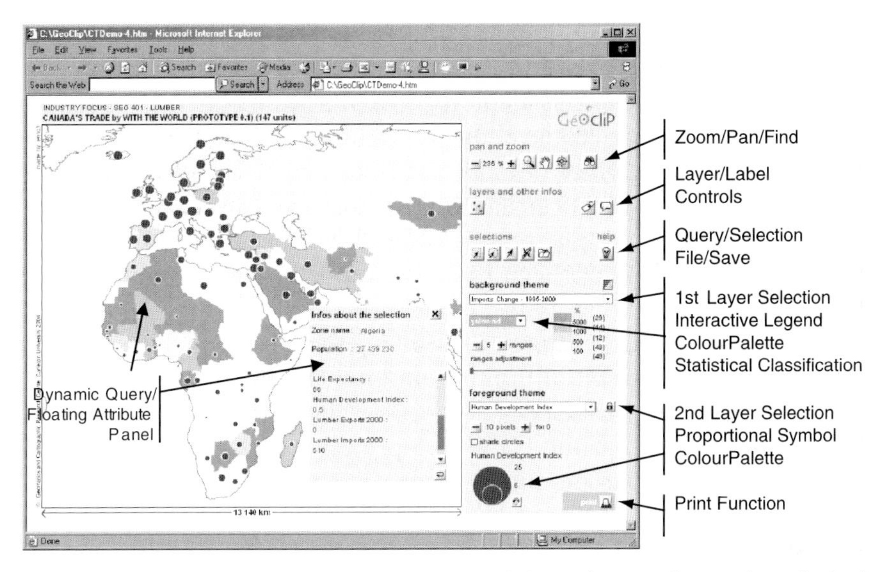

FIG. 22.5. Example beta-phase prototype of CTW interactive and analytical functionality. (Note: this prototype was developed using GeoClip technology, see www.geoclip.net).

Fig. 22.5 is a function that allows users to switch to a "distribution plot" of the data from which they can then set their legend intervals visually. The background theme presents statistical data in a choropleth format, and the foreground theme is presented as proportional symbols (Fig. 22.5).

Figure 22.5 shows the percentage change in Canada's lumber exports to Africa and Europe between 1995 and 2000 against the Human Development Index for each country over the same time period. The print function switches the display to a printable map format, where users can add additional text and notes to their output. Many of these features can be expanded and GeoClip is presented as an example of the directions open source technologies with SVG, and Flash type interfaces are taken. The interface provides a number of standard cartographic functions that are very easy to use, even for novice level users, relieving the cartographer from having to mediate individual user preferences, while at the same time, ensuring that users are following commonly accepted cartographic principles, and data access protocols.

6. Conclusions

Chapter 3 explored the concept of cybercartography through the transdisciplinary lens of integral theory, and presented a synthetic, holistic model of the components and dynamics involved in a cybercartographic domain. In this chapter, we explored our practical experience in applying this CHI model during the development of a cyberatlas of Canada's Trade with the World. The CHI model has demonstrated its utility in this cybercartographic endeavor. A macro-level project planning framework that situates cyber, human and interface domain activities as parallel efforts over four general phases of production has been presented. Reyes (in Chapter 4) argues that cybercartography requires a modeling perspective. This chapter describes how such an approach has been applied to the production of the CTW.

Much of what was experienced in applying the CHI model and described in this chapter is the result of inputs from numerous individuals contributing in a variety of ways to the development of the CTW, both on the CANE project scale, and within the CTW sub-project itself. Although the CTW has an intended user base in mind, we also consider all those involved in its development to be the "first users" of the atlas. Subsequent users, including students, teachers, the public, and policy makers are expected to contribute to its ongoing maintenance and adaptation to new user contexts. The "Human Domain," in all of its complexity, is of paramount importance. Applying a transdisciplinary framework that aims to be as inclusive as possible to better understand these complexities has proven to be beneficial in the creation of the CTW, illustrating the value of this element of the cybercartographic paradigm in a concrete way.

One of the key principles of cybercartography emphasizes "process" in relation to "product." The approach taken here is in itself a work in progress. We do not make any definitive conclusions on the benefits in applying a CHI model to other contexts other than to say that it has been helpful in anticipating and understanding some of the complexities encountered in our experience. At the time of writing, we were at mid-point of a 4-year project, about to enter into Gamma phase research and development, and have not yet had the chance to assess the usefulness of the model in these later phases of the project. The usefulness of the CHI model, and its theoretical underpinnings will be determined through the interaction of theory and practice. It has proven useful in our specific context but requires further development and testing by other creators of cybercartographic products. Comments and feedback in this respect are welcome.

We have described some of the problems in putting cybercartographic theory into practice. There are many novel and innovative aspects of this new paradigm, and it is tempting to be seduced by the positive aspects they suggest. The production of the CTW has however, revealed both expected and unexpected realities that affect all aspects of the production process. The promise of the cybercartographic paradigm must be tempered and informed by this experience.

Our experience reiterates that an important distinction between cybercartography and classical or conventional cartography is emphasized in one of the seven elements of cybercartography: that it is "part of an information system" (Taylor 2003). The CHI model looks beyond the conventional cartographic issues at a multitude of other factors that will inevitably influence the kinds of maps and mapping that emerge in an information era. Cybercartographic applications, or a cyber atlas, can be seen as a "tip of an iceberg." Novice users may not concern themselves with what lies beneath, but for those involved in cybercartographic production, which will increasingly include intermediate and advanced users (in dialogical interactions), having an adequate understanding of these aspects of information theory, systems integration, and information production are critically important.

One of the principles of integral theory emphasizes that in order for integration to occur, adequate differentiation of that, which is being integrated, must take place. Our analysis reveals critical contrasts in requirements and constraints between the human and cyber domains, which intersect and interact at the level of the interface. Cybercartography in the information era will be used for real time, or near real time information dissemination about the world at large. Information pathways move from the world, throughout enterprises working in collaboration, and back to society in a recursive fashion. Interface design and development may be where it all comes together, whereas Peterson (1995) argues that cartographers are now taking on the role of "interface designers," we would extend this argument to suggest that cybercartographers are now taking on the role of information systems designers, developers, organizational facilitation, critics, artisans, and philosophers. The challenges involved in integration across these fields extend well beyond the capacities of any single individual or discipline, and require interdisciplinary teams of which cybercartographers are one component, and this may raise questions of professional identity for cartographers. Cybercartography however sees maps as central by definition and the cartographers who make them have an important role to play.

References

Atlas of Canada (2004) Available from www.atlas.gc.ca

Beck, D. and C. Cowan (1996) *Spiral Dynamics: Mastering Values, Leadership and Change*, Blackwell, Oxford.

Eddy, B. and F. McDonnell (2003) *Canada's Trade with the World: Presentation of First Term Research Results*, Presentation given internally at the Geomatics and Cartographic Research Centre (GCRC).

Eddy, B., I. Guttler, O. Pidufala and F. McDonnell (2004) *Cybercartographic Atlas of Canada's Trade with the World:* Proposed Chapters and Draft Text internal Report of the Geomatics and Cartographic Research Centre (GCRC).

GCRC (2003) *Micro-data Research Agreement*, Statistics Canada and the Geomatics and Cartographic Research Centre (GCRC), Internal Project Document.

Industry Canada (2004) *Trade Data Online Internet Site*, accessed 7 February, 2005 from http://strategis.gc.ca/sc_mrkti/tdst/engdoc/tr_homep.html

Khan, S., M. Rasouli, A. Parush and G. Lindgaard (2004) *Requirements Specification*, Human Oriented Technology Lab Ottawa, Canada.

Latour, B. (1999) *Pandora's Hope: Essays on the Reality of Science Studies*, Harvard University Press, Cambridge, Massachusetts.

Lindgaard, G. (2003) "Stakeholders vision statement", *Project Summary and Project Report*, Internal Project Document.

Peterson, M. P. (1995) "Maps and the changing medium" in Peterson, M. P. (ed.), *Interactive and Animated Cartography*, Prentice Hall, pp. 3–9.

Philp, K., M. Rasouli, S. Narasimhan and G. Lindgaard (2004) "User needs analysis report for cybercartography and new economy project", *Canada's Trade with the World Product* (CD-ROMs, Human Oriented Technology (HOTLab), Carleton University, Ottawa, Canada.

Taylor, D. R. F. (1997) "Maps and mapping in the information era", *Keynote address to the 18th ICA Conference, Proceedings*, Vol. 1, Swedish Cartographic Society, Stockholm, Sweden, pp. 1–10.

Taylor, D. R. F. (2003) "The concept of cybercartography" in Peterson, M. P. (ed.), *Maps and the Internet*, Elsevier, Cambridge, pp. 403–418.

US Dept. of Commerce (2004) *Trade Stats Express*, accessed 7 February, 2005 from http://ese.export.gov

Wilber, K. (1999) "Integral psychology", *The Collected Works of Ken Wilber*, Vol. 4, Shambhala, Boston, pp. 423–722.

CHAPTER 23

Remaining Challenges and the Future of Cybercartography

D. R. FRASER TAYLOR

Geomatics and Cartographic Research Centre (GCRC)
and Distinguished Research Professor in International Affairs, Geography
and Environmental Studies, Carleton University, Ottawa, Ontario, Canada

Abstract

This chapter reviews what has been learned about the paradigm of cybercartography, what challenges remain, and what the main directions are for future research. The new paradigm of cybercartography has much to offer though many questions remain to be answered, and much further research is required.

1. Introduction

In Chapter 4, Carmen Reyes poses the question "Are we really at a breakthrough in the development of cartography as Taylor suggests...." (Reyes, Chapter 4: 92). This concluding chapter will address this question by looking at what we have learned about cybercartography, what challenges remain, and what the key research directions are for the future.

Much has been learned since the paradigm of cybercartography was formally introduced in 1997 and the chapters in this book make a major contribution in this respect, but as was pointed out in the introductory chapter and by authors of a number of subsequent chapters, the paradigm is a relatively recent one and many questions on both cybercartographic theory and practice still remain.

2. Major Lessons Learned

Throughout the chapters in the book many interesting and important observations have been made. The objective here is not to repeat these in detail but to summarize some of the key lessons that have emerged from the previous 22 chapters.

2.1. Cybercartography and Society

The relationship between maps and society has always been central to cartography as an applied discipline, and this relationship expands as the paradigm of cybercartography evolves. It is clear from several chapters in the book that cybercartographic products are social products. In traditional cartography the dominant paradigm was a supply-driven approach, where cartographers produced maps in response to perceived societal demand. Over the last two decades, the supply-driven approach of many national mapping organizations has been increasingly modified to a demand-driven approach where users can choose online and access only maps of interest to them, or in some instances, to create their own maps interactively from cartographic databases. Cybercartography takes this still further. Societies can be active participants in the creation of cybercartographic atlases as is illustrated by the cybercartographic atlases produced by CentroGeo in Mexico (Chapter 6). In these cases the entire community can be involved and the cybercartographers are immersed in the context from which the atlas emerges. This qualitative, dynamic, and interactive process is an important departure from past practice. There are echoes of this process in other cybercartographic products described in the book. The division between "mapmaker" and "map-user" is blurred as the "map-user" becomes a "mapmaker" and the "mapmaker" a "map-user." In addition, Web-based cybercartographic atlases are "living" atlases, which continue to evolve and develop.

Cybermaps are leading to increased understanding of complex issues as is the case with the Cybercartographic Atlas of Antarctica, the Cybercartographic Atlas of Canada's Trade with the World, the Georgia Basin Digital Library, and the five cybercartographic atlases including that of Lake Chapala produced by CentroGeo in Mexico, described in this volume. This increased understanding can lead to more informed decision making.

But in addition, there is evidence in the Mexican case that the immersion of cybercartography in society is creating direct social action on environmental issues. Here, the cybercartographic atlases are not only an aid to government decision making but also help to create social and political actions and in this sense contribute to increased democratization and good governance. Cybercartography allows different perspectives to be presented and different voices to be heard both literally and metaphorically through the use of voice commentary and video as an integral part of a cybercartographic atlas. On contentious and complex issues, such as environmental policies, where different stakeholders can have conflicting priorities, cybercartography can present several viewpoints on the same "reality" in interesting new ways. As Caquard and Taylor argue in Chapter 12, cybercartographic products can encourage the user to reflect on a number of different interpretations of the information presented. Caquard goes even further and argues that new forms of artistic cybercartographic "anti-maps" can be created as a form of social criticism, and Greenspan suggests that games can inspire a new interactive cartography.

The variety of new ways in which information can be presented makes cybercartography of particular value in education as outlined by Baulch *et al.* in Chapter 21. In both Mexico and Canada the formal education system has been identified as an important user for cybercartographic products, and CentroGeo has produced a separate Cybercartographic Atlas of Lake Chapala for schools. In addition, cybercartographic atlases will be useful in informing and educating the general public, which for example, was a key factor in developing the Georgia Basin Digital Library. The use of maps in the education system is, of course, not new, but cybercartography that uses the map as part of a much more comprehensive package is potentially much more useful to both the formal and informal education processes as it presents information in different formats and encourages dynamic interactive learning (Baulch *et al.* Chapter 21).

Geospatial awareness is increasing in societies all over the world, and the concept of spatial data infrastructures is gaining in importance as Pulsifer and Taylor point out in Chapter 7. The importance to society of location based goods and services to society is indicated by the following comment in an article by Gewin in the influential journal Nature appropriately entitled "Mapping Opportunities." "Earlier this year, the US Department of Labour identified geotechnology as one of the three most important fields, along with nanotechnology and biotechnology. Job opportunities are growing and diversifying as geospatial technologies prove their value in ever more areas." (Gewin 2004: 376)

A cybercartographic atlas is one form of organizing and presenting both qualitative and quantitative spatial information. In this respect, cybercartographic atlases represent new ways of integrating knowledge. Cybercartographic atlases should not, however, be random collections of spatially referenced information with which a user can interact in multimedia and multimodal formats. The production experiences described in this book suggest that cybercartographic atlases must have structure and purpose if they are to be most useful. Lindgaard and Brown (Chapter 9) argue that multimedia applications can enhance understanding over a single medium but that this must support or supplement material presented in another, or the user's attention may be distracted and both learning and understanding will suffer. A well-designed cybercartographic atlas must tell a story (as Pulsifer suggests, the cybercartographer can act as a "mediator" between an atlas and its users in Chapter 7). A conceptual model is required, and the atlas must respond to the needs of the societal context in which it emerges. Cybercartographic atlases are different from traditional atlases and contain features that are not "atlas-like" at all, but the term atlas has been retained as a metaphor because it suggests an organization of spatially referenced information in a logical structure and for specific purposes as conceived in the model used to conceptualize the atlas.

2.2. Theory and Practice

This book has revealed the important relationship between theory and practice in developing the paradigm of cybercartography. Reyes argues that integration begins with practice and that cybercartographic atlases represent "... the unfolding of a virtual spiral which represents a qualitative dynamic model of the interactions among society, technology, information and knowledge" (Reyes, Chapter 4: 64). Eddy and Taylor (Chapter 3) present the Cybercartographic-Human Interface model as a theoretical framework for cybercartography and illustrate the utility of this framework for the creation of the Cybercartographic Atlas of Canada's Trade with the World (Chapter 22). In this case theory is the starting point for practice. Pulsifer and Taylor (Chapter 7) argue for the importance of the concepts of geographic and cartographic mediation and semantic interoperability, which has been utilized in developing the Cybercartographic Atlas of Antarctica (Chapter 20). Again, the link between theory and practice is emphasized. Martinez and Reyes (Chapter 5) argue that the theory of cybernetics is of value to cybercartography. Throughout the book, the central importance of the dynamic and iterative feedback processes between theory and practice in the development of the paradigm of cybercartography is a persistent theme regardless of the disciplinary background from which the author is writing or the context of the discussion.

2.3. *The Language of Cybercartography*

The language of traditional cartography is primarily a visual one using images and text. Cybercartography adds new elements to these, including sound and touch (Théberge, Chapter 17, Araujo de Almeida and Tsuji, Chapter 18). Although not presented in the book, research is continuing to incorporate smell and taste to complement the other senses so that the language of cybercartography will be truly multisensory. Martinez and Reyes (Chapter 5) make the important observation that this is not just a new language for cartographers but a new language that allows the users of cybercartographic atlases to communicate with each other on topics of interest to their community. The Cybercartographic Atlas of Lake Chapala (Chapter 6) created a new medium of communication for the various communities living around the lake. The grammar and syntax of the new language are also influenced by the communication medium being used, which is interaction with a database using a computer with information appearing on screens of various sizes. Images and text with which traditional cartography has considerable experience are quite different in this dynamic interactive computer environment.

Text is an interesting case in point. Hypertext is an important textual language for computers, and research on its implications is an important element for cybercartography (Taylor 2003; Greenspan Chapter 13, Koskimaa 2002). Hypertexts are nonlinear and allow linkages with other databases and Websites, which make them of considerable interest to cybercartographers. As early as the 16th century cartographers were experimenting with text. Mercator published a treatise in Latin in 1541 (Mercator 1541) on how cartographers should use italic script on maps (Crane 2002). Mercator was the first to introduce italic script to the Low Countries. Crane comments, "Fluidity and composition on the pages were as salient as color and perspective on a canvas. The lettered line connected the imagination of the reader with the intent of the author. While the value of the text was derived from its meaning, its effect was measured by clarity and elegance. To an increasing number of humanists, italicized Latin was the language of knowledge" (Crane 2002: 157–158). Hypertext is an integral part of the emerging language of cybercartography, which in itself is a new language of knowledge integration.

Visualization is an important part of cybercartographic language (Zhou *et al.* 2004) and the topic of a growing body of research (Kraak 2003). Sound is also clearly of considerable importance, as discussed by Théberge in Chapter 17, Almeida and Tsuji in Chapter 18, and Gartner in Chapter 16. The Canadian Broadcasting Corporation radio stations, for example, exhort their listeners to "hear the big picture" and assert that "some pictures require a thousand words." New syntheses of visualization, sound, and touch form the current content of the emerging language of cybercartography. These will be supplemented by smell and eventually by taste to create a cybercartographic language that comes closer to how human beings communicate with each other in the real-world. The use of video allows the potential addition of gestures and other nonverbal body language.

2.4. Navigation and Interaction with Databases

Several chapters in this volume deal with this important topic. Interface design challenges in virtual space are considered by Lindgaard and Brown in Chapter 9. The psychological literature suggests that although multimedia certainly entertain, the learning outcome of utilizing multimedia is much less certain. Great care must be taken when designing multimedia products, such as cybercartographic atlases, to consider cognitive models and learning processes and their implications for design. Without a coherent model in mind, entertainment may take place but little learning. The user can be stimulated and entertained by cybercartographic products, but cognitive overload can add to the distraction mentioned in Section 2.1. Roberts *et al.* in Chapter 10 use a conceptual framework to develop guidelines to support navigation in cyberspace, and Trbovich *et al.* in Chapter 11 look at the role of vision, speech, gesture, sound, and touch. The design of cybercartographic atlases must include a careful consideration of the lessons coming from the field of cognitive and human factors psychology in terms of both use and usability. Cybercartographic design is largely an iterative process, and this must be informed by the User Needs Assessments (UNA) being carried out as an integral part of the Cybercartography and the New Economy project (Rasouli *et al.* 2004). An iterative process of designing, testing, then redesigning is being used in the creation of both the Cybercartographic Atlas of Antarctica and the Cybercartographic Atlas of Canada's Trade with the World and is resulting in the design of new graphic interfaces and built-in aids for navigation such as the use of landmarks. This is also discussed by Roberts *et al.* in Chapter 10.

The gaming metaphor and approach hold out considerable promise as has been suggested by a number of authors (Cartwright 2004; Greenspan, Chapter 13; Cartwright, Chapter 14). As part of the Cybercartography and New Economy project the Games Cluster (Chapter 8) is developing a game module to integrate critical perspectives on environmental issues as part of the Cybercartographic Atlas of Antarctica (Pulsifer and Taylor, Chapter 20; Dormann *et al.* 2005b). The group has modified the popular role-playing game "Never Winter Nights" to look at global warming in

Antarctica and has transformed the game into an educational learning module. The game has been titled "McMurdo on the Rocks: A Game Quest to Explore Antarctica and Global Warming" (Dormann *et al.* 2005a). The violence in the original game has been replaced by a variety of different forms of interaction in order to retain the drama of the game space. Further development and testing is required to determine how this modified computer game can be used for educational purposes, but initial results are very promising. A video clip illustrating the game appears on the CD-ROM accompanying this book.

Ticoll (2004) argues that gaming is defining a new generation and that the average age of the "gaming generation" is now 29 unlike the situation only a few years ago when teenaged boys dominated. Sales of electronic games are now greater than box office receipts from movie theaters. Ticoll notes that when the game Halo 2 from Microsoft's Xbox was released in November 2004 the first day sales were the largest on record "for the entire entertainment industry" including movies, books, and music (Ticoll 2004: B10). It is interesting to note, given the comments in Section 2.1 that Halo 2 follows a storyline and is not just a set of sequential and progressive interactive gaming events.

Cartwright, in making a strong case for use of gaming concepts, argues that:

> Geographic information delivered through the use of New Media is seen as part of popular media, rather than scientific documents. As the Web can be considered as an accepted part of popular media users could consider its use to be done in similar ways as they use television, video, movies, books, journals, radio and CD-ROM. The delivery of cartographic artifacts via the Web to naïve or inexpert users involves different strategies to traditional map delivery and use.
>
> There now exists a new genre of users. They may never have used maps before and they consider geographical information in the same way as any other commodity that they can obtain on the Web. These users are attuned to getting information via interactive on-line resources and they are willing to explore different methods of access other than conventional mapping products. They already have the skills to navigate through information spaces differently. The tool they are accustomed in using, and in many cases, the tool they have ready access to, is a games machine (Cartwright 2004: 1).

Clearly this is a new way of engaging the user, which is one of the seven elements of cybercartography outlined in Chapter 1 but gaming approaches must be used with caution. The Georgia Basin Digital Library (Harrap *et al.* Chapter 19) experimented with the popular "joysticks" of computer games as an interface device. These were situated inside models of helicopters in public spaces. Many of the young users, however, were seeking ways to blow things up rather than interact with the database and rapidly lost interest when they could not do so. In addition, Lindgaard *et al.* (Chapter 9) argue that there is a contrast between "knowledge seekers" who seek challenges and try to increase their understanding and "feature explorers" who play with the functionalities of a game and pay little attention to the content. Poorly designed "edutainment" can lead to very shallow learning, or in some cases, no learning at all (Taylor 2003).

Harris reports that children as young as two or three years of age are now interacting with computer databases and can send e-mails before they can read or write through a software package called Magic Mail. "Magic Mail relies on images and sound rather than text for communication" (Harris 2004: A5) and was being tested in Norway in 2004. Navigation and interaction with databases is taking on new forms, and completely new types of audiovisual interfaces are emerging.

2.5. Design Issues

Monmonier (Chapter 2) indicates that there are no standards for design in cybercartography. This is partially due to the neglect of the map by GIS and GIScience as argued in Chapter 1. Considerable work on scientific and technical standards in GIS has taken place, but design standards or specifications for maps have been virtually ignored in the geospatial data standards and specifications being developed by the International Standards Organization and by the Open Geospatial Consortium. Monmonier argues that there has been a failure to adapt existing cartographic knowledge on graphic quality, graphic logic, and visual effectiveness to GIS products and that design research on most products has been very limited. There has been no shortage of design and market research in the computer gaming world, but little of this has found its way into cybercartography, as the directions of such research are different. There is, however, some congruence, especially on technical issues. Many computer games are essentially spatial in nature, and they are both multimedia and interactive. Research findings in the industry, however, are rarely reported in the academic literature and technical, design and market research is rarely shared for competitive reasons.

In several chapters of this book important design guidelines for cybercartography have been suggested. These perhaps fall short of the design standards, which Monmonier seeks but make a significant contribution to creating a body of design specifications. In addition, the Mexican experience with cybercartography underlines the importance of cultural relevance and aesthetics in the design of cybercartographic atlases. The interfaces used in these atlases are not only highly functional but are also designed to appeal to the culture in which they are situated. The same applies to the choice of the music used in the atlases, which involved considerable debate in the production teams. The aesthetic and artistic element in cybercartography is an important qualitative feature of the new paradigm, which sets it apart from GIS, as argued in Chapter 1. Caquard and Taylor consider the emerging relationship between art and cartography in chapter 12, and Greenspan from a different perspective in Chapter 13.

The contribution of cognitive and human factors psychology to the design of cybercartographic atlases has been significant. The lessons related to effective interaction and navigation with databases have important design implications, and research in this area is ongoing.

2.6. Interdisciplinarity, Team Building and New Partnerships

Important lessons have been learned on this element of the cybercartographic paradigm. Some of these are outlined by Lauriault and Taylor in Chapter 8, but additional observations are appropriate here. Building an effective interdisciplinary team is the single most important factor in the theory and practice of cybercartography. Human networks are much more important than computer networks.

The National Academies' study on the Intersection between Geospatial Information and Information Technology comments:

> To make any significant progress in geospatial applications, the research community must adopt an integrated interdisciplinary approach. One of the greatest hindrances to benefiting from the massive amounts of geospatial data already being collected is the fragmented nature of current research efforts. Most of the research in the accessibility, analysis, and use of geospatial data has been conducted in isolation within single disciplines" (National Academies of Sciences 2003: 1).

This study looked at the geospatial sciences, but the problem is compounded further in cybercartography, where science, social science, and humanities are all involved.

A more recent study by the National Academies on Facilitating Interdisciplinary Research was released in late November 2004 just as this book was going to press. The Committee that produced the report came up with 15 key findings that defined interdisciplinary research, the current situation, and the changes needed to facilitate such research. A number of these are of special relevance to the Cybercartography and the New Economy project. These include:

> 1. Definition – Interdisciplinary Research (IDR) is a mode of research by teams or individuals that integrates information, data, techniques, tools, perspectives, concepts, and/or theories from two or more disciplines or bodies of specialized knowledge to advance fundamental understanding or to solve problems where solutions are beyond the scope of a single discipline or area of research practice.
>
> Current Situation
>
> 2. IDR is pluralistic in method and focus. It may be conducted by individuals or groups and may be driven by scientific curiosity or practical needs.
> 3. Interdisciplinary thinking is rapidly becoming an integral feature of research as a result of four powerful "drivers": the inherent complexity of nature and society, the desire to explore problems and questions that are not confined to a single discipline, the need to solve societal problems, and the power of new technologies...
> 6. ...Leaders with a clear vision and effective communication and team building skills can catalyze the integration of disciplines.
>
> Challenges to Overcome
>
> 7. ....IDR is typically collaborative and includes people of different backgrounds. Thus it may take extra time for building consensus and for learning new methods, languages and cultures.
> 8. Social-science research has not yet fully elucidated the complex social and intellectual processes that make for successful IDR. A deeper understanding of these processes will further enhance the prospects for the creation and management of successful IDR programs....
>
> Lessons Learned from Industry and National Laboratories
>
> 15. Collaborative interdisciplinary research partnerships among universities, industry and government have increased and diversified rapidly. Although such partnerships still face significant barriers, well documented studies provide strong evidence of both their research benefits and their effectiveness in bringing together diverse cultures. (National Academies of Sciences 2004: 2–3)

These all apply to the Cybercartography and the New Economy project. Findings 6 and 7 are of particular relevance. The time taken for team members to understand each other's language and culture was considerably longer than anticipated and took nearly two years. Common terms, such as "navigation" for example, were defined differently by each of the disciplines involved in the project making effective communication difficult, as discussed by Lauriault and Taylor in Chapter 8. The experience of the Cybercartography and the New Economy project substantiates the validity of Finding 6, but the processes required not just vision but also considerable time and ongoing effort. Many of the recommendations of the National Academies of Sciences' report had already been implemented during the Cybercartography and the New Economy project and proved to be effective, especially the adoption of a flexible problem oriented matrix approach (Lauriault and Taylor, Chapter 8).

The Cybercartography and the New Economy project is also a successful example of Finding 15. The partnerships with Statistics Canada, the National Atlas of Canada, Natural Resources Canada, Students on Ice, the Scientific Committee for Antarctic Research (SCAR), the Canadian Committee for Antarctic Research (CCAR), DM Solutions Group, Orbital Media Holdings, the Geomatics Industry Association of Canada, the Open Geospatial Consortium (OGC), and other private sector partners have proved to be very productive. The success of these partnerships owes much to the fact that there are clear mutual benefits to all partners concerned. It also owes much to the personal relationships and friendships among many of the key players involved who have worked together before. These existing social networks were important building blocks.

Interdisciplinarity was also aided by the fact that team members were contributing to the production of the two cybercartographic atlases, which gave a focus to the research. This focus was, however, not without its drawbacks as some team members saw the emphasis on production as detrimental to their own research agendas. In addition, Cybercartography and the New Economy project researchers in psychology, in particular would relate strongly to the comments made in the National Academies of Science study (2004) that many professional journals have yet to accept the value of interdisciplinary research articles.

Virtual and knowledge networks are considered to be of importance in enhancing interdisciplinary research, and as outlined in Chapter 8, the Cybercartography and the New Economy project utilizes these and other computer-based communication techniques. One lesson that has been learned, however, is the importance of human interaction in real space. The majority of the research teams in this project share one floor of a research building allowing face-to-face contact on a daily basis. This facilitates both formal and informal contact. Similarly, in Mexico the researchers of CentroGeo are all based in one building and again, interact formally and informally on a daily basis. There are also regular personal visits between Carleton University and CentroGeo, and key CentroGeo researchers are part of the electronic forum established for the project.

In Mexico considerable importance is given to an individual researcher's ability to interact with and contribute to the work of the group as a whole. This is, in fact, an integral part of the formal evaluation of the performance of individual researchers carried out on an annual basis. Evaluating interdisciplinary research is difficult, as the National Academies of Science (NAS) study points out (NAS 2004). The Mexican approach is an interesting one in this respect. Building the coherence of a group working toward a common purpose is important if interdisciplinary work is to succeed. CentroGeo makes this part of the formal career evaluation process that determines promotion and salary increases, which recognizes its importance in concrete terms.

The situation in the Cybercartography and the New Economy project is different. Canadian cultural systems put less emphasis on group work in the social sciences and humanities and more on individual research contributions, but the importance of building a group, which is dedicated to a common purpose is still important to ensure that interdisciplinary work takes place. Interpersonal relationships and social networks are important elements in this process as are social functions outside of the work place. One of the most successful elements in the Cybercartography and the New Economy project has been the enthusiasm and cohesion of the graduate student group involved in the project that has been built by both formal and informal methods as outlined in Chapter 8.

3. Remaining Challenges and Directions for Future Research

3.1. *The Need for More Practice*

If theory and practice are closely linked, as has been argued in this chapter and elsewhere in the book, then advances in the paradigm of cybercartography will only come if more cybercartographic products are created. The number of such products is currently quite small. There is a need to extend cybercartography into other content areas and also into other cultural contexts in order to give more empirical content from which an increased understanding of cybercartography can be obtained. Carmen Reyes argues that "the methodological framework should be refined, transmitted and compared to other research and development groups so that cybercartographic atlases are enriched and applied in other contexts in terms of content themes, geographic regions and societal groups" (Reyes, Chapter 4: 93). As cybercartographic products move into the public domain in pervasive public map displays as discussed by Peterson (Chapter 15) more empirical evidence will be created.

3.2. *The Need for More Rigorous Theory*

The theoretical approaches to cybercartography outlined in several chapters in this book are an interesting first step, but it is clear that they require further refinement and development. A number of avenues for research in this respect have been identified, but to extend the metaphor of exploration used in Chapter 1 these are little more than sign posts pointing to possible directions for further theoretical research. Each of these avenues for research will have to be explored in a more rigorous way. Some will lead to useful contributions to our theoretical understanding. Others will be false trails that may have to be abandoned. Integral theory, general systems theory, second wave cybernetics, semantic ontologies, communications theory, cybercartographic mediation, the link between mathematics and cartography, dynamic modeling, cognitive theories, learning theories, human factors theories, artificial intelligence, and theories of interdisciplinarity among others have all been presented as topics worthy of further exploration in the various chapters of this book. The paradigm of cybercartography argues for integration of knowledge, but how this integration is to take place and what theories best lend themselves to an increased understanding remain a major research challenge.

3.3. Design Challenges

In an online, multimedia, and multisensory cybercartographic world the design challenges loom large. Design challenges are both social and technical. In social terms there is clearly a need to find better ways of involving users in the process of creating cybercartographic atlases. User Needs Assessments, as discussed in various chapters of the book, have a contribution to make, but these are necessary but not sufficient techniques. It has been argued that cybercartographic atlases are social products, but the challenge is to find more effective ways of involving society in their production. As social products, cybercartographic atlases are culturally specific and this increases the challenges. The choice of what is to be cybermapped and how that cyber mapping is done is obviously a key question that is part of the debate over the power of maps and their role in society. Caquard and Taylor in Chapter 12 discuss this and make a case for cybercartography as a form of social criticism. The role of the cybermap in society takes on new significance when cartographic products are widely distributed over the Web. It is very difficult to determine who the "users" are in situations, where anyone anywhere can access a cybercartographic atlas on the Web. Cartwright has argued "...that once cartographers knew their users" (Cartwright 2004: 1), but that the situation has changed dramatically with the widespread availability of maps on the Web, which poses a number of new design challenges as cybercartography becomes part of the information mainstream. In addition, there has been a substantial increase in pervasive public map displays as Peterson (Chapter 15) points out.

The technical challenges of design are also significant, as several chapters in this book indicate. The extent of these challenges is considerable and only a few will be considered here. Interface design is of key importance, and it is clear that existing interfaces leave much to be desired and new approaches are required. Integrating content information on a wide variety of topics is also challenging, and interoperability of databases, including semantic interoperability, is a key element. The Cybercartography and the New Economy project uses an open source philosophy and utilizes Open Geospatial Consortium specifications. At the same time, use is made of proprietary software where appropriate. The project team has also had to design its own customized software as existing products did not fully meet project needs. The costs and benefits of an open source as opposed to a proprietary software approach are a topic of much debate. A Web-server approach has been used in producing the Cybercartographic Atlas of Antarctica and the Cybercartographic Atlas of Canada's Trade with the World. Such an approach holds much promise, but there are still technological challenges and rapid technological change to a wireless environment with mobile devices also poses many new technical and design problems (Gartner, Chapter 16). In Section 2.4 the importance of exploring a gaming environment was considered. Research on the relationships between cartography and computer games is in its infancy and creates many new challenges (Cartwright 2004).

3.4. Art, the Humanities and Cybercartography

As Caquard and Taylor, Greenspan, and Théberge argue in this book (Chapters 12, 13, and 17), the potential contribution of art, language, literature, and music to cybercartography is considerable both in terms of design and content. Although some progress has been made in integrating a humanities' perspective into the theory and practice of cybercartography that progress is limited, and much more research is required. We have really only scratched the surface of this important topic. Aesthetics are important as is illustrated in the beautiful graphic design of the cybercartographic atlases produced by CentroGeo in Mexico, but the contribution of art goes well beyond this as Caquard and Taylor argue in Chapter 12. How should new forms of cybercartographic "anti-maps" be created and designed and what forms should these take? The debate over cartography as an art or a science is not a new one. Cybercartography is both an art and a science, but the nature and extent of this important relationship require further research.

3.5. The Utility of Cybercartography

Throughout this book claims have been made about the utility of cybercartography as an educational tool and in the case of Mexico as a catalyst for social action. These claims are substantiated by some empirical evidence, but there is a need for more impact and user studies. Do people actually learn more effectively in an educational setting using cybercartographic atlases? Do gaming approaches help? If so why?

If cybercartographic atlases are leading to social action, such as in Mexico, what is the extent of this impact? How and why did cybercartography contribute? Under what conditions and in what contexts was it effective? Was the process of social action sustained over time? These and other important questions cannot be answered until more cybercartographic products are created, and time is an element here.

3.6. Multisensory Research

Cybercartography is multisensory, and Taylor (2003) outlines the state of research in this respect. Considerable progress has been made with visualization, sound, and touch as illustrated in several chapters in this book. Research on smell and taste in cybercartography has been much more limited. The sense of smell is an important one both for communication and memory. It has been argued that people retain a sense of smell in memory for a much longer time than for all the other senses (Watson 1999). In awarding the Nobel Prize in Science to Richard Axel and Lind Buck for their work on smell the Nobel Assembly noted that smell was "the most enigmatic of our senses" (Ritter 2004: 15). One of the more promising approaches lies in the application of the work of the French company Olfacom (www.olfacom.com) to cybercartography. This innovative company has developed an apparatus that can be attached to a PC to diffuse a number of odors from a changeable cartridge. Diffusion is by a sublimation process and less invasive than other approaches. Odors available from Olfacom include scents for the four seasons, scents of the sea and the forest, making possible the approach to including smell in cybercartography as suggested by Taylor (2003). One of the more interesting applications of Olfacom technology was by the artist Alex Sandover, who presented what he called "synaesthesia" in the Peterborough Art Gallery in the United Kingdom in April 2001 (http://www.alexasandover.com/Peterborough.html). This included the combination of smell, photography, video, and sound. The relationship between cybercartography and art is self-evident here. Equally interesting is the "Balades Olfactives" produced in cooperation with France Telecom and the Bureau Interprofessionel des Vins de Bourgogne (http://olfacom.com /pages/actualit.html). This is based on a map of the wine growing regions of the Bourgogne with Web enabled olfaction and multimedia capabilities. Olfacom can produce a wide variety of odors at a reasonable cost, and the experience of linking these with multimedia objects lends itself to cybercartographic application. As with many other cartographic applications the technology of olfaction has been developed for other purposes but can be adapted for cybercartography. Here, the impact of the marketplace is significant. The companies producing the Visual Odor Displays mentioned by Taylor (2003), such as Digiscents, went out of business because of lack of market demand and poor pricing strategies, and as a result, their products were unavailable for purchase by the Cybercartography and the New Economy project when research funds became available. Olfacom appears to be much more robust with more reasonable pricing strategies and much more careful market research.

There are other odor emitting technologies on the market such as the Scent Dome being utilized by Telewest Broadband in the United Kingdom and at the lower end of the market, the Febreze Scentstories Player, and Theme Disks. The Febreze product was being aggressively marketed in North America through television in November/December 2004 as a stand-alone product for home use. Scentstories, as their names suggest, are CD-ROMS, which when inserted in the player release a series of "stories" through smell.

Much of the existing research on smell is being carried out by large telecommunication companies, but in December 2004, Staples reported the entry of the tourist industry in this field (Staples 2004: D2). The offices of the Thomson travel agency are now offering potential tourists a "virtual tour" of some destinations. A virtual reality headset is used, which provides the user with a three-dimensional movie supplemented by appropriate smells. The initial test experience is of Egypt and the multimedia product, including the smells, was financed by the company with support from the Egyptian Tourist Agency. "In the Valley of the Kings, virtual travelers are treated to the musty aroma of decaying mummies. While scrunching their toes into the virtual sand of an Egyptian beach, they smell the 'sea breeze.' Poolside at a resort on the Red Sea 'Riviera' comes the familiar coconut aroma of suntan lotion and so on" (Staples 2004: D2). Other realistic smells, such as spices in a market, are also being used. "Multisensory campaigns have technically been possible since the arrival of digital technology, which renders sounds, smells and visual images into computerized code. The idea has been slow to catch on with all, but the largest companies mainly because such interaction exhibits are expensive compared with less-stimulating traditional visuals. . . . " (Staples 2004: D2). Staples also reports that smells are being used for advertising on the London Underground and by Disney to attract visitors to various areas of the company's theme parks. Museums are also experimenting with adding smell to their virtual displays.

The use of smell is "...part of an emerging advertising trend to lure customers into stores by appealing to more of their five senses" (Staples 2004: D2). As mentioned in an earlier section, market demand and cost will determine whether this new approach that has similarities to cybercartography will be further developed, but the potential is clear.

Opinions on the value of smell to cybercartography vary, and virtually nothing is known of user acceptance and reaction to the addition of smell to cybercartographic products. In addition, further research and development of the technologies themselves is required. Cybercartographic research on smell is at an early stage.

Research on taste is at an even earlier stage, and the likelihood of the effective integration of taste into cybercartographic products in the near future is not great. Much taste research is concentrated on water quality, and one possibility is to represent taste by interactive graphs of chemical content by the location at which the taste sample was taken.

The most recent developments in touch, sound, taste, and smell were reported at the computer graphics and interactivity conference SIGGRAPH 03, held in San Diego from 27–29 July 2003 and SIGGRAPH 04 held in Los Angeles. Another interesting development was reported by Stoerig *et al.* under the provocative title of "Seeing through the Ears." Visual images from a portable video camera are transmitted into sound patterns. The camera is connected to a computer, and the images are "described" by the sound patterns transmitted. Pitch, volume, frequency, and the timing of different sounds help the user to understand and reconstruct the object in question, and users are informed of how these sound patterns are being used. With extended practice users could, in fact, reconstruct the images being described by the sounds (Stoerig *et al.* 2004). The potential of this approach for blind users is worthy of further research. For blind users cybermaps are important bridges to reality. This is quite different from the sighted user for whom a map is an abstraction of reality (Araujo de Almeida and Tsuji, Chapter 18).

Many research challenges remain for cartography in the multisensory area.

3.7. *Archiving Cybercartographic Atlases*

There is an increasing realization of the need to archive interactive digital products given that much of our digital heritage over the last two decades has been lost (Taylor and Lauriault 2004). Archiving cybercartographic products poses many problems. An important step is finding adequate metadata descriptions for these complex multisensory and multimedia products. A number of existing metadata standards for different media exist and are being codified by the International Standards Organization, but no one standard is completely adequate for cybercartographic products. Current research (Zhou *et al.* 2004) is examining combinations of metadata standards building on the TC211 ISO standards and extending the ISO19115 standard to include new elements.

Defining adequate metadata descriptions is only one of a complex set of problems and other questions arise (Taylor and Lauriault 2004). What is to be archived and how is this to be captured for archival purposes? What technological approaches should be used? How is the rapid change in technology to be dealt with, including the important topic of technological obsolescence? What is an authentic digital record? What institutional arrangements are required to store such data, and how can the data be made available to interested users? These topics are being addressed at a number of levels by research projects, such as InterPARES 2 (The International Research on Permanent Authentic Records in Electronic Systems) (www.interpares.org) and by the Archival Working Group of CODATA (Committee on Data for Science and Technology) (CODATA 2004). The Organization for Economic Cooperation and Development (OECD) has produced an interesting study (OECD 2003a). This led to an important statement by the governments of all the OECD countries in January 2004 entitled "Declaration on Access to Research Data from Public Funding" (OECD 2003b). The nature and complexity of the research challenges involved is captured in the documentation for the Canadian National Consultation on Access to Scientific Data Task Force (Task Force November 2004) and an important report on infrastructures, access, and preservation by the Social Sciences and Humanities Research Council of Canada (SSHRC 2002). The Cybercartographic Atlas of Antarctica is a case study for the InterPARES 2 project, but although progress has been made there are still many questions to be answered before cybercartographic atlases can be archived effectively.

4. Conclusions

In answer to the question posed by Carmen Reyes in the introduction to this chapter, the evidence presented by the chapters in this book suggest that we are at a breakthrough point in the development of cartography and that the paradigm of cybercartography is well worth further exploration. There are, however, many questions still to be answered, and much further research is required if cybercartography is to reach its full potential.

References

Balades Olfactives (2002) *Olfacom Internet Site*, accessed 20 November 2004 from http://www.olfacom.com /pages/actualit.html

Cartwright, W. (2004) *Using Games Strategies and Games Machines for Geographical Information Exploration*, Presentation to the Cybercartography and the New Economy Project, Carleton University, Ottawa, September.

Committee on Data for Science and Technology (CODATA) (2004) *The CODATA Task Group on Preservation and Archiving of Scientific and Technical Data*, CODATA, Paris.

Crane, N. (2002) *Mercator: The Man Who Mapped the Planet*, Phoenix, Orion Books, London.

Dormann, C., J.-P. Fiset, S. Caquard, B. Woods, A. Hadziomerovic, E. Whitworth, A. Hayes and R. Biddle (2005a) *McMurdo on the Rocks: A Game Quest to Explore Antarctica and Global Warming*, Video of Computer Game in the CD-ROM accompanying this book.

Dormann, C., J.-P. Fiset, S. Caquard, B. Woods, A. Hadziomerovic, E. Whitworth, A. Hayes and R. Biddle (2005b) *Computer Games as Homework: How to Delight and Instruct*, Home Oriented Informatics and Telematics Conference, April 2005, University of York, United Kingdom (forthcoming), pp. 13–15

Gewin, V. (2004) "Mapping opportunities" *Nature*, November, Vol. 427, January, pp. 376–377.

InterPARES 2, (2004) *International Records on Permanent Authentic Records in Electronic Systems InterPARES 2 Experimental, Interactive and Dynamic Records*, accessed 3 December 2004 from www.interpares.org

Harris, M. (2004) *You're Never Too Young for E-mail: New Software Lets Kids Send Greetings,* The Ottawa Citizen, Sunday, October 17, A5, Ottawa.

Koskimaa, R. (2002) *Visual Structuring of Hyperfiction Narratives, Digital Literature: From Text to Hypertext and Beyond*, accessed 24 November 2004 from http://www.cc.jyu/Tilde goes hereKoskimaa/thesis/chapter5.htm

Kraak, M.-J. (2003) "Geovisualization illustrated", *ISPRS Journal of Remote Sensing*, Vol. 57, pp. 390–397.

Mercator, G. 1540/1541, *Literarum, Latinarium, Cursoriasque Volat, Scribendaru Ratio*, 1549 (edn.), Louvain (British Library), (quoted in Crane 2002).

National Academies of Sciences (2003) *The Intersection between Geospatial Information and Information Technology*, National Academies of Sciences, Washington.

National Academies of Sciences (2004) *Facilitating Interdisciplinary Research,* National Academies of Sciences, Washington.

Organisation of Economic Cooperation and Development (OECD) (2003a) *Promoting Access to Public Research Data for Scientific, Economic and Social Data, Final Report*, OECD, Paris.

Organizations of Economic Cooperation and Development (OECD) (2003b) *Declaration on Access to Research Data form Public Funding*, 30 June OECD, Paris.

Olfacom, (2004) *Home Page*, accessed 1 December 2004. www.olfacom.com

Rasouli, M., K. Philp, S. Khan, G. Dunn, G. Lindgaard and M. Parush (2004) *User-centred Design of Educational Cybercartographic Atlases*, Presentation to the Canadian Association of Geographers Ontario Meeting, Waterloo.

Ritter, M. (2004) "US scientists win nobel prize by a nose", *The Globe and Mail,* Tuesday, 5 October A15, Toronto.

Sandover, A. (2002) *Synaesthesis*, accessed 1 December, 2004 from http://www. alexsandover. com/peterborough.html

SIGGRAPH (2004) *SIGGRAPH Los Angeles, Conference Proceedings*, accessed from www.siggraph.org/s2004/

SSHRC (Social Sciences and Humanities Research Council of Canada) (2002) *National Data Archive Consultation: Building Infrastructure for Access to and Preservation of Research Data*, SSHRC, Ottawa.

Staples, S. (2004) "Let's take aroma holiday", *The Ottawa Citizen*, 16 December D2, Ottawa, Canada.

Stoerig, P, E. Ludewig, T. Mierdorf, A. Oros-Pesquens, J. N Shah and A. Pascual Leone (2004) *Seeing Through the Ears? Identification of Images Converted to Sounds Improves with Practice*, Program No.177.15, Abstract Viewer/Itinerary Planner, Society for Neuroscience, Washington, DC.

Ticoll, D. (2004) "Electronic gaming is defining the new generation", *The Globe and Mail*, Thursday, 18 November, B10, Toronto.

Task Force on National Consultation on Access to Scientific Data (2004) *Enhanced Access to Scientific Research Data: Framework of for a Vision for 2010, CNADRS*, Government of Canada, November.

Taylor, D. R. F. (2003) "The concept of cybercartography" in M. P. Peterson (ed.), *Maps and the Internet*, Elsevier, Amsterdam.

Taylor, D. R. F. and T. P. Lauriault (2004) *The Cybercartographic Atlas of Antarctica and Related Archival Issues*, Paper to the 19th International CODATA Conference, 7–10 November, Berlin.

Watson, L. (1999) *Jacobson's Organ and the Remarkable Nature of Smell*, Allen Lane, The Penguin Press, Middlesex, UK.

Zhou, Y, Liu, Xiuxia, B. Woods, P. L Pulsifer and D. R. F. Taylor (2004) *Visualization in Cybercartography*, Presentation to the Canadian Association of Geographers Ontario Meeting, Waterloo.

Subject Index

H